D0023523

AN INTRODUCTION TO
HYDRODYNAMICS
AND WATER WAVES

HYDRODYNAMICA

SIVE

DE VIRIBUS ET MOTIBUS FLUIDORUM COMMENTARII

'Remember, when discoursing about water, to induce first experience, then reason.'

—Leonardo da Vinci

AN INTRODUCTION TO HYDRODYNAMICS AND WATER WAVES

Bernard Le Méhauté

SPRINGER-VERLAG
New York Heidelberg Berlin
1976

Bernard Le Méhaute
Senior Vice President
Tetra Tech, Inc.
630 North Rosemead Boulevard
Pasadena, California 91107

This volume has been totally revised from an earlier version published by the U.S. Government Printing Office as ESSA Technical Reports ERL 118-POL 3-1 and 3-2.

Printed in the United States of America.

Library of Congress Cataloging in Publication Data

Le Méhauté, Bernard, 1927–
An introduction to hydrodynamics and water waves.

Includes bibliographical references and index.
1. Hydrodynamics. 2. Water waves. I. Title.
QA911.L39 532′.5 75-12754
ISBN 0-387-07232-2

ISBN 0-387-07232-2 Springer-Verlag New York

ISBN 3-540-07232-2 Springer-Verlag Heidelberg Berlin

Preface

Hydrodynamics is the science which deals with the motion of liquid in the macroscopic sense. It is essentially a field which is regarded as applied mathematics because it deals with the mathematical treatments of basic equations for a fluid continuum obtained on a purely Newtonian basis. It is also the foundation of hydraulics, which, as an art, has to compromise with the rigorous mathematical treatments because of nonlinear effects, inherent instability, turbulence, and the complexity of "boundary conditions" encountered in engineering practice. Therefore, this book can be considered as the text for a course in basic hydrodynamics, as well as for a course in the fundamentals of hydraulic and related engineering disciplines.

In the first case, the students learn how to make use of their mathematical knowledge in a field of physics particularly suitable to mathematical treatments. Since they may have some difficulty in representing a physical phenomenon by a mathematical model, a great emphasis has been given to the physical concepts of hydrodynamics. For students with an undergraduate training in engineering, the difficulty may be a lack of appropriate mathematical tools. Their first contact with hydraulics has been on an essentially practical basis. They may be discouraged in attempting the study of such books as *Hydrodynamics*, by Lamb, which remains the bible of hydrodynamicists. Hence, mathematical intricacies have been introduced slowly and progressively. Also, the emphasis on the physical approach has made it possible to avoid mathematical abstractions so that a concrete support may be given to equations.

Finally, the author has tried to make this book self-contained in the sense that a practicing engineer who wants to improve his theoretical background can study hydrodynamics by himself without attending lectures. Too often articles in scientific journals present some discouraging aspects to practicing engineers and the most valuable messages can only reach a few specialists. It is felt that the learning of some basic theories will help hydraulic engineers to keep abreast of and participate in new developments proposed by theorists.

Considering that a good assimilation of the basis is essential before further study, great care has been taken to develop a clear understanding, both mathematically and physically, of the fundamental concepts of theoretical hydraulics. The introduction of mathematical simplifications and assumptions, often based on physical considerations, has also been developed by examples. The mathematical difficulties have been cleared up by introducing them progressively and by developing all the intermediate calculations. Also, all the abstract concepts of theoretical hydraulics have been explained as concretely as possible by use of examples. It will appear that the first chapter is the easiest to understand, and it is assumed that the mathematical background increases as the student progresses toward the end of the book. However, it is taken for granted that the student already has some notion of elementary hydraulics.

Finally, the succession of the various chapters have been chosen in order to build up a structure as logical and as deductive as possible in order to avoid that the various subjects appear as a succession of different mathematical recipes rather than as a unique and logical subject.

Part One deals with the establishment of the fundamental differential equations governing the flow motion in all possible cases. The possible approximations are also indicated. Part Two deals with general methods of integrations and the mathematical treatments of these equations. Integrations of general interest, and integrations in some typical particular cases are presented. Part Three is devoted to water wave theories, as one of the most important topics of hydrodynamics.

It is pointed out that the emphasis of the book is on water waves. Therefore the treatment of motion of compressible fluids has been judged beyond the scope of this book, with a few exceptions. Also, almost all the calculations are presented in a Cartesian (or cylindrical) system of coordinates. Vectorial and tensorial operations have been minimized in order to reduce the necessary mathematical background. However, vectorial and tensorial notations are slowly introduced for sake of recognition in the literature.

It is hoped that this book will entice students gifted in mathematics to apply their capabilities to the study of fluid motion and dynamical oceanography. It is hoped also that it will instill in engineering students the desire for further study in hydrodynamics and mathematics. It is also hoped that the book will be of great help to students in hydraulics, civil and coastal engineering, naval architecture, as well as in physical oceanography, marine geology, and sedimentology, who want to learn or revise one of the theoretical aspect of their future profession.

ACKNOWLEDGMENTS

The author wishes to express his deep gratitude to Nicholas Boratynski, President of Tetra Tech, Inc. Without his encouragement and support, these lecture notes would not have been revised and published.

Many valuable suggestions and contributions are credited to the senior engineering and scientific staff of Tetra Tech.

The author would also like to acknowledge Dr. Viviane Rupert of the Lawrence Livermore Laboratory for her help in editing the book.

Bernard Le Méhauté
Pasadena, California

Contents

PART ONE

Establishing the Basic Equations that Govern Flow Motion

PART TWO

Some Mathematical Treatments of the Basic Equations

PART THREE

Water Wave Theories

PART ONE

Establishing the Basic Equations that Govern Flow Motion

Chapter 1

Basic Concepts and Principles

1-1 Basic Concepts of Hydrodynamics

1-1.1 Definition of an Elementary Particle of Fluid

Studies of theoretical fluid mechanics are based on the concept of an elementary mass or particle of fluid. This particle has no well-defined existence. It may be considered as a *corpus alienum*, a foreign body in the mechanics of a continuum. It is an aid toward the understanding of the physical meaning of the differential equations governing the flow motion.

Just as the fundamental concepts of the theoretical mechanics of solid matter are based on the mechanics of a so-called "material point," the basis of theoretical fluid mechanics rests on the mechanics of an elementary mass of fluid. Such an elementary mass of fluid, in common with the material point in the kinematics of a solid body, is assumed to be either infinitely small or small enough that all parts of the element can be considered to have the same velocity of translation $\mathbf{V}$ and the same density ρ. This elementary fluid particle is assumed to be homogeneous, isotropic, and continuous in the macroscopic sense. The molecular pattern and the molecular and Brownian motions within the particle, a subject dealt with in the kinetic theory of fluids, are not taken into account.

1-1.2 Theoretical Approach

The laws of mechanics of a solid body system (a rotating disk, for example) are obtained by the integration of the laws of mechanics for a "material point" with respect to the area or the volume of the system under consideration. Similarly, the laws of fluid mechanics used in engineering practice are obtained by integration—exact or approximate—of the laws governing the behavior of a fluid particle along a line or throughout an area or a volume. Hence, studies in hydrodynamics may be divided into two different parts.

1-1.2.1 The first part consists of establishing the general differential equations which govern the motion of an elementary particle of fluid. The fluid may be assumed either perfect (without friction forces) or real. In the latter case, the flow may be either laminar or turbulent.

1-1.2.2 The second step involves the study of different mathematical methods used to integrate these basic differential equations. Practical general relationships, such as the well-known Bernoulli equation, may thereby be deduced. Solutions, valid for special cases, can also be obtained by direct integration.

1-1.3 Relations between Fluid Particles: Friction Forces

In a solid material, points in a system (on a disk, for example) do not change their relative position (except for elastic deformations which are described by well-defined laws). On the other hand, fluid particles may be deformed and each particle may have a particular motion which differs quite markedly from the motion of other particles. The forces exerted between fluid particles are the pressure forces and the friction forces.

The friction force per unit area in a given direction, called the shear stress τ, is assumed to be either zero ("ideal" or perfect fluid), or proportional to the coefficient of viscosity μ (viscous fluid). The shear stress τ is a scalar. The set of shear stresses at a point constitutes a tensor. The significance of this statement is developed in Chapter 5. For now it is sufficient to know that the shearing stress, at any point of a plane parallel to a unidirectional flow is

$$\tau = \mu \frac{d\mathbf{V}}{dn}$$

where n is the perpendicular direction to the flow moving with velocity $\mathbf{V}$.

Hydrodynamics is primarily concerned with a "Newtonian fluid," that is, its viscous stress tensor depends linearly, isotropically, and covariantly (Chapter 5) on the rate of strain or derivatives of the velocity components. It does not deal with "plastic" fluids where the coefficient μ is replaced by a function of the intensity or duration of the shear.

1-2 Streamline, Path, Streakline, and Stream Tube

1-2.1 Notation

Consider the point $A(x,y,z)$ in a Cartesian system of coordinates. The axes OX, OY, OZ are mutually perpen- pendicular (see Fig. 1-1). Consider an infinitely small rectangular element of fluid with the point as a corner. The edges of this element are dx, dy, dz. Its volume is $dx\,dy\,dz$ and its weight is $\varpi\,dx\,dy\,dz$ or $\rho g\,dx\,dy\,dz$. ϖ is the specific weight and g is the acceleration due to gravity.

The pressure at point A is a scalar quantity which is completely specified by its magnitude. The pressure is always exerted perpendicular to the considered surface (see Section 5-3.1). The corresponding force is a vector quantity, which is specified by its magnitude and direction. The magnitude of the pressure p is a function of the space coordinates of A and time t; i.e., $p = f(x,y,z,t)$. Its direction is normal to the area on which the pressure is exerted. The

Figure 1-1 *Notation in Cartesian coordinates.*

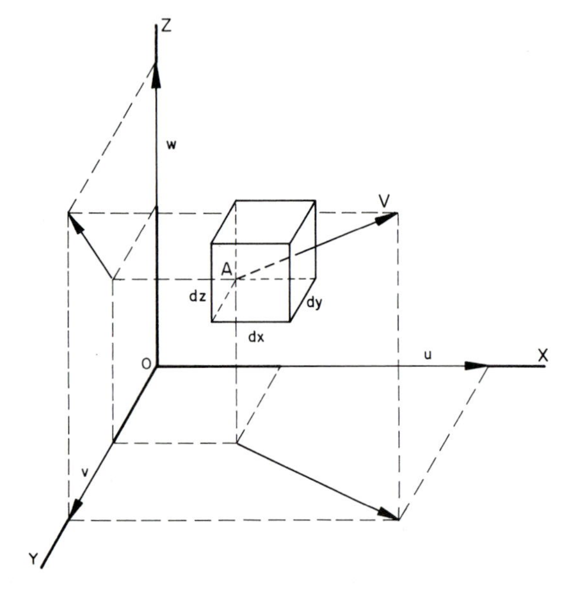

gradient of p (**grad** p or ∇p), its derivative with respect to space, is also a vector quantity. The components of **grad** p along the three coordinate axes OX, OY, OZ, are given by the derivative of p with respect to x, y, z, respectively; i.e., $\partial p/\partial x$, $\partial p/\partial y$, $\partial p/\partial z$.

The velocity of fluid particles at A is **V**. The components of **V** along the three Cartesian coordinate axes OX, OY, OZ, are u, v, and w, respectively. If **i**, **j**, **k** are unit vectors along the axes OX, OY, OZ respectively, then: $\mathbf{V} = \mathbf{i}u + \mathbf{j}v + \mathbf{k}w$. Since the system of reference is rectangular, the magnitude of the velocity is given by $V = [u^2 + v^2 + w^2]^{1/2}$. V is a scalar quantity and therefore completely defined by its magnitude, like the pressure p. **V** is a vector quantity and is specified by its direction and magnitude. Since **V** and its components u, v, and w are functions of the space coordinates of A and the time t, they can be written in the form $\mathbf{V}(x,y,z,t)$.

1-2.2 *Definitions*

1-2.2.1 The displacement **dS** of a fluid particle is defined by the vector equation, $\mathbf{dS} = \mathbf{V}\,dt$, which is valid for both magnitude and direction. This equation may be written more specifically in terms of the displacements in each of the three Cartesian coordinate directions as follows:

$$dx = u\,dt$$
$$dy = v\,dt$$
$$dz = w\,dt$$

1-2.2.2 A *streamline* is defined as a line which is tangential at every point to the velocity vector at a given time t_0. A device for visualizing streamlines is to imagine a number of small bright particles distributed at random in the fluid, and then to photograph them with a short exposure (Fig. 1-2). Every particle photographs as a small line segment. Each line which is drawn tangentially to these small segments is a streamline.

At time t_0, the equations $dx = u\,dt$, $dy = v\,dt$, and $dz = w\,dt$ become:

$$\frac{dx}{u(x,y,z,t_0)} = \frac{dy}{v(x,y,z,t_0)} = \frac{dz}{w(x,y,z,t_0)}$$

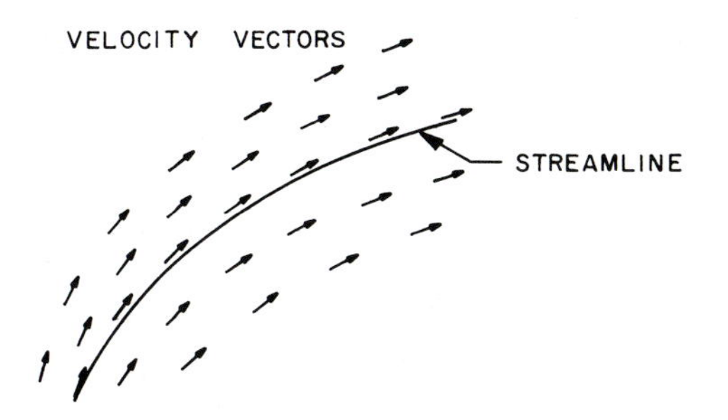

Figure 1-2 *Streamlines observed by short-exposure photography of various particles.*

This is the mathematical definition of a streamline. These equalities express the fact that the velocity is tangential to the displacement of the particle at time t_0. Figure 1-3 illustrates this fact in the case of a two-dimensional motion. In this case $dx/u = dy/v$, which implies $v\,dx - u\,dy = 0$.

Streamlines do not cross, except at point of theoretically infinite velocity (see Figs. 11-6 and 11-7) and at stagnation and separation points of a body where the velocity is zero. Fixed solid boundaries and steady free surfaces are streamlines. Moving boundaries, such as propeller blades, and unsteady free surfaces are not streamlines.

1-2.2.3 The *path* of a specific particle of the fluid is defined by its position as a function of time. It may be

Figure 1-3 *Definition of a streamline in a two-dimensional motion.*

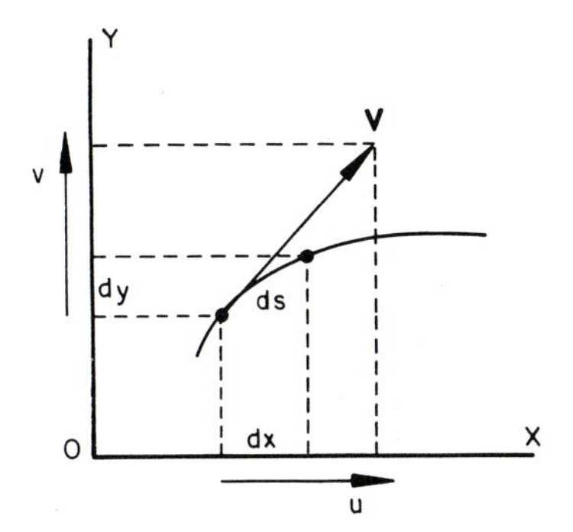

determined by photographing a bright particle with a long exposure. The path line is tangential to the streamline at a given time t_0. However, the time has to be included as a variable for defining a path. Hence, the path lines are defined mathematically as

$$\frac{dx}{u(x,y,z,t)} = \frac{dy}{v(x,y,z,t)} = \frac{dz}{w(x,y,z,t)} = dt$$

1-2.2.4 A *streakline* is given by an instantaneous shot photographing a number of small bright particles in suspension which were introduced into the fluid at the same point at regular intervals of time (Fig. 1-4).

1-2.2.5 An elementary flow channel bounded by an infinite number of streamlines crossing a closed curve is known as a *stream tube* (Fig. 1-5).

1-2.3 Steady and Unsteady Flow

1-2.3.1 For steady flows defined by time-independent quantities, streamlines, streaklines, and particle paths are identical. However, for unsteady or time-dependent flows, these lines are different and a clear understanding of their generation is necessary to properly interpret the results of a given experiment. For example, if dye is injected at a given point of a fluid flow, the dye pattern will be a streakline; if the successive location of a neutrally buoyant small ball are determined, a particle path can be traced; finally, if a large number of short threads are attached to a body, the instantaneous direction of these threads will yield a streamline pattern. All these methods are commonly used in fluid flow experimental studies.

Streamlines, paths, streaklines, and stream tubes are different in unsteady flow, that is, flow changing with respect to time. Turbulent flow is always an unsteady flow; however, it will be seen in that case that the mean motion with respect to time of a turbulent flow may be considered as steady. Then streamlines, paths, and streaklines of the mean motion are the same (see Chapter 7). Figures 1-6 and 1-7 illustrate these definitions in some cases of unsteady motion.

1-2.3.2 In some cases of unsteady flow (a body moving at constant velocity in a still fluid, a steady wave profile such as those due to a periodic wave or a solitary wave) it is possible to transform an unsteady motion into a steady motion relative to a coordinate system which moves with the body or the wave velocity. The construction of a steady pattern is then obtained by subtracting the velocity of the body from the velocity of the fluid. This is the *Galilean transformation*. Steady streamlines can then be defined relative to a moving observer who travels with the body or with the wave (see Fig. 1-8).

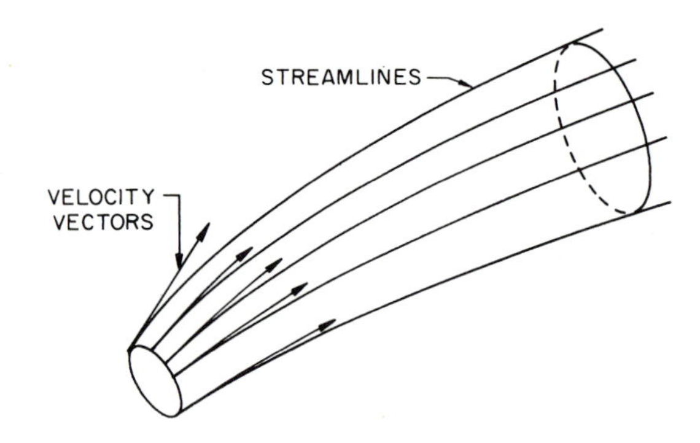

Figure 1-5 *Stream tube.*

Figure 1-4 *Streakline obtained by instantaneous photography of various particles coming from the same point.*

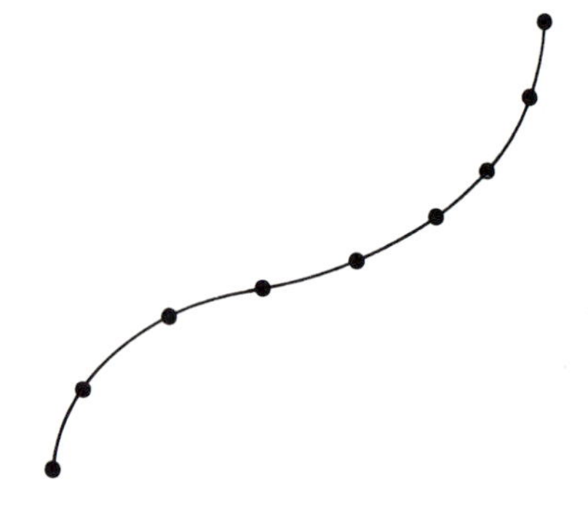

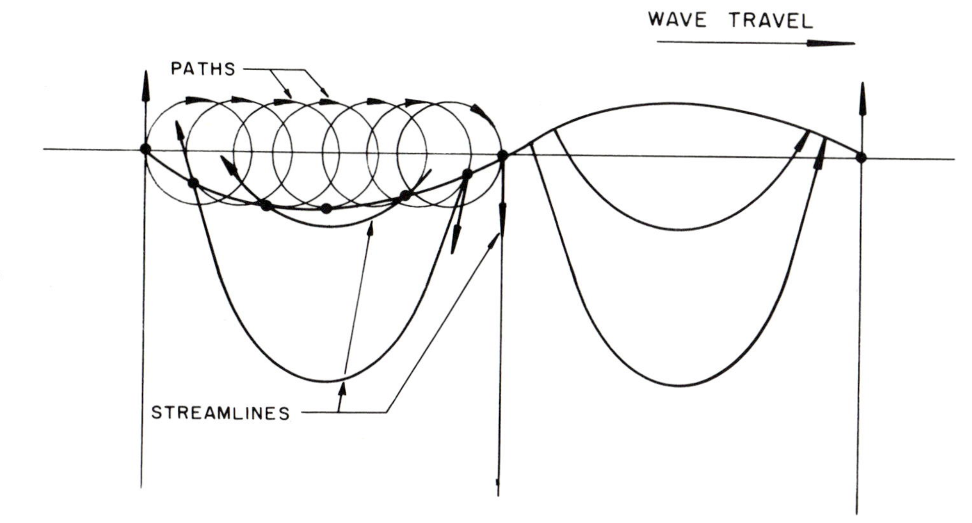

Figure 1-6

Periodic gravity wave in deep water.

1-3 Methods of Study

The motion of a fluid can be studied either by the method of Lagrange or the method of Euler.

1-3.1 Lagrangian Method

The Lagrangian method may be used to answer the question: What occurs to a given particle of fluid as it moves along its own path? This method consists of following the fluid particles during the course of time and giving the paths, velocities, and pressures in terms of the original position of the particles and the time elapsed since the particles occupied their original position. In the case of a compressible fluid, densities and temperatures are also given in terms of the original position and the elapsed time.

If the initial position of a given particle at time t_0 is x_0, y_0, z_0, a Lagrangian system of equations gives the position x, y, z, at the instant t as:

$$x = F_1(x_0, y_0, z_0, t - t_0)$$
$$y = F_2(x_0, y_0, z_0, t - t_0)$$
$$z = F_3(x_0, y_0, z_0, t - t_0)$$

In practice this method is seldom used in hydrodynamics. Lagrangian coordinates are, however, often used in theories relative to periodical gravity waves. The velocity and

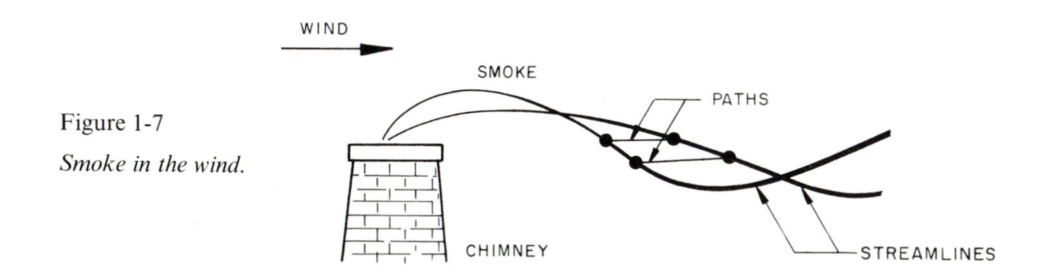

Figure 1-7

Smoke in the wind.

Figure 1-8 (Top) *Streamlines, paths, streaklines for a steady flow around a fixed body.*
(Middle) *Streamlines, paths for an unsteady flow around a body moving at constant velocity in a still fluid.*
(Bottom) *Vectorial relationship between the two kinds of motion: Galilean transformation.*

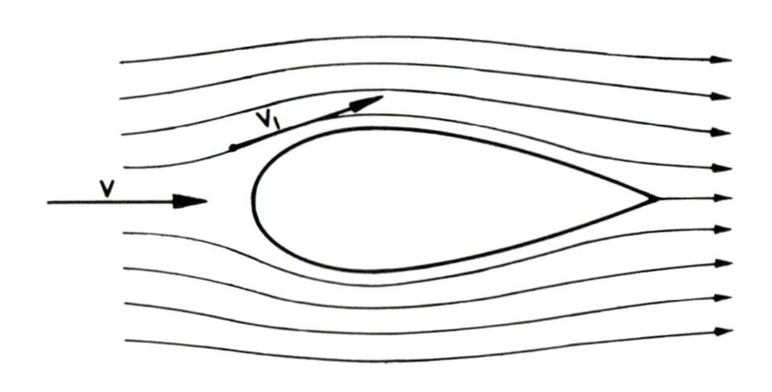

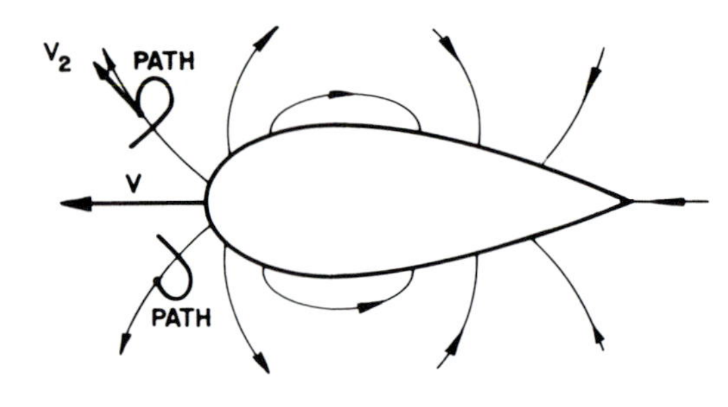

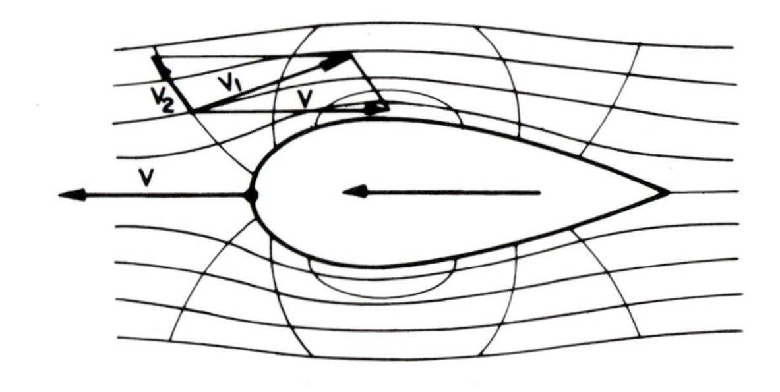

acceleration components at point (x_0,y_0,z_0) are then obtained by a simple partial differentiation with respect to time, such that

$$u = \left.\frac{\partial x}{\partial t}\right|_{x_0,y_0,z_0}$$

$$v = \left.\frac{\partial y}{\partial t}\right|_{x_0,y_0,z_0}$$

$$w = \left.\frac{\partial z}{\partial t}\right|_{x_0,y_0,z_0}$$

Similarly, the acceleration components are $\partial^2 x/\partial t^2$, $\partial^2 y/\partial t^2$, $\partial^2 z/\partial t^2$.

1-3.2 *Eulerian Method*

The Eulerian method may be used to answer the question: What occurs at a given point in a space occupied by a fluid in motion? This is the most frequent form of problem encountered in hydrodynamics. This method gives, at a given point $A(x,y,z)$, the velocity $\mathbf{V}(u,v,w)$ and the pressure p (and, in the case of a compressible fluid, density and temperature) as functions of time t. Since

$$\mathbf{V} = \mathbf{F}(x,y,z,t)$$

then

$$u = f_1(x,y,z,t)$$

$$v = f_2(x,y,z,t)$$

$$w = f_3(x,y,z,t)$$

and

$$p = F_1(x,y,z,t)$$

The Eulerian system of equations is found by a total differentiation of u, v, and w with respect to t and by consideration of the pressure components. In the following example the Eulerian system of coordinates is used.

1-3.3 An Example of Flow Pattern

Let us consider an Eulerian system of coordinates where the two-dimensional wave motion is represented by the velocity components:

$$u = f_1(x,z,t) = \frac{dx}{dt} = \frac{H}{2} ke^{mz} \cos (kt - mx)$$

$$w = f_3(x,z,t) = \frac{dz}{dt} = -\frac{H}{2} ke^{mz} \sin (kt - mx)$$

The equations for the streamlines are obtained from the differential equation

$$\frac{dx}{u(x,z,t_0)} = \frac{dz}{w(x,z,t_0)}$$

Thus,

$$\frac{dx}{k(H/2)e^{mz} \cos (kt_0 - mx)} = \frac{dz}{-k(H/2)e^{mz} \sin (kt_0 - mx)}$$

or

$$dz = -\tan (kt_0 - mx)\, dx$$

If t_0 is taken as 0, this equation becomes:

$$dz = -\tan (-mx)\, dx = \tan mx\, dx$$

The integration of this equation gives

$$e^{mz} \cos mx = \text{constant}$$

By varying the value of the constant the streamlines form the general pattern illustrated in Fig. 1-7.

The paths (or particle orbits) are defined by the differential equation:

$$\frac{dx}{u(x,z,t)} = \frac{dz}{w(x,z,t)} = dt$$

where t is a variable. If it can be assumed that x and z differ little from some given values x_0 and z_0, the differential equation, to a first approximation becomes:

$$dx = k \frac{H}{2} e^{mz_0} \cos (kt - mx_0)\, dt$$

Hence,

$$x - x_i = \frac{H}{2} e^{mz_0} \sin (kt - mx_0)$$

$(z - z_i)$ is found by a similar procedure. It is

$$z - z_i = \frac{H}{2} e^{mz_0} \cos (kt - mx_0).$$

In order to eliminate t, square the equations for $(x - x_i)$ and $(z - z_i)$ and add the results. This gives:

$$(x - x_i)^2 + (z - z_i)^2 = \left[\frac{H}{2} e^{mz_0}\right]^2$$

This is the equation of a circle of radius $(H/2)e^{mz_0}$. It is seen that the paths are circular and the radius tends to zero as $z_0 \rightarrow -\infty$. It will be seen in the linear wave theory that, at a first approximation, one has $x_i \cong x_0$, $z_i \cong z_0$ (Section 16-1), and x_0, z_0 can be considered the location of the fluid particle at rest.

1-4 Basic Equations

1-4.1 The Unknowns in Fluid Mechanics Problems

In general, the density of a liquid is assumed constant so that equations are needed only for velocity and pressure. Hence, in the Eulerian system of coordinates, the motions are completely known at a given point x, y, z if one is able to express $\mathbf{V}$ and p as functions of space and time: $\mathbf{V} = \mathbf{F}(x,y,z,t)$ and $p = F_1(x,y,z,t)$. Therefore, to solve problems in hydrodynamics two equations are necessary, one of them being vectorial. If **V** is expressed by its components u, v, and w, four scalar or ordinary equations are necessary.

In free surface flow problems, the free surface elevation $\eta(x,y,z,t)$ around the still water level, or the water depth $h(x,y,z,t)$, is unknown and a kinematic condition is also required. However, in that case the pressure p is known and in general is equal to the atmospheric pressure.

For gases, two more unknowns need to be considered, namely, the density ρ and the absolute temperature T. Hence, to solve problems in the most general cases of fluid mechanics, four equations are necessary. If **V** is expressed by u, v, and w, then six ordinary equations are needed.

In hydrodynamics, basic equations are given by the physical principles of continuity and conservation of momentum. The equation of state and the principle of the conservation of energy must be added in the case of compressible fluid.

The reduction of a problem to such a small number of variables (2 in hydrodynamics and 4 in gas dynamics), does not occur for trivial reasons, but as a result of several important arguments and assumptions. A number of phenomenological functions are assumed to be known. For example, it is assumed that the fluid is Newtonian and either perfect or viscous, which defines the stress tensor. The fluid obeys Fourier's law of conduction. Also, a number of coefficients, such as heat conductivity, specific heat, and viscosity, are supposed to be known functions of the other unknown variables, such as density and/or temperature.

1-4.2 Principle of Continuity

The continuity principle expresses the conservation of matter, i.e., fluid matter in a given space cannot be created or destroyed. In the case of an incompressible homogeneous fluid, the principle of continuity is expressed by the conservation of volume, except in the special case of cavitation where partial voids appear.

The continuity principle gives a relationship between the velocity **V**, the density ρ, and the space coordinates and time. If ρ is constant (in the case of an incompressible fluid), it gives a relationship between the components of **V** and the space coordinates, which are x, y, z. The equation of continuity then becomes

$$\frac{\partial u}{\partial x} + \frac{\partial v}{\partial y} + \frac{\partial w}{\partial z} = 0$$

as it is demonstrated in Section 3-2.

It will be seen that **V** may be found in some cases of flow under pressure, independent of the absolute value for p, from the principle of continuity alone, but p will always be a function of **V** except at the free surface.

1-4.3 The Momentum Principle

The momentum principle expresses the relationship between the applied forces **F** on a unit volume of matter of density ρ and the inertia forces $d(\rho\mathbf{V})/dt$ of this unit volume of matter in motion. The inertia forces are due to the natural tendency of bodies to resist any change in their motion. It is Newton's first law that "every body continues in its state of rest or uniform motion via a straight line unless it is compelled by an external force to change that state." The well-known Newtonian relationship is derived from his second law: "The rate of change of momentum is proportional to the applied force and takes place in the direction in which the force acts." $\mathbf{F} = d(m\mathbf{V})/dt$.

In fluid mechanics this equation takes particular forms which take into account the fact that the fluid particle may be deformed. These equations will be studied in detail. For an incompressible fluid, the integration of the momentum equation with respect to distance gives an equality of work and energy, expressing a form of the conservation of energy principle.

If **V** is expressed by u, v, w, then Newton's second law has to be expressed along the three coordinate axes. This gives the three equations

$$F_x = \rho\frac{du}{dt} \qquad F_y = \rho\frac{dv}{dt} \qquad F_z = \rho\frac{dw}{dt}$$

where ρ is assumed constant and F_x, F_y, F_z are the components of **F** along the three coordinate axes, respectively.

1-4.4 *Equation of State*

When considering a compressible fluid, two other equations are required in addition to the equations expressing the two above principles. These are the equation of state and the equation expressing the conservation of energy.

The equation of state expresses the relationship which always exists between pressure p, density ρ, and absolute temperature T. For a perfect gas, this equation has the very simple form

$$\frac{p}{\rho g R T} = 1 \qquad \text{or} \qquad \frac{p}{\varpi R T} = 1$$

where R is the universal gas constant ($R = 53.3$ ft/°R for air*) and ϖ is the specific weight.

In a more general case of a real gas, it may take the form $p/\rho g R T = 1 + f_1(T)\rho + f_2(T)\rho^2 + \cdots$ where f_1 and f_2 are functions of the absolute temperature T only. In the case of an incompressible fluid, the equation of state is simply $\rho =$ constant. The temperature can then be treated as an independent variable having an (experimentally) known significant influence on the coefficient of viscosity only.

1-4.5 *Principle of Conservation of Energy*

The next equation expresses the conservation of the total energy (internal energy and mechanical energy). It is the first law of thermodynamics.

The following equation is derived from this law in the particular case of an adiabatic flow—that is, where no heat is added or removed from the fluid mass. In that case: $p/\rho^k =$ constant, where k is the adiabatic constant defined as the ratio of the specific heat at constant pressure C_p to the specific heat at constant volume C_v.

* °R are degrees *Rankine*, equal to degrees Fahrenheit plus 459.58°.

In the case of isothermal flow at constant temperature which may necessitate the removal or addition of heat from/to the fluid mass, $p/\rho =$ constant.

Inasmuch as hydrodynamic problems alone are being considered in this book, it is not necessary to further consider the equation of state and the equation of conservation of total energy. The density ρ will be assumed known and constant and the temperature T will be a variable without influence upon the phenomenon under consideration. However, it is evident that the dissipation of energy by viscous forces may create a (small) elevation of temperature which in turn modifies the characteristics of the fluid. In general, these effects are of secondary importance in hydrodynamics, and in particular, the coefficient of viscosity μ is considered as a known constant.

1-4.6 *Boundary Conditions*

1-4.6.1 It is evident that a general solution of the system of equations described above does not exist, but in many particular cases solutions can be found when the boundary conditions are specified. There are three main kinds of boundary conditions:

1. At a free surface where the pressure is known and generally equal to atmospheric pressure. The cases of wind–water wave interaction and impulses on the free surface are special cases where the variation of the pressure at free surface is taken into account.
2. At a solid boundary, since the fluid cannot pass through or escape from the boundary.
3. At infinity when the motion tends to a known value. In such a case, the known conditions at infinity are considered as "boundary" conditions.

1-4.6.2 At the *free surface* the pressure is known, but the location of this free surface with respect to horizontal datum level is unknown in general. So two conditions must be specified: a dynamic condition, stating the value of pressure, and a kinematic condition, stating that the particle at the free surface remains at the free surface.

Since p is normally a constant at any time, the total differential of $p(x,y,z,t)$ is zero; that is,

$$dp = \frac{\partial p}{\partial x}dx + \frac{\partial p}{dy}dy + \frac{\partial p}{\partial z}dz + \frac{\partial p}{\partial t}dt = 0$$

The total derivative of p with respect to t is given as:

$$\frac{dp}{dt} = \frac{\partial p}{\partial t} + \frac{\partial p}{\partial x}\frac{dx}{dt} + \frac{\partial p}{\partial y}\frac{dy}{dt} + \frac{\partial p}{\partial z}\frac{dz}{dt} = 0$$

If the variables $u = dx/dt$, $v = dy/dt$, and $w = dz/dt$, are used (see Section 1-2.2), the free surface dynamic condition becomes in the most general case:

$$\frac{\partial p}{\partial t} + u\frac{\partial p}{\partial x} + v\frac{\partial p}{\partial y} + w\frac{\partial p}{\partial z} = 0$$

This condition, involving a force, has to be used with the equation expressing the momentum principle.

The kinematic condition will be developed in Section 16-1.3.2. For the time being, it is sufficient to know that if

$$z = \eta(x,y,t)$$

is the equation of the free surface, the kinematic condition is:

$$w = \frac{\partial \eta}{\partial t} + u\frac{\partial \eta}{\partial x} + v\frac{\partial \eta}{\partial y}$$

1-4.6.3 At *fixed solid boundaries*, friction reduces the velocity to zero, so that $\mathbf{V} = 0$. This condition is used in the continuity equation, and since a friction force is involved, it must also be used in the momentum equation. If the fluid is assumed to be perfect (or ideal), only the component perpendicular to the boundary is zero. The component of the velocity $\mathbf{V}$ which is tangential to the boundary remains. It is used primarily with the continuity relationship. It does not involve a force but a continuity statement: the fluid cannot pass through or escape from the boundary (unless there is cavitation).

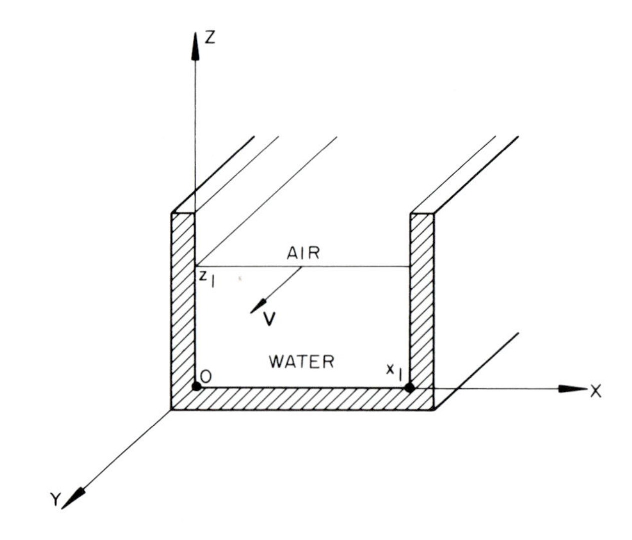

Figure 1-9 *Uniform flow in a rectangular channel.*

For instance, the boundary conditions in the case shown in Fig. 1-9 are:

$$u = 0 \text{ for } x = 0 \text{ and } x = x_1$$
$$w = 0 \text{ for } z = 0$$
$$p = \text{constant for } z = z_1$$

More generally, if $F(x,y,z) = \text{constant}$ is the equation of the boundary, the following boundary condition expresses the fact that the surface F and $\mathbf{V}$ are tangential at any point,

$$u\frac{\partial F}{\partial x} + v\frac{\partial F}{\partial y} + w\frac{\partial F}{\partial z} = 0$$

i.e.,

$$\mathbf{V} \cdot \mathbf{grad}\, F = 0$$

1-4.6.4 At *movable solid boundaries* (wheel of turbine, wave paddle, etc.) the boundary condition expresses the fact that the fluid follows the boundary (see Fig. 1-10). Thus, the velocity component of the fluid perpendicular to the boundary is equal to the corresponding component

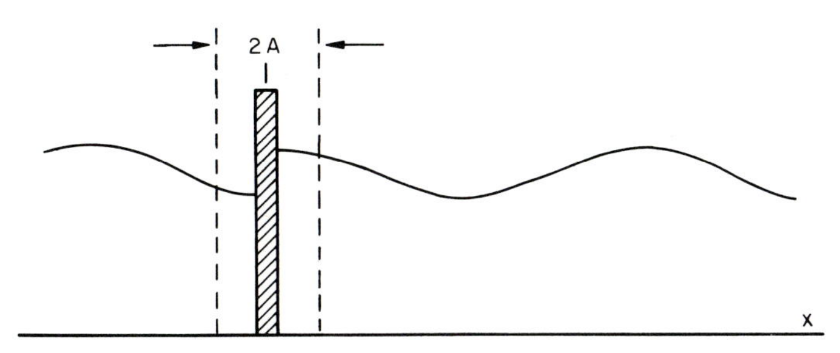

Figure 1-10 *A piston wave paddle gives a movable boundary condition.*

of the boundary itself. The other component follows the corresponding boundary motion component for a real fluid only.

If $F(x,y,z,t)$ = constant is the equation of the movable boundary, the following boundary condition expresses the fact that the fluid remains at the boundary:

$$\frac{\partial F}{\partial t} + u\frac{\partial F}{\partial x} + v\frac{\partial F}{\partial y} + w\frac{\partial F}{\partial z} = 0$$

1-4.6.5 An *infinite distance* can give a boundary condition if the motion tends to a well known value far from the studied space. For example, consider the diagram shown in Fig. 1-11. The motion is known at infinity and can be written (as far as friction effect is negligible) $\mathbf{V}$ = constant for $x \to \pm\infty$.

As another example, it is well known that the motion of swell in deep water is limited to a zone near the free

Figure 1-11 *Flow in a pipe past a diaphragm:* $V = V_0$ *when* $x \to \pm\infty$.

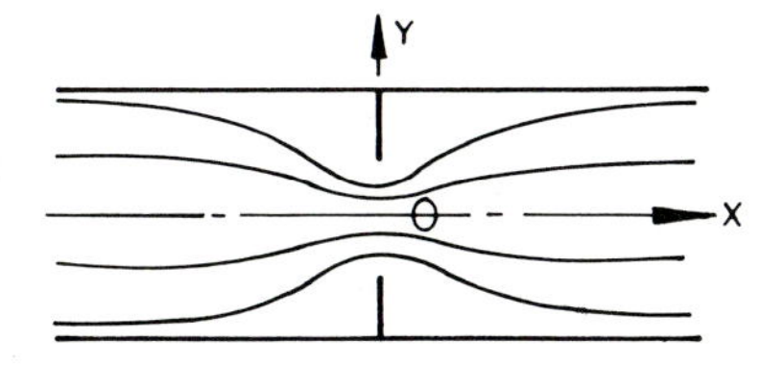

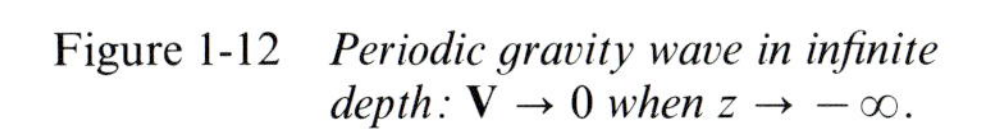

Figure 1-12 *Periodic gravity wave in infinite depth:* $\mathbf{V} \to 0$ *when* $z \to -\infty$.

surface. Hence, the periodic gravity wave theory in infinite depth is based on the boundary condition $\mathbf{V} \to 0$ when the distance from the free surface tends to infinity: $z \to -\infty$ (Fig. 1-12).

PROBLEMS

1.1 Consider a two-dimensional flow motion defined by the velocity components:

$$u = A + Bt \qquad v = C$$

where A, B, and C are constant parameters. Demonstrate that the streamlines are straight lines and that the particle paths are parabolas.

1.2 A disk of radius R rolls without slipping on a horizontal plane at a constant angular velocity k. Demonstrate that the "streamlines" are circular and that the paths are trochoidal.

1.3 Consider a fixed cylinder in a uniform current of constant velocity. It will be assumed that there is no separation. Sketch the streamlines, the paths, and the streaklines intuitively. Consider now a cylinder moving at constant velocity in still water, and sketch the streamlines, the paths, and the streaklines. Explain the differences between the two cases, considered as a steady and an unsteady motion, respectively.

1.4 Devise a general graphical method of construction to determine a steady flow pattern around a body moving at constant speed in a still fluid from the pattern of streamlines defined with respect to a fixed coordinates system and vice versa.

1.5 A two-dimensional flow motion (linear periodic gravity waves in water depth, d) is defined in a Lagrangian system of coordinates by the equations:

$$x = x_i + \frac{H}{2}\frac{\cosh m(d+z)}{\sinh md}\sin(kt - mx)$$

$$z = z_i + \frac{H}{2}\frac{\sinh m(d+z)}{\sinh md}\cos(kt - mx)$$

where H is the wave height; m, k, and d are constants. ($m = 2\pi/L$; L is the wavelength; $k = 2\pi/T$, T is the wave period; and d is the water depth.) Find the equations for the streamlines, and sketch them. Assuming $|x - x_0|$ and $|z - z_0|$ are small, find the approximate equation for the particle paths. Sketch these paths.

1.6 Express mathematically the boundary conditions for any kind of flow motion taking place between the boundaries defined by Fig. 1-13. A hinged paddle will be assumed to have a small sinusoidal motion of amplitude e at the free surface.

1.7 Consider a two-dimensional body moving at velocity U in the negative X direction. The nose of this body can be defined by a curve such that $y = x^{1/3}$, and u and v are the components of velocity along the body. Establish the relationship between u, v, U, and y.

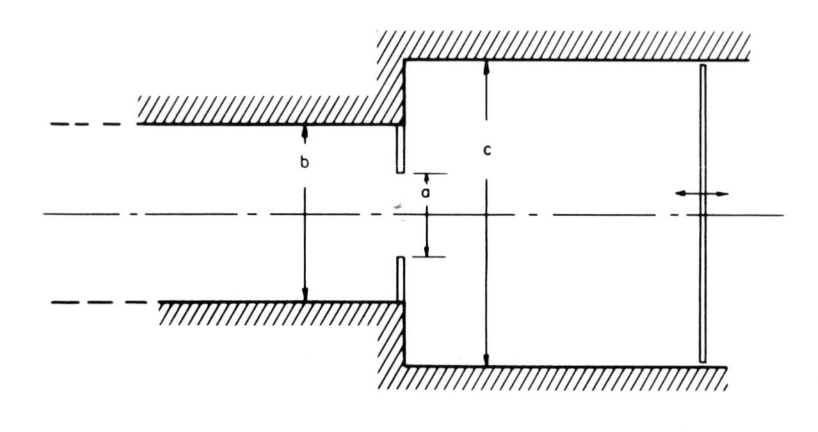

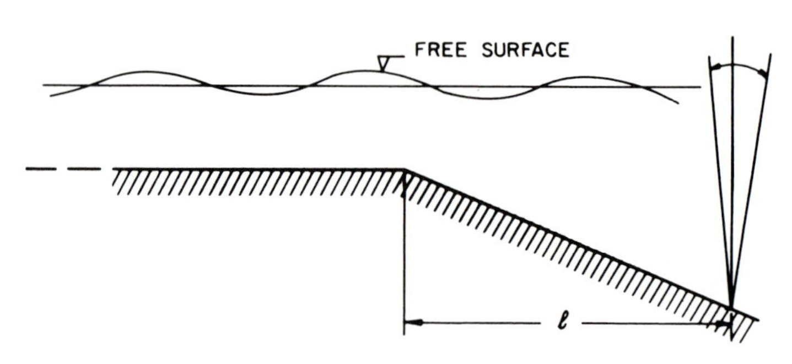

Figure 1-13

Wave basin, top and side view.

Then consider the case where the body is fixed and the fluid is moving at a velocity U.

1.8 Consider a translatory wave in a channel moving without deformation at a constant velocity C in the negative X direction. At a given time t the wave profile is defined approximately by the relationship $z = Ax^{1/2}$ where A is a constant. Demonstrate that the free surface velocity components u_s and w_s are related by the equation

$$w_s = (u_s - C)\frac{A^2}{2z}$$

1.9 A sphere of radius R is moving at a velocity $\mathbf{U}(u_s,v_s,w_s)$ through a fluid at rest. Establish the equation for the boundary condition in the case of a perfect fluid.

1.10 Draw the streamlines and paths of monochromatic periodic waves as given in Problem 1.5, relative to a system of Cartesian coordinates moving at speed $C = k/m$ in the wave direction. The free surface is defined by $z \cong 0$ and the bottom by $z = -d$. Are they streamlines?

Chapter 2

Motions of Fluid Elements; Rotational and Irrotational Flow

2-1 Introduction to the Different Kinds of Motion

In mathematical terms, the motion of the fluid elements along their own paths is considered as the superposition of different kinds of primary motions. The physical interpretation of these motions is given first by considering the simple case of a two-dimensional fluid element, where all velocities are parallel to the OX axis and depend only on y (like a laminar flow between two parallel planes).

Consider the infinitesimal square element $ABCD$ of area $dx\,dy$ at time t and the same element at time $t + dt$: $A_1B_1C_1D_1$ (Fig. 2-1).

The velocity of A and D is u, and the velocity of B and C is $u + du = u + (\partial u/\partial y)\,dy$ since $AB = dy$, and u in this case is a function of y only.

It is possible to go from $ABCD$ to $A_1B_1C_1D_1$ in three successive steps

1. A translatory motion which gives $A_1B_2C_2D_1$; the speed of translation is u
2. A rotational motion which turns the diagonals A_1C_2 and D_1B_2 to A_1C_3 and D_1B_3, respectively
3. A deformation which displaces C_3 to C_1 and B_3 to B_1

If in the limit dt tends to zero, C_1C_2 tends to zero. If this occurs the angle $C_2C_1C_3$ will tend to 45° when $dx = dy$. Hence

$$C_2C_3 = \frac{C_1C_2}{\sqrt{2}} = \frac{(\partial u/\partial y)\,dy\,dt}{\sqrt{2}}$$

The rate of angular rotation is:

$$\frac{dr}{dt} = \frac{d}{dt}\left\{\frac{\text{segment}}{\text{radius}}\right\} \simeq \frac{d}{dt}\frac{C_2C_3}{A_1C_2} = \frac{d}{dt}\frac{C_2C_3}{\sqrt{2}dy}$$

Introducing the value C_2C_3 previously given, it is found that the rate of angular rotation is:

$$\frac{dr}{dt} = \frac{1}{2}\frac{\partial u}{\partial y}$$

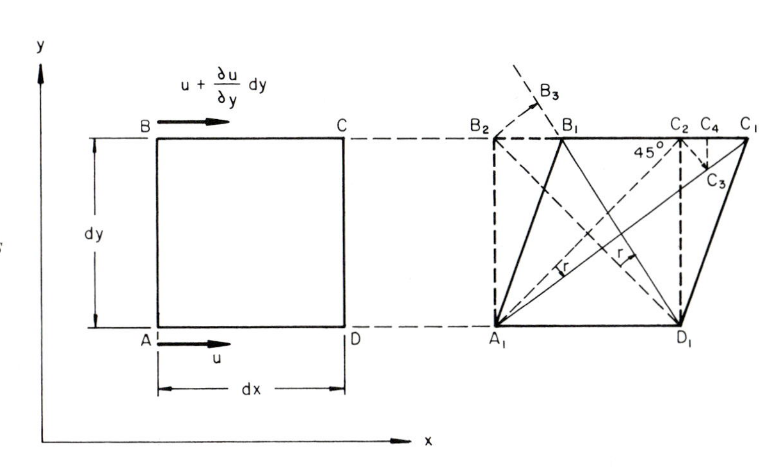

Figure 2-1

Elementary analysis of different kinds of motion of a fluid particle.

Similarly, the rate of deformation would be found to be equal to:

$$\frac{\partial}{\partial t}\left(\frac{C_3C_1}{A_1C_3}\right) = \frac{1}{2}\frac{\partial u}{\partial y}$$

In the general case, there are three major constituents of particle motions and deformations. They are:

1. The velocity components $\mathbf{V}(u,v,w)$: translation
2. The variation of velocity components in their own direction: dilatation
3. The variation of velocity components with respect to a direction normal to their own direction: rotation and angular deformation

These three constituents are successively discussed in the following sections.

2-2 Translatory Motion

Consider the particle at the point $A(x,y,z)$ at time t. The point is a corner of a small rectangular element, the edges of which are parallel to the three axes OX, OY, OZ, respectively (Fig. 2-2). When the particle moves so that the edges of the rectangular elements remain parallel to these axes, and maintain a constant length, it is a translatory motion only. This implies no space dependence of the velocity components. The translation can be along a straight line or a curved line.

If x, y, and z are the coordinates of A at time t, then $x + \Delta x$, $y + \Delta y$, and $z + \Delta z$ are the coordinates at time $t + \Delta t$. The translatory motion is defined by the equations

$$\Delta x = u\,\Delta t \qquad\qquad dx = u\,dt$$
$$\Delta y = v\,\Delta t \qquad \text{or} \qquad dy = v\,dt$$
$$\Delta z = w\,\Delta t \qquad\qquad dz = w\,dt$$

The flow of particles along parallel and straight streamlines with a constant velocity (so-called uniform flow) is a case of translatory motion only (Fig. 2-3).

The translatory motion may be defined more rigorously as the motion of the center of the rectangular element

Figure 2-2 *Translatory motion.*

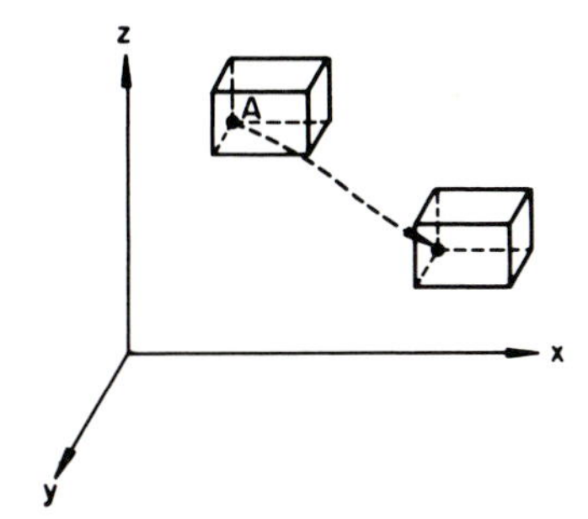

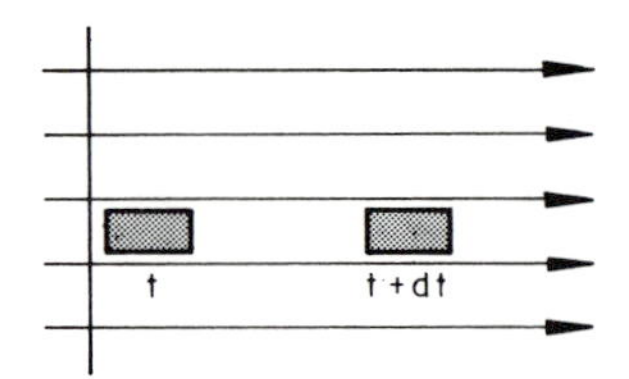

Figure 2-3 *An example of translatory motion: uniform flow.*

instead of the motion of the corner of the element. However, this change complicates slightly the development of figures and equations and gives, finally, the same result. Hence in the following discussion, translatory motion will be defined as the motion of a corner.

In the following, the physical meanings and the corresponding mathematical expressions are studied in the case of a two-dimensional motion at first, then they are generalized for a three-dimensional motion.

2-3 Deformation

It is easier to explain this kind of motion with the aid of an example. Two kinds of deformation have to be distinguished: dilatational deformation and angular deformation.

2-3.1 Dilatational or Linear Deformation

In a converging flow, the velocity has a tendency to increase along the paths of particles. Therefore, the velocities of the edges perpendicular to vector **V** (or to the streamlines) are not the same (Fig. 2-4). The particle becomes longer and thinner. It is a case of dilatational or linear deformation superimposed on a translation only provided the angles between the edges do not change.

Figure 2-4 *Dilatational deformation of fluid particle in a convergent.*

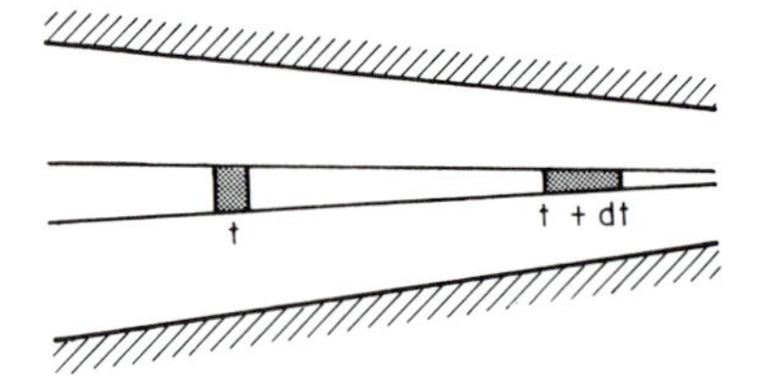

Now consider the two-dimensional particle $ABCD$ of which the velocity in the x direction of the edge AB is u, and the velocity of the edge CD is $u + du = u + (\partial u/\partial x)\,dx$, since $AD = dx$ (Fig. 2-5). Similarly, the velocity of AD in the y direction is v, and the velocity of BC is: $v + (\partial v/\partial y)\,dy$. Note that here the derivatives of u with y or v with x are not considered and the derivatives of velocity $(\partial u/\partial x)\,dx$ and $(\partial v/\partial y)\,dy$, do not depend upon time. The velocities of dilatational deformations are $(\partial u/\partial x)\,dx$ and $(\partial v/\partial y)\,dy$. After a time dt, BC becomes $B'C'$, the length BB' being equal to the product of the change of velocity and the time, that is, $BB' = (\partial v/\partial y)\,dy\,dt$. [The velocity $(\partial v/\partial y)\,dy$ is negative in the case of Fig. 2-5.] CD becomes $C'D'$ and similarly DD' is equal to: $DD' = (\partial u/\partial x)\,dx\,dt$.

The velocities of dilatational deformation are per unit length:

$$\frac{(\partial u/\partial x)\,dx}{dx} = \frac{\partial u}{\partial x} \qquad \frac{(\partial v/\partial y)\,dy}{dy} = \frac{\partial v}{\partial y}$$

The sum $\partial u/\partial x + \partial v/\partial y$ is the total rate of dilatational deformation, i.e., the rate of change of area per unit area. Areas $BCEB'$ and $D'C'ED$ must be equal in the case of an

Figure 2-5 *Components of dilatational deformation.*

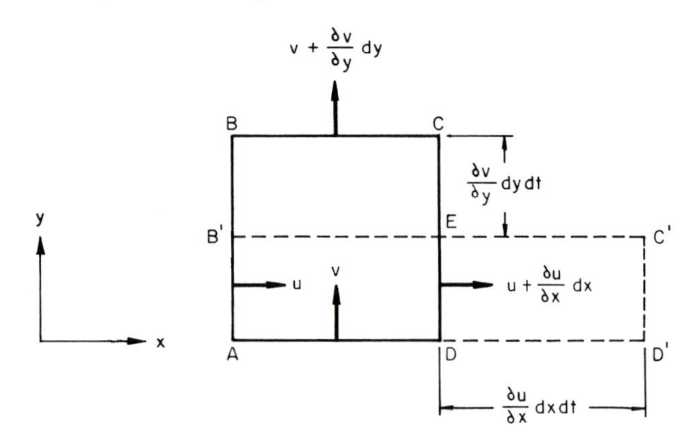

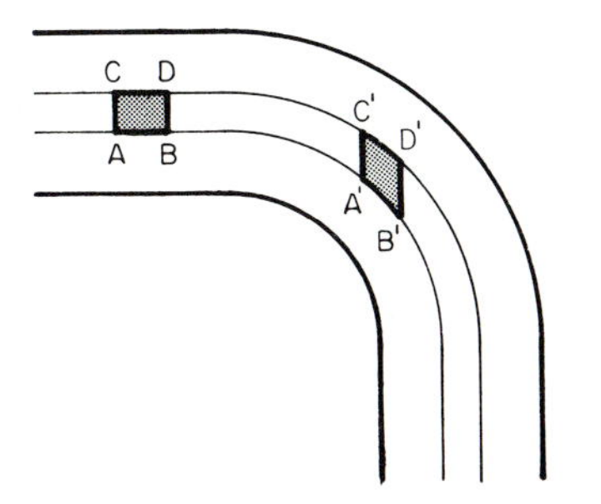

Figure 2-6 *Shear deformation in a bend.*

incompressible fluid. Their difference gives the rate of expansion or compression in the case of a compressible fluid.

2-3.2 Angular Deformation or Shear Strain

Angular deformation may be illustrated by the behavior of a fluid particle flowing without friction around a bend. It is a matter of common observation that it is windier at a corner than it is in the middle of a street. In the similar case of fluid flow around a bend, neglecting the effects of friction, the velocity has a tendency to be greater on the inside than it is on the outside of the bend. The law $V \times R =$ constant may be approximately applied where V is the velocity and R is the radius of curvature of the paths. Hence, if the particle A is the corner of the rectangle $ABCD$, the edge AB of the rectangle moves at a greater velocity than the edge CD and there is angular deformation (Fig. 2-6). This angular deformation is proportional to the difference of velocity between AB and CD.

Now considering, for example, the case presented in Fig. 2-7, in which the velocity of AB is u, and the velocity of CD is $u + du = u + (\partial u/\partial y)\,dy$, then the distance CC' (or DD') after a time dt is $(\partial u/\partial y)\,dy\,dt$. The angular velocity is

$$\frac{(\partial u/\partial y)\,dy}{dy} = \frac{\partial u}{\partial y}$$

Note that in contrast to the case of dilatational deformation, the derivatives of u with y and v with x are retained here. The derivative of the velocity $(\partial u/\partial y)\,dy$ does not depend upon time.

Similarly BB' (or DD'') is equal to $(\partial v/\partial x)\,dx\,dt$. When these two deformations exist at the same time, the sum of the angular velocities $(\partial u/\partial y) + (\partial v/\partial x)$ is the rate of angular deformation.

Note that $\partial u/\partial y$ was chosen equal to $\partial v/\partial x$ in Figure 2-7, and the bisectors of the angles made by the edges of the square element tend to remain parallel to their initial

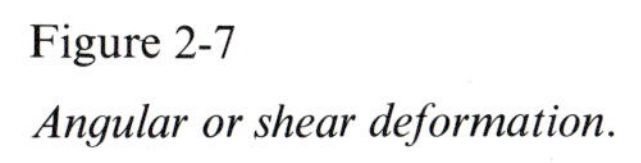

Figure 2-7

Angular or shear deformation.

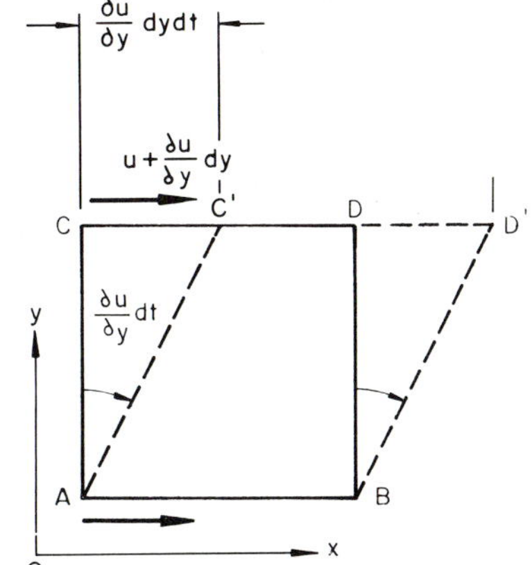

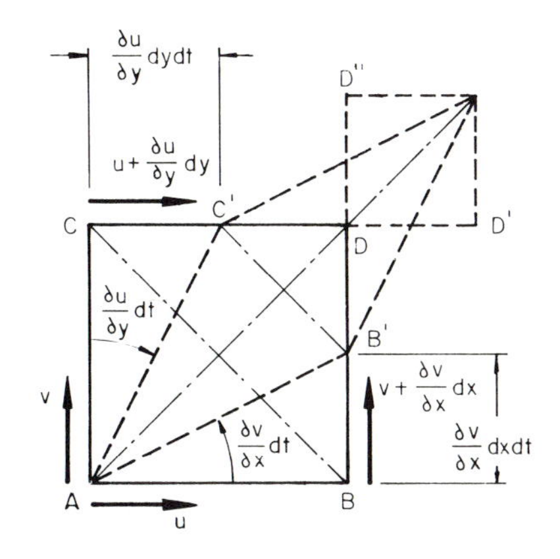

positions during the angular deformation. When the bisectors do not remain parallel to their initial positions, the motion is said to be rotational.

2-4 Rotation

Although flow motions can be classified in various ways according to some of their typical characteristics (such as laminar or turbulent, frictionless or viscous, with or without friction, steady or unsteady), one of the most important divisions in hydrodynamics consists of considering whether a flow is rotational or irrotational. Hence, the abstract concept of irrotationality is fully developed in the following sections.

2-4.1 Mathematical Definitions

For a two-dimensional motion, it has been shown that the angular velocities of deformation are $\partial u/\partial y$ and $\partial v/\partial x$. The rotation of a particle is proportional to the difference between these components. Indeed, if $\partial u/\partial y = \partial v/\partial x$, there is angular deformation without rotation and the bisectors do not rotate (Fig. 2-7). But if $\partial u/\partial y \neq \partial v/\partial x$, the bisectors change their direction, and there is either both rotation and angular deformation, or rotation only (Fig. 2-8).

The difference $(\partial u/\partial y) - (\partial v/\partial x)$ defines the rate of rotation, and therefore, a two-dimensional irrotational motion is defined mathematically by $(\partial u/\partial y) - (\partial v/\partial x) = 0$.

Angular deformation can be considered without rotation when $(\partial u/\partial y) - (\partial v/\partial x) = 0$ and $(\partial u/\partial y) + (\partial v/\partial x) \neq 0$, and theoretically, rotation can exist without deformation when $(\partial u/\partial y) - (\partial v/\partial x) \neq 0$ and $(\partial u/\partial y) + (\partial v/\partial x) = 0$. This case is rare in practice, since rotation generally occurs with angular deformation in physical situations. A forced vortex, such as that schematically shown on Fig. 2-9, is a rare case in which particles rotate without deformation. However, this can be considered more as a special case of hydrostatics where the centrifugal force is added to gravity, rather than a real rotational flow.

2-4.2 Velocity Potential Function: Definition

The concept of irrotational motion is very important in hydrodynamics since many real flows are nearly irrotational.

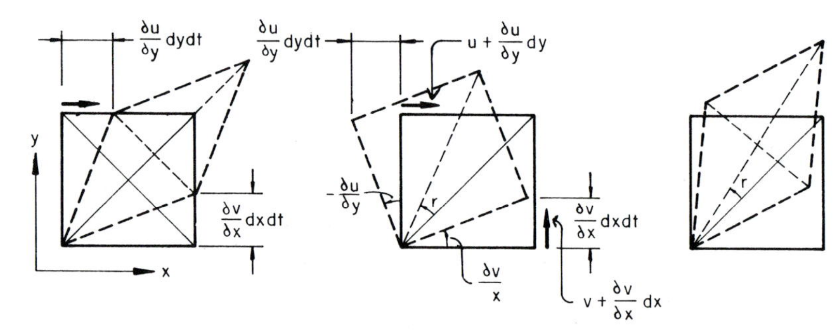

Figure 2-8
Rotation and deformation.

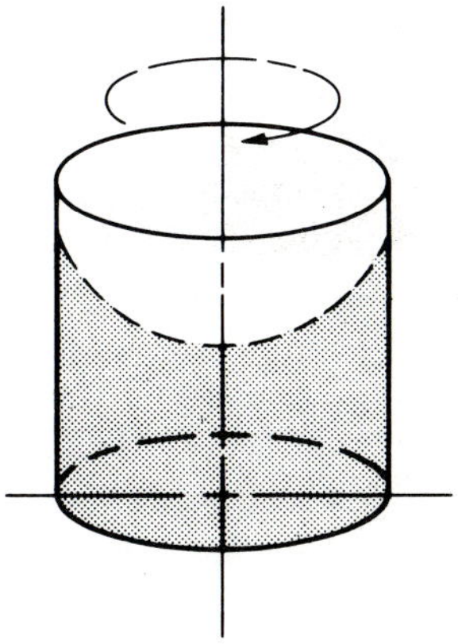

Figure 2-9 *Forced vortex ($V = KR$), rotation without deformation.*

The properties of irrotational motion lead to a number of simple and powerful analytical, graphical, or analog methods which can be used in the solution of hydraulic problems. Most of these methods result from the existence of a special function, the velocity potential.

The velocity potential is defined as a single-valued function ϕ such that $u = (\partial\phi/\partial x)$ and $v = (\partial\phi/\partial y)$ [or alternately $u = -(\partial\phi/\partial x)$, $v = -(\partial\phi/\partial y)$]. If the functions u and v are continuous, this function will satisfy the irrotationality condition, which in two dimensions is $(\partial u/\partial y) - (\partial v/\partial x) = 0$. When the expressions for u and v are substituted into the condition for irrotationality, the results are

$$\frac{\partial^2\phi}{\partial y\,\partial x} - \frac{\partial^2\phi}{\partial x\,\partial y} \equiv 0$$

since the differentiation with respect to two variables is independent of the order in which the differentiation is done. The velocity potential ϕ will be shown to exist for three-dimensional motion as well.

The value of the velocity **V** in terms of velocity potential function ϕ is

$$\mathbf{V} = \mathbf{grad}\,\phi = \mathbf{i}\frac{\partial\phi}{\partial x} + \mathbf{j}\frac{\partial\phi}{\partial y}$$

where **i** and **j** are the unit vectors along the X and Y axis, respectively. The magnitude of the velocity becomes

$$V = \left[\left(\frac{\partial\phi}{\partial x}\right)^2 + \left(\frac{\partial\phi}{\partial y}\right)^2\right]^{1/2}$$

2-4.3 Theoretical Remark on Irrotational Flow

It is useful to study the characteristics of an irrotational flow. For this purpose, the previous example of a flow without friction in a bend, or of a free vortex motion defined by the equation $VR = K$, is analyzed (see Fig. 2-10).

Consider the rectangular fluid element $ABCD$ between two path lines defined by their radius of curvature R_1 and R_2 such that $R_2 = R_1 + dR$, dR being infinitely small.

After an interval of time dt, $ABCD$ becomes $A'B'C'D'$, and

$$AA' = CC' = V_1\,dt = K\frac{dt}{R_1}$$

$$BB' = DD' = V_2\,dt = K\frac{dt}{R_2}$$

Figure 2-10 *The case of an infinitely small displacement in an irrotational flow.*

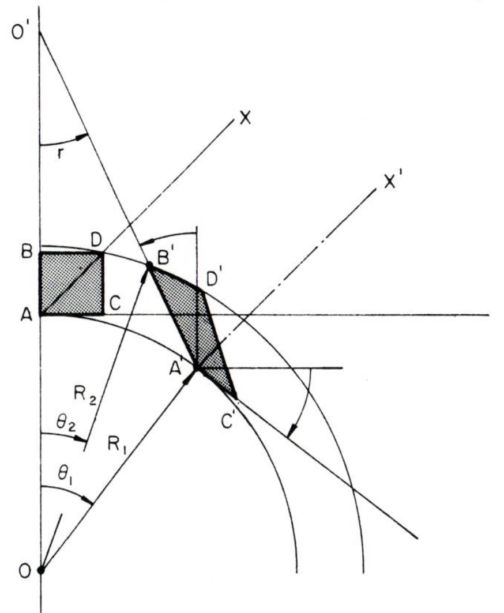

The side AB rotates to $A'B'$ by an infinitely small range r such that

$$r \simeq \tan r \simeq \frac{BB'}{O'B} \simeq \frac{K\,dt}{R_2\,O'B}$$

and

$$r \simeq \tan r \simeq \frac{AA'}{O'A} \simeq \frac{K\,dt}{R_1\,O'A}$$

Equating these last expressions leads to:

$$O'A - O'B = R_2 - R_1 \qquad \frac{R_2}{R_1} = \frac{O'A}{O'B}$$

or $O'B = R_1$ and $O'A = R_2$. When these values are substituted into the equation for r the result is: $r = (K\,dt/R_1R_2)$. Since dR is small, $R_2 \sim R_1$ and the equation can be written as: $r = (K\,dt/R_1^2)$. Since θ_1 is small, $\sin\theta_1 \simeq \theta_1$, and $\theta_1 = (AA'/R_1) = (K\,dt/R_1^2)$, hence $r = \theta_1$.

The side AC rotates into $A'C'$ through the angle θ_1. Since the two sides AB and AC rotate by the same quantity θ_1, but in opposite directions, the bisector AX remains parallel to the bisector $A'X'$. The orientation of this median line remains unchanged, which is the condition for the motion to be irrotational.

It must be emphasized that the previous demonstration holds true only when an infinitely small displacement is considered. It does not hold true for a finite displacement, since the two bisectors have a tendency to rotate in the same direction.

Both the angle of rotation of the bisectors and the angle of angular deformation have finite values for a finite displacement of the element. They both tend to be infinitesimal when the displacement tends to zero. However, in an irrotational motion, the angle of rotation is an infinitesimal of higher order than the angle of deformation. In a real flow, the irrotationality of the motion cannot be determined by observing the deformation of a "particle" in motion along its path since this characteristic is essentially local.

2-5 Practical Limit of Validity of Irrotationality

2-5.1 Rotation Caused by Friction: The Kelvin Theorem

2-5.1.1 In practice it is very important to know when the motion of the fluid particles can be considered as an irrotational motion. Only if the assumption of irrotationality is valid can the powerful methods of calculation based on velocity potential, conformal mapping, relaxation methods, flow nets, electrical analogy, etc, be applied successfully.

The concept of irrotationality is essentially mathematical [$(\partial u/\partial y) - (\partial v/\partial x) = 0$ in the case of two-dimensional motion]. The difficulty arises when one tries to establish some simple practical rules for assessing the validity of this assumption. Indeed, rotation is often caused by viscous forces, but a rotational solution also exists for a perfect fluid, and irrotational flows exist in a viscous fluid.

For example, let us consider a reservoir, where the flow velocity is practically zero, and a connected duct. Initially the fluid is irrotational, but viscous stresses eventually cause the flow to become rotational at the entrance of the duct; here friction forces cause rotation. This experimental fact is translated mathematically by the Kelvin theorem which applies in the case of a viscous fluid of constant density, under a constant gravity force. An exact demonstration of the theorem is beyond the scope of this book but a physical introduction to rotation will suffice in the following.

2-5.1.2 It is easy to see whether a motion is physically rotational or irrotational by a consideration of friction effects. A physical understanding can be gained by the following considerations. Near a fixed boundary, along which particle velocities are zero, particles on adjacent paths have significantly different velocities; hence, a line joining at one time two particles on the same path will rotate much less than that of a line joining two particles on adjacent paths. The difference of direction of the friction forces acting on the two opposite sides of the particle,

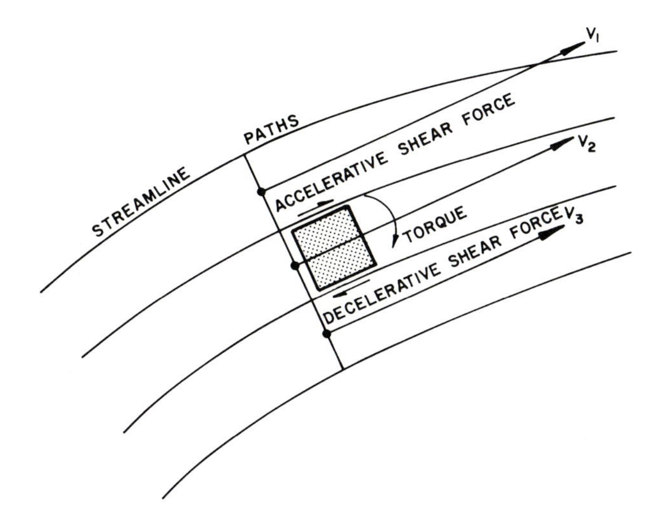

Figure 2-11 *The variation of velocity in a direction perpendicular to the flow induces a difference in the direction of the friction forces and a torque resulting in a rotational motion.*

causes a torque resulting in a net rotation (Fig. 2-11). Figures 2-12 and 2-13 illustrate some cases where it is possible to see easily whether the assumption of irrotational motion is permissible.

In general, the motion can be assumed to be irrotational when the velocity gradient is small (such as in periodic gravity wave), when streamlines converge rapidly, and when the velocity distribution depends on the shape of the boundaries and not on their roughness. Motion is rotational in the neighborhood of boundaries or in a diverging flow.

As mentioned previously, close to boundaries the large velocity differentials between particles on adjacent paths cause the motion to be rotational. A region of high velocity gradients may be adjacent to a region where the velocity gradients are sufficiently small for the motion to be treated mathematically as irrotational. The region of high gradients is called, for instance, a boundary layer if it occurs near a solid boundary or between fluids of different nature (liquid–gas interface) or shear layer between two fluid streams. A motion may be considered irrotational only if the boundary layer is of little importance, i.e., relatively thin. Figure 2-14 illustrates the case of a weir where the boundary layer thickness increases downstream. The motion is irrotational only near the top.

2-5.2 Rotational Solutions in a Perfect Fluid

It is seen that the rotation may physically be due to the friction. Physical considerations of friction effects have resulted in practical rules to follow.

However, there exist mathematical solutions of rotational motion where the friction forces are neglected. The classical Bernoulli equation of elementary hydraulics is valid only along a streamline when motion is rotational without friction (see Chapter 10). One specific example of nondissipative (i.e., without friction) rotational motion is

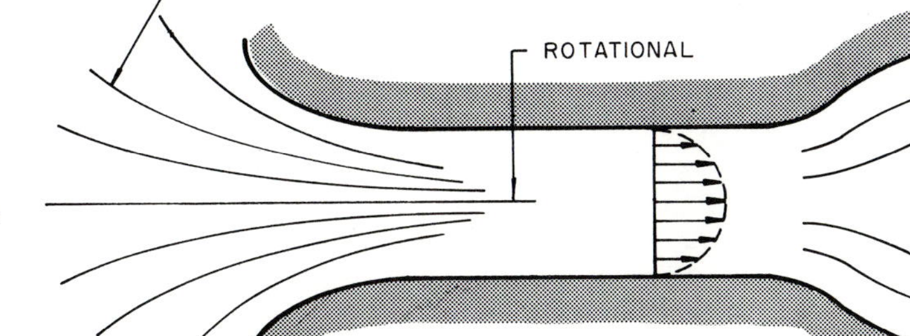

Figure 2-12
Examples of rotational and irrotational motion.

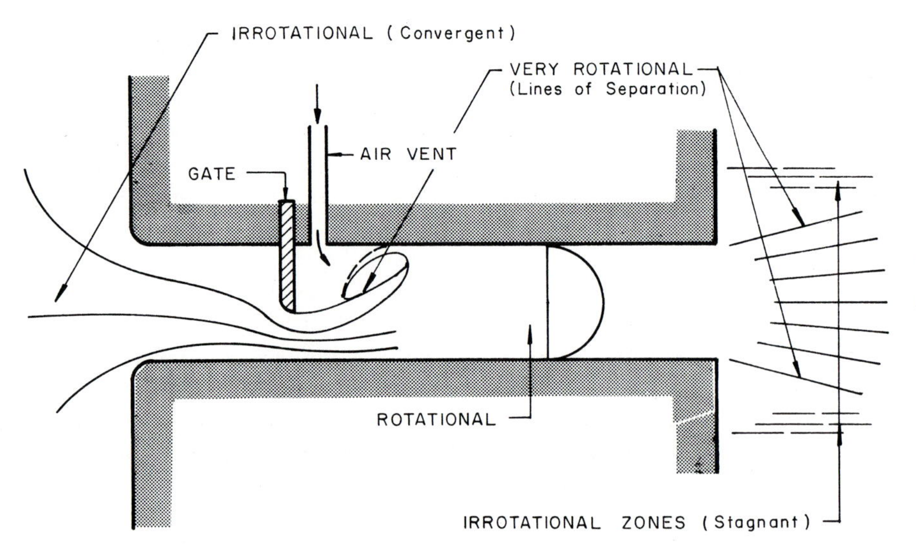

Figure 2-13
Examples of rotational and irrotational motion.

the Gerstner's theory on periodic gravity waves. In this theory the paths of fluid particles describe circles. The particles also rotate about themselves in the opposite direction (Fig. 2-15). The results are expressed by an exact mathematical solution of the basic equations in which the friction terms have been neglected, but in which the inertial rotational terms are taken into account exactly (see Section 17-1.4).

2-5.3 *Irrotational Solutions in Viscous Fluids*

One also finds dissipative motions, which are considered as irrotational. For example, friction forces have a predominant effect on such phenomena as the damping of a

Figure 2-14 *Effect of the boundary layer.*

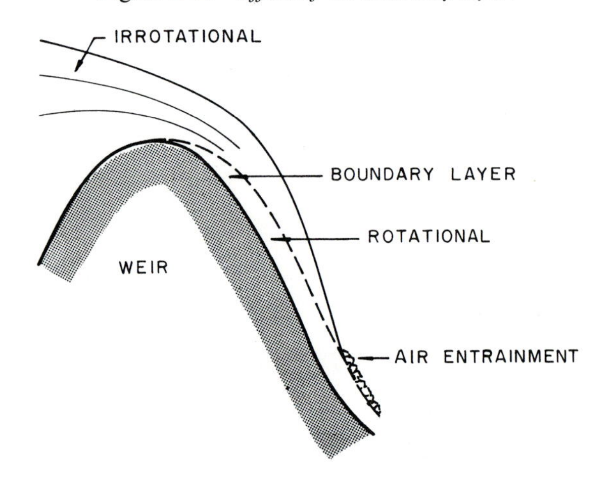

Figure 2-15 *Path and rotation of a fluid particle in a Gerstner wave.*

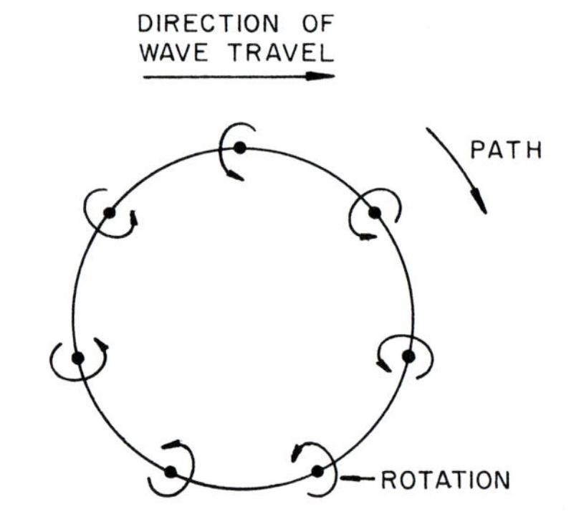

gravity wave though a filter and the flow through a porous medium. However, in these cases, only the mean velocity with respect to space is considered. The actual system of complicated rotational motions through the porous medium is studied as an average motion with respect to space which is irrotational at low Reynolds number (see Chapter 9). Similarly, turbulent flow is strongly rotational but the mean motion with respect to time may often be considered as irrotational (see Chapter 8).

It may also happen that the flow is irrotational when the sum of all the viscous terms which appear in the momentum equation equal zero, although each term individually is different from zero. Such kinds of motion are dissipative and irrotational. A specific example of such a case is the motion generated by a circular cylinder rotating steadily about its axis in an unbounded viscous incompressible fluid. The velocity gradient normal to streamlines can be large near the cylinder. The motion is still irrotational.

The motion of a free vortex is the same whether one considers the fluid perfect or viscous. The solution to the momentum equation for a perfect fluid (VR = constant) makes the sum of all viscous terms of the momentum equation equal to zero.

2-5.4 Energy Dissipation, Shear Deformation, and Rotationality

The fact that the motion is rotational does not necessarily mean that it is dissipative. A motion is dissipative when there are linear and/or angular deformations associated with a non-negligible viscous coefficient. So an irrotational free vortex can be dissipative, while a rotational forced vortex is not dissipative.

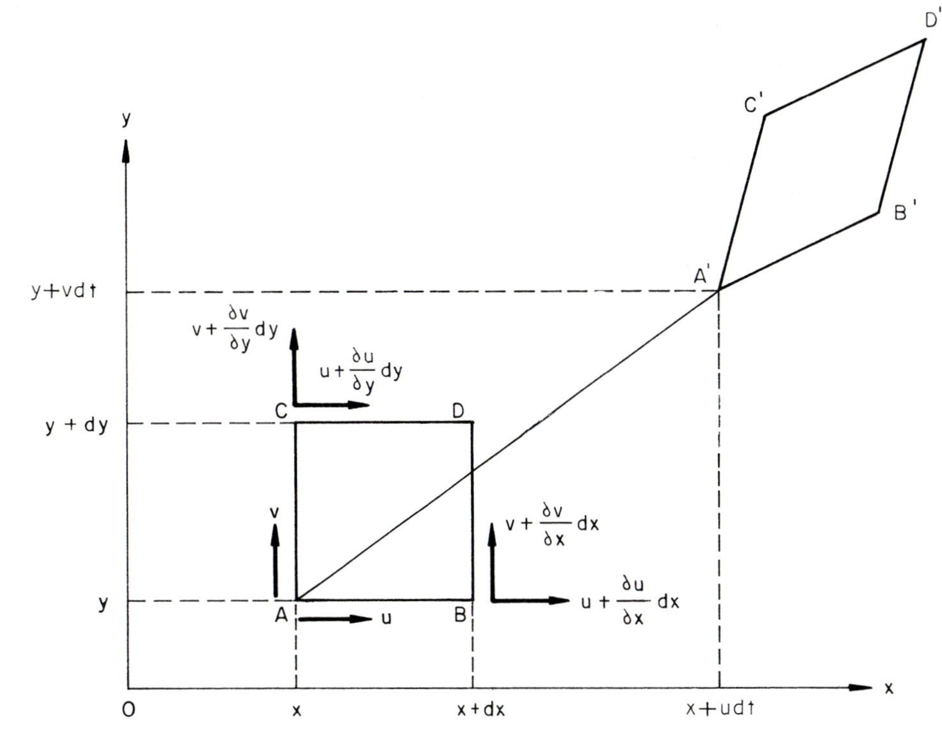

Figure 2-16
Two-dimensional coordinate system.

$$
D'\begin{cases} x + dx \quad + \quad u\,dt \quad + \quad \dfrac{\partial u}{\partial x}\,dx\,dt + \dfrac{1}{2}\left(\dfrac{\partial u}{\partial y} + \dfrac{\partial v}{\partial x}\right) dy\,dt - \dfrac{1}{2}\left(\dfrac{\partial v}{\partial x} - \dfrac{\partial u}{\partial y}\right) dy\,dt \\[2ex] y + dy \quad + \quad v\,dt \quad + \quad \dfrac{\partial v}{\partial y}\,dy\,dt + \dfrac{1}{2}\left(\dfrac{\partial u}{\partial y} + \dfrac{\partial v}{\partial x}\right) dx\,dt + \dfrac{1}{2}\left(\dfrac{\partial v}{\partial x} - \dfrac{\partial u}{\partial y}\right) dx\,dt \end{cases} \tag{2-2}
$$

Initial coordinates | *Translation* | *Dilatational or linear deformation* | *Rate of angular deformation* | *Rate of rotation*

Angular or shear deformation (rate of angular deformation) | *Rotation* (rate of rotation)

Indeed, it will be seen in Section 5-5.3.2 that the viscous stresses are proportional to the coefficients of linear and angular deformations presented in this chapter. Hence, the viscous stresses are dependent on the existence of deformation and not rotationality.

2-6 Mathematical Expressions Defining the Motion of a Fluid Particle

2-6.1 Two-Dimensional Motion

Consider a fluid element $ABCD$ at time t (Fig. 2-16). The velocity components u and v are functions of x and y such that $du = (\partial u/\partial x)\,dx + (\partial u/\partial y)\,dy$ and $dv = (\partial u/\partial x)\,dx + (\partial v/\partial y)\,dy$. At time t the space coordinates of A are x, y and of D are $x + dx$, $y + dy$.

The coordinates of A and D at time $t + dt$ are given in Equation 2-1.

$$
\begin{aligned}
&A'\begin{cases} x + u\,dt \\ y + v\,dt \end{cases} \\
&D'\begin{cases} x + dx + (u + du)\,dt \\ y + dy + (v + dv)\,dt \end{cases}
\end{aligned} \tag{2-1}
$$

or

$$
D'\begin{cases} x + dx + u\,dt + \left(\dfrac{\partial u}{\partial x}\,dx + \dfrac{\partial u}{\partial y}\,dy\right) dt \\[2ex] y + dy + v\,dt + \left(\dfrac{\partial v}{\partial x}\,dx + \dfrac{\partial v}{\partial y}\,dy\right) dt \end{cases}
$$

Adding and subtracting $\frac{1}{2}(\partial v/\partial x)\,dy\,dt$ to the x coordinate and $\frac{1}{2}(\partial u/\partial y)\,dx\,dt$ to the y coordinate leads to the form for the coordinates of D' expressed in Equation 2-2. The physical meaning of the terms becomes apparent by reference to the previous paragraphs.

2-6.2 Three-Dimensional Motion: Definition of the Vorticity

Similar to the two-dimensional case, the coordinates of a point $D'(x + dx,\ y + dy,\ z + dz)$ of a three-dimensional fluid element after a time dt become Equation 2-3.

$$
\begin{aligned}
&x + dx + u\,dt + \left(\frac{\partial u}{\partial x}\,dx + \frac{\partial u}{\partial y}\,dy + \frac{\partial u}{\partial z}\,dz\right) dt \\
&y + dy + v\,dt + \left(\frac{\partial v}{dx}\,dx + \frac{\partial v}{\partial y}\,dy + \frac{\partial v}{\partial z}\,dz\right) dt \\
&z + dz + w\,dt + \left(\frac{\partial w}{\partial x}\,dx + \frac{\partial w}{\partial y}\,dy + \frac{\partial w}{\partial z}\,dz\right) dt
\end{aligned} \tag{2-3}
$$

$$\begin{aligned}
&x + dx + u\,dt + \frac{\partial u}{\partial x}dx\,dt + \left[\frac{1}{2}\left(\frac{\partial v}{\partial x} + \frac{\partial u}{\partial y}\right)dy + \frac{1}{2}\left(\frac{\partial u}{\partial z} + \frac{\partial w}{\partial x}\right)dz + \frac{1}{2}\left(\frac{\partial u}{\partial z} - \frac{\partial w}{\partial x}\right)dz - \frac{1}{2}\left(\frac{\partial v}{\partial x} - \frac{\partial u}{\partial y}\right)dy\right]dt \\
&y + dy + v\,dt + \frac{\partial v}{\partial y}dy\,dt + \left[\frac{1}{2}\left(\frac{\partial w}{\partial y} + \frac{\partial v}{\partial z}\right)dz + \frac{1}{2}\left(\frac{\partial v}{\partial x} + \frac{\partial u}{\partial y}\right)dx + \frac{1}{2}\left(\frac{\partial v}{\partial x} - \frac{\partial u}{\partial y}\right)dx - \frac{1}{2}\left(\frac{\partial w}{\partial y} - \frac{\partial v}{\partial z}\right)dz\right]dt \qquad (2\text{-}4) \\
&z + dz + w\,dt + \frac{\partial w}{\partial z}dz\,dt + \left[\frac{1}{2}\left(\frac{\partial u}{\partial z} + \frac{\partial w}{\partial x}\right)dx + \frac{1}{2}\left(\frac{\partial w}{\partial y} + \frac{\partial v}{\partial z}\right)dy + \frac{1}{2}\left(\frac{\partial w}{\partial y} - \frac{\partial v}{\partial z}\right)dy - \frac{1}{2}\left(\frac{\partial u}{\partial z} - \frac{\partial w}{\partial x}\right)dx\right]dt
\end{aligned}$$

Adding and subtracting

$$\frac{1}{2}\frac{\partial v}{\partial x}dy\,dt \quad \text{and} \quad \frac{1}{2}\frac{\partial w}{\partial x}dz\,dt$$

to the first line;

$$\frac{1}{2}\frac{\partial w}{\partial y}dz\,dt \quad \text{and} \quad \frac{1}{2}\frac{\partial u}{\partial y}dx\,dt$$

to the second line; and

$$\frac{1}{2}\frac{\partial u}{\partial z}dx\,dt \quad \text{and} \quad \frac{1}{2}\frac{\partial v}{\partial z}dy\,dt$$

to the third line leads to Equation 2-4.

The following notations will be used: The coefficients of *dilatational deformation* will be defined as

$$a = \frac{\partial u}{\partial x} \qquad b = \frac{\partial v}{\partial y} \qquad c = \frac{\partial w}{\partial z}$$

The coefficients of *shear deformation* will be defined as

$$f = \frac{1}{2}\left(\frac{\partial w}{\partial y} + \frac{\partial v}{\partial z}\right) \qquad g = \frac{1}{2}\left(\frac{\partial u}{\partial z} + \frac{\partial w}{\partial x}\right)$$

$$h = \frac{1}{2}\left(\frac{\partial v}{\partial x} + \frac{\partial u}{\partial y}\right)$$

The coefficients of *rotation* will be defined as

$$\xi = \frac{1}{2}\left(\frac{\partial w}{\partial y} - \frac{\partial v}{\partial z}\right) \qquad \eta = \frac{1}{2}\left(\frac{\partial u}{\partial z} - \frac{\partial w}{\partial x}\right)$$

$$\zeta = \frac{1}{2}\left(\frac{\partial v}{\partial x} - \frac{\partial u}{\partial y}\right)$$

The coordinates of the point D' are now written as Equation 2-5, in which 2ξ, 2η, and 2ζ are the components of a vector

$$\begin{aligned}
&\underbrace{x + dx}_{\text{Initial coordinates}} + \underbrace{u\,dt}_{\text{Translation}} + \underbrace{a\,dx\,dt}_{\text{Dilatational deformation}} + \underbrace{(h\,dy + g\,dz)\,dt}_{\text{Angular deformation}} + \underbrace{(\eta\,dz - \zeta\,dy)\,dt}_{\text{Rotation}} \\
&y + dy + v\,dt + b\,dy\,dt + (f\,dz + h\,dx)\,dt + (\zeta\,dx - \xi\,dz)\,dt \qquad (2\text{-}5) \\
&z + dz + w\,dt + c\,dz\,dt + (g\,dx + f\,dy)\,dt + (\xi\,dy - \eta\,dx)\,dt
\end{aligned}$$

which represents the vorticity of the fluid at a point. A three-dimensional irrotational motion is defined by $\xi = 0$, $\eta = 0$, and $\zeta = 0$; that is,

$$\frac{\partial w}{\partial y} = \frac{\partial v}{\partial z}, \qquad \frac{\partial u}{\partial z} = \frac{\partial w}{\partial x}, \qquad \frac{\partial v}{\partial x} = \frac{\partial u}{\partial y}$$

2-6.3 *Velocity Potential Function in the Case of a Three-Dimensional Motion*

The velocity potential function is defined in three dimensions by

$$u = \frac{\partial \phi}{\partial x} \qquad v = \frac{\partial \phi}{\partial y} \qquad w = \frac{\partial \phi}{\partial z}$$

This may be written vectorially as $\mathbf{V} = \mathbf{grad}\ \phi$.

When the values of the velocity potential are substituted into the equations for irrotational motion, the results are:

$$\frac{\partial^2 \phi}{\partial z\, \partial y} = \frac{\partial^2 \phi}{\partial y\, \partial z} \qquad \frac{\partial^2 \phi}{\partial x\, \partial z} = \frac{\partial^2 \phi}{\partial z\, \partial x} \qquad \frac{\partial^2 \phi}{\partial x\, \partial y} = \frac{\partial^2 \phi}{\partial y\, \partial x}$$

This substantiates the definition of ϕ since ϕ always satisfies the conditions for an irrotational flow. In other words, the existence of ϕ implies that the flow is irrotational.

The above equations would still hold, even if the velocity potential had been negative, so the velocity potential can also be defined by $\mathbf{V} = -\mathbf{grad}\ \phi$.

2-6.4 *Stokes Analogy: Experience of Shaw*

A three-dimensional rotational motion may be a two-dimensional irrotational motion when the rotation is always in the same plane. For example, a thin layer of water flowing on a horizontal glass plate in which the thickness of the layer is very small in comparison with the other dimensions, has a rotational motion in vertical planes only (Fig. 2-17). If seen in the plane, the motion would appear as two-dimensional irrotational motion.

In the case of Fig. 2-17, motion in the vertical planes XOZ and YOZ is rotational, ξ and $\eta \neq 0$, while the motion in the horizontal plane XOY is irrotational and $\zeta = \frac{1}{2}(\partial u/\partial y - \partial v/\partial x) = 0$. It may be demonstrated that the average velocity with respect to a vertical verifies similar conditions of irrotationality.

The streamlines seen in the plane are simply shown by the injection of dyes. The same result is obtained from a flow between two vertical parallel planes. This method is often used to determine the flow pattern of two-dimensional or

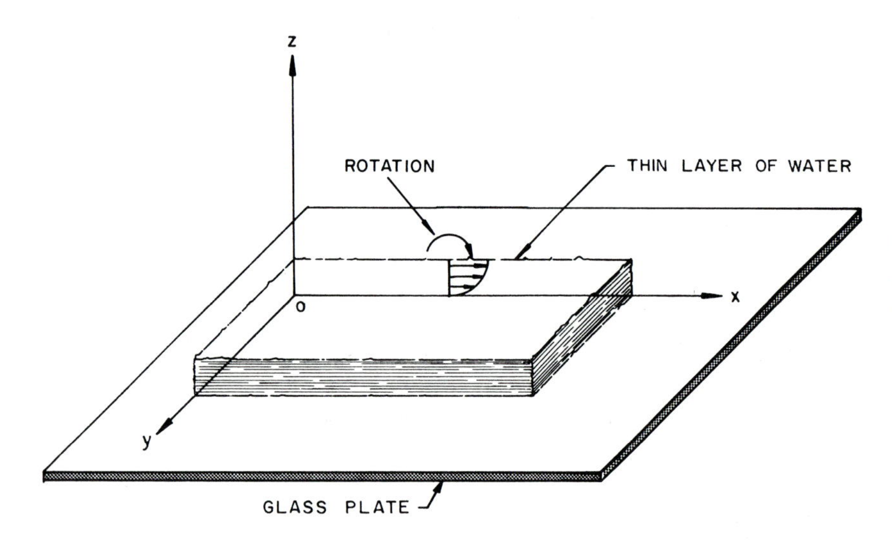

Figure 2-17

In a thin flow of water, rotation exists only in vertical planes.

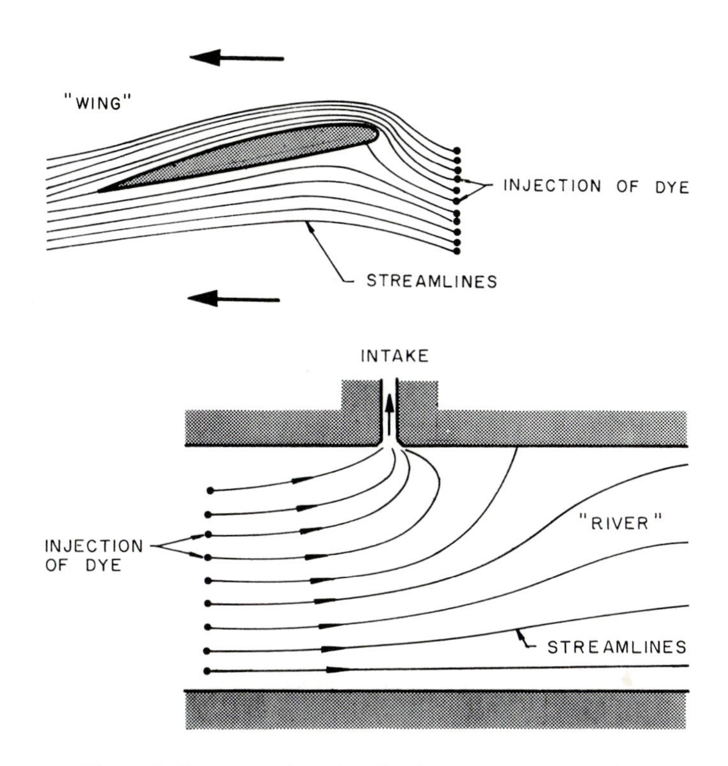

Figure 2-18 *Examples of studies based on Stokes analogy.*

almost two-dimensional motion. Some examples are: flow pattern around a wing, influence of an intake on the flow of a wide and shallow river (Fig. 2-18).

PROBLEMS

2.1 Consider a two-dimensional convergent section as shown by Fig. 2-19. Determine the coefficient of linear deformation at point $x = 0$, $y = 0$ where $V = u = L/t$ (L is a unit length, t is a unit time).

2.2 Indicate the domains of Fig. 2-20, where the flow can be considered as irrotational and the domains where the flow is rotational. Give the reasons which prevail in your choice.

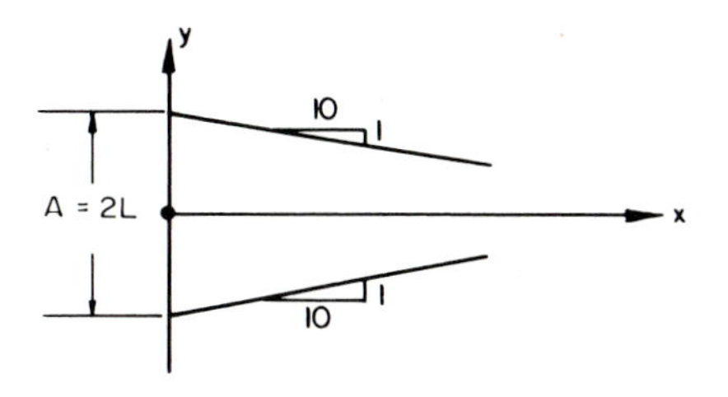

Figure 2-19

2.3 Determine the coefficients of dilatational and shear deformation and rotation for a flow between two parallel planes separated by a distance $d = 0.01L$. One of the planes is assumed to be fixed, the other one moves at a speed $V = 0.1L/t$. The velocity distribution between the two planes is linear.

2.4 The velocity distribution of a laminar flow between two parallel planes is given by the equation

$$u = \frac{1}{2\mu}\alpha y^2 - \frac{2}{2\mu}\left(\alpha - \frac{2\mu V}{e^2}\right)y$$

where μ is the coefficient of viscosity, e is the distance between the two planes, α is a constant equal to the head loss or decrease of pressure per unit length: $\alpha = dp/dx$.

Figure 2-20

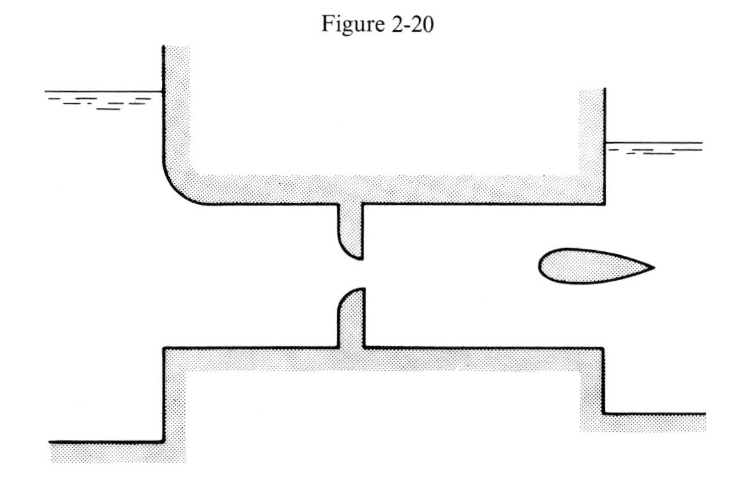

V is the velocity of one of the planes, the other one being assumed to be fixed.

Determine the coefficient of dilatational and shear deformation and rotation as a function of y. Consider the two cases where $\alpha = 0$ and $V \neq 0$ on one hand, and $\alpha \neq 0$ and $V = 0$, on the other hand, as two particular cases, and explain their significance.

2.5 Express the velocity components as a function of a potential function ϕ in a cylindrical (r,θ,z) and spherical (r,Φ,θ) system of coordinates.

2.6 Derive the expression for irrotationality in polar system of coordinates (r, θ). The components of velocity are: radial velocity v_r, tangential velocity v_θ.

2.7 Consider two coaxial cylinders of radius R_1 and R_2 which are rotating at angular velocity ω_1 and ω_2, respectively. The velocity distribution of the fluid between these two cylinders is given as a function of r by the expression $(R_1 < r < R_2)$

$$v_\theta(r) = \left[\frac{r}{R_2^2 - R_1^2}\right]\left[(\omega_2 R_2^2 - \omega_1 R_1^2) - \frac{R_1^2 R_2^2}{r^2}(\omega_2 - \omega_1)\right]$$

Determine the value of the coefficients of rotation and the relationship between ω_1 and ω_2 which makes the flow motion irrotational.

2.8 The equations for an average viscous flow through a porous medium are defined by

$$\frac{\partial p}{\partial x} = Ku$$

$$\frac{\partial p}{\partial y} = Kv$$

where p is the hydrostatic pressure. Demonstrate that such a flow is irrotational. The equations for an average fully turbulent flow through rocks are

$$\frac{\partial p}{\partial x} = Ku^2$$

$$\frac{\partial p}{\partial y} = Kv^2$$

Demonstrate that such a flow is in general rotational.

2.9 Demonstrate that at a given location in a two-dimensional flow, the value of angular rotation is independent of the axis system of references, i.e.,

$$\frac{1}{2}\left(\frac{\partial v}{\partial x} - \frac{\partial u}{\partial y}\right) = \frac{1}{2}\left(\frac{\partial v'}{\partial x'} - \frac{\partial u'}{\partial y'}\right)$$

u' and v' being the velocity components along the x' axis and the y' axis, respectively.

2.10 Demonstrate that

$$\frac{\partial \xi}{\partial x} + \frac{\partial \eta}{\partial y} + \frac{\partial \zeta}{\partial z} \equiv 0$$

Chapter 3

The Continuity Principle

3-1 Elementary Relationships

3-1.1 The Continuity in a Pipe

The principle of continuity expresses the conservation of mass in a given space occupied by a fluid. The simplest, well-known form of the continuity relationship in elementary fluid mechanics expresses that the discharge for steady flow in a pipe is constant; that is, ρVA = constant, where A is the cross-sectional area of the pipe and V is the mean velocity.

In the case of an incompressible fluid (ρ = constant) in a uniform pipe (A = constant), the continuity relationship becomes simply: V = constant. If the X axis is taken as the axis of the pipe, then $V = u$, and the continuity principle expressed in differential form becomes:

$$\frac{dV}{dx} \equiv \frac{\partial u}{\partial x} = 0$$

The case of a two-dimensional motion of an incompressible fluid is now given. The general case will be treated later.

3-1.2 Two-Dimensional Motion in an Incompressible Fluid

Since no fluid is being added or subtracted during the motion, the quantity of fluid involved is constant. This may be expressed mathematically in the case of two-dimensional incompressible motion. The development of the mathematics follows.

Consider a rectangular element in two-dimensional fluid motion as shown in Fig. 3-1. The rectangular boundaries have sides of length a and b and are considered to be fixed with respect to the axes. It is not a moving fluid element.

The volume of fluid entering the left-hand boundary line by unit of time is au_1, and at the same instant, the amount leaving the right-hand boundary line is au_2. The difference in amount in the OX direction is thus: $a(u_2 - u_1) = a\,\Delta u$.

Similarly, the difference in amount in the OY direction is: $b(v_2 - v_1) = b\,\Delta v$.

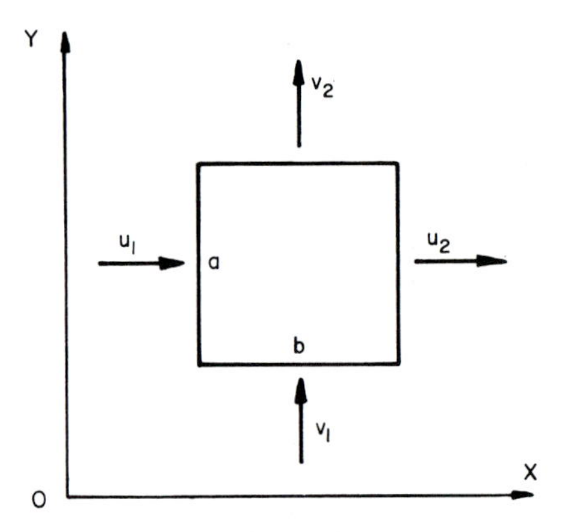

Figure 3-1 *Rectangular element of an incompressible fluid.*

Since the total mass of fluid within the boundaries is constant, the total loss must be zero: $a\,\Delta u + b\,\Delta v = 0$; that is, $(\Delta u/b) + (\Delta v/a) = 0$.

In the limit, when b and a approach zero, one obtains $(\partial u/\partial x) + (\partial v/\partial y) = 0$. This differential form is permitted because of the assumption of a continuous fluid. It should be noted that $\partial u/\partial x$ and $\partial v/\partial y$ are the rates of linear deformation of a fluid particle; hence, in an incompressible fluid, the total sum of linear deformation is nil, as has been previously noted.

3-2 The Continuity Relationship in the General Case

3-2.1 Establishment of the Continuity Relationship

Consider a fixed volume of fluid of which the edges dx, dy, dz are parallel to the axes OX, OY, OZ, respectively (Fig. 3-2). The continuity relationship is obtained by considering that the change of fluid mass inside the volume $dx\,dy\,dz$ during the time dt is equal to the difference between the rates of influx into and efflux out of the considered volume during the same interval of time.

The fluid mass at the time t is: $\rho\,dx\,dy\,dz$.

After a time dt, because of the change of density with respect to time, the quantity of fluid mass becomes

$$\left(\rho + \frac{\partial \rho}{\partial t}\,dt\right) dx\,dy\,dz$$

Hence the change of fluid mass in a time dt is

$$\rho\,dx\,dy\,dz - \left(\rho + \frac{\partial \rho}{\partial t}\,dt\right) dx\,dy\,dz = -\frac{\partial \rho}{\partial t}\,dt\,dx\,dy\,dz \tag{3-1}$$

On the other hand, if one takes into account the change in velocity and in density with respect to space coordinates, the quantity of fluid mass entering through the section $ABCD$ during a time dt, parallel to the OX axis, is the product ρu times the area perpendicular to OX ($ABCD$) and the time dt. Since $ABCD = dy\,dz$, the quantity of fluid mass entering is $\rho u\,dy\,dz\,dt$. The derivative of u along AB and AD with respect to dz and dy is of an infinitely small order and can be neglected. Now the quantity of fluid mass coming out

Figure 3-2 *Coordinate system for continuity equation.*

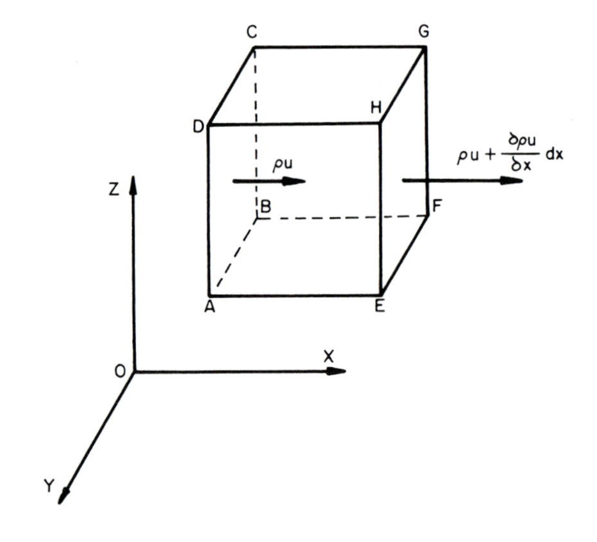

during the same interval of time through the section *EFGH* is:

$$\left[\rho u + \frac{\partial(\rho u)}{\partial x} dx\right] dy\, dz\, dt$$

In the general case, both the density ρ and velocity u are assumed to be changed along dx. Hence the difference is

$$\left[\rho u + \frac{\partial(\rho u)}{\partial x} dx\right] dy\, dz\, dt - \rho u\, dy\, dz\, dt = \frac{\partial(\rho u)}{\partial x} dx\, dy\, dz\, dt$$

Similarly, the difference due to the components of motion parallel to the OY and OZ axes are, respectively,

$\frac{\partial(\rho v)}{\partial y} dx\, dy\, dz\, dt$ due to the difference of discharge across the sections $BFGC$ and $AEHD$ $(dx\, dz)$

$\frac{\partial(\rho w)}{\partial z} dx\, dy\, dz\, dt$ due to the difference of discharge across the sections $AEFB$ and $DHGC$ $(dx\, dy)$

The total change of mass contained within the elementary region during the time dt is

$$\left[\frac{\partial(\rho u)}{\partial x} + \frac{\partial(\rho v)}{\partial y} + \frac{\partial(\rho w)}{\partial z}\right] dx\, dy\, dz\, dt \qquad (3\text{-}2)$$

Equating Equations 3-1 and 3-2 yields

$$-\frac{\partial \rho}{\partial t} dx\, dy\, dz\, dt = \left[\frac{\partial(\rho u)}{\partial x} + \frac{\partial(\rho v)}{\partial y} + \frac{\partial(\rho w)}{\partial z}\right] dx\, dy\, dz\, dt$$

Dividing both sides by $dx\, dy\, dz\, dt$ yields

$$\frac{\partial \rho}{\partial t} + \frac{\partial(\rho u)}{\partial x} + \frac{\partial(\rho v)}{\partial y} + \frac{\partial(\rho w)}{\partial z} = 0$$

Since $[\partial(\rho u)]/\partial x = \rho(\partial u/\partial x) + u(\partial \rho/\partial x)$, and similarly for the terms $\partial(\rho v)/\partial y$ and $\partial(\rho w)/\partial z$, the continuity relationship becomes

$$\frac{\partial \rho}{\partial t} + \rho\left(\frac{\partial u}{\partial x} + \frac{\partial v}{\partial y} + \frac{\partial w}{\partial z}\right) + u\frac{\partial \rho}{\partial x} + v\frac{\partial \rho}{\partial y} + w\frac{\partial \rho}{\partial z} = 0$$

These continuity relationships can be written in a shorter way as follows:

$$\frac{\partial \rho}{\partial t} + \operatorname{div} \rho\, \mathbf{V} = 0$$

or

$$\frac{\partial \rho}{\partial t} + \rho \operatorname{div} \mathbf{V} + \mathbf{V} \cdot \mathbf{grad}\, \rho = 0$$

3-2.2 *Physical Meaning and Approximations*

Consider, respectively, the three groups of terms:

$$\frac{\partial \rho}{\partial t}$$

$$\rho\left(\frac{\partial u}{\partial x} + \frac{\partial v}{\partial y} + \frac{\partial w}{\partial z}\right) \qquad \text{or} \qquad \rho \operatorname{div} \mathbf{V}$$

$$u\frac{\partial \rho}{\partial x} + v\frac{\partial \rho}{\partial y} + w\frac{\partial \rho}{\partial z} \qquad \text{or} \qquad \mathbf{V} \cdot \mathbf{grad}\, \rho$$

3-2.2.1 The first term, $\partial\rho/\partial t$, is the derivative of the density with time at a given point. This term is nil in the case of (1) incompressible fluid, since ρ is a constant; and (2) a steady motion of a compressible fluid. This term has to be considered when sound, water hammer, shock waves, etc. are studied.

3-2.2.2 The second group of terms is proportional to the derivative of speed in the direction of motion at a given time. In the simple case of a three-dimensional motion of an incompressible fluid

$$\frac{\partial u}{\partial x} + \frac{\partial v}{\partial y} + \frac{\partial w}{\partial z} = 0 \qquad \text{or} \qquad \operatorname{div} \mathbf{V} = 0$$

When $\operatorname{div} \mathbf{V} > 0$, an expansion of the fluid is indicated, and conversely, $\operatorname{div} \mathbf{V} < 0$ signifies a compression.

3-2.2.3 The third group of terms is proportional to the derivative of density with respect to the space coordinates at a given time. This derivative is usually negligible in comparison with the others.

For example, consider a unidimensional sinusoidal acoustic wave. The density is given as a function of time and distance by the relationship $\rho = A \sin (Ct + x)$. The derivative of ρ with respect to time t is $\partial\rho/\partial t = AC \cos (Ct + x)$, and with respect to the distance x along the OX axis, $\partial\rho/\partial x = A \cos (Ct + x)$. Hence,

$$\frac{\partial\rho/\partial x}{\partial\rho/\partial t} = \frac{1}{C}$$

where C is the wave velocity.

Since C is usually large compared to the particle velocity u, $u(\partial\rho/\partial x)$ is usually negligible by comparison with $\partial\rho/\partial t$.

However, there is an exception to this rule: the case of shock waves where the variation of ρ with respect to space is theoretically infinite at the front of the wave.

Table 3-1

Uniform (one-dimensional) flow of an incompressible fluid	$\frac{\partial u}{\partial x} = 0$
Two-dimensional flow of an incompressible fluid	$\frac{\partial u}{\partial x} + \frac{\partial v}{\partial y} = 0$
Three-dimensional flow of an incompressible fluid	$\frac{\partial u}{\partial x} + \frac{\partial v}{\partial y} + \frac{\partial w}{\partial z} = 0$ $\operatorname{div} \mathbf{V} = 0$
Unsteady motion in a compressible fluid at usual speed (acoustic wave, water hammer)	$\frac{\partial\rho}{\partial t} + \rho\left(\frac{\partial u}{\partial x} + \frac{\partial v}{\partial y} + \frac{\partial w}{\partial z}\right) = 0$ $\frac{\partial\rho}{\partial t} + \rho \operatorname{div} \mathbf{V} = 0$
Unsteady motion in a compressible fluid at high speed (shock wave)	$\frac{\partial\rho}{\partial t} + \frac{\partial(\rho u)}{\partial x} + \frac{\partial(\rho v)}{\partial y} + \frac{\partial(\rho w)}{\partial z} = 0$ $\frac{\partial\rho}{\partial t} + \operatorname{div} (\rho\mathbf{V}) = 0$

This phenomenon occurs when a large variation of pressure results in the celerity of the wave exceeding the velocity of sound. A shock wave travels at a higher speed than usual pressure waves such as acoustic sound or water hammer. Hence, when a supersonic flow or the effect of an underwater explosion is studied, the derivative of the density with respect to the distance has to be taken into account in the continuity relationship.

3-2.2.4 Table 3-1 summarizes these considerations. Unless otherwise specified, incompressible fluids only are considered in the rest of this text.

3-3 Some Particular Cases of the Continuity Relationship

The continuity relationship is often used in other forms in hydraulics. These forms are not so general but more adapted to integration for the phenomena to be studied. Some examples of the different forms used are provided by the case of unsteady flow, mainly unidimensional either at a free surface (channel, river) or under pressure (pipe, gallery).

An unsteady free surface flow resulting in a change of level with respect to time and space and caused by gravity action is called a gravity wave. Some examples are flood waves in a river, bore (translatory waves), tsunami waves due to earthquake, tides, harbor oscillation, and seiche. An unsteady flow under pressure resulting in a change of pressure with respect to time and space, caused by the pressure gradient is called a pressure wave. Two examples are water hammer and acoustic wave. Such gravity waves and pressure waves are studied from special forms of the continuity relationships. They are valid when the distribution of velocity in a cross section can be assumed to be a constant; then the gravity waves are called long waves.

Here, the continuity relationship is obtained by stating that the change of volume of water during the interval of time dt between two finite cross sections separated by the infinitely small distance dx is equal to the difference between the influx and efflux from the considered volume during the same interval of time

3-3.1 *Translatory Gravity Waves*

Consider the volume defined by the two cross sections x and $x + dx$ and the free surface at time t (Fig. 3-3). The volume of influx, during a time dt, into the considered volume at x is $q\,dt$ or $hu\,dt$, where q is the discharge, h the depth, and u the horizontal velocity component.

The efflux out of the volume at $x + dx$ is

$$\left(q + \frac{\partial q}{\partial x} dx\right) dt \qquad \text{or} \qquad \left[hu + \frac{\partial(hu)}{\partial x} dx\right] dt$$

Hence the change of volume ΔV between these cross sections x and $x + dx$ is a difference

$$\Delta V = -\frac{\partial q}{\partial x} dx\,dt \qquad \text{or} \qquad \Delta V = -\frac{\partial(hu)}{\partial x} dx\,dt \qquad (3\text{-}3)$$

On the other hand, the volume at the time t is $h\,dx$, and at the time $t + dt$, since the free surface level changes the volume is $[h + (\partial h/\partial t)\,dt]\,dx$. Hence the change of volume during time dt is:

$$\Delta V = \frac{\partial h}{\partial t} dt\,dx \qquad (3\text{-}4)$$

Equating Equations 3-3 and 3-4 and dividing by $dx\,dt$, one obtains $(\partial h/\partial t) + (\partial q/\partial x) = 0$, $(\partial h/\partial t) + (\partial hu/\partial x) = 0$, or

$$\frac{\partial h}{\partial t} + u\frac{\partial h}{\partial x} + h\frac{\partial u}{\partial x} = 0$$

3-3.2 *Irrotational Flow*

If the density ρ is a constant, the continuity relationship has been seen to be $(\partial u/\partial x) + (\partial v/\partial y) + (\partial w/\partial z) = 0$.

In the case of irrotational motion, a velocity potential function ϕ has been defined by the relationships

$$u = \frac{\partial \phi}{\partial x} \qquad v = \frac{\partial \phi}{\partial y} \qquad w = \frac{\partial \phi}{\partial z}$$

Hence, introducing these expressions into the continuity relationship yields

$$\frac{\partial^2 \phi}{\partial x^2} + \frac{\partial^2 \phi}{\partial y^2} + \frac{\partial^2 \phi}{\partial z^2} = 0$$

which can be written $\nabla^2 \phi = 0$. This is the well-known Laplace equation which has been subjected to extensive research in mathematical physics.

3-3.3 *Lagrangian System of Coordinates*

It has been explained in Chapter 1 that it is possible to study problems in hydraulics either in Eulerian coordinates or in Lagrangian coordinates (See Section 1-3.1). Since this last system of coordinates is rarely used, the continuity

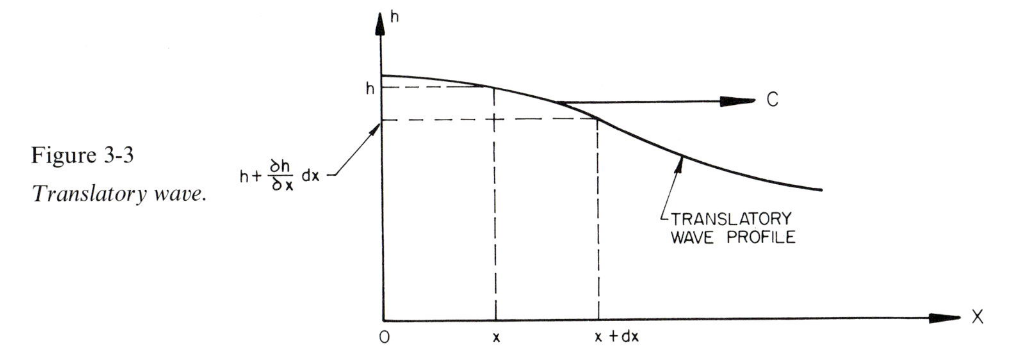

Figure 3-3
Translatory wave.

relationship is given here without any comment for the simple purpose of recognition in literature:

$$\frac{\partial(x,y,z)}{\partial(x_0,y_0,z_0)} = 1$$

where x_0, y_0 and z_0 are the coordinates of the considered particle at time t_0 and x, y, z at time t.

PROBLEMS

3.1 Demonstrate that the continuity equation for a stream tube can be written as:

$$\frac{\partial(\rho A)}{\partial t} + \frac{\partial(\rho A V)}{\partial s} = 0$$

where A is the cross section of the stream, and ds an element of streamline.

3.2 Consider the two-dimensional motions defined by their velocity components $u = A$ and $v = 0$ on one hand, and $u = Ax + B$ and $v = 0$ on the other hand, where A and B are different from zero. Calculate the divergence, and tell in which case the fluid is compressible.

3.3 Establish the continuity equation in a cylindrical system of coordinates (r,θ,z). A polar element of volume $r\,dr\,d\theta\,dz$ will be used. The velocity components will be v_r along a radius, v_θ perpendicular to v_r, and v_z along the axis OZ.

3.4 Express the Laplace equation $\nabla^2\phi = 0$ in a cylindrical system of coordinates (r,θ,z).

3.5 Establish the continuity equation for a stratified fluid of density varying with z as $\rho = k/z$ ($\partial\rho/\partial t$ will be assumed to be zero).

3.6 Verify that the motion defined by the potential function

$$\phi = -a\frac{k}{m}\frac{\cosh m(d+z)}{\cosh md}\cos(kt - mx)$$

(1) is irrotational; (2) satisfies the continuity equation; (3) is such that $\partial\phi/\partial z = 0$ for $z = -d$.

3.7 Derive the continuity equation and $\nabla^2\phi$ for an incompressible liquid in spherical polar coordinates (r,θ,Φ) by considering a small volume bounded by the surface: $r, r + dr; \theta, \theta + d\theta; \Phi, \Phi + d\Phi$.

3.8 Consider a uniform cylindrical pipe of diameter D, wall thickness e, and elasticity modulus E. The bulk modulus of elasticity of the fluid is K. It will be assumed that under the influence of a variation of pressure $\partial p/\partial x$, the pipe is elastic and the fluid compressible. Demonstrate that the continuity relationship is

$$\frac{\partial p}{\partial t} = \left(\frac{1}{K} + \frac{D}{eE}\right)^{-1}\frac{\partial u}{\partial x}$$

3.9 Demonstrate that the continuity relationship for long wave (tidal motion) in a spherical system of coordinates is

$$\frac{\partial\eta}{\partial t} + \frac{1}{a\sin\Phi}\left\{\frac{\partial}{\partial\Phi}[h(\theta,\Phi)U_E\sin\Phi] + \frac{\partial}{\partial\theta}[h(\theta,\Phi)U_S]\right\} = 0$$

where a is the mean earth radius, η the free surface elevation around the still water level, U_E is the component of the current velocity in the direction of increasing θ, U_S is the component of velocity in the direction of increasing Φ, θ is the longitude, Φ is the latitude, $h(\theta,\Phi)$ is the water depth.

Chapter 4

Inertia Forces

4-1 Mass, Inertia, and Acceleration

4-1.1 The Newton Equation

To cause the motion of a constant mass M, or, more generally, to change the state of an existing motion, it is necessary to apply to this mass a force $\mathbf{F}$, which causes an acceleration $d\mathbf{V}/dt$ such that $\mathbf{F} = M(d\mathbf{V}/dt)$. This is a vector relationship, i.e., true for both magnitude and direction. The product $M(d\mathbf{V}/dt)$ is the inertia force, which characterizes the natural resistance of matter to any change in its state of motion.

The considered mass M is the mass of a unit of volume of fluid

$$M = \rho \cdot (\text{unit of volume}) = \rho$$

where ρ is the density. Hence the fundamental equation of momentum has the form $\mathbf{F} = \rho(d\mathbf{V}/dt)$. Its three components along the three coordinate axes OX, OY, OZ are $\rho(du/dt)$, $\rho(dv/dt)$, and $\rho(dw/dt)$, respectively.

4-1.2 Relationships between the Elementary Motions of a Fluid Particle and the Inertia Terms

To each kind of motion of the fluid particles (Chapter 2) there corresponds an inertia force. The relationship between the kind of motion described and the corresponding inertia force is straightforward.

The elementary components of velocity of a fluid particle as given in Chapter 2 are, in the case of a two-dimensional motion,

Translation	u, v	
Dilatational deformation	$\dfrac{\partial u}{\partial x}\, dx$	$\dfrac{\partial v}{\partial y}\, dy$
Shear deformation	$\dfrac{1}{2}\left(\dfrac{\partial u}{\partial y} + \dfrac{\partial v}{\partial x}\right) dy$	
	$\dfrac{1}{2}\left(\dfrac{\partial u}{\partial y} + \dfrac{\partial v}{\partial x}\right) dx$	

Rotation $$-\frac{1}{2}\left(\frac{\partial v}{\partial x} - \frac{\partial u}{\partial y}\right) dy$$
$$\frac{1}{2}\left(\frac{\partial v}{\partial x} - \frac{\partial u}{\partial y}\right) dx$$

To each of these velocity components corresponds a component of acceleration, which multiplied by ρ, yields a component of inertia force.

Two types of inertia forces may be distinguished, depending on the type of acceleration or elementary motion considered. These are:

1. Local acceleration—corresponding to a variation of the velocity of translation or the derivative of velocity with respect to time.
2. Convective acceleration—corresponding to a variation of velocity of deformation and rotation or derivative of velocity with respect to space.

The physical meaning of these accelerations and the corresponding inertia forces is first examined; then their mathematical expression is demonstrated. Chapter 5 deals with the applied forces **F** which have to be equated to these inertia forces to obtain the momentum equation.

4-2 Local Acceleration

Local acceleration characterizes any unsteady motion, i.e., motion where the velocity at a given point changes with respect to time. Local acceleration results from a change in the translatory motion of a fluid particle imposed by external forces **F**.

4-2.1 Examples of Flow with Local Inertia

Local acceleration occurs in the following cases.

4-2.1.1 In the first case, the velocity stays in the same direction along a straight line and changes in magnitude. If the velocity increases at a given point, which involves a positive local acceleration, the inertia of the mass of fluid in motion tends to slow it down.

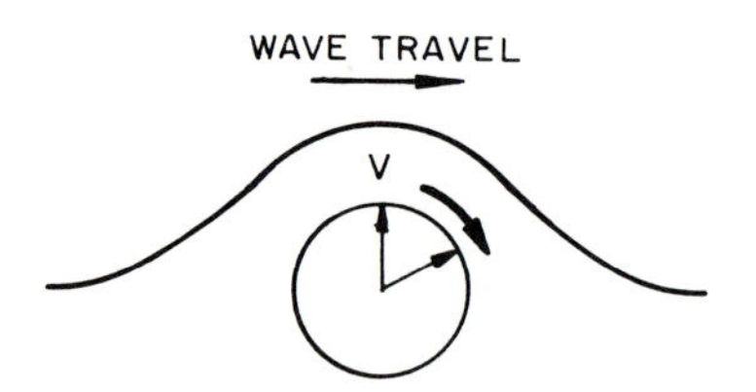

Figure 4-1 *Periodic gravity wave in deep water. The velocity at a given point changes in direction only.*

Alternatively, if the velocity decreases, which corresponds to a negative local acceleration, the inertia of the mass of fluid in motion tends to maintain the original velocity.

This type of phenomenon is observed in pipes or tunnels, where a fluid stops or starts or balances because of a gate movement. Local inertia has to be taken into account in hydraulic engineering applications such as surge tanks, water hammer, and locks.

4-2.1.2 In the second case, the velocity maintains the same magnitude, but changes its direction. In this case the inertia force is due to the centrifugal acceleration. For example, in a periodic gravity wave in infinite depth, the magnitude of the velocity at a given point is a constant but its direction revolves continuously at all points (Fig. 4-1).

4-2.1.3 In the third case, the velocity changes at a given point both in magnitude and direction. Some examples of this case are turbulent flow (this important case is fully developed in Chapters 7 and 8); alternate vortices; displacement caused by a ship in motion; bore and tide in an estuary; and periodic gravity waves in shallow water.

4-2.2 Mathematical Expression of Local Inertia

The mathematical expression of the inertia forces caused by a local acceleration is given by the change in the velocity of the translatory motion with respect to time only. The corresponding inertia force is equal to $\rho(\partial \mathbf{V}/\partial t)$ of which the components along the three axes are, respectively: $\rho(\partial u/\partial t)$, $\rho(\partial v/\partial t)$, and $\rho(\partial w/\partial t)$. The derivatives with respect to space are not taken into account.

4-3 Convective Acceleration

Convective acceleration characterizes any nonuniform flow, i.e., when the velocity at a given time changes with respect to distance. It is sometimes called field acceleration. Convective acceleration results from any linear or angular deformation, or from a change in the rotation of fluid particles, imposed by external forces **F**.

4-3.1 The Case of Linear Deformation

4-3.1.1 In a convergent pipe, it has been seen that the velocity of a fluid particle, although constant with time at a fixed location, tends to increase along the converging streamlines. The velocity of the fluid particle increases with respect to space. This is a positive convective acceleration. The fluid tends to resist this acceleration by convective inertia.

In a divergent conduit, the velocity decreases and the fluid tends to continue its motion with the same velocity because of its inertia. The applied forces cause a negative convective acceleration.

Expansion or contraction of a compressible fluid is the sum of linear deformations and also results in corresponding inertia forces.

4-3.1.2 It has been seen that the linear deformation velocity components are those given in Equation 4-1.

$$\left.\begin{matrix}\left.\begin{matrix}\dfrac{\partial u}{\partial x}dx \\ \\ \dfrac{\partial v}{\partial y}dy\end{matrix}\right\} \text{Two-dimensional motion} \\ \\ \dfrac{\partial w}{\partial z}dz\end{matrix}\right\} \text{Three-dimensional motion} \qquad (4\text{-}1)$$

The expressions $\partial u/\partial x$, $\partial v/\partial y$, and $\partial w/\partial z$ are taken at a given time, as seen in section 2-3.1. The corresponding acceleration is

$$\frac{d}{dt}\left[\frac{\partial u}{\partial x}dx\right] = \frac{\partial u}{\partial x}\frac{dx}{dt}$$

Two similar expressions result for w and v. If $u = dx/dt$, $v = dy/dt$, and $w = dz/dt$, are substituted in these expressions and the result is multiplied by the density, the inertia forces are obtained. They are

$$\rho u\frac{\partial u}{\partial x} = \frac{1}{2}\rho\frac{\partial(u^2)}{\partial x}$$

$$\rho v\frac{\partial v}{\partial y} = \frac{1}{2}\rho\frac{\partial(v^2)}{\partial y}$$

$$\rho w\frac{\partial w}{\partial z} = \frac{1}{2}\rho\frac{\partial(w^2)}{\partial z}$$

It should be noticed that the last group of expressions may be written as $(\partial/\partial x)(\rho u^2/2)$. This shows that the inertia force is equal to the derivative of the kinetic energy with respect to space along the three direction axes OX, OY, and OZ, respectively.

4-3.2 The Case of Shear Deformation

4-3.2.1 In a bend, where the fluid particles are angularly deformed, the fluid paths are curved and because of its inertia, the fluid tends to continue along a straight line. This causes a centrifugal force proportional to the change of direction which is imposed by the applied forces.

It is possible for the velocity of a fluid particle to keep the same magnitude along its path, but with a change in direction. This is the case of free vortex motion.

4-3.2.2 It has been seen that the velocity components of angular deformation for a two-dimensional motion are

$$\frac{1}{2}\left(\frac{\partial u}{\partial y} + \frac{\partial v}{\partial x}\right)dy$$

$$\frac{1}{2}\left(\frac{\partial u}{\partial y} + \frac{\partial v}{\partial x}\right)dx$$

Hence, as in the previous case, using the substitutions $u = dx/dt$ and $v = dy/dt$, the corresponding inertia forces become:

$$\frac{1}{2}\rho v\left(\frac{\partial u}{\partial y} + \frac{\partial v}{\partial x}\right)$$

$$\frac{1}{2}\rho u\left(\frac{\partial u}{\partial y} + \frac{\partial v}{\partial x}\right)$$

4-3.3 *The Case of a Change of Rotation*

4-3.3.1 In the entrance to a pipe (Fig. 4-2), because of the change in friction forces, there is a variation of rotation of the fluid particles. Hence there are inertia forces corresponding to the natural resistance of the fluid to change its rotational motion. In a uniform pipe, the rotation of particles exists but there is no change in rotational magnitude and the corresponding acceleration is zero.

4-3.3.2 As in the two previous cases, since

$$-\frac{1}{2}\left(\frac{\partial v}{\partial x} - \frac{\partial u}{\partial y}\right)dy$$

$$\frac{1}{2}\left(\frac{\partial v}{\partial x} - \frac{\partial u}{\partial y}\right)dx$$

Figure 4-2 *Zone of acceleration of rotation.*

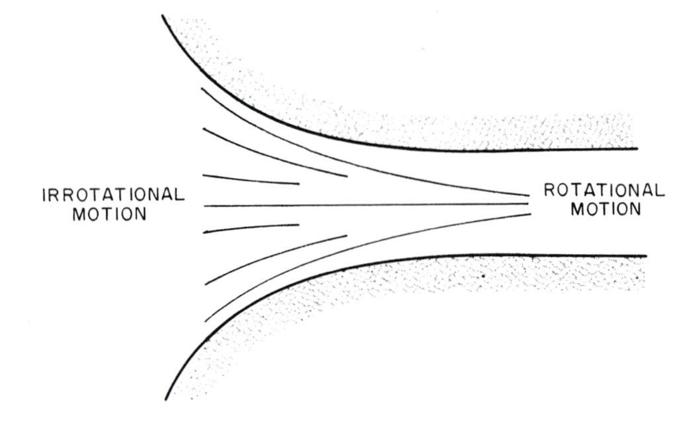

are the velocities of the components of rotation in a two-dimensional motion, the corresponding inertia forces obtained are

$$-\frac{1}{2}\rho v\left(\frac{\partial v}{\partial x} - \frac{\partial u}{\partial y}\right)$$

$$\frac{1}{2}\rho u\left(\frac{\partial v}{\partial x} - \frac{\partial u}{\partial y}\right)$$

It has been shown that it is possible to assume that the motion is irrotational when friction effects are negligible. It is evident that the same conditions lead to neglect of rotational inertia forces.

4-4 General Mathematical Expressions of Inertia Forces

4-4.1 *Local and Convective Acceleration*

In the general case both local acceleration and convective acceleration occur at the same time. A simple example is when a fluid oscillates in a nonuniform curved pipe. Hence, in the general case, **V** and its components u, v, and w are functions of both time and space coordinates. For example, $u(x,y,z,t)$. The total differential of u is

$$du = \frac{\partial u}{\partial t}dt + \frac{\partial u}{\partial x}dx + \frac{\partial u}{\partial y}dy + \frac{\partial u}{\partial z}dz$$

The acceleration in the x direction is thus given by the total differential of u, with respect to time:

$$\frac{du}{dt} = \frac{\partial u}{\partial t} + \frac{\partial u}{\partial x}\frac{dx}{dt} + \frac{\partial u}{\partial y}\frac{dy}{dt} + \frac{\partial u}{\partial z}\frac{dz}{dt}$$

Similar expressions occur for dv/dt and dw/dt.

Substituting $u = dx/dt$, $v = dy/dt$, and $w = dz/dt$, and multiplying by the density ρ, the inertia forces given by Equation 4-2 are obtained.

$$\begin{aligned}
&\rho\left(\frac{\partial u}{\partial t} + u\frac{\partial u}{\partial x} + v\frac{\partial u}{\partial y} + w\frac{\partial u}{\partial z}\right)\\
&\rho\left(\frac{\partial v}{\partial t} + u\frac{\partial v}{\partial x} + v\frac{\partial v}{\partial y} + w\frac{\partial v}{\partial z}\right)\\
&\rho\left(\underbrace{\frac{\partial w}{\partial t}}_{\text{Local acceleration terms}} + \underbrace{u\frac{\partial w}{\partial x} + v\frac{\partial w}{\partial y} + w\frac{\partial w}{\partial z}}_{\text{Convective acceleration terms}}\right)
\end{aligned} \qquad (4\text{-}2)$$

4-4.2 Elementary Acceleration Components

Following a procedure similar to that used in the study of the elementary motions of fluid particles (Section 2-6.2), that is, adding and subtracting $\frac{1}{2}\rho v(\partial v/\partial x)$ and $\frac{1}{2}\rho w(\partial w/\partial x)$ to the first line above, gives Equation 4-3, which emphasize the previous physical considerations.

Similar forms can be obtained for the y and z components of the forces.

4-4.3 Separation of Rotational Terms

It is often useful to transform the acceleration terms to a form which emphasizes both the kinetic energy terms and the rotational terms. Adding and subtracting $\rho[v(\partial v/\partial x) + w(\partial w/\partial x)]$ to the first line, gives the following expression, valid along the OX axis:

$$\rho\left[\frac{\partial u}{\partial t} + \left(u\frac{\partial u}{\partial x} + v\frac{\partial v}{\partial x} + w\frac{\partial w}{\partial x}\right) + v\left(\frac{\partial u}{\partial y} - \frac{\partial v}{\partial x}\right) + w\left(\frac{\partial u}{\partial z} - \frac{\partial w}{\partial x}\right)\right]$$

But

$$u\frac{\partial u}{\partial x} + v\frac{\partial v}{\partial x} + w\frac{\partial w}{\partial x} = \frac{1}{2}\frac{\partial}{\partial x}(u^2 + v^2 + w^2) = \frac{\partial}{\partial x}\frac{V^2}{2}$$

When the coefficients of the rotational vector

$$2\eta = \left(\frac{\partial u}{\partial z} - \frac{\partial w}{\partial x}\right)$$

$$2\zeta = \left(\frac{\partial v}{\partial x} - \frac{\partial u}{\partial y}\right)$$

are introduced, the following expression for the inertia forces along the OX axis results

$$\rho\left[\frac{\partial u}{\partial t} + \frac{\partial}{\partial x}\left(\frac{V^2}{2}\right) + 2(\eta w - \zeta v)\right]$$

Similarly, it may be found that the inertia forces along the OY and OZ axes are

$$\rho\left[\frac{\partial v}{\partial t} + \frac{\partial}{\partial y}\left(\frac{V^2}{2}\right) + 2(\zeta u - \xi w)\right]$$

$$\rho\left[\frac{\partial w}{\partial t} + \frac{\partial}{\partial z}\left(\frac{V^2}{2}\right) + 2(\xi v - \eta u)\right].$$

$$\rho\Bigg[\underbrace{\frac{\partial u}{\partial t}}_{\substack{\text{Local acceleration}\\ \text{resulting in a change in}\\ \text{translatory motion}}} + \overbrace{\underbrace{u\frac{\partial u}{\partial x}}_{\substack{\text{Acceleration}\\ \text{in linear}\\ \text{deformation}}} + \underbrace{\frac{1}{2}v\left(\frac{\partial v}{\partial x} + \frac{\partial u}{\partial y}\right) + \frac{1}{2}w\left(\frac{\partial u}{\partial z} + \frac{\partial w}{\partial x}\right)}_{\substack{\text{Acceleration}\\ \text{in angular}\\ \text{deformation}}} + \underbrace{\frac{1}{2}w\left(\frac{\partial u}{\partial z} - \frac{\partial w}{\partial x}\right) - \frac{1}{2}v\left(\frac{\partial v}{\partial x} - \frac{\partial u}{\partial y}\right)}_{\substack{\text{Acceleration in}\\ \text{rotation}}}}^{\text{Convective acceleration terms}}\Bigg] \qquad (4\text{-}3)$$

These three expressions may be written more concisely in vector notation as shown in Equation 4-4.

$$\rho\Bigg(\underbrace{\frac{\partial \mathbf{V}}{\partial t}}_{\substack{\textit{Local}\\ \textit{acceleration}}} + \underbrace{\underbrace{\mathbf{grad}\,\frac{V^2}{2}}_{\substack{\textit{Kinetic energy}\\ \textit{term}}} + \underbrace{\mathbf{curl}\,\mathbf{V}\times\mathbf{V}}_{\substack{\textit{Rotational}\\ \textit{term}}}}_{\textit{Convective acceleration}}\Bigg). \tag{4-4}$$

It has to be noticed that the convective inertia term

$$\rho\,\mathbf{grad}\,\frac{V^2}{2} \equiv \rho\left[\mathbf{i}\frac{\partial}{\partial x}\left(\frac{V^2}{2}\right) + \mathbf{j}\frac{\partial}{\partial y}\left(\frac{V^2}{2}\right) + \mathbf{k}\frac{\partial}{\partial z}\left(\frac{V^2}{2}\right)\right]$$

is, in fact, the derivative with respect to space of the kinetic energy, $\rho V^2/2$, of the particle.

4-5 On Some Approximations

4-5.1 Cases Where the Local Acceleration Is Neglected

4-5.1.1 A rigorous steady motion never exists. There is always a beginning and an end. However, many motions in hydraulics are actually very close to being steady during a given interval of time. In this case, since $\mathbf{V}$ does not vary with time, the corresponding inertia term $\partial \mathbf{V}/\partial t$ is zero. (The very important case of turbulent motion will be studied in Chapters 7 and 8.)

However, there are many unsteady motions in hydraulic engineering in which the local acceleration and the corresponding inertia terms are neglected.

This occurs when the velocities are slow and their variations with time are very slow. For instance, in the case of a periodic motion in which the period T is very long: $\partial \mathbf{V}/\partial t \approx \mathbf{V}/T$. Hence $\rho(\partial \mathbf{V}/\partial t)$ would be negligible in comparison with other forces.

Some particular cases where this approximation is valid are (1) flow in a porous medium: variation of the ground water table with respect to time; (2) flood wave in a river; (3) variation of level in a reservoir because of the variation of the upstream river flux, the spillway and bottom outlet control, and turbined discharge; and (4) emptying of a basin by a small valve. In all these cases, the flow is considered as a succession of steady motions and calculated as such without taking account of local inertia.

4-5.1.2 As an example of unsteady motion analyzed as a succession of steady motions, the variation of level in a basin is studied.

Consider the emptying of a rectangular basin of horizontal cross-sectional area A. The volume of water above a small hole of area S is A times z. The variation of z as a function of time is given by the differential equation $A\,dz = Q\,dt$ where $Q = C_d S V$, C_d being the coefficient of discharge and V being given by the formula of Torricelli: $V = (2gz)^{1/2}$. The formula of Torricelli is well known in elementary hydraulics, but it is important to point out herewith that this formula is only valid when the local inertia is negligible.

Introducing the value of $Q = C_d S(2gz)^{1/2}$ in the above equation and integrating gives the total time required to empty such a basin, (z_0 is the initial depth)

$$T = \frac{2A}{C_d S(2g)^{1/2}}\,(z_0)^{1/2}$$

If S were large and A small, it would be necessary to take account of the local inertia to calculate T.

Another similar example, previously cited, is that of the variation of level in a reservoir of horizontal section $S(z)$ because of the variation of the upstream river flux. The corresponding calculation of the economical height of the dam, the number of turbines and the spillway capacity are deduced from Equation 4-5.

4-5.2 Cases Where the Convective Acceleration Is Neglected or Approximated

4-5.2.1 The local inertia term is proportional to the velocity $\mathbf{V}$ and thus, it is a linear term. The convective acceleration terms are quadratic and are proportional to V^2 (or a product u^2, v^2, w^2, uv, uw, vw). Since the convective

$$\underbrace{Q(t)\,\Delta t}_{\substack{\text{Upstream}\\ \text{influx of}\\ \text{the river}}} = \underbrace{S(z)\,\Delta z}_{\substack{\text{Change of}\\ \text{level in}\\ \text{reservoir}}} + \underbrace{Q_t\,\Delta t}_{\substack{\text{Turbined}\\ \text{volume}}} + \underbrace{Q_s(z)\,\Delta t}_{\substack{\text{Volume}\\ \text{over the}\\ \text{spillway}}} + \underbrace{f(S)\,\Delta t}_{\substack{\text{Loss by}\\ \text{evaporation}}} \qquad (4\text{-}5)$$

acceleration introduces a quadratic term, the general equation of momentum is nonlinear.

It is well known that it is easy to mathematically solve many linear differential systems of equations. But it is often difficult to solve a nonlinear system. This is the chief cause of difficulty in fluid mechanics. For this reason, it is helpful to know when it is possible to neglect this quadratic term.

When V tends to zero, a quadratic term proportional to V^2 tends to zero more rapidly than a linear term proportional to V. Hence, in practice, when V is small, V^2 is negligible and the convective inertia term is negligible in comparison with the other terms expressing the local inertia and applied forces. Such motions are called "slow motions."

Some examples of such motion are: (1) periodic gravity wave theory (first order of approximation); (2) flow in a porous medium, which obeys the linear law of Darcy (such motion is defined only by an equality of applied forces since the local acceleration is negligible); and (3) motion of a small sphere in a viscous fluid (Stokes' formula).

4-5.2.2 Sometimes a partial effect of the convective acceleration is taken into account by the use of an approximate solution given by a number of terms of a series. (Example: gravity wave theories to the second order, third order, etc. of approximation; laminar boundary layer theory; etc.)

4-5.2.3 Another method to take account of a partial effect of the convective acceleration is by the assumption of irrotationality when the friction effects are negligible as seen previously.

4-5.2.4 Sometimes only some terms of the convective acceleration may be neglected. The case of a two-dimensional boundary layer on a flat plate is given here as an example (Fig. 4-3). This example is particularly helpful in understanding how the mathematical simplifications may be based on physical considerations. Hypothesis: u is large in comparison with v; the derivatives with respect to y are large compared to the derivatives with respect to x. The continuity equation for two-dimensional motion shows also that $\partial u/\partial x$ and $\partial v/\partial y$ are of the same order. (Section 3-1.2). Hence, the OY components of the convective inertia terms are negligible because

$$u\frac{\partial u}{\partial x} + v\frac{\partial u}{\partial y} \gg u\frac{\partial v}{\partial x} + v\frac{\partial v}{\partial y}$$

Similar approximations are made to analyze the development of a jet, and in the nonlinear long wave theory (Chapter 18).

Figure 4-3 *Introduction of simplifying assumptions in the theory of development of a boundary layer.*

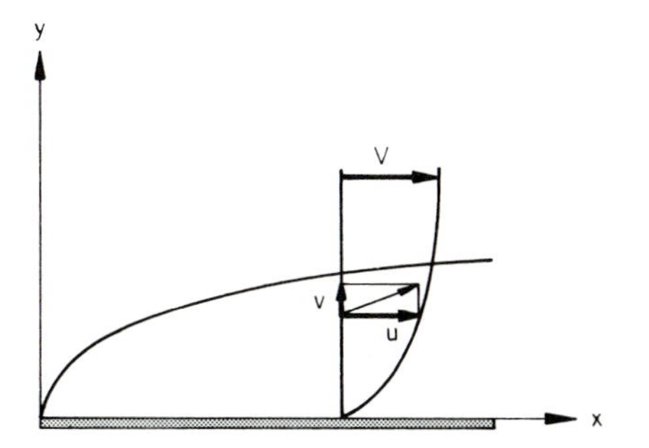

4-5.2.5 Linearizing the quadratic terms consists, for instance, of substituting for the quadratic terms, $y = AV^2$ or $dy = A_1V\,dV$, the linear terms, $y_1 \cong BV$ or $dy_2 \cong B_1V$ or $dy_2 \cong B_2\,dV$, such that B = mean value of AV, B_1 = mean value of $A_1\,dV$, and B_2 = mean value of A_1V.

This method is applicable when V or dV varies within a small range; otherwise, the value of B, B_1, B_2 must be changed and the problems solved step by step. Surge tank stability calculations, laminar flow stability study and motion of a sphere in a viscous fluid (Oseen's theory) are some examples where this method is used. The Oseen theory is an attempt to improve the Stokian theory for the flow around a sphere by taking into account some linearized effects of convective inertia.

4-6 Geostrophic Acceleration

An additional inertial force due to the rotation of the earth is now considered. This force is important in the study of tidal motion, oceanic circulation, and storm surge. (The storm surge phenomena is the piling up of seawater along the coast due to the wind stress generated by storm and hurricane.) It is called the Coriolis effect, which, in this case, is due to the geostrophic acceleration.

4-6.1 The Coriolis Effect

4-6.1.1 Different parts of the earth rotate at different speeds, depending upon their distance from the earth's axis. Hence a fluid particle in the northern hemisphere moving south, toward the equator, finds itself over a portion of the earth which rotates faster than where it comes from. Therefore, it appears for an observer fixed to the earth to fall behind the rotating earth and the seabottom underneath. Since the earth rotates from west to east, the deflection is to the west, i.e., to the right of the direction of motion with respect to the earth. The fluid particle will describe a path westward (Fig. 4-4). Similarly, a fluid particle moving northward from the equator has, by inertia, an excess of eastward velocity relative to the seabottom underneath, which is moving slower. Consequently, its path turns

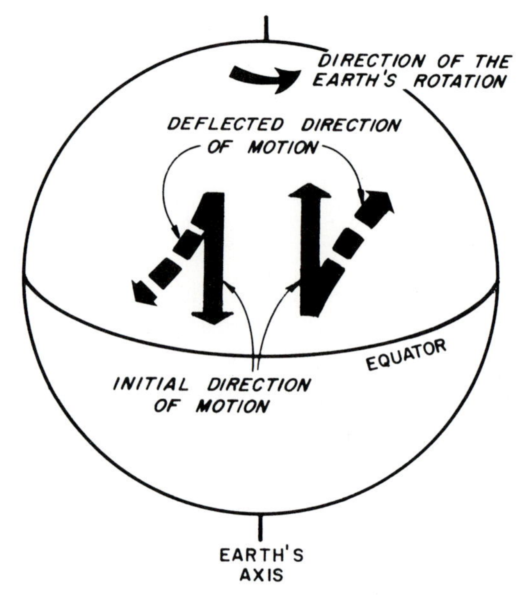

Figure 4-4 *The Coriolis effect in the northern hemisphere.*

eastward and the deflection is again to the right of the direction of motion. In the southern hemisphere, the same reasoning applies, and the deflection is always found to the left of the direction of motion (Fig. 4-5).

4-6.1.2 Let us now consider a fluid particle which rotates with the earth. Its velocity about the earth's axis creates a centrifugal force. Since the fluid particle moves at the same speed as the earth, this force just balances the component of gravity force which is perpendicular to the earth's axis. If now it moves eastward with respect to the earth along a parallel in the northern hemisphere, the particle experiences an additional centrifugal force. Since the particle is moving faster than the speed for which the balance is possible, the particle tends to move further away from the earth's axis. Therefore, it is deflected towards the equator, where even though the distance from the earth's center is the same, the distance to the earth's axis is increased. This deviation tends to restore the balance

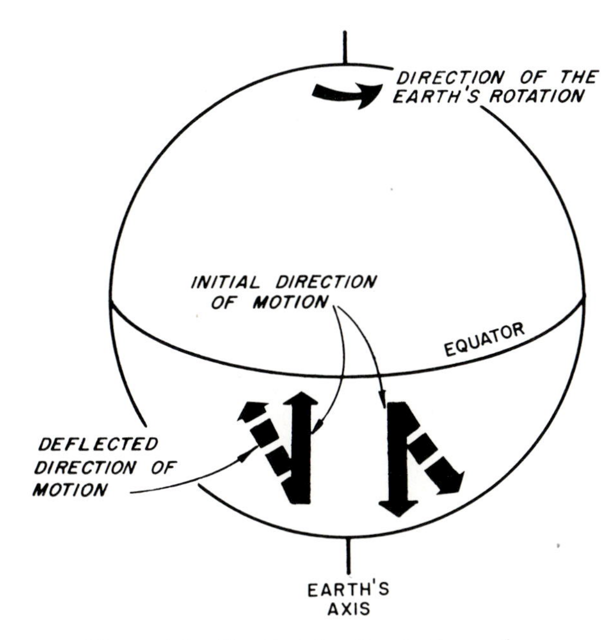

Figure 4-5 *The Coriolis effect in the southern hemisphere.*

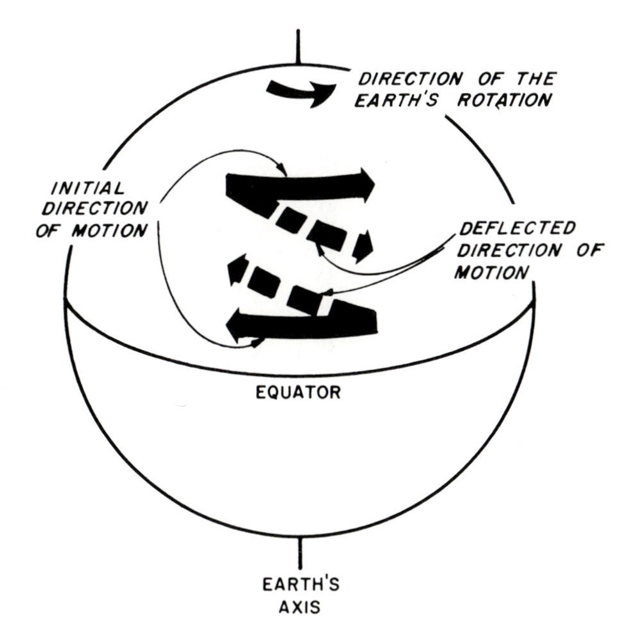

Figure 4-6 *The Coriolis effect in the eastward and westward motion in the northern hemisphere.*

of forces (Fig. 4-6). For a fluid particle moving westward the centrifugal force is not sufficient to balance the component of gravity force. Therefore, the fluid particle is deflected toward the Pole.

In the northern hemisphere, the deflection is to the right, and in the southern hemisphere, the deflection is to the left.

4-6.1.3 In the case where a component of motion is perpendicular to the earth's axis, a similar effect is observed. Angular momentum is conserved. Therefore, if the particle goes downward, towards the center of the earth, it will be rotating faster than an earth fixed particle at the same location and therefore it will deflect eastward. If the particle goes upward, away from the center of the earth, it will not be rotating sufficiently fast, and therefore will deflect westward. Since most oceanic motions have very small velocity components, this effect is most often neglected. All these deviations are called the Coriolis effect.

4-6.2 Geostrophic Inertial Components

4-6.2.1 Let us consider fluid particle A at a latitude Φ. The horizontal velocity component U has a southerly component U_S and an easterly component U_E. The vertical component is W (Fig. 4-7). U_S may be resolved into $U_S \sin \Phi$, along a perpendicular AB to the earth rotation axis NS, and a component $U_S \cos \Phi$ parallel to the earth's axis and directed toward the south. The vertical component W, can also be resolved into a component $W \cos \Phi$, along AB, and a component $W \sin \Phi$ parallel to NS and directed toward the south. The components $U_S \cos \Phi$, $W \sin \Phi$, being parallel to the earth's axis, do not contribute to modifying the distance of the fluid particle from the earth's axis, and therefore, do not result in a Coriolis effect. The components U_E, $U_S \sin \Phi$, and $W \cos \Phi$ lie in a plane which is perpendicular to the earth's axis, and therefore contribute to deflect the path of the fluid particle.

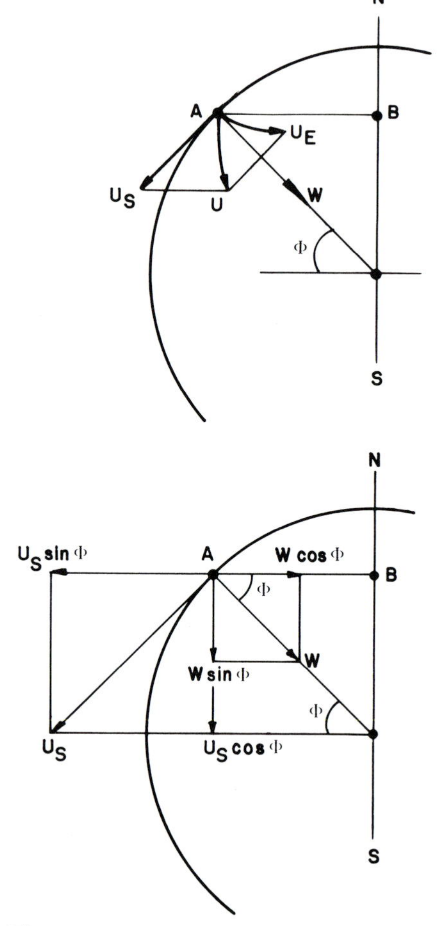

Figure 4-7 *Velocity components of motion.*

4-6.2.2 Let us now consider the components of motion perpendicular to the earth's axis, U_E, $U_S \sin \Phi$, and $W \cos \Phi$. This is equivalent to considering a particle A which moves in a plane perpendicular to the earth's axis. This plane cuts the earth's axis NS at B, and let $BA = r$ (Fig. 4-8). Let M be a fixed point on this plane, which rotates with the earth. The position of A is defined by the latitude Φ, $BA = r$, and the angle $MBA = \theta$. It is assumed that θ is positive when A is to the east of M. Owing to the earth's rotation alone, the fluid particle A has a velocity $r\omega$ toward the east. (ω is the angular rotation of the earth.) The velocity components relative to the earth are $U_E = r(d\theta/dt)$ toward the east, and $(U_S \sin \Phi - W \cos \Phi) = dr/dt$ along BA. Therefore, the resultant velocity has an easterly component $r[\omega + (d\theta/dt)]$ and a radial component dr/dt.

4-6.2.3 Consider a vector defined by the two components $X = R \cos \alpha$ and $Y = R \sin \alpha$. The derivatives of this vector with respect to time are given by the derivatives of its components, i.e.,

$$\frac{dX}{dt} = \frac{dR}{dt} \cos \alpha - R \sin \alpha \frac{d\alpha}{dt}$$

$$\frac{dY}{dt} = \frac{dR}{dt} \sin \alpha + R \cos \alpha \frac{d\alpha}{dt}$$

The derivative of this vector is thus given by a sum. The first part, in the direction of the vector, has the magnitude dR/dt. The second part has a magnitude $R(d\alpha/dt)$.

Since

$$-\sin \alpha = \cos\left(\alpha + \frac{\pi}{2}\right) \quad \text{and} \quad \cos \alpha = \sin\left(\alpha + \frac{\pi}{2}\right),$$

the second part is oriented perpendicular to the vector (X, Y) and directed toward the left of this vector.

We are concerned with the vector components in a plane XBY, perpendicular to the earth axis NS. It is seen

Figure 4-8 *Plane perpendicular to the earths axis.*

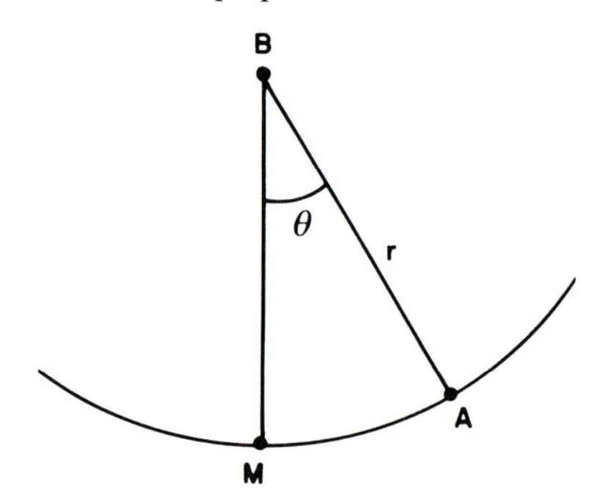

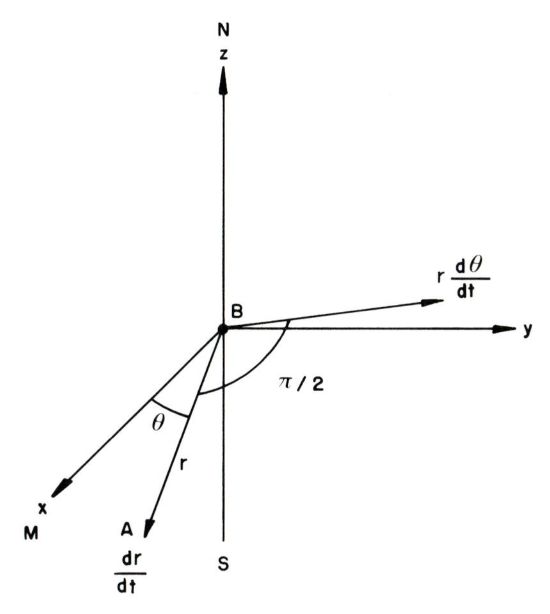

Figure 4-9 *Components of the derivative of a vector r.*

from Fig. 4-9 that the component $r(d\theta/dt)$, perpendicular to the vector (X, Y) is actually to the east of this vector.

4-6.2.4 The previous results will be applied to both the radial component, dr/dt, and the easterly component, $r[\omega + (d\theta/dt)]$, of the velocity vector. The acceleration or time rate of increase of dr/dt has a component d^2r/dt^2 in the same radial direction which is due to centrifugal acceleration, and an easterly component perpendicular to BA which is equal to

$$\frac{dr}{dt}\left(\omega + \frac{d\theta}{dt}\right)$$

Similarly, the acceleration or time rate of increase of the easterly component $\{r[\omega + (d\theta/dt)]\}$ has a component,

$$\frac{dr}{dt}\left(\omega + \frac{d\theta}{dt}\right) + r\frac{d^2\theta}{dt^2}$$

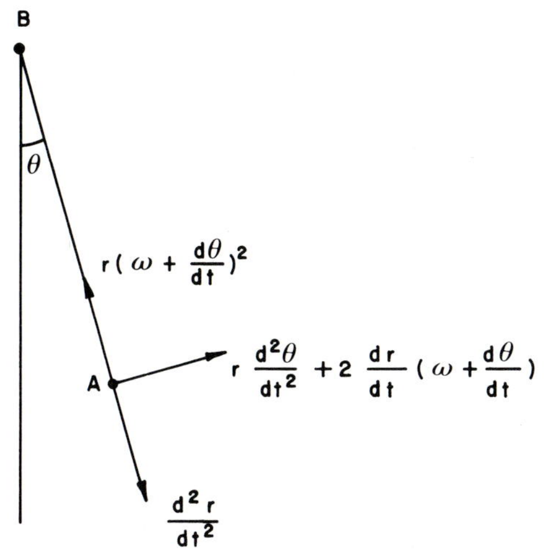

Figure 4-10 *Components of acceleration.*

in the easterly direction, and a radial component equal to

$$r\left(\omega + \frac{d\theta}{dt}\right)^2$$

The total acceleration is composed of these four components, as shown on Fig. 4-10.

If the radial component $r[\omega + (d\theta/dt)]^2$ is written as $r[\omega^2 + 2\omega(d\theta/dt) + (d\theta/dt)^2]$, further observations can be made.

1. The terms which are independent from ω

$$\frac{d^2r}{dt^2} \qquad 2\left(\frac{dr}{dt}\right)\left(\frac{d\theta}{dt}\right) + r\frac{d^2\theta}{dt^2} \quad \text{and} \quad -r\left(\frac{d\theta}{dt}\right)^2$$

   are the acceleration of the particle relative to the earth.
2. The term $r\omega^2$, directed toward B, is the centrifugal acceleration.
3. The two terms $2\omega(dr/dt)$ and $2\omega r(d\theta/dt)$ are perpendicular to and to the left of the velocity components

$$\frac{dr}{dt} = U_S \sin\phi - W\cos\phi \qquad \text{and} \qquad r\frac{d\theta}{dt} = U_E$$

   respectively and are the geostrophic acceleration terms.

4-6.2.5 The corresponding inertial components are attained by multiplying these accelerations by the density ρ. Therefore the corresponding geostrophic inertial forces are

$$2\rho\omega(U_S \sin \Phi - W \cos \Phi)$$

directed horizontally toward the east, and

$$2\rho\omega U_E$$

directed along BA. Now, $(2\rho\omega U_E)$ has a horizontal component $(2\rho\omega U_E \sin \Phi)$ and a vertical downward component $(2\rho\omega U_E \cos \Phi)$. $\mathbf{U}_E + \mathbf{U}_S = \mathbf{U}$ is the vector sum of the velocity components. The vector sum of $(2\omega U_S \sin \phi)$ and $(2\omega U_E \sin \phi)$ is $(2\omega U \sin \phi)$ along a horizontal perpendicular to the direction of $\mathbf{U}$ and to the left of $\mathbf{U}$. Finally, the geostrophic inertial force is composed of three terms:

1. A horizontal component $2\rho\omega U \sin \Phi$ perpendicular to and to the left of U. In the southern hemisphere, ϕ is negative, and the horizontal component $2\rho\omega U \sin \Phi$ is directed to the right of U.
2. A horizontal component $2\rho\omega W \cos \Phi$ directed toward the west (due to the negative sign appearing in the previous expression), and generally negligible as the motions are generally so nearly horizontal that W/U is very small.
3. A vertical component $2\rho\omega U_E \cos \Phi$ directed downward (or upward if the sign of U_E is negative) which is added or subtracted to the gravity, and which is also negligible compared to gravity.

4-6.3 Total Inertial Components for a Nearly Horizontal Motion

In the case of nearly horizontal motion, only the term $(2\rho\omega U \sin \Phi)$ is to be considered. $\Omega = 2\omega \sin \Phi$ is the Coriolis parameter. Let us now have another two-dimensional axis system, X, Y, fixed with respect to the earth's surface and horizontal. The velocity components of U are defined as u and v, respectively. Since the Coriolis force acts perpendicularly and to the left of U, the two geostrophic terms in this system of coordinates are: $-\Omega v$ along the OX axis and $+\Omega u$ along the OY axis. Adding the local and convective forces, with the vertical component w neglected, the total inertial forces are

$$\rho\left(\frac{\partial u}{\partial t} + u\frac{\partial u}{\partial x} + v\frac{\partial u}{\partial y} - \Omega v\right)$$

along the OX axis, and

$$\rho\left(\frac{\partial v}{\partial t} + u\frac{\partial v}{\partial x} + v\frac{\partial v}{\partial y} + \Omega u\right)$$

along the OY axis.

In the case of a periodic motion of frequency k, the local acceleration term has an amplitude kU_0, U_0 being the maximum horizontal velocity. The Coriolis acceleration amplitude is ΩU_0. Therefore, it appears that the Coriolis effect is important when Ω/k is large, i.e., when the wave period is of the order of magnitude of a day, which is the case of tidal motion and storm surge.

PROBLEMS

4.1 Consider an unsteady two-dimensional flow where the velocity components of $x = L$, $y = L$ at time $t = 0$ are $u = L/t_0$, $v = 2L/t_0$ and at time $t = t_0$ are $u = 2L/t_0$, $v = 3L/t_0$.

Moreover, at time $t = 0$, the velocity components at point $x = 2L$, $y = L$, are $u = 1.2L/t_0$ and $v = 2.4L/t_0$ and at point $x = L$, $y = 2L$ they are $u = 1.1L/t_0$, $v = 1.8L/t_0$. Calculate the value of the total acceleration at $x = L$, $y = L$ by assuming that the variations of velocity with time and distance are linear.

4.2 Calculate the total variation of temperature of a train which travels 300 miles a day in the northern direction. The mean daily variation of temperature is $-2°$F per 1000 miles. The daily variation at a given location is $4 \sin (2\pi t/T)°$F where $T = 24$ hr.

4.3 In the case of a progressive acoustic wave in a pipe, such that

$$u = a \sin \frac{2\pi}{L}(x - Ct)$$

calculate the ratio of convective inertia to local inertia.

4.4 Consider a basin such as shown in Fig. 4-11, where the particle velocity at the orifice is $V = (2g\eta)^{1/2}$, $z = \eta$ is the height of the free surface above the orifice. The horizontal area of the basin being $A = 4$ ft^2, (3716 cm^2), determine:

1. The variation of the free surface with respect to time, i.e., the function $\eta(t)$. At time $t = 0$, $\eta_0 = 20$ ft (609 cm). The local inertia will be neglected for this calculation.
2. The local inertia at point $x = 3$ ft (91.44 cm) $z = 0$, i.e., in the converging section.
3. The convective inertia at the same point.
4. Is the neglect of the local inertia a valid assumption? Explain. The friction will be neglected.

Figure 4-11

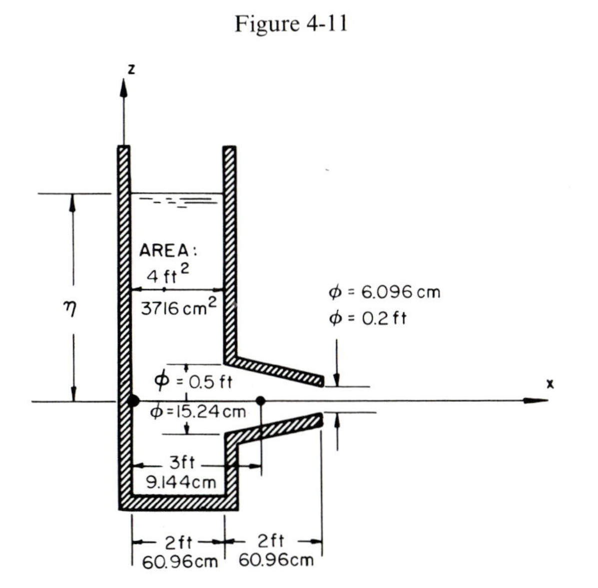

5. When a constant discharge $q_0 = 1$ ft^3/sec (28,317 cm^3/sec) is poured into the tank, establish the function $\eta(t)$ ($\eta_0 = 20$ ft, 609 cm) and determine η when $t \to \infty$.

The discharge coefficient C_D will be assumed to be unity.

4.5 Determine the convective inertia terms which can be neglected in a jet.

4.6 The influx of discharge into a reservoir is defined by the equation

$$Q(t) = 10{,}000\,(1.5 - \sin kt) \text{ ft}^3/\text{sec}$$
$$= [283\,(1.5 - \sin kt) \text{ m}^3/\text{sec}]$$

where $k = 2\pi/T$ and T is a period of 1 year. The horizontal area of the reservoir is defined by

$$A(z) = 10{,}000z^2 \text{ ft}^2 \quad (929z^2 \text{ m}^2)$$

The top of the spillway for flood discharge is located at an elevation $z_1 = 100$ ft (30.48 m) and has a discharge capacity

$$Q_S = Cl[z - z_1][2g(z - z_1)]^{1/2}$$

where the coefficient of discharge $C = 0.5$, and the length of the spillway $l = 100$ ft (30.48 m). The turbined discharge is constant and equal to 7000 ft^3/sec (198 m^3/sec).

Determine the variation of the level of the free surface in the reservoir as a function of time and the maximum discharge over the spillway for each year following time $t = 0$. The maximum possible discharge over the spillways will also be determined. At time $t = 0$, one will take the free surface elevation at $z = 30$ ft (9.144 m).

4.7 Consider a periodic two-dimensional oscillation in a rectangular tank (seiche) of length l and depth $d(l \gg d)$ (Fig. 4-12). The period of oscillation is $T = 2l/(gd)^{1/2}$. The horizontal velocity component u is assumed to be a constant along a vertical and is a function of time only. The equation of the free surface is ($a \ll d$):

$$h = d + a \cos mx \cos kt$$
$$m = 2\pi/L \qquad 2l = L = T(gd)^{1/2} \qquad k = 2\pi/T$$

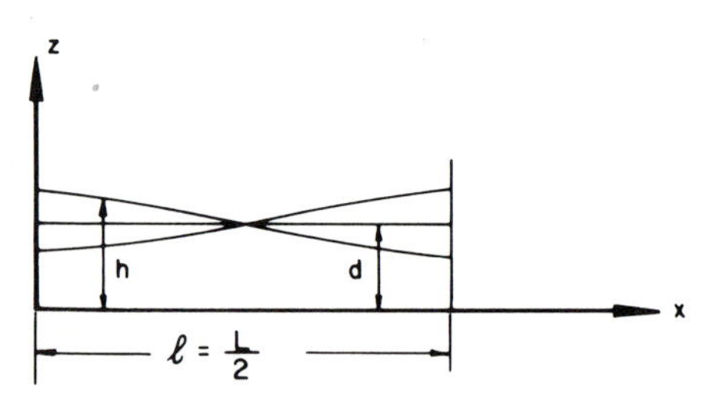

Figure 4-12

Determine the maximum value of u and w and the location where they are maximum. Some simplifying assumptions will be accepted for these calculations. Determine the expression for the local inertia and the convective inertia, their maximum values, and the ratio of the maximum value of convective inertia to the maximum value of the local inertia. Present a criterion permitting the evaluation when the convective inertia is negligible.

4.8 Demonstrate that

$$\frac{du}{dt} = \frac{\partial u}{\partial t} + 2(w\eta - v\zeta) + \frac{\partial}{\partial x}\left(\frac{V^2}{2}\right)$$

and find similar expressions for dv/dt and dw/dt.

4.9 Express the components of acceleration A_r, A_θ, A_z in a cylindrical system of coordinates.

4.10 Define the components of velocity u, v, w in a natural system of coordinates, i.e., the axes OX and OY are defined from a given point along and perpendicularly, respectively, to a streamline. Give the components of acceleration. In a second case, where it is assumed that paths and streamlines are different, the axis will then be defined at a given point with respect to paths. Then, give also the components of acceleration (R is the radius of curvature of the path).

4.11 Demonstrate that in a spherical system of coordinates (r, θ, Φ) related to the earth, the operator d/dt is:

$$\frac{d}{dt} = \frac{\partial}{dt} + W\frac{\partial}{\partial r} + \frac{U_E}{r \sin \Phi}\frac{\partial}{\partial \theta} + \frac{U_S}{r}\frac{\partial}{\partial \Phi}$$

where W is the vertical velocity component, U_E is the east-west velocity component positive to the east, U_S is the north-south velocity component positive to the north, r is the distance from the center of the earth, Φ is the latitude, and θ is the longitude.

4.12 Compare the vertical component of geostrophic acceleration with the gravity, in the case where $U_E \cos \Phi = 1$ knot (1.688 ft/sec; 51.4 cm/sec).

Chapter 5

Applied Forces

5-1 Internal and External Forces

The applied forces on an elementary mass of fluid consist of internal forces and external forces.

5-1.1 Internal Forces

Internal forces result from the interaction of the interior points of the considered mass of fluid. According to the principle that action equals reaction, these internal forces balance in pairs and their sum is zero. Their total torque is also zero. However, the work of these internal forces is not zero. It is for this reason that it is important to mention their existence. For example, the headloss in a pipe is the result of the work of internal viscous forces.

5-1.2 External Forces

The forces on the boundaries of the considered particle of fluid, called surface forces, and the forces which are always acting in the same direction on its mass, called body or volume forces, are not balanced. These are the external forces.

5-1.2.1 *Surface forces* result from forces acting on the outside of the considered volume. They are caused by molecular attraction. They decrease very quickly away from the boundaries of the particle of fluid, and their action is limited to a very thin layer. In practice, if the fluid is a continuous medium, this layer can be considered infinitesimally thin and blended with the surface of the particle. These surface forces can be divided into (1) normal forces—due to pressure; and (2) shearing forces—due to viscosity. These two kinds of forces also exist within the particle, but are always balanced in pairs and their sum is zero, as previously noted.

5-1.2.2 *Body forces* result from an external field (such as gravity or a magnetic field) which acts on each element of the considered volume in a given direction. For this reason, they are called body or volume forces. Except for some rare cases, for example the study of the motion of a fluid metal in a magnetic pump, only gravitational force has to be considered in fluid mechanics. In general, this gravitational force can be considered as acting in the same fixed direction. However in some studies, such as tidal motion and oceanic circulation, the gravity acceleration must be considered radial.

5-1.2.3 Other applied external forces are:

1. The capillary forces due to the difference of molecular attraction between two media are of particular importance in free surface flow through a porous medium and in small gravity waves such as generated by a gentle breeze. These are called capillary waves.
2. The geostrophic force caused by the Coriolis acceleration due to the earth's rotation is sometimes considered as a body force similar to the gravity force, even though it is really an inertial force.

5-2 Gravity Forces

Similar to the inertia forces, the volume forces are proportional to the mass of the fluid and to the acceleration caused by an external field. In the case of gravity action, the volume force per unit of volume is simply equal to the fluid weight: $\varpi = \rho g$, where g is the acceleration due to gravity. This force is independent of the motion. It is the same whether the fluid is static or in any viscous or turbulent motion.

The components X, Y, Z of the gravity force, expressed in a differential form in the three-axis system OX, OY, OZ are given below. The vertical axis OZ is considered as positive upward along the normal to the earth's surface. The components are $X = 0$; $Y = 0$; and $Z = -\rho g = -\partial/\partial z(\rho g z)$. In vector form, they are $-\mathbf{grad}(\rho g z)$, since $X = -\partial/\partial x(\rho g z) = 0$ and $Y = -\partial/\partial y(\rho g z) = 0$.

The gravity force is often neglected in gas dynamics except, for example, in meteorology or in the calculation for chimneys and ventilation openings, where the phenomena are influenced by the variations of gravity forces due to density changes.

5-3 Pressure Forces

Pressure forces result from the normal components of the molecular forces near the boundary of the considered volume. The magnitude of the pressure at a point is obtained by dividing the normal force against an infinitely small area by the area.

5-3.1 Pressure Magnitude, Pressure Force, and Direction

The pressure magnitude is a scalar quantity that is independent of the orientation of the area on which the force is applied. This may be demonstrated by considering a triangular two-dimensional element in a fluid at rest (Fig. 5-1).

Since there is no motion, inertia forces and viscous forces are zero, and the only forces are gravity and pressure. The projections of these forces along the OX and OY axes yield the equalities

$$p_x\, dy - p\, ds \sin \alpha = 0$$

$$p_y\, dx - p\, ds \cos \alpha = \rho g \frac{dx\, dy}{2}$$

Introducing $dy = ds \sin \alpha$, $dx = ds \cos \alpha$, and neglecting the second-order term, $\rho g(dx\, dy/2)$, one obtains $p = p_x$, $p = p_y$. Hence, $p = p_x = p_y$. Since α is any arbitrary angle,

Figure 5-1 *Pressure magnitude is independent of orientation.*

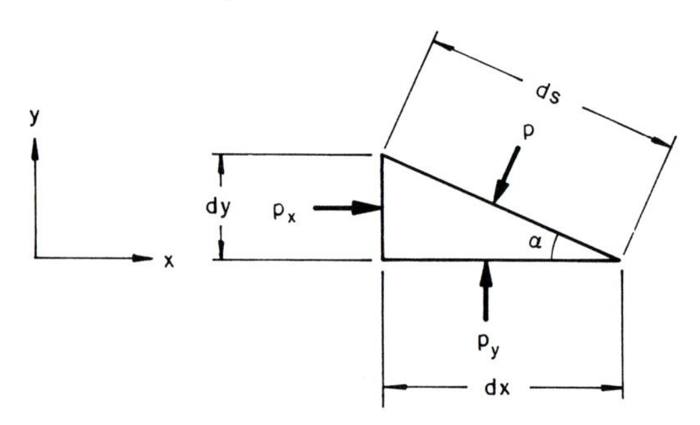

the pressure is seen to be the same in all directions. A similar demonstration is possible in three dimensions. However, it is evident that the gradient of pressure force (which is a vector) changes with direction. In the same way, the force caused by pressure against an area (which is a vector) changes direction as the normal to the considered area changes direction.

5-3.2 *Rate of Pressure Force Per Unit of Volume*

Consider an elementary fluid particle ($dx\ dy\ dz$) (Fig. 5-2). The pressure force due to the external adjacent fluid particle acting against the side $ABCD$ is p(area $ABCD$) = $p\ dy\ dz$. The pressure force against the other side acts in the opposite direction and may be written:

$$-\left(p + \frac{\partial p}{\partial x}dx\right)(\text{area } EFGH) = -\left(p + \frac{\partial p}{\partial x}dx\right)dy\ dz$$

Hence, the difference of pressure forces acting in opposite directions is

$$p\ dy\ dz - \left(p + \frac{\partial p}{\partial x}dx\right)dy\ dz = -\frac{\partial p}{\partial x}dx\ dy\ dz$$

Similarly, the pressure force differences acting in the OY and OZ directions are $-(\partial p/\partial y)\ dx\ dy\ dz$ and $-(\partial p/\partial z)\ dx\ dy\ dz$.

Figure 5-2 *Difference in pressure forces in a unit of volume.*

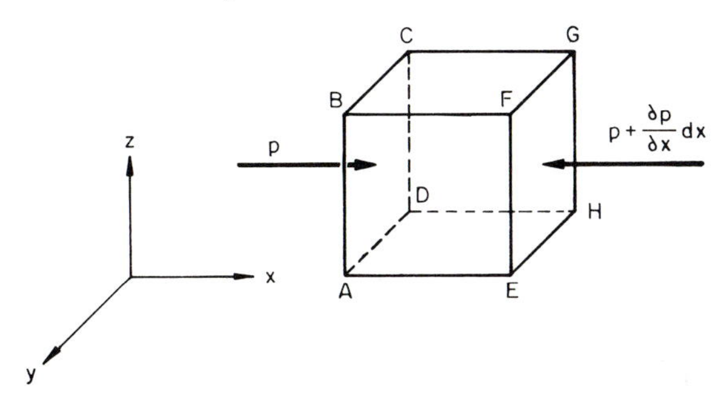

Hence, the rate of change of pressure force per unit of volume is given by the three components $-\partial p/\partial x$, $-\partial p/\partial y$, and $-\partial p/\partial z$, which can be written vectorially: $-$**grad** (p).

5-3.3 *Fluid Motion and Gradient of Pressure*

It is interesting to note that the motion of the fluid particle does not depend upon the absolute value of p, but only upon the gradient of p. Consider the motion in a tunnel. The motion depends upon the difference in pressure levels between the upstream and the downstream sections. Therefore, it is possible to operate a scale model experiment at any convenient arbitrary absolute pressure, provided the pressure gradient is kept in similitude. However, if the pressure level drops below a critical value, cavitation occurs and, for similitude, the scale model must be operated under partial vacuum (see Appendix B).

5-3.4 *Pressure and Gravity*

The total force due to the pressure force and gravity force per unit volume is

$$\mathbf{grad}\ p + \mathbf{grad}\ \rho g z = \mathbf{grad}\ (p + \rho g z)$$

The sum of these two linear quantities ($p + \rho g z$) is a constant in hydrostatics since $p - p_a = -\rho g z$ where p_a is a constant external pressure (atmospheric). This property is also verified in a cross section of a uniform flow as in a channel or in a pipe, or more generally when the curvature of the paths is negligible or the motion is very slow (see Section 10-2.1.1). Hence the sum ($p + \rho g z$) may often be conveniently replaced by the single term p^*: $p^* = p + \rho g z$. In hydrostatics p^* = constant. Whereas $p/\rho g$ is known as the *pressure head*, $p^*/\rho g$ is called the *piezometric head.*

5-4 Viscous Forces

5-4.1 *Mathematical Expression for the Viscous Forces*

Shear stresses are present because of fluid viscosity and are caused by the transfer of molecular momentum. The friction force τ is assumed to be proportional to the

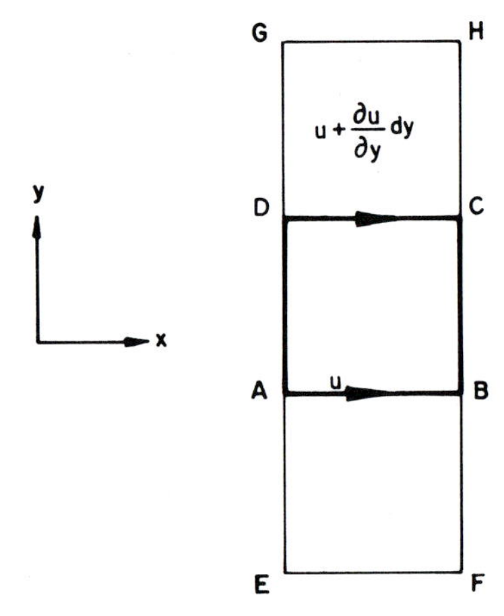

Figure 5-3 *Two-dimensional element of fluid.*

coefficient of viscosity μ and to the rate of angular deformation.

Consider a two-dimensional element of fluid (Fig. 5-3). The friction force on the side AB of length dx is: $\tau\, dx = \mu(\partial u/\partial y)\, dx$. Since the velocity at C is $(u + (\partial u/\partial y)\, dy)$, the friction force on the side CD is

$$\left(\tau + \frac{\partial \tau}{\partial y} dy\right) dx = \mu \frac{\partial}{\partial y}\left(u + \frac{\partial u}{\partial y} dy\right) dx$$

$$= \mu \frac{\partial u}{\partial y} dx + \mu \frac{\partial^2 u}{\partial y^2} dy\, dx.$$

These forces act in opposite directions. If the force due to the particle $GHCD$ acts in the OX direction on the side CD of $ABCD$, the force due to the particle $ABCD$ will act in the same direction of the AB side of a $ABFE$ and by reaction $ABFE$ will cause a force in the $-OX$ direction on $ABCD$. The total shear force thus becomes:

$$\frac{\partial \tau}{\partial y} dx\, dy = \mu \frac{\partial^2 u}{\partial y^2} dx\, dy$$

Dividing by $dx\, dy$, the friction force per unit of area is:

$$\frac{\partial \tau}{\partial y} = \mu \frac{\partial^2 u}{\partial y^2}$$

More generally, for a three-dimensional incompressible fluid, it is possible to demonstrate that the friction force components per unit of volume are:

$$\mu\left(\frac{\partial^2 u}{\partial x^2} + \frac{\partial^2 u}{\partial y^2} + \frac{\partial^2 u}{\partial z^2}\right) = \mu \nabla^2 u$$

$$\mu\left(\frac{\partial^2 v}{\partial x^2} + \frac{\partial^2 v}{\partial y^2} + \frac{\partial^2 v}{\partial z^2}\right) = \mu \nabla^2 v$$

$$\mu\left(\frac{\partial^2 w}{\partial x^2} + \frac{\partial^2 w}{\partial y^2} + \frac{\partial^2 w}{\partial z^2}\right) = \mu \nabla^2 w$$

They are written vectorially:

$$\mu\left(\frac{\partial^2 \mathbf{V}}{\partial x^2} + \frac{\partial^2 \mathbf{V}}{\partial y^2} + \frac{\partial^2 \mathbf{V}}{\partial z^2}\right) = \mu \nabla^2 \mathbf{V}$$

5-4.2 *Approximations Made on Viscous Forces*

It has been shown experimentally that it is sometimes possible to consider friction effects as negligible. The preceding expression for the friction forces show that they can be neglected when the Laplacian of the velocity components ($\nabla^2 \mathbf{V}$) is small. This is often true outside the boundary layer where the fluid motion is similar to that of a perfect fluid.

Sometimes it is possible to neglect only one part of the viscous friction terms. For example, as similarly explained in Section 4-5.2.4, in a two-dimensional laminar boundary layer or in a jet, $\partial^2 v/\partial x^2$ and $\partial^2 v/\partial y^2$ and also $\partial^2 u/\partial x^2$ may be neglected in comparison to $\partial^2 u/\partial y^2$, and only the friction force $\mu(\partial^2 u/\partial y^2)$ has to be taken into account.

5-5 Some Theoretical Considerations of Surface Forces

5-5.1 *A General Expression for Surface Forces*

Surface forces, as previously seen, consist of pressure force and friction force. These surface forces may be introduced without consideration of their physical nature. The advantage of so expressing the surface forces lies in its applicability to any kind of motion, e.g., perfect, viscous or turbulent, compressible or incompressible. However, the values of these surface forces are expressed differently when their physical nature is taken into account.

5-5.2 The Nine Components of the External Forces: Components of Lamé

5-5.2.1 Consider an elementary mass of fluid in the form of a cube; its edges are parallel to the three coordinate axes OX, OY, and OZ, as shown in Fig. 5-4.

On each side of this elementary cube, surface forces may be completely defined by three components parallel to the three coordinate axes. Two of these components are shear stresses while the third is a normal stress.

Since a cube has six sides, 18 components have to be considered. These components are defined with the help of two subscripts. σ are the normal forces and τ are the shearing forces. The first subscript x, y, or z refers to the axis normal to the considered side. The second subscript x, y, or z refers to the direction in which the force acts.

The pairs of parallel forces acting on two opposite sides of the cube act in opposite directions, and their difference is obtained by a simple partial derivative with respect to the distance between the two considered sides.

Hence, the external forces may be defined by a tensor of rank two:

$$\begin{vmatrix} \sigma_{xx} & \tau_{yx} & \tau_{zx} \\ \tau_{xy} & \sigma_{yy} & \tau_{zy} \\ \tau_{xz} & \tau_{yz} & \sigma_{zz} \end{vmatrix}$$

These forces are completely given by Table 5-1. Now, the addition of all the forces per unit volume acting in the same direction yields

In the OX direction:

$$\left(\frac{\partial \sigma_{xx}}{\partial x} + \frac{\partial \tau_{yx}}{\partial y} + \frac{\partial \tau_{zx}}{\partial z}\right)$$

In the OY direction:

$$\left(\frac{\partial \tau_{xy}}{\partial x} + \frac{\partial \sigma_{yy}}{\partial y} + \frac{\partial \tau_{zy}}{\partial z}\right)$$

In the OZ direction:

$$\left(\frac{\partial \tau_{xz}}{\partial x} + \frac{\partial \tau_{yz}}{\partial y} + \frac{\partial \sigma_{zz}}{\partial z}\right)$$

5-5.2.2 On the other hand, consider the torque of a fluid particle about one point (for example, A in Fig. 5-5).

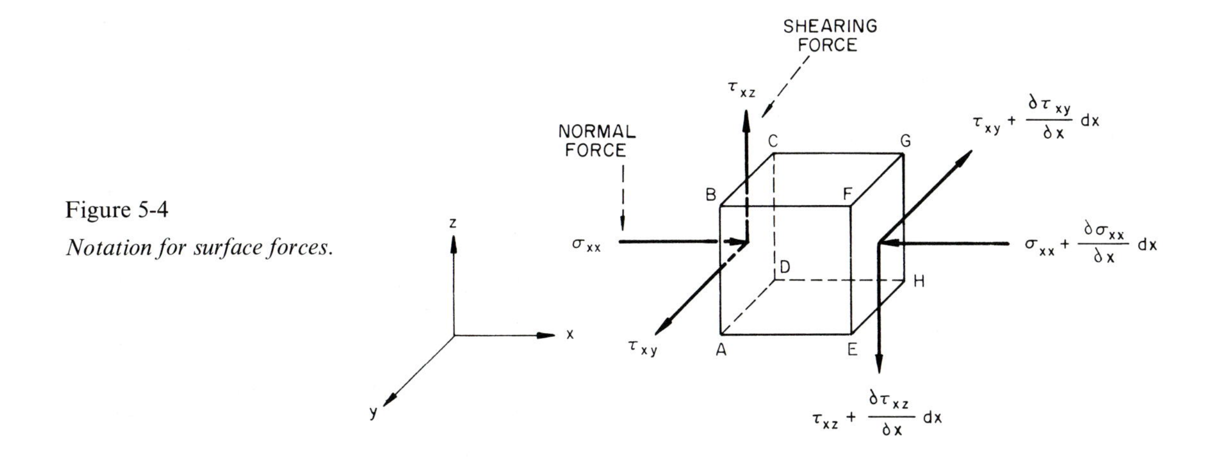

Figure 5-4
Notation for surface forces.

Table 5-1 *External forces*

		Stresses applied to the side normal to the axis					
		OX		*OY*		*OZ*	
On the side of area (see Fig. 5-4)		*ABCD* $dy\,dz$	*EFGH* $dy\,dz$	*DHGC* $dx\,dz$	*AEFB* $dx\,dz$	*AEHD* $dx\,dy$	*BFGC* $dx\,dy$
In the direction of	*OX*	σ_{xx}	$\left(\sigma_{xx} + \frac{\partial \sigma_{xx}}{\partial x} dx\right)$	τ_{yx}	$\left(\tau_{yx} + \frac{\partial \tau_{yx}}{\partial y} dy\right)$	τ_{zx}	$\left(\tau_{zx} + \frac{\partial \tau_{zx}}{\partial z} dz\right)$
	OY	τ_{xy}	$\left(\tau_{xy} + \frac{\partial \tau_{xy}}{\partial x} dx\right)$	σ_{yy}	$\left(\sigma_{yy} + \frac{\partial \sigma_{yy}}{\partial y} dy\right)$	τ_{zy}	$\left(\tau_{zy} + \frac{\partial \tau_{zy}}{\partial z} dz\right)$
	OZ	τ_{xz}	$\left(\tau_{xz} + \frac{\partial \tau_{xz}}{\partial x} dx\right)$	τ_{yz}	$\left(\tau_{yz} + \frac{\partial \tau_{yz}}{\partial y} dy\right)$	σ_{zz}	$\left(\sigma_{zz} + \frac{\partial \sigma_{zz}}{\partial z} dz\right)$

The total sum of the torque caused by the shearing stresses is

$$\tau_{xy}(dy\,dz)\,dx - \tau_{yx}(dz\,dx)\,dy$$

This torque is equal to the mass times the square of the radius of gyration $(dR)^2$, times the square of the angular velocity (ω^2), which may be written $\rho\,dx\,dy\,dz(dR)^2\omega^2$.

Since dR is infinitesimally small, having the same order as dx, dy, and dz, $(dR)^2$ is of the second order and the speed of gyration becomes infinite, which is physically impossible. Hence, the total torque must be zero. This condition is possible only when $\tau_{xy} = \tau_{yx}$.

Figure 5-5 *Torque about the point A.*

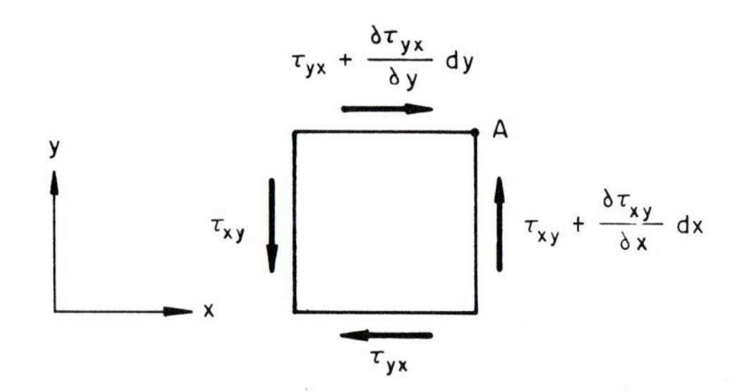

Similarly, it may be shown that $\tau_{yz} = \tau_{zy}$ and $\tau_{xz} = \tau_{zx}$. Hence, the nine components of the tensor of external force are reduced to the six so-called components of Lamé.

$$\begin{vmatrix} \sigma_{xx} & \tau_{xy} & \tau_{xz} \\ \tau_{xy} & \sigma_{yy} & \tau_{yz} \\ \tau_{xz} & \tau_{yz} & \sigma_{zz} \end{vmatrix}$$

5-5.3 *Value of the Lamé Components in Some Particular Cases*

5-5.3.1 In the case of a *perfect fluid*, the shearing stresses are zero and the normal forces become simply the pressure forces:

$$\sigma_{xx} = \sigma_{yy} = \sigma_{zz} = -p$$
$$\tau_{xy} = \tau_{yz} = \tau_{zx} = 0$$

5-5.3.2 In a *viscous incompressible fluid*, it is possible to demonstrate that the normal forces (σ) are the sum of the pressure force and a viscous force proportional to the coefficients of linear deformation:

$$\sigma_{xx} = -p + 2\mu \frac{\partial u}{\partial x}$$

Similar expressions exist for σ_{yy} and σ_{zz}.

The shearing stresses τ are functions of the coefficients of angular deformation:

$$\tau_{yz} = \mu\left(\frac{\partial w}{\partial y} + \frac{\partial v}{\partial z}\right)$$

and similar expressions for τ_{xz} and τ_{xy}.

Now, introducing these values in the sum of forces acting in the same direction as, for example, in the OX direction, it is easy to verify that

$$\left(\frac{\partial \sigma_{xx}}{\partial x} + \frac{\partial \tau_{xy}}{\partial y} + \frac{\partial \tau_{xz}}{\partial z}\right) = -\frac{\partial p}{\partial x} + \mu\nabla^2 u$$

5-5.3.3 In the case of a *viscous compressible fluid*, the shearing stresses are the same as in the above case, but the normal forces have to take into account the change of volume of the fluid particle. It may be seen that:

$$\sigma_{xx} = -p + \lambda\left(\frac{\partial u}{\partial x} + \frac{\partial v}{\partial y} + \frac{\partial w}{\partial z}\right) + 2\mu\frac{\partial u}{\partial x}$$

Two similar relationships are easily deduced for σ_{yy} and σ_{zz}. λ is a second coefficient of viscosity for a gas. From the kinetic theory of gases, it may be shown that for a monoatomic gas $3\lambda + 2\mu = 0$. In practice this relationship is considered accurate enough for any kind of gas.

5-5.4 Dissipation Function

The energy transformed into heat either by change of volume or by friction may be obtained by adding the work done by all the external forces. This work is equal to the external forces times their displacement.

For instance, in the OX direction, the work of pressure forces is:

$$p\,dy\,dz\,u\,dt - \left(p + \frac{\partial p}{\partial x}dx\right)dy\,dz\left(u + \frac{\partial u}{\partial x}dx\right)dt$$

The work of all the forces in the OX direction is:

$$\sigma_{xx}\,dy\,dz\,u\,dt - \left(\sigma_{xx} + \frac{\partial \sigma_{xx}}{\partial x}dx\right)dy\,dz\left(u + \frac{\partial u}{\partial x}dx\right)dt$$

$$+ \tau_{xy}\,dy\,dz\,u\,dt - \left(\tau_{xy} + \frac{\partial \tau_{xy}}{\partial x}dx\right)dy\,dz\left(u + \frac{\partial u}{\partial x}dx\right)dt$$

$$+ \tau_{xz}\,dy\,dz\,u\,dt - \left(\tau_{xz} + \frac{\partial \tau_{xz}}{\partial x}dx\right)dy\,dz\left(u + \frac{\partial u}{\partial x}dx\right)dt$$

The total work per unit volume changed into heat in a unit period of time is called the "dissipation function" Φ. It is a function of the linear and angular rates of deformation and is found by substituting the values of σ and τ.

$$\Phi = \lambda(\text{div } \mathbf{V})^2 + \mu\left[2\left(\frac{\partial u}{\partial x}\right)^2 + 2\left(\frac{\partial v}{\partial y}\right)^2 + 2\left(\frac{\partial w}{\partial z}\right)^2 + \left(\frac{\partial w}{\partial y} + \frac{\partial v}{\partial z}\right)^2 + \left(\frac{\partial u}{\partial z} + \frac{\partial w}{\partial x}\right)^2 + \left(\frac{\partial v}{\partial x} + \frac{\partial u}{\partial y}\right)^2\right]$$

$\lambda(\text{div } \mathbf{V})^2$ is equal to zero in an incompressible fluid.

This function can be used, for example, in the calculation of head loss of a viscous flow in a pipe and the damping of gravity waves.

PROBLEMS

5.1 Demonstrate that the viscous forces acting on an element of incompressible fluid of volume unity can be expressed in terms of the rotation by the following expression:

$$-2\mu\left(\frac{\partial \zeta}{\partial y} - \frac{\partial \eta}{\partial z}\right)$$

The two other expressions are obtained by permutation.

5.2 Demonstrate that in an irrotational flow of an incompressible fluid, the sum of the viscous forces is theoretically zero.

5.3 Calculate the viscous force acting on a cubic element of water of volume $10^{-3}L^3$ and located between $y = \frac{1}{10}L$ and $y = \frac{2}{10}L$ in a two-dimensional flow defined by the velocity components $u = 10^{-4}(g/\nu)(4 - y)y$, $v = 0$, and $w = 0$. Calculate the values of σ and τ acting on each side of this cube, and the rate of dissipation of energy in the cube. ($\nu = 1.076 \times 10^{-5}$ ft^2/sec; 0.01 cm^2/sec.)

5.4 Express $\mu\nabla^2\mathbf{V}$ in a cylindrical system of coordinates.

5.5 Express the stresses σ and τ in a cylindrical system of coordinates for an incompressible fluid.

Chapter 6

Forms of the Momentum Equation: Equations of Euler and Navier–Stokes

6-1 Main Differential Forms of the Momentum Equation

The momentum equation is obtained by equating the applied forces to the inertia force for a unit volume of the fluid. The physical meaning and the mathematical expressions of these forces have been developed in Chapters 4 and 5.

The different forms of the momentum equation corresponding to a number of cases encountered in hydrodynamics are now presented.

6-1.1 Perfect Fluid: Equations of Euler

6-1.1.1 The first major approximation is to assume that the fluid is perfect. In this case the friction forces are zero and the applied forces consist of gravity and pressure only. The momentum equation is obtained directly from the expressions developed in Chapters 4 and 5, in the three-axis system OX, OY, OZ, where OZ is assumed to be vertical (see Table 6-1). When the expressions du/dt and p^* are

Table 6-1 *The momentum equation*

Inertia forces per unit of volume (see Section 4-1.1)		*Pressure and gravity forces*† *per unit of volume of fluid (see Section 5-3.4)*
$\rho \frac{du}{dt}$	$=$	$-\frac{\partial p^*}{\partial x}$
$\rho \frac{dv}{dt}$	$=$	$-\frac{\partial p^*}{\partial y}$
$\rho \frac{dw}{dt}$	$=$	$-\frac{\partial p^*}{\partial z}$

Written in vector notation, these become

$$\rho \frac{d\mathbf{V}}{dt} + \mathbf{grad}\, p^* = 0$$

† Recall ($p^* = p + \rho g z$).

expanded (see Section 4-4.1), the momentum equation takes the form along the OX axis given by Equation 6-1.

$$\underbrace{\underbrace{\rho\frac{\partial u}{\partial t}}_{\text{Local inertia}} + \underbrace{\rho\left(u\frac{\partial u}{\partial x} + v\frac{\partial u}{\partial y} + w\frac{\partial u}{\partial z}\right)}_{\text{Convective inertia}}}_{\text{Inertia forces}} = \underbrace{-\frac{\partial}{\partial x}(p + \rho g z)}_{\text{Applied forces: Pressure, Gravity}} \qquad (6\text{-}1)$$

Two similar equations may be written in the OY and OZ directions. These are called the equations of Euler.

Such a system of equations associated with the continuity relationship $\partial u/\partial x + \partial v/\partial y + \partial w/\partial z = 0$ forms the basis of the largest part of the hydrodynamics dealing with a perfect incompressible fluid. These equations are mathematically of the first order but are nonlinear (more specifically quadratic) because of the convective inertia terms. This quadratic term is the cause of most mathematical difficulties encountered in hydrodynamics.

6-1.1.2 It has been explained in Chapter 1 that it is possible to study hydrodynamic problems either in Eulerian coordinates or in Lagrangian coordinates. It is recalled that the Lagrangian method consists of following particles along their paths instead of dealing with particles at a given point. This method is used, for example, in some studies of periodic gravity waves over a horizontal bottom. The corresponding equations are given here only for the purpose of recognition in the literature and will not be developed.

If X, Y, Z are the volume or body forces, i.e., gravity, the Lagrangian equation along the OX axis is written:

$$\frac{1}{\rho}\frac{\partial p}{\partial x_0} = \left(X - \frac{\partial^2 x}{\partial t^2}\right)\frac{\partial x}{\partial x_0} + \left(Y - \frac{\partial^2 y}{\partial t^2}\right)\frac{\partial y}{\partial y_0} + \left(Z - \frac{\partial^2 z}{\partial t^2}\right)\frac{\partial z}{\partial z_0}$$

Two similar equations give the value of $\partial p/\partial y_0$ and $\partial p/\partial z_0$ by permutation of x_0, y_0, z_0, which are the coordinates of a particle at time $t = t_0$. These are called the equations of Lagrange.

6-1.2 Viscous Fluid and the Navier–Stokes Equations

6-1.2.1 If the friction forces are introduced in the Eulerian equations, the Navier–Stokes equations are obtained (see Section 5-4.1), as shown in Equation 6-2.

The Navier–Stokes equations are the basis of most problems in fluid mechanics dealing with liquid. They are second-order differential equations because of the friction terms, and nonlinear because of the convective inertia terms.

6-1.2.2 These Navier–Stokes equations are written in a very concise manner with the aid of tensorial notation. Although a knowledge of tensorial calculus is not required, it is given here as a guide to further reading on the subject.

Use is made of two subscripts, i and j, which indicate when an operation is to be systematically repeated and which component of a vector quantity (such as **V**) is being considered. When an index is repeated in a term, the considered quantity has to be summed over the possible components. For example, the continuity equation $\partial u/\partial x + \partial v/\partial y + \partial w/\partial z = 0$ is tensorially written: $\partial u_i/\partial x_i = 0$, since the subscript i indicates that the quantity (here **V**) has to be summed over the three components OX, OY, OZ.

The three previous Navier–Stokes equations, may be written simply as:

$$\rho\left[\frac{\partial u_i}{\partial t} + u_j\frac{\partial u_i}{\partial x_j}\right] = -\frac{\partial(p + \rho g z)}{\partial x_i} + \mu\frac{\partial^2 u_i}{\partial x_j\,\partial x_j}$$

Here, the subscript i is called "free index" and indicates the component being considered; the subscript j, called "dummy index," indicates repeated operations.

$$\underbrace{\underset{\text{Local inertia}}{}\quad\underset{\text{Convective inertia}}{}}_{\text{Inertia forces}} = \underbrace{\underset{\text{Pressure}}{}\quad\underset{\text{Gravity}}{}\quad\underset{\text{Friction}}{}}_{\text{Applied forces}}$$

$$\begin{aligned}
\rho\left(\frac{\partial u}{\partial t} + u\frac{\partial u}{\partial x} + v\frac{\partial u}{\partial y} + w\frac{\partial u}{\partial z}\right) &= -\frac{\partial p}{\partial x} + \mu\left(\frac{\partial^2 u}{\partial x^2} + \frac{\partial^2 u}{\partial y^2} + \frac{\partial^2 u}{\partial z^2}\right)\\
\rho\left(\frac{\partial v}{\partial t} + u\frac{\partial v}{\partial x} + v\frac{\partial v}{\partial y} + w\frac{\partial v}{\partial z}\right) &= -\frac{\partial p}{\partial y} + \mu\left(\frac{\partial^2 v}{\partial x^2} + \frac{\partial^2 v}{\partial y^2} + \frac{\partial^2 v}{\partial z^2}\right)\\
\rho\left(\frac{\partial w}{\partial t} + u\frac{\partial w}{\partial x} + v\frac{\partial w}{\partial y} + w\frac{\partial w}{\partial z}\right) &= -\frac{\partial(p + \rho g z)}{\partial z} + \mu\left(\frac{\partial^2 w}{\partial x^2} + \frac{\partial^2 w}{\partial y^2} + \frac{\partial^2 w}{\partial z^2}\right)
\end{aligned} \tag{6-2}$$

6-1.2.3 These Navier–Stokes equations are often written in another way in order to emphasize the role of the rotational component of motion. It is sufficient in this case to use the expression of the inertia force demonstrated in Chapter 4, which yields (see Section 4-4.3) Equation 6-3.

6-1.2.4 The three components of Equation 6-3 are more concisely written in the vector form of Equation 6-4, which may be transformed as

$$\mathbf{grad}\left(\rho\frac{V^2}{2} + p + \rho g z\right) = -\rho\frac{\partial \mathbf{V}}{\partial t} - \rho(\mathbf{curl}\,\mathbf{V}) \times \mathbf{V} + \mu\nabla^2\mathbf{V}$$

6-1.2.5 In the case of a steady ($\partial\mathbf{V}/\partial t = 0$) irrotational flow ($\mathbf{curl}\ \mathbf{V} = 0$) of a perfect fluid ($\mu = 0$), the above equation gives at once:

$$\mathbf{grad}\left(\rho\frac{V^2}{2} + p + \rho g z\right) = 0$$

Since the derivative of the sum in parentheses is zero in all directions, one obtains

$$\rho\frac{V^2}{2} + p + \rho g z = \text{constant}$$

which is the well-known Bernoulli equation, fully developed in Chapter 10.

6-1.3 The General Form of the Momentum Equation

It has been shown that the applied forces may be expressed independently of their physical nature with the help of the tensor of rank two:

$$\begin{vmatrix} \sigma_{xx} & \tau_{xy} & \tau_{xz} \\ \tau_{xy} & \sigma_{yy} & \tau_{yz} \\ \tau_{xz} & \tau_{yz} & \sigma_{zz} \end{vmatrix}$$

The main advantage of such a notation is that it is valid for any kind of fluid—perfect or real—and any kind of

Inertia forces = *Applied forces*

Inertia forces: *Local inertia*; *Convective inertia* (*Caused by variation of kinetic energy*; *Caused by rotation*)

Applied forces: *Pressure*; *Gravity*; *Friction*

$$\rho\left[\frac{\partial u}{\partial t} + \frac{\partial}{\partial x}\left(\frac{V^2}{2}\right) + 2(w\eta - v\zeta)\right] = -\frac{\partial(p + \rho g z)}{\partial x} + \mu\nabla^2 u$$

$$\rho\left[\frac{\partial v}{\partial t} + \frac{\partial}{\partial y}\left(\frac{V^2}{2}\right) + 2(u\zeta - w\xi)\right] = -\frac{\partial(p + \rho g z)}{\partial y} + \mu\nabla^2 v \qquad (6\text{-}3)$$

$$\rho\left[\frac{\partial w}{\partial t} + \frac{\partial}{\partial z}\left(\frac{V^2}{2}\right) + 2(v\xi - u\eta)\right] = -\frac{\partial(p + \rho g z)}{\partial z} + \mu\nabla^2 w$$

Inertia forces = *Applied forces*

Inertia forces: *Local inertia*; *Convective inertia* (*Caused by variation of kinetic energy*; *Caused by rotation*)

Applied forces: *Pressure*; *Gravity*; *Friction*

$$\rho\left[\frac{\partial \mathbf{V}}{\partial t} + \mathbf{grad}\left(\frac{V^2}{2}\right) + (\mathbf{curl}\ \mathbf{V}) \times \mathbf{V}\right] = -\mathbf{grad}\,(p + \rho g z) + \mu\nabla^2\mathbf{V} \qquad (6\text{-}4)$$

$$\underbrace{\text{Inertia forces}} = \overbrace{\underbrace{\text{Volume forces}} \quad \underbrace{\text{Surface forces}}}^{\text{Applied forces}}$$

$$\begin{aligned}
\rho \frac{du}{dt} &= X + \left(\frac{\partial \sigma_{xx}}{\partial x} + \frac{\partial \tau_{xy}}{\partial y} + \frac{\partial \tau_{xz}}{\partial z}\right) \\
\rho \frac{dv}{dt} &= Y + \left(\frac{\partial \tau_{xy}}{\partial x} + \frac{\partial \sigma_{yy}}{\partial y} + \frac{\partial \tau_{yz}}{\partial z}\right) \qquad (6\text{-}5) \\
\rho \frac{dw}{dt} &= Z + \left(\frac{\partial \tau_{xz}}{\partial x} + \frac{\partial \tau_{yz}}{\partial y} + \frac{\partial \sigma_{zz}}{\partial z}\right)
\end{aligned}$$

motion—laminar or turbulent. It will be shown that if in the momentum equation the real values u, v, w, and p are replaced by the average values $\bar{u}$, $\bar{v}$, $\bar{w}$, and $\bar{p}$ in a turbulent flow, the surface forces σ and τ include additional components due to the turbulent fluctuations.

Hence, the advantage of using the notations σ and τ exists in expressing general equations which are independent of the nature of the flow. Equating the inertia forces to the applied forces expressed in the manner shown in Chapter 5 yields Equation 6-5.

In practice, if OZ is vertical upward, $X = 0$, $Y = 0$, $Z = -\rho g = -(\partial/\partial z)(\rho g z)$.

6.1-4 Synthesis of the Most Usual Approximations

Tables 6-2 and 6-3 recall the physical meaning of different terms and possible approximations accepted in the studies of flow motions, which may be investigated in the following. Complex disordered and random motions, even though also obeying the Navier–Stokes equation, cannot be analyzed on a purely Newtonian deterministic approach. The motion is averaged and the friction term $\mu\nabla^2\mathbf{V}$ is replaced by an empirical functional relationship proportional to $\mathbf{V}$ in the case of flow trough porous medium, or to V^2 in the case of fully turbulent motion. These two points are analyzed in Chapters 8 and 9.

6-2 Exact Integration vs Numerical Solutions

6-2.1 An Example of an Exact Solution of Navier–Stokes Equations. Flow on a Sloped Plane

6-2.1.1 It is to be expected that a general solution of the system of differential equations given by the continuity and momentum principle does not exist. However, some exact solutions can be obtained if the boundary conditions are simple. Examples where exact solutions may be obtained include flow between parallel plates (i.e., the Couette flow, the Poiseuille flow), flow due to a rotating disk,

Table 6-2

	$\rho\left[\frac{\partial \mathbf{V}}{\partial t}\right.$	$+\ \mathbf{grad}\left(\frac{V^2}{2}\right)$	$+\ (\mathbf{curl\ V}) \times \mathbf{V}\left.\right]$	$= -\ \mathbf{grad}(p$	$+\ \rho g z)$	$+\ \mu \nabla^2 \mathbf{V}$
Physical meaning	Local inertia	Variation of kinetic energy with space	Rotational term	Pressure force	Gravity force	Friction force
		Convective inertia		Applied forces		
Mathematical characteristics	First-order linear term	Nonlinear (quadratic) term		First-order linear term	Constant term	Second-order linear term
Approximation	=0 In a steady flow		=0 for irrotational motion; solution given by a harmonic function		≅0 in a gas (with exceptions)	=0 in an ideal fluid
		=0 for slow motion	=0 for slow motion			

uniform unsteady flow over an infinite flat plate (see Section 13-2.3.3), etc.

6-2.1.2 The very simple example of a two-dimensional steady uniform flow on an inclined plane of infinite dimensions is given here as an example (Fig. 6-1); the Navier–Stokes equation given in Section 6-1.2.1 may be simplified in the following manner:

Since the motion is steady, $\partial u/\partial t = 0$ and $\partial v/\partial t = 0$. Since the motion is two-dimensional, $w = 0$, and all derivatives with respect to z are zero. Since the motion is uniform and parallel to the axis OX, v and all its derivatives are zero. All derivatives with respect to x are also zero. The components of the gravitational force are $X = \rho g \sin \alpha$ and $Y = -\rho g \cos \alpha$. Since the flow is uniform, $v = 0$, and the continuity equation is reduced to $\partial u/\partial x = 0$.

The Navier–Stokes equations are reduced to:

$$\rho g \sin \alpha + \mu\left(\frac{\partial^2 u}{\partial y^2}\right) = 0$$

$$0 = -\frac{\partial p}{\partial y} - \rho g \cos \alpha$$

The second equation yields

$$p = p_a - \rho g y \cos \alpha,$$

where p_a is the atmospheric pressure. Hence the lines of equal pressure are parallel to the OX axis.

The boundary conditions are $u = 0$ for $y = -d$ on the plane, and $du/dy = 0$ for $y = 0$ at the free surface. Taking into account these boundary conditions, the integration of

$$\frac{\partial^2 u}{\partial y^2} = -\frac{\rho g}{\mu} \sin \alpha$$

Table 6-3

Local inertia	*Convective inertia*	*Friction*	*Equations*	*Some applications*
Steady motion or motion considered as a succession of steady motions	Slow motion	Without friction	$\mathbf{grad}\,(p + \rho g z) = 0$	Hydrostatics
		With friction	$-\mathbf{grad}\,(p + \rho g z) + \mu \nabla^2 \mathbf{V} = 0$	Steady uniform flow Flow in a porous medium
	Irrotational motion	Without friction	$\mathbf{grad}\left(\rho \frac{V^2}{2} + p + \rho g z\right) = 0$	Nonuniform (convergent) Steady flow at a constant total energy. Calculation of pressure in a two-dimensional flow net
	Rotational motion	With friction	$\mathbf{grad}\left(\rho \frac{V^2}{2} + p + \rho g z)\right)$ $= -\rho(\mathbf{curl}\ \mathbf{V}) \times \mathbf{V} + \mu \nabla^2 \mathbf{V}$	General case of steady motion; laminar boundary layer theories
Unsteady motion	Slow motion	Without friction	$\rho \frac{\partial \mathbf{V}}{\partial t} + \mathbf{grad}\,(p + \rho g z) = 0$	Gravity wave (first-order theory); water hammer theory
		With friction	$\rho \frac{\partial \mathbf{V}}{\partial t} + \mathbf{grad}\,(p + \rho g z) - \mu \nabla^2 \mathbf{V} = 0$	Gravity wave damping
	Irrotational motion	Without friction	$\rho \frac{\partial \mathbf{V}}{\partial t} + \mathbf{grad}\left(\rho \frac{V^2}{2} + p + \rho g z\right) = 0$	Most nonlinear wave theories
	Rotational motion	Without friction	$\mathbf{grad}\left(\rho \frac{V^2}{2} + p + \rho g z\right)$ $+ \rho \frac{\partial \mathbf{V}}{\partial t} + \rho(\mathbf{curl}\ \mathbf{V}) \times \mathbf{V} = 0$	Gravity wave theory of Gerstner
		With friction	General case	Tidal wave in an estuary

gives successively,

$$\frac{\partial u}{\partial y} = -\frac{g \sin \alpha}{\nu} y \qquad \left(\nu = \frac{\mu}{\rho}\right)$$

and

$$u = \frac{g \sin \alpha}{2\nu}(d^2 - y^2)$$

which is the equation of a parabola.

The discharge per unit of width is:

$$q = \int_{-d}^{0} u\, dy = \frac{g \sin \alpha}{2\nu} \int_{-d}^{0} (d^2 - y^2)\, dy$$

$$= \frac{g \sin \alpha}{3\nu} d^3$$

The loss of energy per unit length may be given by the dissipation function Φ, which in this case is $\mu(\partial u/\partial y)^2$.

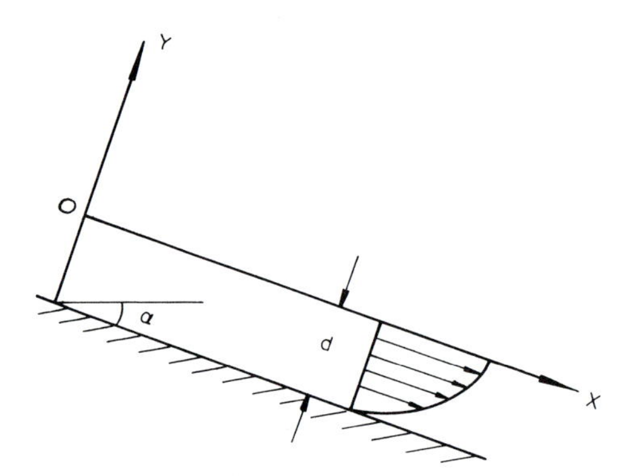

Figure 6-1 *Laminar flow on an inclined plane.*

Hence the loss of energy per unit length is

$$\int_0^d \Phi \, dy = \mu \int_0^d \left(\frac{\partial u}{\partial y}\right)^2 dy = \frac{(\rho g \sin \alpha)^2 d^3}{3\mu}$$

This can also be obtained by determining the work done by friction forces F_f as follow;

$$\int_0^d F_f \, du = \mu \int_0^d \frac{\partial u}{\partial y} du = \mu \int_0^d \left(\frac{\partial u}{\partial y}\right)^2 dy$$

6-2.2 *Numerical Treatments of the Navier–Stokes Equations*

It is now possible, thanks to the development of high-speed computers, to treat the Navier–Stokes equations directly by finite differences. This permits the study of complex flow motions beyond the usual limits of analytical solutions.

Among many possible methods which have been developed, one must mention the MAC (markers and cells) method for two-dimensional or axially symmetric incompressible fluid, and the PIC (particle in cell) method for two-dimensional compressible fluid.

In brief, these methods consist in solving time dependent flow motion at successive intervals of time from a given set of boundary conditions and the knowledge of the flow motion at time $t = 0$. The space intervals define a square mesh or a grid. Considering one (or two) particle(s) at the center of each of these squares at time $t = 0$, it is then possible to calculate the paths of these particles at successive intervals of time. The results are printed directly by the computer and give a Lagrangian representation of the flow pattern as a function of time. It is also possible to obtain and print velocity vectors and pressure distributions (isobars) directly.

It is easily realized that this method is extremely powerful, as is evident from Fig. 6-2. This figure represents the flow patterns which will be obtained by the sudden release of a vertical wall of water (this is the dam-break problem) and hitting an obstacle. However, this method, as any numerical method, also has its limitations. The accuracy of the results is rapidly limited by the error which is made by replacing differential terms by finite difference. These are the truncation errors. A round-off error is also added, as will be explained in further detail in Section 18-3.1.2.

Any calculation also requires a preliminary analysis of stability conditions so that the cumulative error does not increase out of proportion. This method is costly due to computing time. Nevertheless, it is to be expected that these kinds of methods will be used more and more for solving problems of increasing complexity. There are theoretical limits, however, to this method, since the phenomena of turbulence (which is three-dimensional) could not be analyzed in a two-dimensional finite difference scheme. Also, the viscosity coefficient μ is taken arbitrarily larger than its real value for the sake of numerical stability.

6-3 The Stability of Laminar Flow

6-3.1 *The Natural Tendency for Fluid Flow to Be Unstable*

Consider two layers of fluid moving with different velocities because of the effect of friction (Fig. 6-3). If for any reason a small undulation exists between these two layers, the velocity of layer 2 decreases; hence, according to the Bernoulli equation, the pressure tends to increase.

On the other hand, the velocity of layer 1 tends to increase; hence, the pressure tends to decrease. The pressure action being in the same direction as the inertial forces

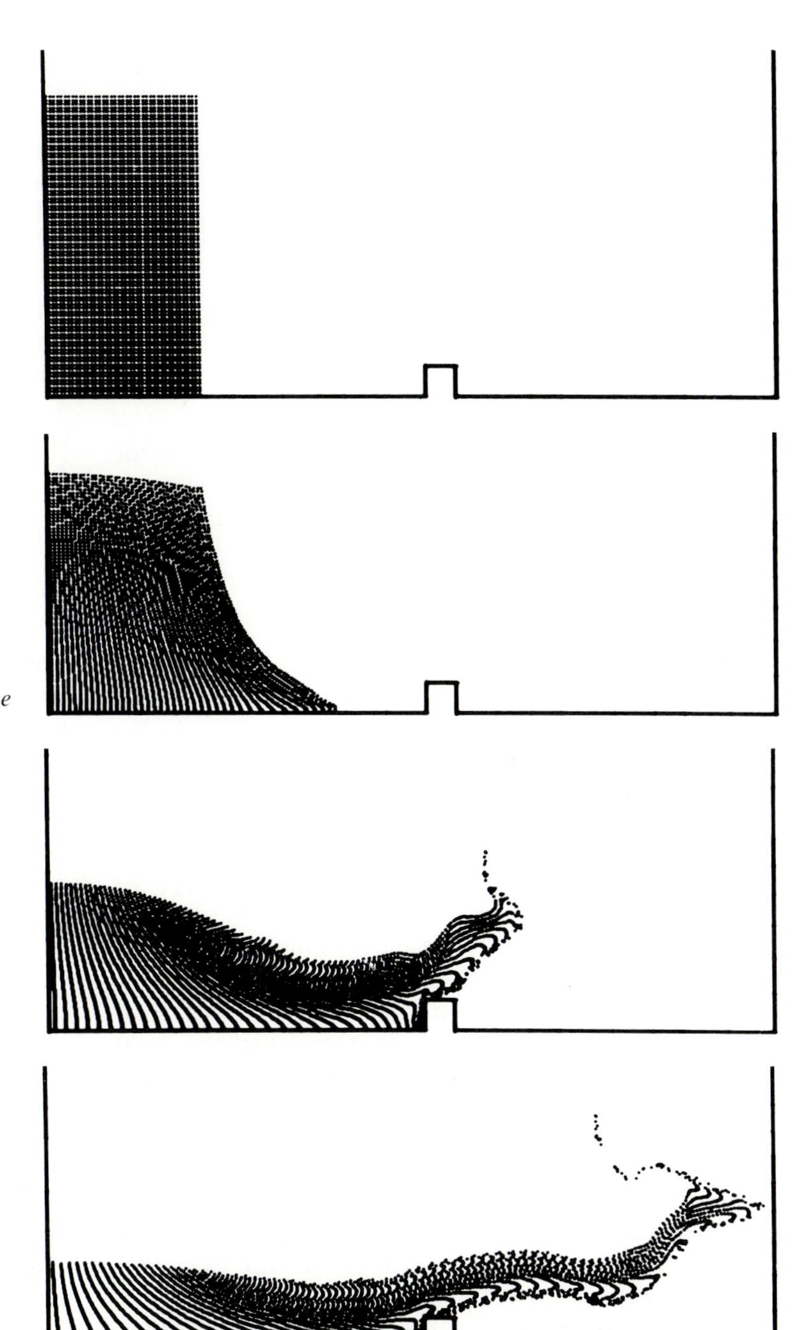

Figure 6-2

An example of an application of numerical treatment of the Navier–Stokes equation. (Courtesy of Dr. F. Harlow.)

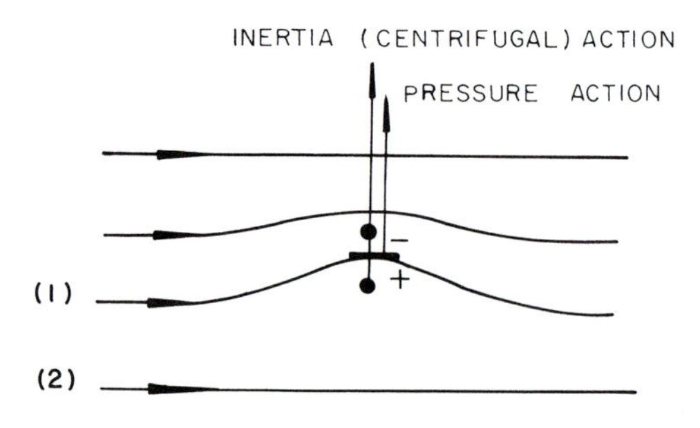

Figure 6-3 *Fluid flow is fundamentally unstable.*

(by centrifugal action), the undulation has a natural tendency to increase in amplitude. However, this increase in path length of the particles in motion causes an increase in friction effect, which in turn has a tendency to dampen such an undulation. This disturbance would tend to be dampened out by friction unless there is a transfer of energy (or a transfer of momentum by convective inertia forces) from the primary motion to this disturbance. Hence, in a turbulent motion, the rate of turbulence depends on the rate of energy which is transmitted from the primary flow to be finally absorbed entirely by friction.

6-3.2 *Free Turbulence, Effects of Wall Roughness*

Instabilities initially occur either within the fluid or at a boundary. In the first case the phenomenon is called "free turbulence." The lumps of turbulence come initially from the zone where the gradient of kinetic energy is a maximum, as for example from the boundary of a submerged jet. Hydraulic jumps, breaking waves on a beach, and whitecaps generated at sea under wind action, are also cases of free turbulence. Initial instabilities are most often caused by roughness of a fixed boundary. Indeed, any roughness causes a local increase of velocity which consequently produces a local strong gradient of kinetic energy, causing an instability (Fig. 6-4). This instability also arises between two fluids of different density. For example, the wind blowing on a liquid causes ripples. These ripples are due to an instability between the air flow and water.

The very interesting question concerning the origin of turbulence will not be studied in detail here. It is simply emphasized that the Navier–Stokes equations give unstable solutions which represent exact motions only at low Reynolds numbers, i.e., when the friction forces are large in comparison with the kinetic energy gradient. Turbulence will be studied further in Chapters 7 and 8.

Since exact motions occur only at low Reynolds numbers, the stability of laminar flow depends upon the ratio of the gradient of the kinetic energy (dimensionally equal to the convective inertia forces) to the viscous forces. This ratio is dimensionally equivalent to the Reynolds number ($R_e = VL/\nu$), where V is a velocity, and L is a characteristic length.

Because of the definite instability of the flow, a disturbance caused by external forces (such as that created by a roughness at the boundary) grows exponentially if the disturbance is large enough. If the disturbance is small, the friction forces cause its damping. But if the ratio of the gradient of kinetic energy (dimensionally equal to a convective inertia force) to the viscous force is large enough, even an infinitely small disturbance is able to cause instability. Hence, laminar flow is naturally and basically unstable at large Reynolds numbers. But even at low

Figure 6-4 *Instability of a laminar flow because of a roughness.*

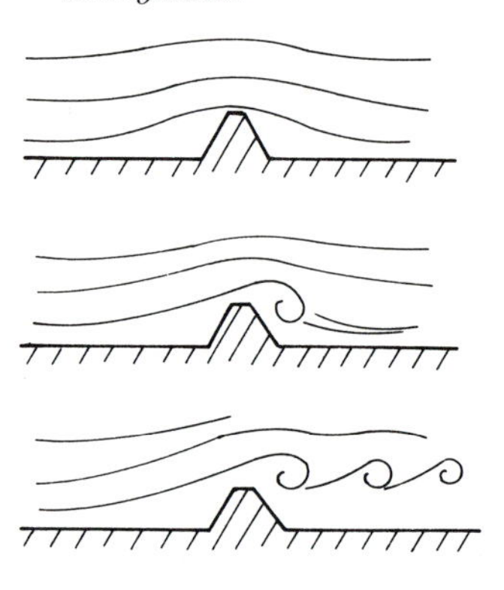

Reynolds numbers, laminar flow is unstable if the disturbance is large enough. It is possible, with many precautions, to obtain a laminar flow in a very smooth pipe up to a Reynolds number of 40,000, although under normal conditions the critical value of Reynolds number for a pipe is 2,000. A disturbance superimposed on the primary motion causes, as previously seen, a large local increase in the convective inertia forces.

PROBLEMS

6.1 Consider successively a circular pipe and a square pipe rotating around their own axes at an angular velocity varying suddenly from $\omega = 0$ at time $t = 0$ to $\omega = \omega_1$ at time $t = \varepsilon$ (ω_1 small) and $\omega = \omega_2$ at time $t = t_1$ (ω_2 large). These two pipes are successively half filled and fully filled with liquid. Describe qualitatively the liquid motion in the two cases where (1) the fluid is perfect, and (2) the fluid is viscous.

6.2 Demonstrate that the velocity distribution for a flow between two parallel planes, one of them being fixed and the other one moving at a constant velocity U, is $u = Uy/e$ where e is the distance between the two planes, and under a gradient of pressure dp/dx, the flow between two fixed parallel planes is

$$u = -\frac{dp}{dx}\frac{1}{2\mu}(ey - y^2)$$

6.3 Write a Navier–Stokes equation for an unsteady flow between two parallel planes in which one of the planes is fixed, while the other one is moving at a speed $u(t)$.

Then write the Navier–Stokes equation for a two-dimensional steady flow between two planes almost parallel; one plane is fixed, and the other plane is moving at constant velocity U. Do the simplifying approximations that you think are permissible for analyzing the flow motion.

6.4 Calculate the two-dimensional velocity distribution $u(z)$ between two parallel horizontal planes between which there are two layers of fluid of thickness e_1 and e_2, viscosity μ_1 and μ_2, and density ρ_1 and ρ_2 ($\rho_1 > \rho_2$), respectively. One plane is fixed and the upper plane moves at constant velocity U.

6.5 Consider a two-dimensional flow between two fixed parallel horizontal planes separated by a distance $2h$.

1. Write the continuity relationship, the Navier–Stokes equation, and the boundary condition. The flow motion will be assumed to be in the OX direction and OZ is perpendicular to the plane.
2. j being the head loss defined by $dp/dx = -\rho g j$, calculate the velocity distribution $u = f(j,z)$ by two successive integrations, and the total discharge per unit of width $Q = f(j,h)$.
3. Calculate the mean velocity $\bar{u} = f(j,h)$ and express u as a function of $\bar{u}$, z, and e.
4. Calculate d^2u/dz^2 and $dp/dx = f(\bar{u},h)$.
5. Calculate the rotational coefficients ξ, η, ζ as functions of j, z, h.
6. Calculate the loss of energy per unit length of the direction of the flow: $\rho g j Q = f(j,h)$ and the value of j as a function of Q and h.
7. Should an obstacle be inserted between the two planes, demonstrate that the mean motion with respect to the vertical OZ is irrotational, i.e., $(\partial\bar{u}/\partial y) - (\partial\bar{v}/\partial x) = 0$. (It is the *Hele–Shaw* analogy.) Express the potential function as a function of p, h, and μ.

6.6 Calculate the ratio of inertial force to viscous forces in the case of a laminar steady uniform flow. Discuss the statement which consists of saying that the Reynolds number is a significant dimensionless parameter giving the relative importance of the inertial force to viscous force. Is the ratio of the gradient of kinetic energy to the viscous force a more significant definition?

6.7 The following dimensionless quantities are defined:

$$x^* = \frac{x}{L} \qquad y^* = \frac{y}{L} \qquad z^* = \frac{z}{L}$$

$$t^* = \frac{t}{T} \qquad \mathbf{V}^*(u^*,v^*,w^*) = \frac{\mathbf{V}(u,v,w)}{U}$$

$$p^* = \frac{p}{\rho U^2} \qquad \mathbf{F}^* = \frac{\mathbf{F}(F_x, F_y, F_z)}{g}$$

where L, T, U are an arbitrary typical length, time, and velocity, and $\mathbf{F}$ is the gravity force. Then demonstrate that the Navier–Stokes equations can be written in dimensionless form as:

$$\left(\frac{L}{UT}\right)\frac{\partial u^*}{\partial t^*} + \frac{\partial}{\partial x^*}\left(p^* + \frac{1}{2}u^{*2}\right) + v^*\frac{\partial u^*}{\partial y^*} + w^*\frac{\partial u^*}{\partial z^*}$$
$$= \left(\frac{gL}{U^2}\right)F_x^* + \left(\frac{\nu}{UL}\right)\nabla^2 u^*$$

and two other similar equations. Explain the physical significance of the parameters:

$\dfrac{UT}{L}$ (sometimes called reduced frequency)

$\dfrac{U^2}{gL}$ (Froude number)

$\dfrac{UL}{\nu}$ (Reynolds number)

6.8 Demonstrate that in a flow defined by $v = w = 0$ and $u = f(y,z) \neq 0$, one has

$$\rho\frac{\partial \eta}{\partial t} = \mu\nabla^2\eta$$

$$\rho\frac{\partial \zeta}{\partial t} = \mu\nabla^2\zeta$$

6.9 Demonstrate that the Navier–Stokes equation can still be written:

$$\rho\frac{d\xi}{dt} - \mu\nabla^2\xi = \rho\left[\xi\frac{\partial u}{\partial x} + \eta\frac{\partial u}{\partial y} + \zeta\frac{\partial u}{\partial z}\right]$$

Obtain the two other equations by circular permutation.

6.10 Establish the equation of flow for a laminar flow in an horizontal cylindrical pipe of radius R. Determine the velocity distribution $u(r)$ and the relationship between the total discharge Q and the head-loss. The head-loss is defined by $j = (1/\rho g)(\partial p/\partial x)$ where $\partial p/\partial x$ is the pressure gradient along the pipe.

6.11 Establish the equation of motion for a horizontal steady flow in an unbounded ocean subjected to a wind stress τ and Coriolis acceleration. It will be assumed that the fluid is viscous, the free surface horizontal, and the pressure gradient at the free surface negligible ($\partial p/\partial x, \partial p/\partial y = 0$). The wind stress τ is in the OY direction, the axis OX and OY are horizontal, and the axis OZ is positive downward from the free surface. Draw three-dimensional diagram of the velocity vector $\mathbf{V}$ as a function of depth (called *Ekman spiral*).

Chapter 7

Turbulence: Mean and Fluctuating Components of Motion

7-1 The Definition of Mean Motion and Mean Forces

7-1.1 Characteristics of Mean Motion vs Actual Motion

In the previous chapters, theory was sometimes illustrated by examples in which the real motion was actually turbulent, despite the fact that we were dealing only with an ideal fluid.

In these examples, it was implied but not specified that only the average values of the velocity and the pressure were considered. In a turbulent motion, the true velocity and pressure vary in a disorderly manner. In fact, a turbulent motion is always unsteady, since at a given point the velocity changes continuously in a very irregular way. It is also nonuniform, since the velocity changes from point to point at a given time, and rotational, since the friction forces, proportional to $\nabla^2 \mathbf{V}$, are important. These characteristics are true as far as the actual motion is concerned. However, by splitting the motion into mean and fluctuating components, the average motion may often be considered steady, uniform (in a pipe), or irrotational (over a weir). The previously considered examples were relative to the average values.

Now this method has to be justified and the differences between the motion of an ideal fluid or a viscous flow, and a mean turbulent motion have to be further considered. This is the purpose of this chapter.

7-2.1 Validity of the Navier–Stokes Equation for Turbulent Motion

Equalities between the inertia forces and the applied forces on an elementary fluid particle are valid even if the motion is turbulent. Hence, the basic Navier–Stokes equations and continuity relationships are also theoretically valid in the study of turbulent motion.

It has been seen that it is sometimes possible to calculate a laminar solution where the boundary conditions are simple. It is also possible to determine theoretically the stability of the solution, that is to determine whether a small disturbance will increase or be damped out by friction. However, a fully deterministic approach is no longer possible in the case of turbulent motion, because of the random nature of these turbulent fluctuations.

On the other hand, in engineering practice, it is not always necessary to know the exact fine structure of the flow. Only the average values and the overall and statistical effects of turbulent fluctuations have to be studied.

7-1.3 *Definitions of the Mean Values in a Turbulent Flow*

In a turbulent motion, as in the case of a viscous flow, velocity and pressure have to be known as functions of the space coordinates and time.

The instantaneous velocity $\mathbf{V}$ at a fixed point is the vectorial sum of the mean velocity $\bar{\mathbf{V}}$ with respect to time (referring to the basic primary movement) and the fluctuation velocity $\mathbf{V}'$ which varies rapidly with time both in intensity and direction. This can be expressed by the relationships $\mathbf{V} = \bar{\mathbf{V}} + \mathbf{V}'$ where, by definition,

$$\bar{\mathbf{V}} = \frac{1}{T}\int_0^T \mathbf{V}\,dt$$

and

$$\bar{\mathbf{V}}' = \frac{1}{T}\int_0^T \mathbf{V}'\,dt = 0$$

where T is a time interval. Similarly, the instantaneous components of velocity are defined as follows:

Real velocity		*Mean velocity*		*Fluctuation velocity*
u	$=$	$\bar{u}$	$+$	u'
v	$=$	$\bar{v}$	$+$	v'
w	$=$	$\bar{w}$	$+$	w'

and $\overline{u'} = \overline{v'} = \overline{w'} = 0$.

Similarly, the instantaneous pressure p is the scalar sum of the mean pressure $\bar{p}$ and a fluctuation term p' such that $p = \bar{p} + p'$ where

$$\bar{p} = \frac{1}{T}\int_0^T p\,dt \quad \text{and} \quad \bar{p}' = \frac{1}{T}\int_0^T p'\,dt = 0$$

Hence, turbulent motion may be considered as the superposition of a mean motion and a fluctuating and disorderly motion, random in nature, which obeys statistical laws.

7-1.4 *Steady and Unsteady Mean Turbulent Flows*

The mean value is defined for intervals of time T, which are large compared to the time scale of turbulent fluctuations but small compared to the time scale of the mean motion.

If, for example, one considers the oscillation of water in a tunnel between a surge tank and a reservoir, the instantaneous velocity at a fixed point may vary quickly because of the turbulence. The average velocity defined for a relatively short interval of time varies also with respect to time, but its change is slow.

The real motion is always unsteady because of turbulence and in this case, the mean motion is also unsteady (Fig. 7-1).

In the following discussion, a motion is called unsteady only if the mean value of the velocity varies. The interval of time, which permits a realistic definition of the mean motion, is relative to the frequency of turbulent fluctuations. It varies with the phenomenon to be studied. For example, it is long for the meteorologist who deals with atmospheric motion, and it is short for the aerodynamist who deals with the turbulence effects in the boundary layer along a wing.

7-1.5 *Mean Forces*

Since the real value of the inertia forces is always equal to the sum of the real values of the applied forces in any kind of motion (laminar or turbulent), the mean value of the inertia forces with respect to time is equal to the mean value of the applied forces with respect to time. This may be expressed as shown in Equation 7-1. Since

Figure 7-1
The steadiness of a turbulent flow is defined by the mean velocity only.

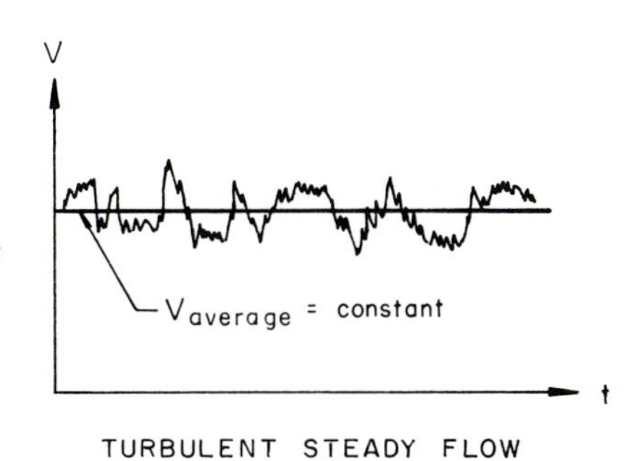

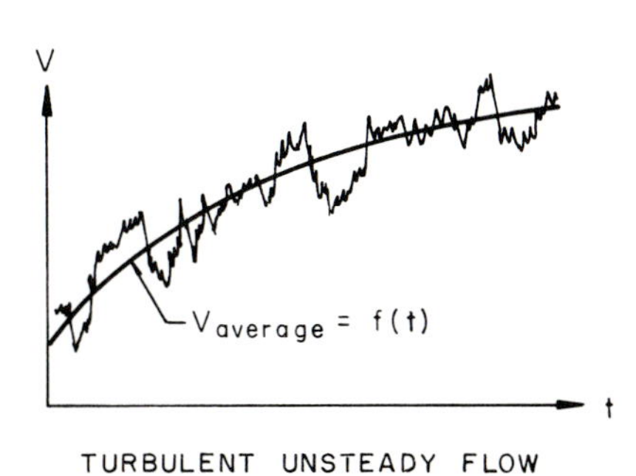

Local inertia forces + *Convective inertia force* + *Pressure force* + *Gravity force* + *Friction force* = 0

one also has:

Mean value with respect to time of [*Local inertia force* + *Convective inertia force* + *Gravity force* + *Pressure force* + *Friction force*] = 0 (7-1)

Equation 7-1 is expressed mathematically, along the OX axis, as (see Section 6-1.2.1)

$$\frac{1}{T}\int_0^T \rho\left(\frac{\partial u}{\partial t} + u\frac{\partial u}{\partial x} + v\frac{\partial u}{\partial y} + w\frac{\partial u}{\partial z}\right) dt = \frac{1}{T}\int_0^T \left(-\frac{\partial(p + \rho g z)}{\partial x} + \mu\nabla^2 u\right) dt$$

or, using the σ and τ notations and the rotational coefficients η, ζ, and ξ (see Sections 6-1.2.2 and 6-1.3)

$$\frac{1}{T}\int_0^T \rho\left(\frac{\partial u}{\partial t} + \frac{\partial}{\partial x}\left(\frac{V^2}{2}\right) + 2(w\eta - v\zeta)\right) dt = +\frac{1}{T}\int_0^T \left(-\frac{\partial}{\partial x}\rho g z + \frac{\partial}{\partial x}\sigma_{xx} + \frac{\partial}{\partial y}\tau_{xy} + \frac{\partial}{\partial z}\tau_{xz}\right) dt$$

Similar equations are found for the OY and OZ axes.

Now each of these mean forces has to be expressed as a function of the mean values and fluctuating values of the velocity and the pressure. For this purpose, one considers:

1. The constant forces	Gravity force
2. The linear forces	Pressure force, linear function of p
	Local inertia force, linear function of $\mathbf{V}$
	Friction force, linear function of $\mathbf{V}$
3. The quadratic force	Convective inertia, function of V^2 or product of two components of velocity: u^2, v^2, w^2, uv, uw, vw

7-2 Calculation of the Mean Forces

The mean forces are calculated as a function of the mean values of velocities and pressure. For these calculations, it is assumed that the order of mathematical operations has no effect on the final result. In particular, integration during an interval of time T and derivatives with respect to time or space can be interchanged.

7-2.1 The Constant Force

The gravity force depends only on the density of the elementary particle. The fluctuations of pressure are too small to have a significant effect on the density. Hence, the gravity force is the same for laminar and turbulent motion.

The mean value of the gravity force is equal to this constant gravity force. Mathematically, this may be expressed

$$\overline{\rho g} = \frac{1}{T}\int_0^T \rho g\, dt = \rho g \frac{1}{T}\int_0^T dt = \rho g$$

Since ρg is a constant with respect to time, then

$$\overline{\frac{\partial}{\partial z}(\rho g z)} = \frac{\partial}{\partial z}(\rho g z)$$

7-2.2 Linear Forces

7-2.2.1 The mean value of the local inertia force, may be obtained by considering the mean value of any of its components. For example,

$$\overline{\rho \frac{\partial u}{\partial t}} = \frac{1}{T}\int_0^T \rho \frac{\partial u}{\partial t}\, dt = \rho \frac{\partial}{\partial t}\frac{1}{T}\int_0^T u\, dt$$

Introduce $u = \bar{u} + u'$, where u' is a fluctuation term, into this equation.

$$\overline{\rho \frac{\partial u}{\partial t}} = \rho \frac{\partial}{\partial t}\frac{1}{T}\int_0^T (\bar{u} + u')\, dt$$

$$= \rho \frac{\partial}{\partial t}\frac{1}{T}\int_0^T \bar{u}\, dt + \rho \frac{\partial}{\partial t}\frac{1}{T}\int_0^T u'\, dt$$

Since

$$\frac{1}{T}\int_0^T \bar{u}\, dt = \bar{u} \quad \text{and} \quad \overline{u'} = \frac{1}{T}\int_0^T u'\, dt = 0$$

the mean value is given by:

$$\overline{\rho \frac{\partial u}{\partial t}} = \rho \frac{\partial \bar{u}}{\partial t}$$

Hence, the mean value of the local inertia force with respect to time is equal to the inertia force caused by the change of value of the mean velocity alone.

7-2.2.2 Similarly averaging the pressure forces yields, for example:

$$-\overline{\frac{\partial p}{\partial x}} = -\frac{\partial \bar{p}}{\partial x}$$

7-2.2.3 The viscous force has a mean value which may be calculated by considering, for example, one of the second-order terms, such as $\mu(\partial^2 u/\partial x^2)$. Averaging this term leads also to

$$\overline{\mu \frac{\partial^2 u}{\partial x^2}} = \mu \frac{\partial^2 \bar{u}}{\partial x^2}$$

The mean viscous force is equal to the viscous force due to the mean velocity alone, and is mathematically expressed in the same way as the actual motion.

All the linear forces involved in the mean motion are mathematically written in the same way for both mean turbulent flow and actual motion, turbulent or laminar.

7-2.3 The Quadratic Forces

Consider the component $u = \bar{u} + u'$. Squaring and averaging u leads successively to

$$\overline{u^2} = \frac{1}{T}\int_0^T u^2\, dt = \frac{1}{T}\int_0^T (\bar{u}^2 + 2\bar{u}u' + u'^2)\, dt = \bar{u}^2 + \overline{u'^2}$$

This is a result of the following intermediate steps. Since $\bar{u}$ is a constant in the interval of time T:

$$\frac{1}{T}\int_0^T \bar{u}^2\, dt = \bar{u}^2$$

Since $\overline{u'} = 0$:

$$\frac{1}{T}\int_0^T 2\bar{u}u'\, dt = 2\bar{u}\frac{1}{T}\int_0^T u'\, dt = 0$$

Since u' may be positive or negative but u'^2 is always positive and its mean value is different from zero:

$$\frac{1}{T}\int_0^T u'^2\, dt = \overline{u'^2}$$

Similarly, the mean value of uv is

$$\overline{uv} = \bar{u}\,\bar{v} + \overline{u'v'}$$

and the mean values of $\bar{u}v'$ and $u'\bar{v}$ are zero.

Now, consider the mean value of any term of convective inertia, such as $\rho u(\partial u/\partial x)$. One has successively,

$$\overline{\rho u \frac{\partial u}{\partial x}} = \frac{1}{T}\int_0^T \rho u \frac{\partial u}{\partial x}\, dt = \frac{1}{T}\int_0^T \rho(\bar{u} + u')\frac{\partial}{\partial x}(\bar{u} + u')\, dt$$

$$= \frac{1}{T}\int_0^T \left(\bar{u}\frac{\partial \bar{u}}{\partial x} + \bar{u}\frac{\partial u'}{\partial x} + u'\frac{\partial \bar{u}}{\partial x} + u'\frac{\partial u'}{\partial x}\right) dt$$

Considering each of these terms independently one obtains, since $\bar{u}$ and $\partial\bar{u}/\partial x$ are constant with respect to time,

$$\frac{1}{T}\int_0^T \bar{u}\frac{\partial \bar{u}}{\partial x}\, dt = \bar{u}\frac{\partial \bar{u}}{\partial x}$$

$$\frac{1}{T}\int_0^T \bar{u}\frac{\partial u'}{\partial x}\, dt = \bar{u}\frac{1}{T}\frac{\partial}{\partial x}\int_0^T u'\, dt = 0$$

$$\frac{1}{T}\int_0^T u'\frac{\partial \bar{u}}{\partial x}\, dt = \frac{1}{T}\frac{\partial \bar{u}}{\partial x}\int_0^T u'\, dt = 0$$

$$\frac{1}{T}\int_0^T u'\frac{\partial u'}{\partial x}\, dt = \frac{1}{T}\frac{\partial}{\partial x}\int_0^T \frac{u'^2}{2}\, dt = \frac{\partial}{\partial x}\frac{\overline{u'^2}}{2} = \overline{u'\frac{\partial u'}{\partial x}} \neq 0$$

Introducing these values yields

$$\overline{\rho u \frac{\partial u}{\partial x}} = \rho\left(\bar{u}\frac{\partial \bar{u}}{\partial x} + \overline{u'\frac{\partial u'}{\partial x}}\right)$$

Similarly, it is found that

$$\overline{\rho u \frac{\partial v}{\partial x}} = \rho\left(\bar{u}\frac{\partial \bar{v}}{\partial x} + \overline{u'\frac{\partial v'}{\partial x}}\right)$$

and so on.

Hence, the mean value of a convective inertia force with respect to time is equal to the sum of the convective inertia caused by the mean velocity and the mean convective inertia caused by the turbulent fluctuations. As far as the mean value of the velocity alone is concerned, the convective inertia terms have the same mathematical form as for the case of a laminar motion.

7-3 The Continuity Relationship

In the simple case of an incompressible fluid, the continuity relationship is written (see Section 3-2):

$$\frac{\partial u}{\partial x} + \frac{\partial v}{\partial y} + \frac{\partial w}{\partial x} = 0$$

This relationship, expressed as a function of the mean components of velocity and turbulent fluctuations, becomes

$$\frac{\partial}{\partial x}(\bar{u} + u') + \frac{\partial}{\partial y}(\bar{v} + v') + \frac{\partial}{\partial z}(\bar{w} + w') = 0$$

or

$$\frac{\partial \bar{u}}{\partial x} + \frac{\partial \bar{v}}{\partial y} + \frac{\partial \bar{w}}{\partial z} + \frac{\partial u'}{\partial x} + \frac{\partial v'}{\partial y} + \frac{\partial w'}{\partial z} = 0$$

The averaging process, applied to $\partial\bar{u}/\partial x$, gives

$$\overline{\frac{\partial \bar{u}}{\partial x}} = \frac{\partial \bar{u}}{\partial x}$$

and applied to $\partial u'/\partial x$ gives

$$\overline{\frac{\partial u'}{\partial x}} = \frac{\partial \overline{u'}}{\partial x} = 0$$

Then the continuity relationship for the mean motion becomes

$$\frac{\partial \bar{u}}{\partial x} + \frac{\partial \bar{v}}{\partial y} + \frac{\partial \bar{w}}{\partial z} = 0$$

Consequently,

$$\frac{\partial u'}{\partial x} + \frac{\partial v'}{\partial y} + \frac{\partial w'}{\partial z} = 0$$

The mathematical form of the continuity relationship is the same for the mean motion as for the actual motion.

7-4 The Characteristics of the Mean Motion of a Turbulent Flow

Insofar as the mean velocity and the mean pressure alone are concerned, the basic momentum equation and the continuity relationship have exactly the same mathematical form as the corresponding equations for the actual motion. However, other forces exist and have to be added. These new forces are caused by the convective inertia of the turbulent fluctuations. If these "new" forces may be neglected, or as long as only the forces which are functions of the mean velocity and mean pressure are dealt with, the solutions of problems concerning turbulent motion have the same mathematical form as the solutions given by the Navier–Stokes equations. For example, a mean motion which is steady and irrotational and for which the viscous forces $\mu\nabla^2\overline{\mathbf{V}}$ are neglected obeys the Bernoulli equation:

$$\rho\frac{\overline{V}^2}{2} + \bar{p} + \rho g z = \text{constant}$$

as found in Section 6-1.2.5 where the velocity $\overline{V}$ and the pressure $\bar{p}$ now designate the average values. When applying this equation, the assumptions must be made relative to the mean motion; i.e., the mean motion must be steady, irrotational, and without viscous friction (despite the fact that the actual turbulent motion is always unsteady, rotational, and dissipative).

In practice the fluctuations of pressure p' are very small by comparison with the real pressure p, so that $p \approx \bar{p}$. On the other hand, the viscous forces $\mu\nabla^2\overline{\mathbf{V}}$ caused by the mean motion are generally small in comparison with the other forces, in particular with the convective inertia forces caused by the turbulent fluctuations. The viscous forces can often be neglected except, for example, in a laminar boundary sublayer between a smooth wall and a turbulent boundary layer.

Now the effects of the convective fluctuating forces on the mean motion have to be studied. Then a relationship between the value of the mean velocity and the fluctuating velocity has to be established. Indeed since another unknown $\mathbf{V}'(u',v',w')$ has been added, another relationship is necessary. This will be the purpose of the next chapter.

7-5 Reynolds Equations

7-5.1 *Purpose of the Reynolds Equations*

Expressing each force in the Navier–Stokes equation as a function of the mean values $\overline{\mathbf{V}}(\bar{u},\bar{v},\bar{w})$ and the fluctuating values $\mathbf{V}'(u', v', w')$ and averaging, leads to the Reynolds equation. The Reynolds equation is the form of the Newton or momentum equation for turbulent motion.

Since each of the mean forces has been calculated in the previous sections, it is possible to obtain directly the Reynolds equations. To do this, the sum of these mean forces is equated to zero. Recall that each force has the same mathematical form as in the Navier–Stokes equation, but it is expressed as a function of the mean values of velocity or pressure. However, additional convective inertia forces exist, caused by the fluctuating terms. For example, the mean value of the quadratic inertia term, $\overline{\rho u(\partial u/\partial x)}$, is

$$\rho\bar{u}\frac{\partial\bar{u}}{\partial x} + \overline{\rho u'\frac{\partial u'}{\partial x}}$$

Hence, the momentum equation valid for the average motion in the OX direction may be written directly as Equation 7-2. (Since the calculation method is identical in the OY and OZ directions, only the momentum equation along the OX axis is studied.)

7-5.2 *Reynolds Stresses*

The convective inertia caused by the fluctuating velocity components is given in Section 7-2.3 as

$$\rho\left(\overline{u'\frac{\partial u'}{\partial x}} + \overline{v'\frac{\partial u'}{\partial y}} + \overline{w'\frac{\partial u'}{\partial z}}\right)$$

$$\rho\left(\frac{\partial \bar{u}}{\partial t} + \bar{u}\frac{\partial \bar{u}}{\partial x} + \bar{v}\frac{\partial \bar{u}}{\partial y} + \bar{w}\frac{\partial \bar{u}}{\partial z} + \overline{u'\frac{\partial u'}{\partial x}} + \overline{v'\frac{\partial u'}{\partial y}} + \overline{w'\frac{\partial u'}{\partial z}}\right) = -\frac{\partial}{\partial x}(\bar{p} + \rho g z) + \mu\nabla^2\bar{u} \tag{7-2}$$

Local inertia — *Convective inertia caused by the mean velocities* — *Convective inertia caused by the fluctuation velocities* — *Pressure and gravity forces* — *Viscous force*

The value of

$$\overline{\rho u'\left(\frac{\partial u'}{\partial x} + \frac{\partial v'}{\partial y} + \frac{\partial w'}{\partial z}\right)}$$

was shown to be zero by the continuity relationship in Section 7-3. Thus, this expression can be added to the convective inertia without changing the value. When these two expressions are added and grouped in pairs, the result is:

$$\rho\left(\overline{u'\frac{\partial u'}{\partial x}} + \overline{u'\frac{\partial u'}{\partial x}} + \overline{v'\frac{\partial u'}{\partial y'}} + \overline{u'\frac{\partial v'}{\partial y}} + \overline{w'\frac{\partial u'}{\partial z}} + \overline{u'\frac{\partial w'}{\partial z}}\right)$$

This can be written as

$$\rho\left(\frac{\partial \overline{u'^2}}{\partial x} + \frac{\partial \overline{u'v'}}{\partial y} + \frac{\partial \overline{u'w'}}{\partial z}\right)$$

Now, by introducing these terms (and two similar terms obtained for the OY and OZ directions) into the general momentum equation, the Reynolds equations (Equation 7-3) are obtained.

It is seen, indeed, that these Reynolds equations are very similar to the Navier–Stokes equations shown previously. The difference is in the convective inertia forces caused by the turbulent fluctuations and in the fact that the other forces are expressed as functions of the mean value of the velocity or pressure.

The turbulent fluctuation forces, are called "Reynolds stresses."

7-5.3 *Value of the Lamé Components in a Turbulent Motion*

The applied forces have been expressed independently of their physical nature, as is shown in Section 6-1.3. It is recalled, for example, that the applied forces along the OX axis are expressed by

$$X + \left(\frac{\partial \sigma_{xx}}{\partial x} + \frac{\partial \tau_{xy}}{\partial y} + \frac{\partial \tau_{xz}}{\partial z}\right)$$

The averaging process applied to these terms (which are either constant, such as X, or linear) gives for the applied forces:

$$\frac{1}{T}\int_0^T \left[X + \left(\frac{\partial}{\partial x}\sigma_{xx} + \frac{\partial}{\partial y}\tau_{xy} + \frac{\partial}{\partial z}\tau_{xz}\right)\right] dt$$
$$= X + \frac{\partial}{\partial x}\bar{\sigma}_{xx} + \frac{\partial}{\partial y}\bar{\tau}_{xy} + \frac{\partial}{\partial z}\bar{\tau}_{xz}$$

Introducing this above expression in the Reynolds equation yields

$$\rho\frac{d\bar{u}}{dt} = X + \frac{\partial}{\partial x}(\bar{\sigma}_{xx} - \rho\overline{u'^2})$$
$$+ \frac{\partial}{\partial y}(\bar{\tau}_{xy} - \rho\overline{u'v'}) + \frac{\partial}{\partial z}(\bar{\tau}_{xz} - \rho\overline{u'w'})$$

From this equation it is easily deduced that the fluctuation terms may be considered as external forces which are added to the other forces defined by normal forces $\bar{\sigma}$

$$
\begin{aligned}
&\rho\left(\frac{\partial \bar{u}}{\partial t} + \bar{u}\frac{\partial \bar{u}}{\partial x} + \bar{v}\frac{\partial \bar{u}}{\partial y} + \bar{w}\frac{\partial \bar{u}}{\partial z}\right) = -\frac{\partial}{\partial x}(\bar{p} + \rho g z) + \mu\nabla^2\bar{u} - \rho\left(\frac{\partial \overline{u'^2}}{\partial x} + \frac{\partial \overline{u'v'}}{\partial y} + \frac{\partial \overline{u'w'}}{\partial z}\right)\\
&\rho\left(\frac{\partial \bar{v}}{\partial t} + \bar{u}\frac{\partial \bar{v}}{\partial x} + \bar{v}\frac{\partial \bar{v}}{\partial y} + \bar{w}\frac{\partial \bar{v}}{\partial z}\right) = -\frac{\partial}{\partial y}(\bar{p} + \rho g z) + \mu\nabla^2\bar{v} - \rho\left(\frac{\partial \overline{u'v'}}{\partial x} + \frac{\partial \overline{v'^2}}{\partial y} + \frac{\partial \overline{v'w'}}{\partial z}\right)\\
&\rho\left(\frac{\partial \bar{w}}{\partial t} + \bar{u}\frac{\partial \bar{w}}{\partial x} + \bar{v}\frac{\partial \bar{w}}{\partial y} + \bar{w}\frac{\partial \bar{w}}{\partial z}\right) = -\frac{\partial}{\partial z}(\bar{p} + \rho g z) + \mu\nabla^2\bar{w} - \rho\left(\frac{\partial \overline{u'w'}}{\partial x} + \frac{\partial \overline{v'w'}}{\partial y} + \frac{\partial \overline{w'^2}}{\partial z}\right)
\end{aligned}
\tag{7-3}
$$

Local inertia — *Convective inertia* — *Pressure and gravity forces* — *Viscous forces* — *Turbulent fluctuation forces*

and shear stresses $\bar{\tau}$. Hence, these new external forces to be dealt with are:

Normal force: $[\sigma_{xx}] = \bar{\sigma}_{xx} - \rho\overline{u'^2}$

$$= -\bar{p} + 2\mu\frac{\partial \bar{u}}{\partial x} - \rho\overline{u'^2}$$

Shear stress: $[\tau_{xy}] = \bar{\tau}_{xy} - \rho\overline{u'v'}$

$$= -\bar{p} + \mu\left(\frac{\partial \bar{u}}{\partial y} + \frac{\partial \bar{v}}{\partial x}\right) - \rho\overline{u'v'}$$

and so on. These new total external forces may also be defined by a tensor of rank two similar to the first tensor defined in Section 5-5.2.

In practice the viscous forces caused by the mean velocity are very often negligible in turbulent flow in comparison with the other forces, and particularly in comparison with the shear stresses caused by the fluctuation terms $\rho\overline{u'v'}$, $\rho\overline{u'w'}$, and $\rho\overline{v'w'}$.

7-5.4 Correlation Coefficients and Isotropic Turbulence

By definition, in isotropic turbulence the mean value of any function of the fluctuating velocity components and their space derivatives is unaltered by a change in the axes of reference. In particular:

$$\overline{u'^2} = \overline{v'^2} = \overline{w'^2} \qquad \overline{u'v'} = \overline{u'w'} = \overline{v'w'} = 0$$

It is evident that isotropy introduces a great simplification in the calculations. However, because of the boundaries, the turbulence is not isotropic and the products $\overline{u'v'}$, $\overline{u'w'}$ and $\overline{v'w'}$ may differ from each other. There exists a correlation between u' and v', u' and w', and v' and w', defined by the coefficients:

$$\frac{\overline{u'v'}}{(\overline{u'^2}\,\overline{v'^2})^{1/2}}, \quad \frac{\overline{u'w'}}{(\overline{u'^2}\,\overline{w'^2})^{1/2}}, \quad \frac{\overline{v'w'}}{(\overline{v'^2}\,\overline{w'^2})^{1/2}}$$

These coefficients are equal to zero in the case of isotropic turbulence. Since the convective inertia forces caused by the fluctuation terms are functions of $\overline{u'^2}$, $\overline{v'^2}$, $\overline{w'^2}$, $\overline{u'v'}$, $\overline{u'w'}$, $\overline{v'w'}$, they may be expressed directly as functions of the coefficients of correlation which are dimensionless.

PROBLEMS

7.1 Express $\overline{\mathbf{grad}\, V^2/2}$ and $\overline{\mathbf{V} \times \mathbf{curl}\, \mathbf{V}}$ in terms of u, u', v, v', w, w' for a turbulent flow.

7.2 Express the average ratio of dilatational and shear deformation in terms of mean and fluctuating velocity components for a turbulent flow.

7.3 Draw a line $u(t)$ at random on graph paper and determine $\bar{u}$ and $(\overline{u'^2})^{1/2}$. On the same graph, draw another line $v(t)$ at random and determine $\bar{v}$ and $(\overline{v'^2})^{1/2}$ and the value of the correlation coefficient

$$\frac{\overline{u'v'}}{(\overline{u'^2}\,\overline{v'^2})^{1/2}}$$

7.4 Demonstrate that the Reynolds equations can still be written:

$$\frac{\partial \bar{u}}{\partial t} + \bar{u}\frac{\partial \bar{u}}{\partial x} + \bar{v}\frac{\partial \bar{u}}{\partial y} + \bar{w}\frac{\partial \bar{u}}{\partial z} = \frac{\partial}{\partial x}\left[\frac{p^*}{\rho} + 2\nu\frac{\partial \bar{u}}{\partial x} - \overline{u'^2}\right] + \frac{\partial}{\partial y}\left[\nu\left(\frac{\partial \bar{u}}{\partial y} + \frac{\partial \bar{v}}{\partial x}\right) - \overline{u'v'}\right] + \frac{\partial}{\partial z}\left[\nu\left(\frac{\partial \bar{u}}{\partial z} + \frac{\partial \bar{w}}{\partial x}\right) - \overline{u'w'}\right]$$

and two other equations which will be determined. Indicate the advantage of this form of the Reynolds equation.

7.5 Write the Reynolds equation in the case of a mean two-dimensional motion. Write the Reynolds equation in the case of isotropic turbulence $[\overline{u'^2} = \overline{v'^2}, \overline{u'v'} = 0]$.

7.6 Write the Reynolds equation for a flow in a straight circular pipe and demonstrate that the pressure is smaller on the axis of the pipe than on the wall.

Chapter 8

Turbulence Effects: Modern Theories

8-1 Some Physical Effects of Turbulent Fluctuations

8-1.1 Velocity Distribution: Effects of Reynolds Stresses

8-1.1.1 The velocity distribution depends upon the total shearing stresses, which are a function of the viscous force and Reynolds stresses:

$$\rho\overline{u'v'} \qquad \rho\overline{u'w'} \qquad \rho\overline{v'w'}$$

In the case of a viscous flow, it has been shown (see Section 6-2.1.2), that the velocity distribution over a sloped plane is parabolic. In the case of a turbulent flow, the velocity distribution is influenced by the Reynolds stresses and therefore different from a parabola. The effects of the Reynolds stresses on the velocity distribution are analyzed qualitatively in the following section.

8-1.1.2 Consider two fluid layers defined by the mean motion, i.e. separated by streamlines tangential to the vector "mean velocity" (Fig. 8-1). Let $\bar{\mathbf{V}}_1$ and $\bar{\mathbf{V}}_2$ be the mean velocities of these two layers in a given cross section. The instantaneous velocity $\mathbf{V}_1$ is the sum of the mean velocity $\bar{\mathbf{V}}_1$ and a fluctuating term $\mathbf{V}'_1$, $\mathbf{V}_1 = \bar{\mathbf{V}}_1 + \mathbf{V}'_1$. $\mathbf{V}'_1$ has two components: v'_1, normal to the mean velocity, and v'_2 in the $\bar{\mathbf{V}}_1$ direction.

Figure 8-1 *Change of momentum by turbulence between two stream tubes.*

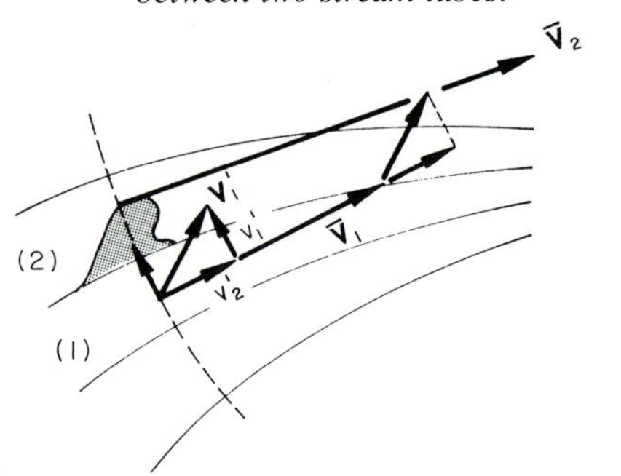

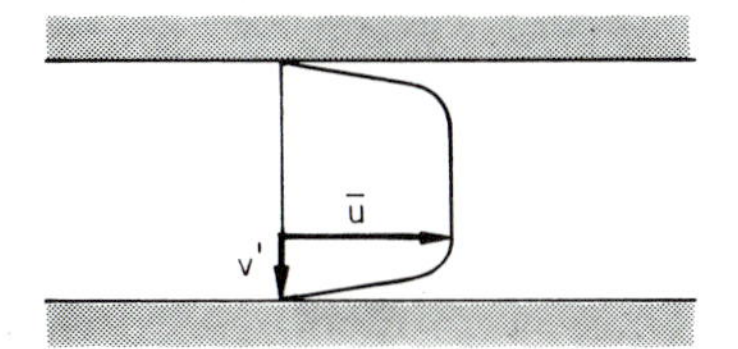

Figure 8-2 *The shearing stresses caused by turbulence decreases near the boundary, while the viscous force increases.*

Because of the normal component v'_1, an amount of fluid moving in the $\bar{\mathbf{V}}_1$ direction at the mean velocity $\bar{\mathbf{V}}_1$ penetrates from layer 1 into layer 2, and since its mean velocity $\bar{\mathbf{V}}_1$ is smaller than the mean velocity $\bar{\mathbf{V}}_2$ of layer 2, this amount of fluid tends to slow down the speed of layer 2.

Similarly, because of the fluctuating components of the velocity, the amounts of fluid penetrating from layer 2 into layer 1 tend to increase the velocity of layer 1. In a word because of the turbulence, the mean velocities of the two layers tend to become equal. It is seen that the fluctuating velocity forces act physically as a shearing stress between the two layers of fluid.

8-1.1.3 In a turbulent flow these shearing stresses caused by turbulence are usually more active than the shearing stresses caused by viscosity. Therefore, adjacent layers tend to have similar velocity as in the case of an ideal fluid. However, at a boundary layer, the terms of the form $\rho\overline{u'v'}$ tend to zero since the velocity normal to the boundary must vanish (Fig. 8-2). Conversely, the viscous term $\mu(\partial^2\bar{u}/\partial y^2)$ increases near the boundary and becomes particularly important in the case of a smooth boundary.

The mean velocity distributions in a pipe, given by Fig. 8-3 and corresponding to different assumptions made on the shearing stresses, illustrate these previous considerations.

The quantitative study of the velocity distribution in a turbulent flow depends upon the assumption made on the distribution of the value of the shearing stress τ. This is the subject of the following sections.

8-1.2 Irrotational Motion

A turbulent motion is strongly rotational since the actual friction forces have an important effect. However, rotational motion occurs at random, like the turbulent fluctuations.

In the case of isotropic turbulence the mean motion could be considered as irrotational. Where the turbulence is nonisotropic, the mean flow is rotational, but out of the boundary layer the turbulence is nearly isotropic in a first approximation. Hence, a number of methods of calculation which give the flow pattern in an ideal fluid may be successfully applied in a turbulent flow, as long as the boundary layer is thin with respect to the main flow.

It is evident that such an assumption is of particular importance in engineering practice since it permits a knowledge of the flow pattern of the mean motion in any convergent short structure, such as a bellmouth gallery or a spillway (Figs. 2-12 and 2-14).

These considerations are also illustrated by Fig. 8-4.

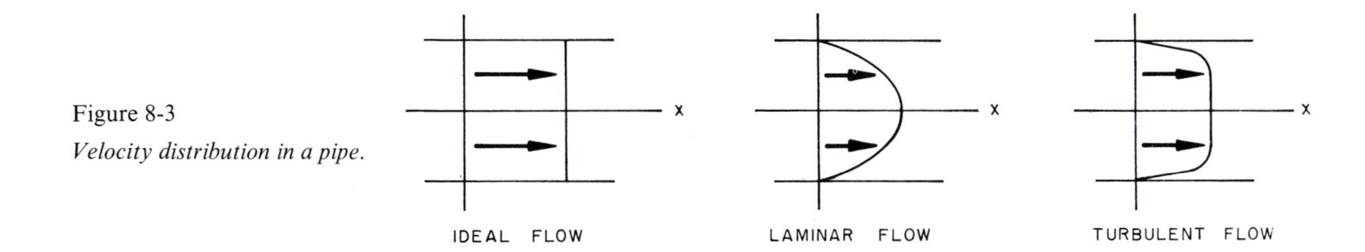

Figure 8-3
Velocity distribution in a pipe.

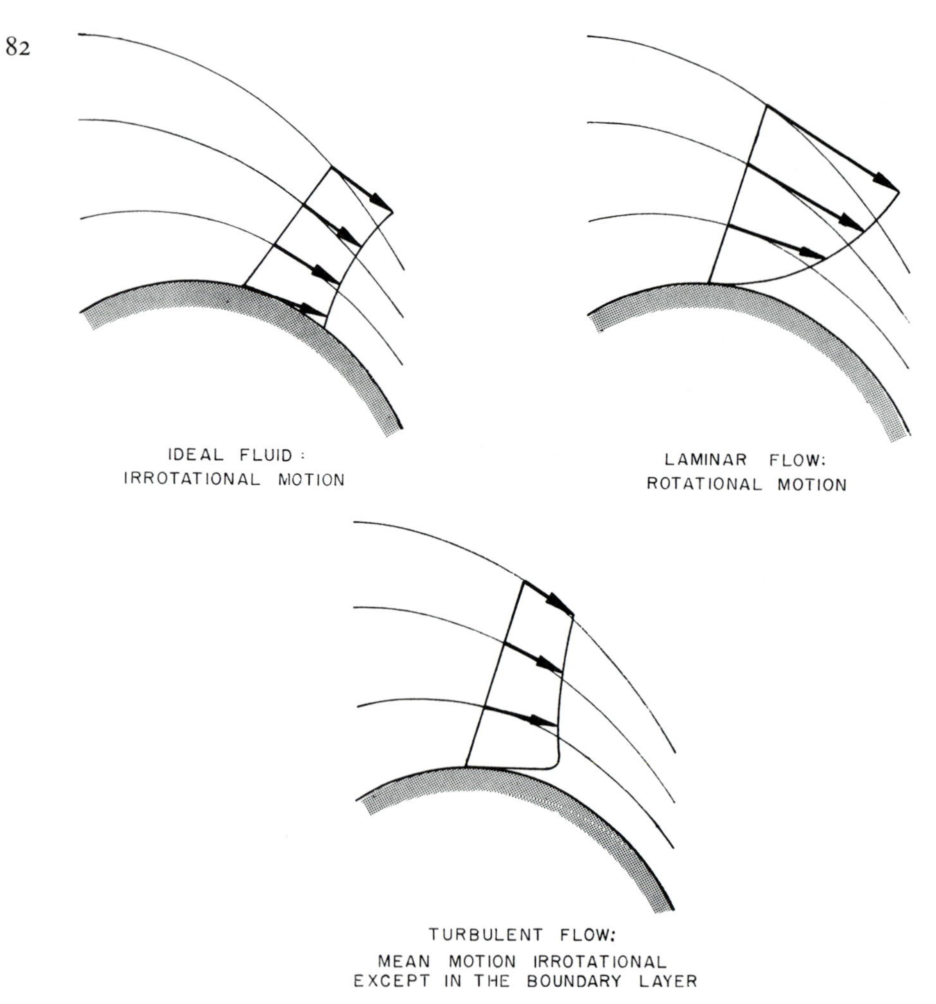

Figure 8-4

A turbulent flow may often be considered as irrotational.

8-2 Pressure Force. Secondary Effects

8-2.1 Paradox of Du Buat

Considering the normal forces

$$[\sigma] = -p + 2\mu \frac{\partial \bar{u}}{\partial x} - \rho \overline{u'^2}$$

It is seen that an additional force has been added to the pressure force. This results in an increase of the average pressure value. For example, it is seen in elementary hydraulics that the hydraulic jump theory is developed by equating the variation of momentum to the external forces, i.e., the difference of pressure forces before and after the jump (Fig. 8-5). To be more exact, it would be necessary to add to the pressure forces the difference in $\int_0^h (\rho \overline{u'^2})\, dh$ (h is the water depth). This term is actually negligible. However, this same factor is sufficient to explain why the resistance of a body moving with velocity V in calm water

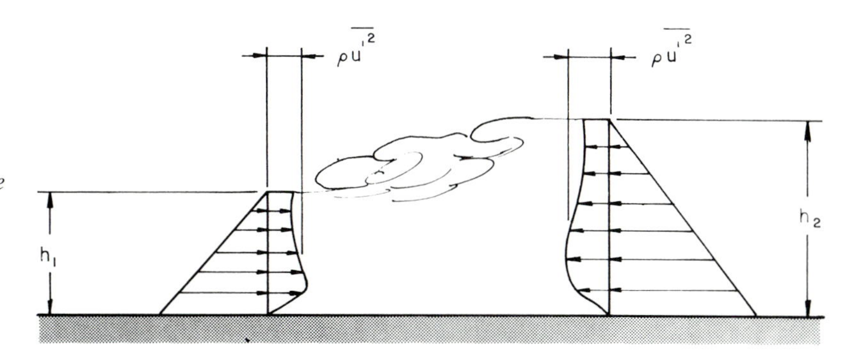

Figure 8-5

Turbulent fluctuation terms have to be added to the pressure forces.

is different from the resistance of this same body when stationary in a turbulent flow of the same mean velocity $\overline{V}$. This is the paradox of Du Buat. This phenomenon is caused by the difference of impulse of the turbulent convective inertia $\rho\overline{u'^2}$ acting against the body in a manner similar to pressure forces.

8-2.2 Turbulent Flow between Two Parallel Planes

8-2.2.1 Consider the simple case of uniform steady turbulent motion between two horizontal parallel planes, as shown by Fig. 8-6. Since the mean motion is steady the local inertia forces are zero: $\rho(\partial\bar{u}/\partial t) = 0$; $\rho(\partial\bar{v}/\partial t) = 0$; $\rho(\partial\overline{w}/\partial t) = 0$. Assuming that the mean velocity vector is parallel to the two planes in the OX direction, the components $\bar{v}$ and $\overline{w}$ along the OY and OZ axis respectively are zero and all the terms of the Reynolds equations where those quantities appear are zero (see Section 7-5.2). Since the mean motion is uniform, $\partial\bar{u}/\partial x = 0$. It is then seen that all the convective inertia terms are zero as in any case of uniform flow.

Now consider the fluctuation terms. The derivatives of $\overline{u'^2}$, $\overline{v'^2}$, $\overline{w'^2}$, $\overline{u'v'}$, $\overline{u'w'}$, $\overline{v'w'}$ with respect to x are zero since the motion is uniform. The derivatives with respect to y are also zero since the two planes are assumed to be infinite, and the motion is two-dimensional.

Finally, the Reynolds equations are reduced to:

$$0 = -\frac{\partial\overline{p^*}}{\partial x} + \mu\frac{\partial^2\bar{u}}{\partial z^2} - \rho\frac{\partial\overline{u'w'}}{\partial z}$$

$$0 = -\frac{\partial\overline{p^*}}{\partial z} - \rho\frac{\partial\overline{w'^2}}{\partial z}$$

where $\overline{p^*} = \bar{p} + \rho g z$. For a laminar flow between two parallel planes the Navier–Stokes equations would be written:

$$0 = -\frac{\partial p^*}{\partial x} + \mu\frac{\partial^2 u}{\partial z^2}$$

$$0 = -\frac{\partial p^*}{\partial z}$$

Figure 8-6 *Turbulent flow between two parallel planes.*

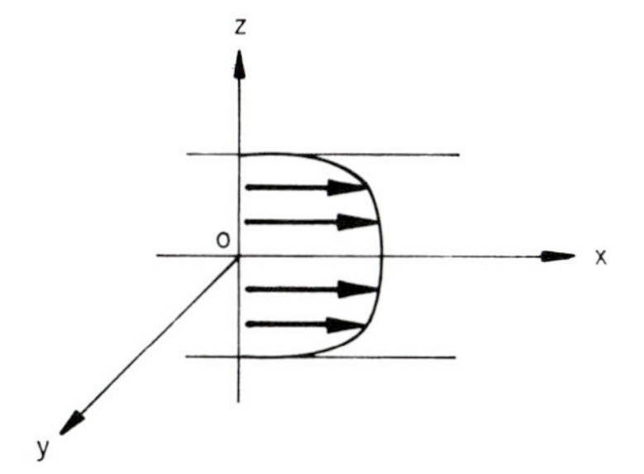

8-2.2.2 Integrating the second of the above equations with respect to z yields

$$\overline{p^*} + \rho\overline{w'^2} = \text{constant}$$

Let the pressure at the boundaries be $\overline{p_0^*}$. Since $w' = 0$ at the boundaries, the mean pressure $\overline{p^*}$ at any point of the flow is smaller than the mean pressure at the boundary $\overline{p_0^*}$ by the quantity $\rho\overline{w'^2}$.

8-2.2.3 If $\overline{U}_0$ is the mean velocity between the two planes, it has been found experimentally that $\overline{u'^2}/\overline{U}_0^2 < 0.01$ and $\overline{v'^2}/\overline{U}_0^2$ and $\overline{w'^2}/\overline{U}_0^2 < 0.0025$. Hence the difference:

$$\frac{\overline{p_0^*} - \overline{p^*}}{(\rho/2)\overline{U}_0^2} = 2\frac{\overline{w'^2}}{\overline{U}_0^2}$$

is always smaller than 0.0050 and is neglected. The pressure distribution in a turbulent uniform flow is hydrostatic (at least within 0.5%).

8-2.3 Secondary Currents

Such variations of pressure caused by the fluctuation terms explain the origin of secondary currents in straight channels and non-circular pipes. Secondary currents take place when nonsymmetrical effects of the turbulent shearing stresses exist in the flow, that is, each time that the boundary is noncircular.

These secondary currents go from the zone of high shearing stress to the zone of lower shearing stresses as shown by Fig. 8-7. They have a tendency to equalize the shearing stresses at the boundary. Hence they are secondary only in name: they partly justify the use by engineers of the concept of hydraulic radius (Section 14-1.1). Indeed, the definition of hydraulic radius is based on the assumption that the shearing stress at the boundary is a constant. The limitations of applicability of the hydraulic radius definition must be known. A change in the secondary current pattern in a flow has an effect on the head loss which is not negligible. However, this effect is neglected in hydraulics because it is not well known yet, and remains small.

Figure 8-7 *Secondary currents in a triangular pipe.*

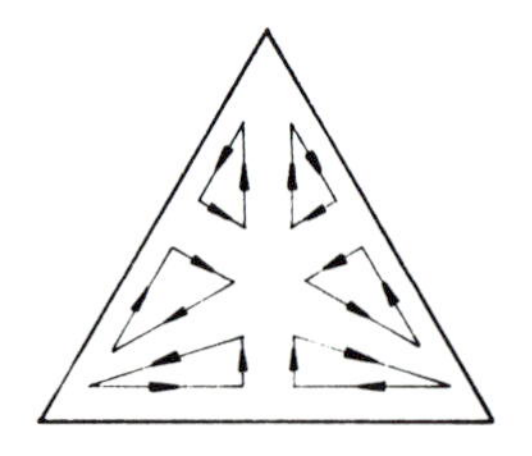

8-3 Modern Theories on Turbulence

8-3.1 The Unknowns in a Turbulent Flow

In Chapter 1, it was seen that problems in hydrodynamics consist of determining the four unknowns, u, v, w, and p. For turbulent flow, the four unknowns are $\bar{u}$, $\bar{v}$, $\bar{w}$, and $\bar{p}$. However, four other unknowns, u', v', w', and p' have been introduced, theoretically requiring four other equations (unless the fluctuation terms may be neglected). The fluctuation values u', v', w', or some function of these values, such as $\overline{u'v'}$, $\overline{v'w'}$, $\overline{u'w'}$, are expressed as functions of the mean values $\bar{u}$, $\bar{v}$, and $\bar{w}$.

It is seen that p' does not appear in the Reynolds equation which governs the mean motion because of the linearity of the pressure forces. Moreover, p' is usually very small in comparison with $\bar{p}$. p' should be taken into account statistically only for some very special problems, such as in the investigation of the growth of wind waves under wind action. Although progress has been made in the statistical theory of turbulence, only isotropic and homogeneous turbulence is well described by use of random functions. However, isotropic turbulence is an idealized case never encountered as is the abstract concept of irrotationality. Further investigation into nonisotropic turbulence is necessary.

8-3.2 Boussinesq Theory

In order to simplify the Reynolds equation, Boussinesq introduced the *turbulent exchange coefficient* ε, with the same dimensions as the coefficient of viscosity μ. In the case of uniform flow parallel to a plane in the OX direction ($\bar{u} = \bar{u}(y)$, $\bar{v} = 0$, $\bar{w} = 0$), ε is defined by the equality $\rho\overline{u'v'} = -\varepsilon(d\bar{u}/dy)$. Then the shearing stress $[\tau]$ becomes $[\tau] = (\mu + \varepsilon)(d\bar{u}/dy)$ instead of $\tau = \mu(du/dy)$. $[\tau]$ is given by a linear relationship. From this relationship it may be seen that the fluctuation term would act similar to the viscous term. They are simply added linearly. However, they are of a different order of magnitude, that is $\varepsilon \gg \mu$ and $[\tau] \cong \varepsilon(d\bar{u}/dy)$.

Such a relationship gives a velocity distribution similar to that obtained in a laminar flow from the Navier–Stokes equations. Since it would be necessary to consider that ε varies with respect to space, the Boussinesq theory is a failure. However, in some cases relative to the motion of atmospheric layer, ε is approximately a constant and the Boussinesq assumption is applied to obtain a result at a first order of approximation.

8-3.3 Prandtl Theory for Mixing Length

The mixing length theory has been introduced by Prandtl by analogy with the mean free path as it is defined in kinetic theory of gases. It is the momentum transfer theory.

Consider the flow $\bar{u} = \bar{u}(y)$, $\bar{v} = 0$, $\bar{w} = 0$ parallel to the OX axis (Fig. 8-8). The mean velocities are $\bar{u}$ and $\bar{u} + d\bar{u}$ at two points on a perpendicular to the boundary defined by $y = 0$.

Figure 8-8 *Flow parallel to the OX axis.*

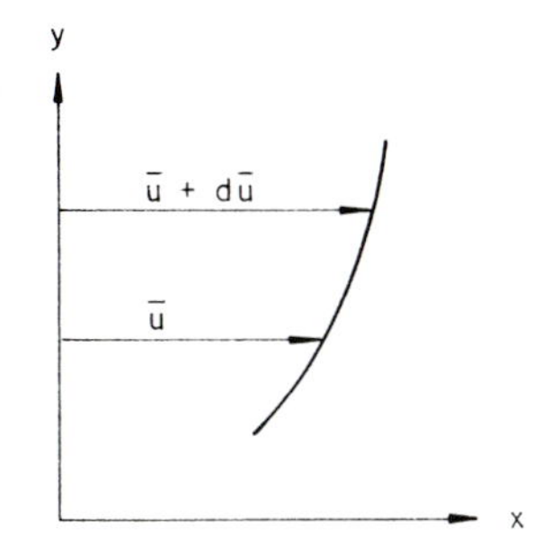

According to Prandtl, it is assumed that the fluctuation terms u' and v' are proportional to the difference in velocity $d\bar{u}$ which is equal to: $d\bar{u} = (d\bar{u}/dy)\ dy$, such that $\overline{u'v'} = -l^2(d\bar{u}/dy)^2$ or $|u'|$ and $|v'| \approx l(d\bar{u}/dy)$. l is the "mixing length." l may be physically considered as the length which may be traversed by a lump of fluid perpendicular to the mean velocity vector **u**. It is evident that according to this definition, l is equal to zero at the boundary since a lump of fluid cannot pass through the boundary.

On the other hand, $\overline{u'v'}$ always has the opposite sign of $d\bar{u}/dy$. If one considers a lump of fluid moving from the boundary to the middle of the flow, $v' > 0$ and it is moving from a layer where $\bar{u}$ is smaller to a layer where $\bar{u}$ is larger. It causes a slowing down of the motion, hence $u' < 0$. Conversely, if one considers a lump of fluid moving toward the boundary, v' is negative while u' is positive. Since $\overline{u'v'}$ is always negative, the shearing stresses caused by turbulence, $\tau = -\rho\overline{u'v'}$, are positive, as is $d\bar{u}/dy$.

If a velocity distribution has been considered such that $d\bar{u}/dy$ is negative, it will be similarly found that $-\rho\overline{u'v'}$ is always negative. Consequently, in any case, $[\tau]$ has the same sign as $d\bar{u}/dy$, which is emphasized by writing:

$$[\tau] = \rho l^2 \left|\frac{d\bar{u}}{dy}\right| \frac{d\bar{u}}{dy}$$

In the general case,

$$[\tau] = \mu\frac{d\bar{u}}{dy} + \rho l^2 \left|\frac{d\bar{u}}{dy}\right| \frac{d\bar{u}}{dy}$$

This function may be linearized with the help of the Boussinesq coefficient ε:

$$\varepsilon = \rho l^2 \left|\frac{d\bar{u}}{dy}\right|$$

Despite a more complex mathematical form, the main advantage of the Prandtl theory over the Boussinesq theory is that it is easier to assume the value of l than the value of ε.

8-3.4 Taylor's Vorticity Transport Theory

Instead of considering the change of momentum from one layer to another as Prandtl did, Taylor considered the change of moment of momentum. This theory sometimes gives the same result. For example, the velocity distributions in a two-dimensional jet given by both theories are the same. However, when the mixing length is a function of the normal distance from the boundary, different results are obtained.

8-3.5 Value of the Mixing Length

Now the value of the mixing length has to be determined. Various formulas are proposed. The first kind of formula for the mixing length is purely empirical and valid only for special cases. Some examples are:

1. At the boundary of a jet l is proportional to the distance from the orifice.
2. Against the wall of a pipe l is assumed to be proportional to the distance y from the boundary: $l = ky$ where k is a constant. This means physically that the amplitude of a turbulent fluid lump is zero at the boundary and increases linearly with the distance from the boundary.

Introducing this value in the Prandtl formulas yields

$$[\tau] = \rho k^2 y^2 \left|\frac{d\bar{u}}{dy}\right| \frac{d\bar{u}}{dy}$$

If $[\tau]$ is considered as a constant, a "universal velocity distribution" is obtained by integrating with respect to y:

$$\bar{u} = \left(\frac{[\tau]}{\rho}\right)^{1/2} \left(\frac{1}{k} \ln y + \text{constant}\right)$$

$[\tau]$ has been considered as a constant by Prandtl in the theory of the boundary layer along a flat plate. The values of the constants are determined by experiment. In a pipe $[\tau]$ is considered as a linear function of the distance from the wall as it is explained in elementary hydraulics. Both cases are theoretically approximate, but give results close to factual measurements.

8-3.6 Von Kármán's Similarity Hypothesis

Von Kármán tried to find a value for l independent of the kind of flow, according to two similarity assumptions:

1. The turbulence mechanism is independent of viscosity (except near a smooth boundary).
2. The turbulent fluctuations are statistically the same at any point but change only in time and length scales.

From this assumption Von Kármán found that

$$[\tau] = \rho l^2 \left|\frac{d\bar{u}}{dy}\right| \frac{d\bar{u}}{dy}$$

as Prandtl, and

$$l = k \frac{d\bar{u}/dy}{d^2\bar{u}/dy^2}$$

Hence

$$[\tau] = \frac{\rho k^2 (d\bar{u}/dy)^4}{(d^2\bar{u}/dy^2)^2}$$

where k is a universal constant, experimentally found to be equal to 0.4.

8-3.7 Other Theories

Other theories were proposed to improve these semi-empirical formulas. Particularly, in order to avoid $\varepsilon = 0$ when $d\bar{u}/dy = 0$, for example, in the middle of a pipe, Prandtl proposed:

$$[\tau] = \rho l^2 \left[\left(\frac{d\bar{u}}{dy}\right)^2 + l'^2 \left(\frac{d^2\bar{u}}{dy^2}\right)^2\right]$$

But it is difficult to know the best value of l'.

These various theories, and particularly the Prandtl and Von Kármán theories, have been very successfully applied in a number of practical cases (wall, pipe, etc.). However, they do not seem so successful when the flow is not uniform (bend, divergent, etc.). It is to be born in mind that the phenomenon of turbulence is random, and therefore non-deterministic. A solution to the problem of mathematical representation of such complexity is found in statistical mechanics, as introduced by Taylor, Von Kármán, and Kampe de Feriet. This subject is beyond the scope of this book.

8-4 Some Considerations on the Loss of Energy in a Uniform Flow

8-4.1 A Review of Elementary Hydraulics

It has been seen in elementary hydraulics that the head loss in a uniform flow is:

1. Proportional to the mean value through a cross section of the velocity V when the flow is laminar.
2. Proportional to its square value V^2 when the flow is turbulent and the boundary is rough.
3. A complex intermediate function of V (V^n) when the flow is turbulent and the boundary is smooth ($1 < n < 2$).

The above result may be partly explained by the following considerations. A part of the kinetic energy of the primary (or mean) motion of a turbulent flow is continuously absorbed to provide energy for the turbulent fluctuations. The kinetic energy of these turbulent fluctuations is a quadratic function of the fluctuating velocities. Since all these fluctuations are finally absorbed by friction, the loss of energy in a turbulent flow is a quadratic function of the fluctuating velocities.

On the other hand, the fluctuating velocities are roughly linear functions of the mean velocities. It has been seen that

$$[\tau] = \mu \frac{d\bar{u}}{dy} - \rho \overline{u'v'} = \mu \frac{d\bar{u}}{dy} + \rho l^2 \left|\frac{d\bar{u}}{dy}\right| \frac{d\bar{u}}{dy}$$

As long as $\mu(d\bar{u}/dy)$ is negligible, $\overline{u'v'}$ is of order $\bar{u}^2$ and the head loss is proportional to $\bar{u}^2$, that is, proportional to V^2.

In the case of a smooth boundary, the term $d\bar{u}/dy$ is no longer negligible in the boundary layer. Hence the head loss is a complex intermediate function of $\bar{u}$, that is, a complex function of V.

In the case of a laminar flow, $[\tau]$ is simply equal to $\tau = \mu(d\bar{u}/dy)$ and the head loss is a linear function of $\bar{u}(\bar{u} = u)$, hence a linear function of V.

8-4.2 Work Done by Turbulent Forces: Dissipation Function

It is evident that because of the turbulence the loss of energy in a turbulent flow is generally much greater than in a laminar flow.

The mean value of the viscous forces per unit of volume has been found to be

$$\mu\overline{\nabla^2 \mathbf{V}} = \mu\nabla^2\bar{\mathbf{V}} + \mu\overline{\nabla^2 \mathbf{V}'}$$

where

$$\mu\overline{\nabla^2 \mathbf{V}'} = \mu\nabla^2\bar{\mathbf{V}}' = 0$$

The term $\mu\nabla^2\bar{\mathbf{V}}$ is small by comparison to the kinematic forces caused by turbulent fluctuations. The mean value of the viscous forces caused by these turbulent fluctuations is zero.

If instead of considering the mean forces, one considers the mean value of the work done by these forces, quite a different result is obtained. Consider, for example, the mean shearing force $\mu\overline{(\partial u/\partial x)}$. This force is equal to $\mu(\partial\bar{u}/\partial x)$ since

$$\mu\overline{\left(\frac{\partial u'}{\partial x}\right)} = \mu\left(\frac{\partial \overline{u'}}{\partial x}\right) = 0.$$

The work done by the force $\mu(\partial u/\partial y)$ in a unit of time is (see Section 5-5.4)

$$\mu\left(\frac{\partial u}{\partial x}\right)^2 dx\, dy\, dz$$

and by unit of volume: $\mu(\partial u/\partial x)^2$. The mean value of this work with respect to time is, successively,

$$W = \mu\overline{\left(\frac{\partial u}{\partial x}\right)^2} = \mu\overline{\left(\frac{\partial \bar{u}}{\partial x} + \frac{\partial u'}{\partial x}\right)^2}$$

The double product is zero:

$$2\overline{\frac{\partial \bar{u}}{\partial x}\frac{\partial u'}{\partial x}} = 2\frac{\partial \bar{u}}{\partial x}\frac{\partial}{\partial x}\frac{1}{T}\int_0^T u'\,dt = 0$$

so that one finally obtains:

$$W = \mu\left(\frac{\partial \bar{u}}{\partial x}\right)^2 + \mu\overline{\left(\frac{\partial u'}{\partial x}\right)^2}$$

The second term is always positive, hence its mean value is not zero.

Moreover, u' is generally smaller than $\bar{u}$, but the derivative of u' with respect to space (in this expression with respect to x) is usually greater than the derivative of $\bar{u}$ with respect to space (in this expression with respect to x). The first term $(\mu)(\partial\bar{u}/\partial x)^2$ may often be neglected and $W \simeq \overline{\mu(\partial u'/\partial x)^2}$. Hence the loss of energy and the head loss are mainly due to the turbulent fluctuations.

A similar result may be obtained by considering all the terms of the dissipation function Φ presented in Section 5-5.4 in which u, v, w are replaced by $\bar{u}$, $\bar{v}$, $\bar{w}$, u', v', w'. Then it is found that the mean value for Φ is the sum of two terms: Φ_m and Φ_t. $\Phi = \Phi_m + \Phi_t$, where Φ_m is a function of the mean values $\bar{u}$, $\bar{v}$, $\bar{w}$ only, and Φ_t, a function of u', v', w' only.

$$\Phi_m = \mu\left[2\left(\frac{\partial \bar{u}}{\partial x}\right)^2 + \cdots + \left(\frac{\partial \bar{u}}{\partial y} + \frac{\partial \bar{v}}{\partial x}\right)^2 + \cdots\right]$$

$$\Phi_t = \mu\left[2\overline{\left(\frac{\partial u'}{\partial x}\right)^2} + \cdots + \overline{\left(\frac{\partial u'}{\partial y} + \frac{\partial v'}{\partial x}\right)^2} + \cdots\right]$$

Φ_t is the part of energy which is absorbed by friction because of the turbulent fluctuations.

PROBLEMS

8.1 Explain why $[\tau]$ is considered as a constant along a perpendicular to the wall in the boundary layer theory and varies linearly with distance from the wall in the case of a uniform flow in a pipe or between two parallel planes. Explain the limitation of these assumptions. What is the criterion for the pressure distribution on which such assumptions are based?

8.2 It has been found experimentally that

$$\bar{u} = \left(\frac{[\tau_0]}{\rho}\right)^{1/2}\left(\frac{y}{D}\right)^{1/6}$$

where D is the distance between two parallel planes. Give the expressions for $[\tau]$ and Prandtl's mixing length l as functions of y.

8.3 Using the Von Kármán similarity rule

$$\tau = \rho k^2 \frac{(du/dy)^4}{(d^2\bar{u}/dy^2)^2}$$

and the relationship $(\partial\bar{p}/\partial x)$ = constant along the centerline, derive the following universal velocity distribution law for a rectangular channel of width $2h$

$$\bar{u} = \overline{u_0} + \frac{1}{k}\left(\frac{\tau_0}{\rho}\right)^{1/2}\left\{\ln\left[1 - \left(\frac{y}{h}\right)^{1/2}\right] + \left(\frac{y}{h}\right)^{1/2}\right\}$$

where $\overline{u_0}$ is the velocity at the centerline $y = 0$ and τ_0 is the shear stress at the wall.

8.4 It will be assumed that the velocity distribution in a cylindrical pipe of radius R is given by the one-seventh power law, i.e.,

$$\frac{u}{U} = \left(\frac{R - r}{R}\right)^{1/7}$$

where U is the maximum velocity on the centerline. Then give an expression for the Prandtl's mixing length as a function of r.

8.5 Determine the expression of the dissipation function due to turbulent fluctuation as a function of $\partial u'/\partial y$ only, in the case of isotropic turbulence.

Chapter 9

Flow in a Porous Medium: Law of Darcy

9-1 Average Motion in a Porous Medium

9-1.1 *Basic Phenomenology*

9-1.1.1 The basic laws to be applied to a flow in a porous medium are again the continuity relationship and the momentum equation. The momentum principle, expressed by the Navier–Stokes equations, is also theoretically valid for this kind of motion. However, because of the complexity of the boundary conditions (since $\mathbf{V} = 0$ at the surface of every grain of the porous medium), this equation is no longer useful in this form. Some approximations and transformations must be performed.

9-1.1.2 First of all, the grains are assumed to be distributed at random. The flow obeys statistical laws. (The case of nonisotropic porous medium, such as varved clays, necessitates the consideration of a coefficient of permeability which varies with direction.) Hence, instead of dealing with the real values of velocity and pressure, varying in a very complex manner, only the mean values need to be considered.

It is evident that such a method considerably simplifies the boundary conditions since these conditions have to be expressed only to the boundaries of the mean flow, i.e., the limits of the porous medium and the free surface.

9-1.1.3 It is known that in a laminar flow the mixing process is very slow since it is caused only by molecular agitation, while in a turbulent flow it is rapid since it is caused by the turbulence fluctuations. In a laminar flow through a porous medium, because of the random nature of the particle distribution, it may be observed that dye diffuses quickly although the flow is laminar (Fig. 9-1). For example, the concentration curve obtained by injecting a colored fluid within a porous medium is given by a Gaussian-shaped distribution.

The angle of the cone of diffusion is a function of the characteristics of the porous medium and is approximately

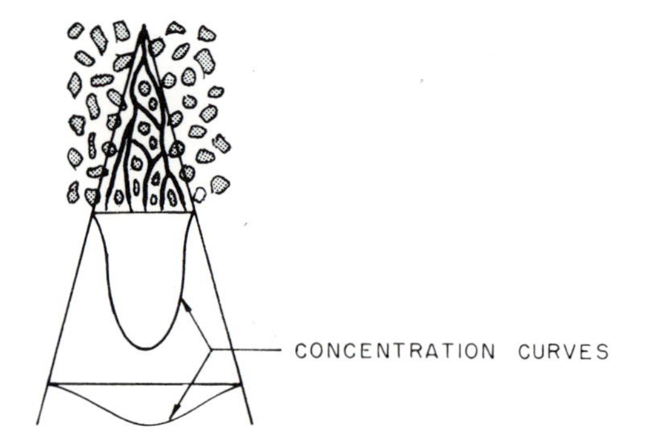

Figure 9-1 *Diffusion through a porous medium.*

6°. This angle increases when the flow becomes turbulent; it is then a function of the Reynolds number as defined in Section 9-3.1.

9-1.2 *Definition of the Mean Motion*

9-1.2.1 The simplest way of defining the mean velocity consists of considering a unidimensional porous medium as shown by Fig. 9-2. The mean velocity or "specific velocity" is the ratio of the discharge Q to the total area A, $V = Q/A$, independent of the void coefficient. Now, if the motion is referred to a three-axis system of coordinates OX, OY, OZ (see Fig. 9-2), the three real velocity components, u, v, w are different from zero. Their mean values in the porous medium are, respectively,

$$\bar{\bar{u}} = \frac{1}{\text{vol}} \iiint_{\text{vol}} u \, d(\text{vol}) = \frac{1}{A} \iint_A u \, dA = V$$

$$\bar{\bar{v}} = \frac{1}{\text{vol}} \iiint_{\text{vol}} v \, d(\text{vol}) = \frac{1}{A} \iint_A v \, dA = 0$$

$$\bar{\bar{w}} = \frac{1}{\text{vol}} \iiint_{\text{vol}} w \, d(\text{vol}) = \frac{1}{A} \iint_A w \, dA = 0$$

where "vol" is the total volume of the porous medium. The mean values with respect to space are written with two bars instead of one to be differentiated from the mean value with respect to time ($\bar{V}$) as used in studies of turbulent motions.

9-1.2.2 For a more complicated pattern of the mean motion, where a variation of the mean value of the velocity with respect to space also exists, the mean value of the

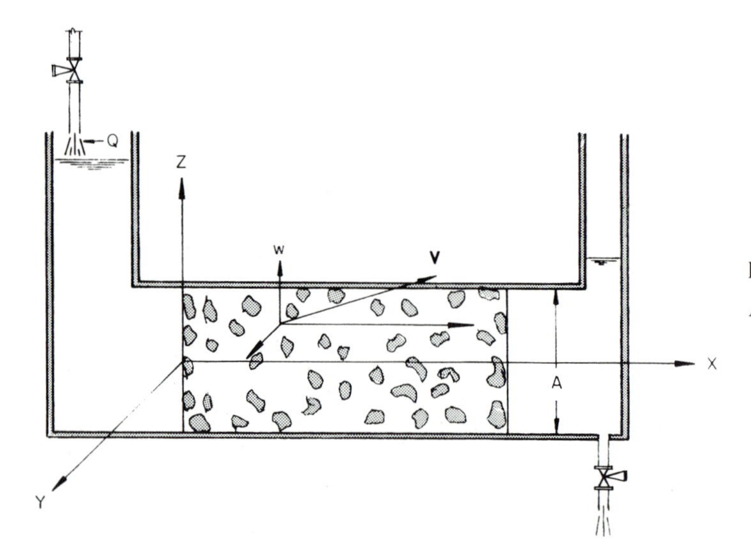

Figure 9-2

Mean uniform flow through porous medium.

velocity vector has to be defined in an elementary volume $\Delta\text{vol} = \Delta x\ \Delta y\ \Delta z$ of porous medium as follows:

$$\bar{\bar{\mathbf{V}}} = \frac{1}{\Delta x\ \Delta y\ \Delta z} \iiint_{\Delta x\ \Delta y\ \Delta z} \mathbf{V}\ dx\ dy\ dz$$

and along three coordinate axes:

$$\begin{bmatrix} \bar{\bar{u}} \\ \bar{\bar{v}} \\ \bar{\bar{w}} \end{bmatrix} = \frac{1}{\Delta x\ \Delta y\ \Delta z} \iiint_{\Delta x\ \Delta y\ \Delta z} \begin{bmatrix} u \\ v \\ w \end{bmatrix} dx\ dy\ dz$$

$\bar{\bar{v}} = 0$ in the case of a mean two-dimensional flow.

Such an elementary volume of porous medium must theoretically be large enough for the averaging process to be valid. Hence, $\Delta x\ \Delta y\ \Delta z$ must be large enough to contain a number of grains distributed at random.

On the other hand, $\Delta x\ \Delta y\ \Delta z$ must theoretically be small enough to be considered as infinitely small $dx\ dy\ dz$ in order to apply the methods of differential calculus.

In other words, $\Delta x\ \Delta y\ \Delta z$ has to be large enough for the averaging process to be valid, but small enough to be considered as infinitely small in the mean motion. For both these opposing conditions to be satisfied the gradient of real velocity has to be much greater than the gradient of the mean velocity. This may be physically translated as: a large flow pattern through relatively small grains or pebbles. A flow pattern around a few large rocks does not obey statistical laws. Therefore it cannot be represented realistically by a mean motion.

9-1.2.3 Similarly the mean pressure is defined by:

$$\bar{\bar{p}} = \frac{1}{\Delta x\ \Delta y\ \Delta z} \iiint_{\Delta x\ \Delta y\ \Delta z} p\ dx\ dy\ dz$$

The variations of p around $\bar{\bar{p}}$ are mainly caused by the curvature of the paths around the grains. Such variations are proportional to the convective inertia, i.e., proportional to the square of the velocity, which is usually negligible.

9-1.2.4 Considering the mean velocities with respect to space passing across the plane sides of a cube defining an elementary volume of a porous medium, it is found by a demonstration similar to that given in Chapter 3 that the continuity relationship is:

$$\frac{\partial \bar{\bar{u}}}{\partial x} + \frac{\partial \bar{\bar{v}}}{\partial y} + \frac{\partial \bar{\bar{w}}}{\partial z} = 0$$

That is, the continuity relationship for the mean motion has the same mathematical form as for other kinds of flow.

9-1.2.5 The boundary conditions are expressed as a function of the mean velocity at the boundary of the porous medium instead of being expressed as a function of the real velocity at the boundary of each grain.

9-1.3 Analogies between Turbulent Flow and Flow through a Porous Medium

9-1.3.1 Interesting theoretical analogies may be made between the methods of studying turbulent flow and flow through a porous medium. In both cases the mean velocity and mean pressure are dealt with because of the random nature of the flows. In the case of turbulence the mean values are defined at a given point with respect to time, while in the case of flow through a porous medium the mean values are defined with respect to space (see Section 7-1.3).

$$\bar{\mathbf{V}} = \frac{1}{T}\int_0^T \mathbf{V}\ dt \qquad \bar{\bar{\mathbf{V}}} = \frac{1}{\text{vol}} \iiint_{\text{vol}} \mathbf{V}\ d(\text{vol})$$

The time T has to be long enough for the averaging process to be valid, but short enough to take account of whether the mean motion is steady or unsteady (see Section 7-1.4). The elementary volume $\Delta x\ \Delta y\ \Delta z$ must obey similar considerations with respect to space as has been discussed in Section 9-1.2.2.

The fluctuation terms u', v', w', p' found in the studies of turbulence exist also with respect to space in the studies of flow through a porous medium, and their mean value is also zero by definition. If one defines $\mathbf{V} = \bar{\bar{\mathbf{V}}} + \mathbf{V}'$, then

$$\frac{1}{\Delta x\ \Delta y\ \Delta z} \iiint_{\Delta x\ \Delta y\ \Delta z} \mathbf{V}'(\text{or } p')\ dx\ dy\ dz = 0$$

The momentum equation of the mean motion in a porous medium is obtained by averaging each of the forces with respect to space, as has been done with respect to time in the study of turbulent flow. This will be the subject of Section 9-2. Both turbulent flow and flow through a porous medium are strongly rotational as far as the real motion is concerned. However, their mean motions may be irrotational (see Sections 8-1.2 and 9-2.2.2). An isotropic turbulent flow may by considered analogous to flow through an isotropic medium. Turbulent flow through a porous medium is studied by considering the mean values with respect to both space and time such that

$$\bar{\bar{\mathbf{V}}} = \frac{1}{T}\int_0^T \frac{1}{\Delta x\,\Delta y\,\Delta z}\iiint_{\Delta x\,\Delta y\,\Delta z} \mathbf{V}\,dx\,dy\,dz\,dt$$

9-2 Law of Darcy

9-2.1 *Approximations*

9-2.1.1 First of all, it must be noted that for certain flows with a free surface through a very fine porous medium, the capillarity forces could have an appreciable effect on the flow pattern and the discharge.

For example, the rise of the free surface in an earth dam of grain size near 0.1 mm is about 1 ft (30 cm) (Fig. 9-3).

9-2.1.2 Insofar as these capillarity effects may be neglected, the momentum equation expressed as a function of the mean values is given by the same averaging operation with respect to space as was done for a turbulent flow with respect to time.

Since the sum of the real value of the different forces involved is always zero,

$$\begin{matrix}\text{Inertia}\\ \text{force}\end{matrix} + \begin{matrix}\text{Gravity}\\ \text{force}\end{matrix} + \begin{matrix}\text{Pressure}\\ \text{force}\end{matrix} + \begin{matrix}\text{Friction}\\ \text{force}\end{matrix} = 0$$

The sum of the forces' mean values with respect to space is also equal to zero

$$\begin{matrix}\text{Mean}\\ \text{value with}\\ \text{respect to}\\ \text{space}\end{matrix}\left[\begin{matrix}\text{Inertia}\\ \text{force}\end{matrix} + \begin{matrix}\text{Gravity}\\ \text{force}\end{matrix} + \begin{matrix}\text{Pressure}\\ \text{force}\end{matrix} + \begin{matrix}\text{Friction}\\ \text{force}\end{matrix}\right] = 0$$

9-2.1.3 For a first approximation, the inertia forces are neglected. The local inertia is neglected because the variation of the ground water table is usually very slow. From this point of view, unsteady motions through a porous medium are usually studied as a succession of steady motions (see Section 4-5.1). However, some special problems require the consideration of the local inertia, for example, perviousness of a rockfill breakwater to periodical gravity waves.

The convective inertia is also neglected. Since the velocity is usually very small, the square of the velocity and terms which are functions of the square of the velocity (such as the convective inertia forces) are of a second order of magnitude in comparison with other terms (see Section 4-5.2.1). The range of validity of such an assumption is studied further.

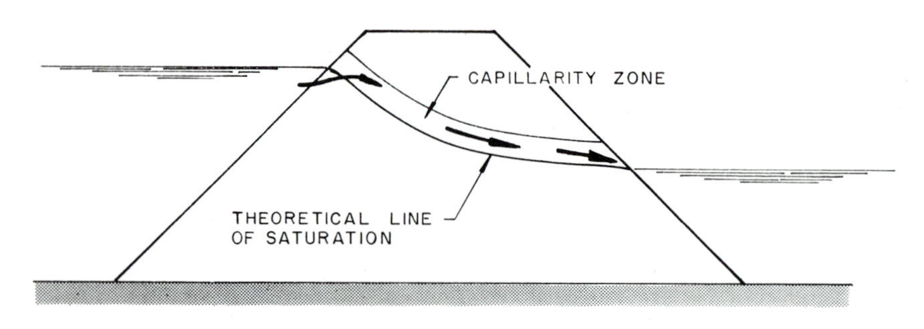

Figure 9-3

Capillarity effects in a flow through porous medium.

9-2.2 *Mean Forces*

9-2.2.1 Finally the momentum equation is reduced to an equality of applied forces:

$$\frac{1}{\Delta x\, \Delta y\, \Delta z} \iiint_{\Delta x\, \Delta y\, \Delta z} \left(-\frac{\partial (p + \rho g z)}{\partial x} + \mu \nabla^2 u \right) dx\, dy\, dz = 0$$

Two similar equations may be written along the two other axes OY and OZ. These three equations are, in vector form:

$$\frac{1}{\Delta x\, \Delta y\, \Delta z} \iiint_{\Delta x\, \Delta y\, \Delta z} [-\textbf{grad}\,(p + \rho g z) + \mu \nabla^2 \mathbf{V}]\, dx\, dy\, dz = 0$$

In calculating these equations as functions of the mean value $\bar{\bar{\mathbf{V}}}$ and $\bar{\bar{p}}$, it is to be noted that they include:

One constant force: Gravity force
Two linear forces: Pressure force and viscous force

Following the same process of integration as that used in turbulent motion (see Section 7-2), it is found that

$$\frac{1}{\Delta x\, \Delta y\, \Delta z} \iiint_{\Delta x\, \Delta y\, \Delta z} -\textbf{grad}\,(p + \rho g z)\, dx\, dy\, dz = -\textbf{grad}\,(\bar{\bar{p}} + \rho g z)$$

Now consider the viscous forces. Since they are linear, it is reasonable to expect that they are proportional to the mean velocity $\bar{\bar{\mathbf{V}}}$ as long as there are no quadratic effects caused by the convective inertia and turbulence. Hence it is written that they are proportional to $\mu \bar{\bar{\mathbf{V}}}$ such that

$$\frac{1}{\Delta x\, \Delta y\, \Delta z} \iiint_{\Delta x\, \Delta y\, \Delta z} \mu \nabla^2 \mathbf{V}\, dx\, dy\, dz = \frac{\mu}{k} \bar{\bar{\mathbf{V}}}$$

where k is the *permeability* of the porous medium, k is an empirical function of the void coefficient and grain size. $K = k/\mu$ is the *hydraulic conductivity* which measures the permeability of the porous medium to the fluid. Hence, one has finally,

$$\bar{\bar{u}} = K \frac{\partial}{\partial x} (\bar{\bar{p}} + \rho g z) = K \frac{\partial \bar{\bar{p}}^*}{\partial x}$$

$$\bar{\bar{v}} = K \frac{\partial}{\partial y} (\bar{\bar{p}} + \rho g z) = K \frac{\partial \bar{\bar{p}}^*}{\partial y}$$

$$\bar{\bar{w}} = K \frac{\partial}{\partial z} (\bar{\bar{p}} + \rho g z) = K \frac{\partial \bar{\bar{p}}^*}{\partial z}$$

or vectorially:

$$\bar{\bar{\mathbf{V}}} = K\, \textbf{grad}\,(\bar{\bar{p}} + \rho g z) = K\, \textbf{grad}\, \bar{\bar{p}}^*$$

This expression defines the "Law of Darcy."

This law states that the mean velocity of the fluid flowing through a porous medium is directly proportional to the pressure gradient acting on the fluid. $\bar{\bar{p}}^* = \bar{\bar{p}} + \rho g z$ is the piezometric head. The simplification of the friction term is of an empirical nature and it seems difficult to justify such a law rigorously. It would be necessary to go through the calculation for a flow as shown in Fig. 9-1. On the other hand, it would seem reasonable to think that the Navier–Stokes equations are no longer valid, since from a microscopic point of view, a flow passing through the very fine channels of a porous medium, like porous china, would probably require a study based on molecular agitation. This subject is relevant to the kinetic theory of liquids.

In a nonisotropic porous medium, K has different values —K_x, K_y, K_z—along the three components axes OX, OY, and OZ, respectively.

9-2.2.2 It is important to note that such a mean motion defined by the law of Darcy is always irrotational.

Introducing the value $\bar{\bar{u}}$, $\bar{\bar{v}}$, $\bar{\bar{w}}$ given above, it is easy to verify that $\partial \bar{\bar{u}}/\partial y - \partial \bar{\bar{v}}/\partial x = 0$, since

$$\frac{\partial}{\partial y}\left[\frac{k}{\mu}\frac{\partial \bar{\bar{p}}^*}{\partial x}\right] - \frac{\partial}{\partial x}\left[\frac{k}{\mu}\frac{\partial \bar{\bar{p}}^*}{\partial y}\right] \equiv 0$$

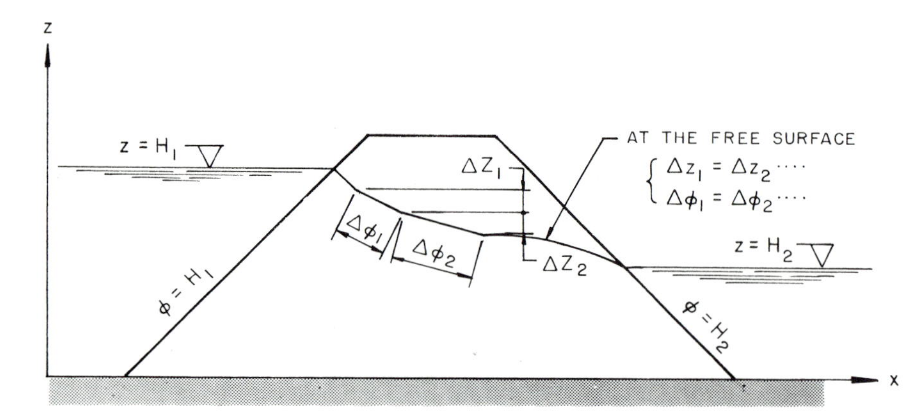

Figure 9-4

Value of the potential function for a free surface flow through porous medium.

Similar demonstrations may be done for the two other conditions given in Section 2-6.2. However, if the flow through the porous medium is turbulent, it is necessarily rotational and hence cannot have its mean motion defined by a potential function.

9-2.2.3 The velocity potential function for a laminar flow through a porous medium is usually defined by $\bar{u} = -K(\partial\phi/\partial x)$, $\bar{v} = -K(\partial\phi/\partial y)$, $\bar{w} = -K(\partial\phi/\partial z)$, or $\bar{\mathbf{V}} = -K\ \mathbf{grad}\ \phi$. Substituting these values into the Darcy equations gives

$$\bar{u} = -K\frac{\partial\phi}{\partial x} = -K\frac{\partial}{\partial x}(\bar{p} + \rho g z)$$

and two similar equations for $\bar{v}$ and $\bar{w}$. From these, it is easy to see that ϕ is equal to the piezometric head: $\overline{\overline{p^*}} = (\bar{p} + \rho g z) = \phi$. Sometimes ϕ is also defined by: $[(\bar{p}/\rho g) + z] = \overline{\overline{p^*}}/\rho g$ and $\bar{u} = -K\rho g(\partial\phi/\partial x)$, etc. Figure 9-4 illustrates the value and the physical meaning of ϕ corresponding to such a definition. The velocity potential function is a constant along the sides of the dike and decreases linearly with z at the free surface inside the dike. Hence a constant Δz corresponds to a constant value for $\Delta\phi$. This property will be used in Section 11-6.3 for devising a graphical method of investigation.

9-3 Range of Validity of the Law of Darcy

9-3.1 Effect of Reynolds Number

9-3.1.1 The permeability k is a function of the porous medium characteristics and the Reynolds number. The Reynolds number of a flow through a porous medium is defined by $\bar{\bar{\mathbf{V}}}\delta/\nu$ where $\bar{\bar{\mathbf{V}}}$ is the mean or specific velocity as previously defined, and δ is the diameter of a grain. The diameter of a grain is easily known while the "diameter of the channels," as used for pipe, would be difficult to define. This process assumes that there is a simple linear relationship between these "channel diameters" and the grain size. However, in a porous medium made of a large grain size distribution, the small particles have a tendency to reduce the size of the "channel." The "channels" have the same order of magnitude as the smallest particles. Hence, it is more exact in this case to define the Reynolds number with the help of the smaller grain sizes. The "characteristic diameter" δ_c may be considered empirically as the average size corresponding to the lowest 10 percent limit.

9-3.1.2 Although the velocity in a porous medium is very small, the derivative of velocity with respect to space is large. It is easy to recognize this since the actual paths in a porous medium are strongly curved. Hence the convective inertia has an appreciable influence on the motion when the Reynolds number is greater than one even before the appearance of turbulence. Since the convective inertia is quadratic, the following law of Forchheimer is more accurate than the Darcy's law.

$$\mathbf{grad}\ \overline{\overline{p^*}} = a\overline{\overline{\mathbf{V}}} + b\overline{\overline{\mathbf{V}}}|\overline{\overline{\mathbf{V}}}|^n$$

where n lies between 0 and 1.

9-3.1.3 At larger Reynolds numbers ($R_e > 100$), the flow becomes turbulent. The above Forchheimer equation may still be applied but the values of the coefficients a and b are changed. At very large Reynolds numbers, the linear term $a\overline{\overline{\mathbf{V}}}$ becomes negligible, and the coefficient n tends to the value 1:

$$\mathbf{grad}\ \overline{\overline{p^*}} = b\overline{\overline{\mathbf{V}}}|\overline{\overline{\mathbf{V}}}|$$

Then the coefficient b for the same void coefficient and the same grain size distribution is a function of the roughness of the rocks. A similar phenomenon has been found in elementary hydraulics in the case of turbulent flow in a rough pipe (see Section 13-4.3).

9-3.2 Permeability Coefficient

The value of the permeability coefficient k is given by dimensional analysis and experimental results. In the general case, it is found to be a function of the Reynolds number $V\delta_c/v$, void coefficient ε, and the Froude number $V^2/g\delta$. Many functions have been proposed, too numerous to be compared and analyzed in this book. Some of them are more or less theoretically justified.

For a first order of approximation, the following equation may be used for both laminar and turbulent flow:

$$\frac{\Delta H}{\Delta L} = \frac{C_x(1 - \varepsilon)^2}{\varepsilon^3} \frac{\overline{\overline{V}}^2}{2g\delta_c}$$

where $\rho g\ \Delta H/\Delta L$ is the gradient of pressure ($= \mu\overline{\overline{V}}/k$) and C_x is the drag coefficient of a rough sphere for the same value of the Reynolds number (Fig. 9-5) (see Section 13-5.1). This empirical law has been established experimentally for a range of Reynolds numbers between 10^2 and 10^5 and for small variations of the void coefficient ε, such as encountered in sand or quarry run used in rubble mound

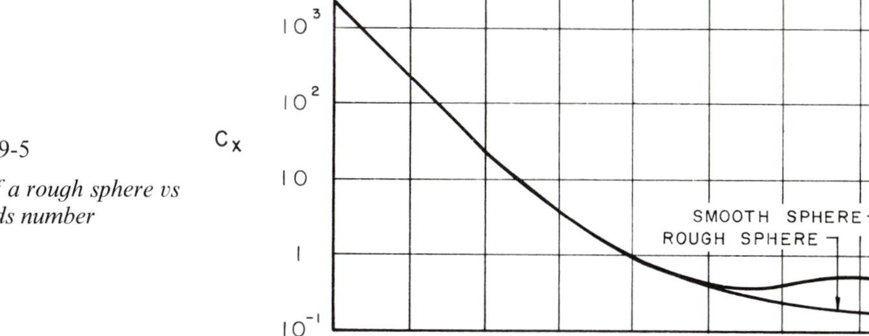

Figure 9-5

Drag of a rough sphere vs Reynolds number

breakwaters and rockfill cofferdams. For a laminar motion with a negligible convective inertia ($R_e < 1$),

$$C_x = \frac{24}{\overline{\overline{V}}\delta_c/\nu}.$$

Hence the hydraulic conductivity is found to be equal to

$$K = \frac{k}{\mu}\frac{\rho g \delta_c^2 \varepsilon^3}{12\mu(1-\varepsilon)^2}$$

9-4 A Comparison Between the Basic Equations of Different Kinds of Flow Motion

9-4.1 *The Unknowns to Be Found*

To solve a problem in hydrodynamics, there are in general two unknowns to be found: the velocity $\mathbf{V}(u,v,w)$ and the pressure p as a function of space coordinates x, y, z, and time t.

However, for turbulent flow, the mean motion with respect to time is dealt with. The two unknowns are $\overline{\mathbf{V}}(\bar{u},\bar{v},\bar{w})$ and $\bar{p}$. The fluctuations of velocity $\mathbf{V}'(u',v',w')$ give rise to some convective inertia forces acting on the mean motion similar to the external forces.

For flow through a porous medium the mean motion with respect to space is dealt with. The two unknowns are $\overline{\overline{\mathbf{V}}}(\bar{\bar{u}},\bar{\bar{v}},\bar{\bar{w}})$ and $\bar{\bar{p}}$.

In all cases (ideal fluid, laminar flow, turbulent flow, flow through porous medium) the two unknowns—velocity and pressure, real or mean with respect to time or with respect to space—are obtained by the continuity relationship and the momentum equation.

9-4.2 *The Continuity Equations*

The continuity relationship has the same mathematical form for four kinds of motion. It is always expressed as a function of the real velocity $\mathbf{V}(u,v,w)$ for an ideal fluid and a laminar flow:

$$\frac{\partial u}{\partial x} + \frac{\partial v}{\partial y} + \frac{\partial w}{\partial z} = 0, \text{ i.e., div } \mathbf{V} = 0$$

and applied as such to ideal fluid and laminar viscous flows. It is also expressed as a function of the mean velocity with respect to time for a turbulent flow:

$$\frac{\partial \bar{u}}{\partial x} + \frac{\partial \bar{v}}{\partial y} + \frac{\partial \bar{w}}{\partial z} = 0$$

	Inertia		*Pressure gravity*		*Viscous friction*		*Convective inertia caused by turbulence*
Perfect fluid: Eulerian equation	$\rho\frac{du}{dt}$	$=$	$-\frac{\partial p^*}{\partial x}$				
Laminar flow: Navier–Stokes equations	$\rho\frac{du}{dt}$	$=$	$-\frac{\partial p^*}{\partial x}$	$+$	$\mu\nabla^2 u$		
Turbulent flow: Reynolds (or Boussinesq) equations	$\rho\frac{d\bar{u}}{dt}$	$=$	$-\frac{\partial \overline{p^*}}{\partial x}$	$+$	$\mu\nabla^2\bar{u}$	$-$	$\rho\left(\frac{\partial \overline{u'^2}}{\partial x} + \frac{\partial \overline{u'v'}}{\partial y} + \frac{\partial \overline{u'w'}}{\partial z}\right)$
Flow through porous medium: Darcy's law when $V\delta/\nu < 1$	0	$=$	$-\frac{\partial \overline{\overline{p^*}}}{\partial x}$	$+$	$\frac{\mu}{k}\bar{\bar{u}}$		

and as a function of the mean velocity with respect to space for a flow through a porous medium:

$$\frac{\partial \bar{\bar{u}}}{\partial x} + \frac{\partial \bar{\bar{v}}}{\partial y} + \frac{\partial \bar{\bar{w}}}{\partial z} = 0$$

9-4.3 *The Momentum Equations*

The momentum equations are written (in the table opposite) along the OX axis only for the four cases to be considered. Since similar terms are found in these four equations, similar methods of integration may be used. Some of them are only valid after some approximations or some assumptions have been used to simplify the basic equations. For example, neglecting the turbulent fluctuation terms and the viscous term, a turbulent flow behaves as a perfect fluid. Hence, in order to simplify the writing and for more generality, only the notation $\mathbf{V}(u,v,w)$ and p are used in the following chapters since it is understood that $\mathbf{V}$ and p means $\bar{\mathbf{V}}$ and $\bar{p}$ for turbulent flow and $\bar{\bar{\mathbf{V}}}$ and $\bar{\bar{p}}$, respectively, for flow through a porous medium.

PROBLEMS

9.1 Calculate the total flow discharge through a porous medium of total cross section $A = 1\ \text{ft}^2$ (929 cm^2) and length in the direction of the flow $l = 3$ ft (91.4 cm) as a function of the head. The significant grain size diameter is $\delta = 0.07$ in (0.3 mm) and the void coefficient is $\varepsilon = 0.40$. One will also make use of Fig. 9-5 for determining the head loss coefficient. Determine the head under which the law of Darcy no longer applies, and the head under which the turbulence appears. Repeat the same calculation when the porous medium is composed of two successive layers: $l = 1.5$ ft (45.7 cm) and $\delta = 0.12$ in (0.5 mm); $l = 1.5$ ft (45.7 cm) and $\delta = 0.024$ in (0.1 mm); and three successive layers of length $l = 1$ ft (30.5 cm) each and $\delta = 0.024$ in (0.1 mm), 0.072 in (0.3 mm), and 0.12 in (0.5 mm) respectively, of same void coefficient (kinematic coefficient of viscosity $\nu = 1.076 \times 10^{-5}\ \text{ft}^2/\text{sec}$ (0.01 cm^2/sec).

9.2 Consider a flow through a porous medium with a cross section $A = 100\ \text{ft}^2$ and a length in the direction of the average flow $L = 100$ ft. One wants to build a scale model of this porous medium at a scale $\lambda = 1/10$ such that $a = 1\ \text{ft}^2$ (929 cm^2) and $l = 10$ ft (305 cm), and with the same void coefficient ε. One wants the discharge to obey the rule of similitude of Froude, i.e., $q = \lambda^{5/2} Q$ under a head $\Delta h_{\text{model}} = \Delta h_{\text{prototype}} \times \lambda$. For this purpose the grain size of the model δ_m will be related to the grain size of the prototype δ_p by the relationship $\delta_m = K\lambda\delta_p$. Determine the value of K in the case where $H = 100$ ft (3047 cm) $\delta = 0.024$ in (1 mm), and $\varepsilon = 0.40$.

REFERENCES FOR PART ONE

Daugherty, R. L., and Ingersoll, A. C., *Fluid Mechanics*, 5th ed. McGraw-Hill, New York, 1947.

Goldstein, S., *Modern Development in Fluid Mechanics*, Vol. I. Oxford University Press, 1938.

Hunter, R., Editor, *Advanced Mechanics of Fluids*. Wiley, New York, 1959.

Lamb, H., *Hydrodynamics*, 6th ed. Dover Publications, New York, 1945.

Landau, L. D., and Lifshitz, L. D., *Fluid Mechanics*. Pergamon Press, London, 1959.

Li, W. H., and Lam, S. H., *Principles of Fluid Mechanics*. Addison-Wesley, Reading, Mass., 1964.

Milne-Thomson, L. M., *Hydrodynamics*. Macmillan, New York, 1960.

Owczarek, J. A., *Introduction to Fluid Mechanics*. International Textbook, Scranton, Pennsylvania, 1968.

Planck, M., *The Mechanics of Deformable Bodies*. Macmillan, London, 1932.

Prandtl, L., and Tietjens, O. G., *Fundamentals of Hydro and Aero-mechanics*. McGraw-Hill, New York, 1934.

Rouse, H., *Elementary Mechanics of Fluids*. Wiley, New York, 1946.

Sabersky, R. H., and Acosta, J. A., *Fluid Flow—A First Course in Fluid Mechanics*. Macmillan, New York, 1964.

Serrin, J., "Mathematical Principles of Classical Fluid Mechanics." In *Encyclopedia of Physics* (S. Fligge, Ed.), Vol. 8-1. Springer, Berlin, 1959.

Streeter, V., Editor, *Handbook of Fluid Mechanics*. McGraw-Hill, New York, 1961.

Streeter, V., *Fluid Dynamics*. McGraw-Hill, New York, 1948.

PART TWO

Some Mathematical Treatments of the Basic Equations

Chapter 10

The Bernoulli Equation

10-1 General Momentum Relationships

10-1.1 Momentum Equation Along a Streamline

The laws that govern the motion of a fluid element have been established in the first part of this book. They are given in differential forms. The purpose of this chapter is to establish general relationships from these equations, the first of which gives the balance of forces along a streamline.

10-1.1.1 Consider first the momentum equations in the Eulerian form:

$$\rho\frac{du}{dt} = -\frac{\partial p^*}{\partial x}$$

$$\rho\frac{dv}{dt} = -\frac{\partial p^*}{\partial y}$$

$$\rho\frac{dw}{dt} = -\frac{\partial p^*}{\partial z}$$

For the sake of simplicity in demonstrating the first relationship, it is assumed that u, v, and w vary with respect to space only, i.e., the motion is steady and the partial derivatives $\partial u/\partial t$, $\partial v/\partial t$, $\partial w/\partial t$, are zero.

If these equations are multiplied by the differentials dx, dy, and dz, respectively, and then summed, the result is

$$\rho\left(\frac{du}{dt}dx + \frac{dv}{dt}dy + \frac{dw}{dt}dz\right) = -\left(\frac{\partial p^*}{\partial x}dx + \frac{\partial p^*}{\partial y}dy + \frac{\partial p^*}{\partial z}dz\right)$$

If it is defined that $u = dx/dt$, $v = dy/dt$, and $w = dz/dt$, where dx, dy, and dz are by definition the components of an element of the streamline $\mathbf{dS}$, then the differentials can be written as $dx = u\,dt$, $dy = v\,dt$, and $dz = w\,dt$. Therefore, the left side of the above equation can be written in the form

$$\rho(u\,du + v\,dv + w\,dw) = \rho d\left(\frac{v^2 + v^2 + w^2}{2}\right) = d\left(\frac{\rho V^2}{2}\right)$$

The right-hand side is the total differential of p^*. Hence, the momentum equation becomes

$$d\left(\rho\frac{V^2}{2}\right) = -dp^*$$

Specifying that the above equation applies along a streamline only and that $p^* = p + \rho gz$, the final result is

$$\frac{d}{dS}\left(\rho\frac{V^2}{2} + p + \rho gz\right) = 0$$

When integrated this becomes

$$\rho\frac{V^2}{2} + p + \rho gz = \text{constant}$$

which is the same equation that was obtained for an irrotational motion (see Section 6-1.2.5). This formula has been obtained without assuming the motion to be irrotational, as the total convective derivatives of u, v, and w, were taken into account exactly. However, the integration of the equation is limited along a streamline because of the substitution of the differential $dx = u\,dt$, etc.

10-1.1.2 When the local inertia terms are introduced and the procedures used for steady motion are applied, the following equations result.

$$\rho\frac{V^2}{2} + p + \rho gz + \rho\int\left(\frac{\partial u}{\partial t}dx + \frac{\partial v}{\partial t}dy + \frac{\partial w}{\partial t}dz\right) = \text{constant}$$

or

$$\rho\frac{V^2}{2} + p + \rho gz + \rho\int\frac{\partial V}{\partial t}dS = \text{constant}$$

Differentiating with respect to S gives

$$\frac{\partial}{\partial S}\left(\frac{\rho V^2}{2} + p + \rho gz\right) + \rho\frac{\partial V}{\partial t} = 0$$

10-1.2 The Case of Irrotational Flow

In the case of an irrotational flow, one has seen (Section 2-6.3) that the velocity vector could be defined by a potential function ϕ such that $\mathbf{V} = \mathbf{grad}\,\phi$, and

$$u = \frac{\partial\phi}{\partial x} \qquad v = \frac{\partial\phi}{\partial y} \qquad w = \frac{\partial\phi}{\partial z} \qquad V = \frac{\partial\phi}{\partial S}$$

Let us consider again the momentum equation in the X direction, and replace the velocity components u, v, w, by their expressions in terms of ϕ (see equation 6-1):

$$\rho\frac{\partial^2\phi}{\partial x\partial t} + \rho\left(\frac{\partial\phi}{\partial x}\frac{\partial^2\phi}{\partial x^2} + \frac{\partial\phi}{\partial y}\frac{\partial^2\phi}{\partial x\partial y} + \frac{\partial\phi}{\partial z}\frac{\partial^2\phi}{\partial x\partial z}\right) = -\frac{\partial p^*}{\partial x}$$

Since

$$\frac{\partial}{\partial x}\left[\frac{1}{2}\left(\frac{\partial\phi}{\partial x}\right)^2\right] = \frac{\partial\phi}{\partial x}\frac{\partial^2\phi}{\partial x^2}$$

$$\frac{\partial}{\partial x}\left[\frac{1}{2}\left(\frac{\partial\phi}{\partial y}\right)^2\right] = \frac{\partial\phi}{\partial y}\frac{\partial^2\phi}{\partial y\partial x}$$

and the same for the z component, the above equation can be written as

$$\frac{\partial}{\partial x}\left\{\rho\left(\frac{\partial\phi}{\partial t}\right) + \frac{\rho}{2}\left[\left(\frac{\partial\phi}{\partial x}\right)^2 + \left(\frac{\partial\phi}{\partial y}\right)^2 + \left(\frac{\partial\phi}{\partial z}\right)^2\right] + p^*\right\} = 0$$

or

$$\frac{\partial}{\partial x}\left(\rho\frac{\partial\phi}{\partial t} + \rho\frac{V^2}{2} + p^*\right) = 0$$

Hence, by integrating,

$$\rho\frac{\partial\phi}{\partial t} + \rho\frac{V^2}{2} + p^* = F(y,z,t)$$

Similar equations are obtained in the Y and Z directions so that the function F is a function of time t only. It is important to point out that this equation is valid everywhere in the fluid and in any direction, particularly along a streamline.

Indeed, taking the derivative with respect to S yields

$$\rho \frac{\partial^2 \phi}{\partial t \partial S} + \frac{\partial}{\partial S}\left(\rho \frac{V^2}{2} + p^*\right) = 0$$

But, since $V = \partial\phi/\partial S$, then this can be written as

$$\rho \frac{\partial V}{\partial t} + \frac{\partial}{\partial S}\left(\rho \frac{V^2}{2} + p^*\right) = 0$$

A number of simple cases are reviewed (*vide infra*), with a physical interpretation.

10-2 Simple Cases and Physical Interpretations

10-2.1 Slow-Steady and Uniform-Steady Motions

10-2.1.1 As the motion is steady, the local inertia terms are zero; because the motion is slow, the convective inertia terms may be neglected; (and because the fluid is perfect, the friction forces are zero). Hence, the momentum equation is reduced to an equality of applied forces—pressure and gravity, mathematically expressed as

$$\frac{\partial p}{\partial x} = 0$$

$$\frac{\partial p}{\partial y} = 0$$

$$\frac{\partial(p + \rho g z)}{\partial z} = 0$$

As the axes OX and OY are horizontal, the first two equations show that the pressure is a constant in a horizontal plane. The third equation gives

$$p + \rho g z = p^* = \text{constant}$$

It is seen that p varies linearly with the distance from the free surface. The law of hydrostatics is recognized; that is, hydrostatics could be considered as a limiting case of steady slow motion of an ideal fluid. Therefore, the same law applies both to steady slow motion and to the case of no motion. However, in the latter case the law of hydrostatics is physically exact, in addition to being mathematically exact, since the case of no motion implies no friction.

10-2.1.2 A similar law is obtained for a uniform steady flow of a real fluid, i.e., with friction forces. Consider the uniform flow studied in Section 6-2.1.1. The pressure distribution obtained was: $p = p_a + \rho g z \cos \alpha$, where p_a is the atmospheric pressure.

The isobars or lines of equal pressure are inclined at an angle α with respect to a horizontal plane. Therefore, buoyancy exerted on any body in such a flow, such as on a rock deposit on the bottom, is exerted at an angle α with the vertical.

For a number of practical cases of flow with a free surface, α is very small and $\cos \alpha$ may be considered equal to unity. Therefore, the pressure distribution in a free surface uniform flow is most often hydrostatic.

This hydrostatics law is considered to be accurate enough even for a nonuniform and nonslow motion when the curvature of the paths is small. The calculation of a backwater curve as seen in elementary hydraulics is usually based on such an assumption, even though this approximation is not always specified.

Figures 10-1 and 10-2 illustrate two cases in which the convective inertia has a nonnegligible influence on the

Figure 10-1 *Pressure distribution is greater than that given by hydrostatic law.*

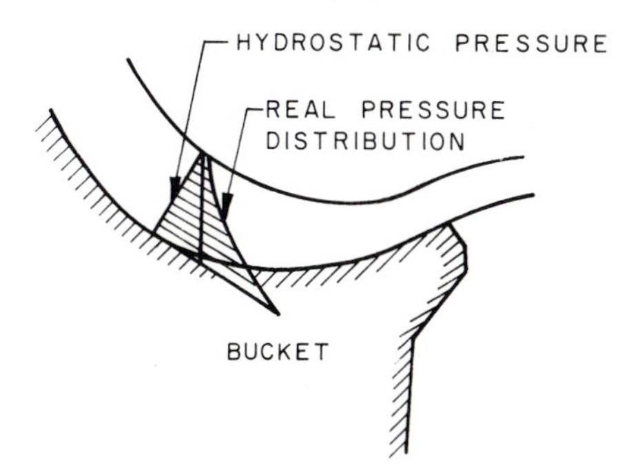

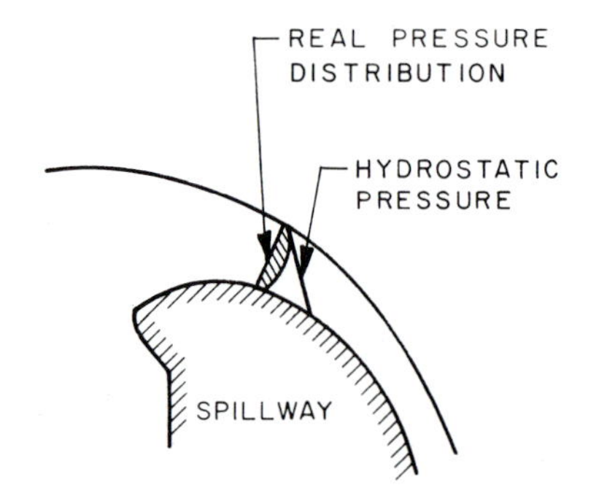

Figure 10-2 *Pressure distribution is smaller than that given by hydrostatic law.*

pressure distribution, and conversely, the pressure distribution has an influence on the flow pattern. The effect of flow curvature is studied in Section 10-2.4.

10-2.2 Slow-Unsteady and Uniform-Unsteady Motions of a Perfect Fluid

The rotational convective inertia terms are nonlinear. Therefore, a slow motion may mathematically be considered irrotational and $\mathbf{V}(u,v,w)$ may be defined by a potential function ϕ. The quadratic term $\rho V^2/2$ may be neglected. Therefore the momentum equation is

$$\rho \frac{\partial \phi}{\partial t} + p + \rho g z = f(t)$$

which is valid in any direction throughout the fluid. The derivative of the function $f(t)$ with respect to space is zero.

10-2.3 Steady Irrotational Motion of a Perfect Fluid

10-2.3.1 The momentum equations in the OX, OY, OZ directions of steady irrotational motion of a perfect fluid give

$$\rho \frac{V^2}{2} + p + \rho g z = \text{constant}$$

(see Section 6-1.2.4). Dividing by $\varpi = \rho g$ yields the total head H, which has the dimension of a length

$$\frac{V^2}{2g} + \frac{p}{\varpi} + z = H$$

The total head is the sum of the *velocity head* $V^2/2g$, the *pressure head* p/ϖ, and the *elevation head* z. The value given by a piezometer, $(p/\varpi) + z$, called the *piezometric head* (see Fig. 10-3).

It is seen that the variation with respect to space of the total head H in an irrotational motion is zero: $\mathbf{grad}\, H = 0$.

Such an equation expresses the conservation of energy of an elementary particle of fluid as a sum of its kinetic energy, pressure energy, and potential energy. It is emphasized that the Bernoulli equation is valid in any direction for an irrotational motion, i.e., along a path as well as along a normal to a path. It is noticed, also, that the velocity V and pressure p refer to the local value of the velocity and do not refer to the mean velocity and mean pressure in a cross section of a stream tube. The cases in which Bernoulli's equation can be approximately applied to stream tube are discussed in the following section.

In engineering practice, the pressure along a boundary may be determined by this equation when the velocity field is given, and the flow is approximately irrotational, as encountered in short, convergent structures (see Section 2-5). In a steady flow, if $\partial H/\partial S$ is positive, it is because of the action along the path dS of an external force such as a pump. If $\partial H/\partial S$ is negative, it is either because of the action along the path dS of an external force, such as a turbine, or because of friction force.

The variation in H along S measures the action of turbo machines or the head losses in the dimension of a length.

10-2.3.2 In a rotational flow H varies from one streamline to another streamline, whereas H is the same for any streamline of an irrotational flow. Figure 10-4 as compared to Fig. 10-3 illustrates this point.

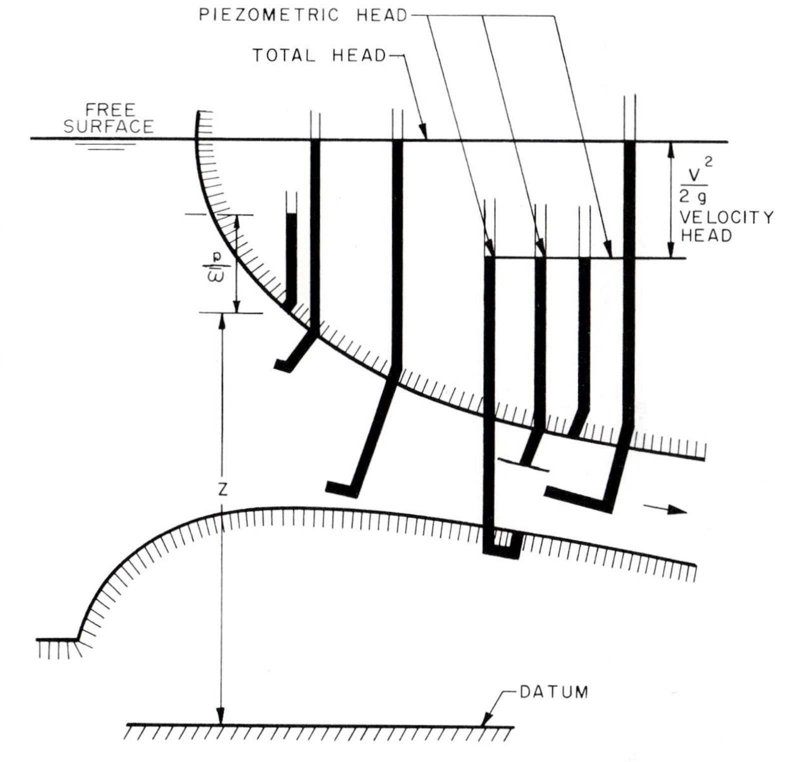

Figure 10-3

In an irrotational flow, the total head is a constant at any point.

10-2.4 Pressure Distribution in a Direction Perpendicular to the Streamlines

10-2.4.1 In an irrotational flow, the derivative of p is known in any direction by applying the Bernoulli equation to the velocity field.

In a rotational flow, the Bernoulli equation gives the derivative of p along a streamline as a function of the derivative of V, but does not give any indication of the derivative of p in a direction perpendicular to the streamlines. However, both of these are of equal importance in engineering practice.

It has been seen that the pressure distribution in a uniform flow is hydrostatic. This hydrostatic law is again valid when the path curvature is small.

10-2.4.2 Now the general case of nonnegligible curvature is studied. Consider an infinitely small, curved, two-dimensional stream tube as shown by Fig. 10-5, and an elementary mass of fluid $\rho\, dR\, dS$ in this stream tube.

Because the motion is in the direction of the stream tube, this elementary mass is in equilibrium in a direction normal to the streamline and is under the action of its inertia and applied forces.

Its inertia gives rise to a centrifugal force equal to $\rho\, dR\, dS\, V^2/R$, where R is the radius of curvature of the streamlines. The applied forces are the difference of pressure forces acting on the two streamlines

$$\left(p + \frac{\partial p}{\partial R}\, dR\right) dS - p\, dS = \frac{\partial p}{\partial R}\, dR\, dS$$

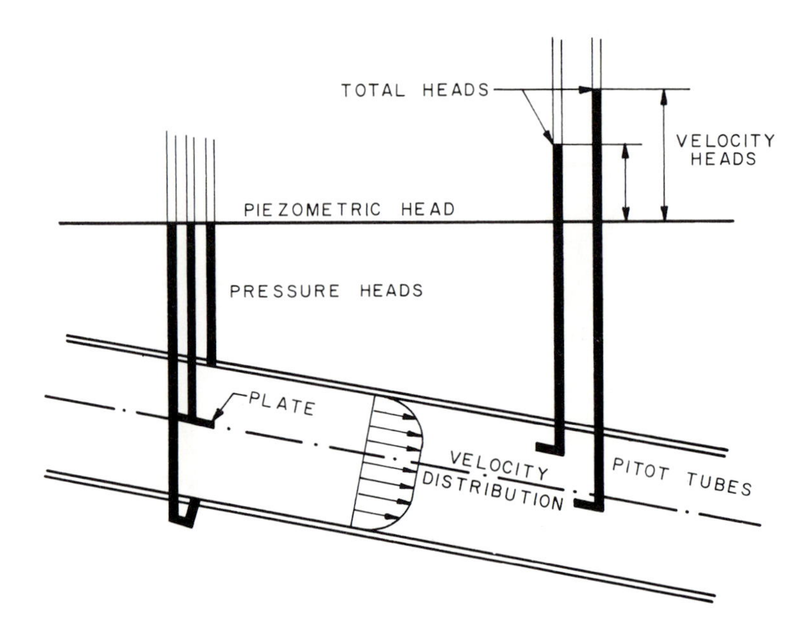

Figure 10-4

In a rotational flow, the total head changes from one streamline to another.

and the gravity

$$\rho g \, dR \, dS \cos \alpha$$

From the diagram it is observed that

$$\cos \alpha = \frac{(\partial z/\partial R) \, dR}{dR} = \frac{\partial z}{\partial R}$$

When the centrifugal force is equated to the applied forces and the resulting equation is divided by the volume $dR \, dS$ it gives

$$\rho \frac{V^2}{R} = \frac{\partial(p + \rho g z)}{\partial R}$$

or

$$\rho \frac{V^2}{R} = \frac{\partial p^*}{\partial R}$$

that is

$$\frac{\partial}{\partial R}\left(\frac{V^2}{g} \ln R - \frac{p^*}{\rho g}\right) = 0$$

Integrating this equation along dR permits the calculation of the pressure distribution from the velocity fields along a curved boundary, as on a bucket of a spillway, for example.

If R tends to infinity, $\partial(p^*/\varpi)/\partial R$ tends to zero and p^* = constant, as has been found in the case of uniform flow.

Figure 10-5 *Notation.*

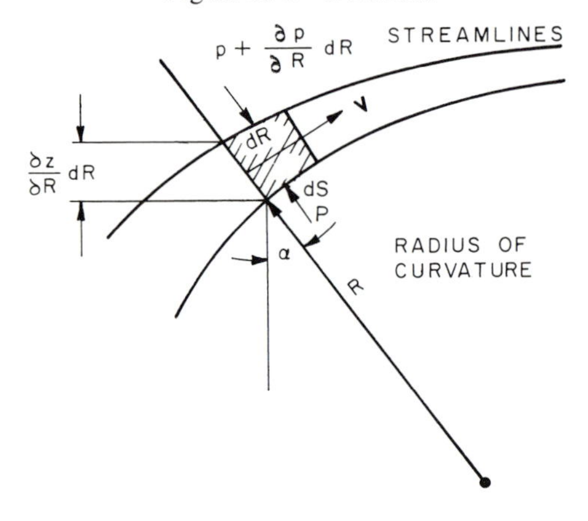

10-2.4.3 The above demonstration does not require the assumption that the flow be irrotational. Hence, it is valid for an irrotational flow as well as for a rotational flow. It has been seen that the pressure distribution in an irrotational flow is also known by the Bernoulli equation.

It is valid for any direction, particularly in a direction perpendicular to the streamlines. Hence, two methods exist for calculating the derivative of the pressure distribution in a direction perpendicular to the streamlines for an irrotational flow. It is evident that the same result must be obtained. This could be demonstrated by combining the above formula with a condition of irrotationality.

The simplest demonstration is that the derivative of the total head $H = (V^2/2g) + (p^*/\varpi)$ along the radius of curvature R is zero, i.e.,

$$\partial H/\partial R = 0$$

Using the relationship demonstrated in the previous section, one has successively

$$\frac{\partial H}{\partial R} = \frac{V}{g}\frac{\partial V}{\partial R} + \frac{\partial(p^*/\varpi)}{\partial R} = \frac{V}{g}\left(\frac{\partial V}{\partial R} + \frac{V}{R}\right)$$

In an irrotational flow $VR =$ constant; $V\,dR + R\,dV = 0$. Hence, $\partial V/\partial R + V/R = 0$, and therefore $\partial H/\partial R = 0$. This means that the constant H is the same for any streamlines of an irrotational flow, as previously shown.

10-2.5 Résumé and Noteworthy Formulas

The following formulas have been found in the case of a perfect fluid:

Hydrostatics
Steady slow motion
$$p^* = p + \rho g z = \text{constant}$$

Steady uniform flow (real fluid). OZ at an angle α with the vertical
$$p + \rho g z \cos\alpha = \text{constant}$$

Unsteady slow motion
$$\rho\frac{\partial \phi}{\partial t} + p^* = f(t)$$

Steady irrotational flow
$$H = \frac{V^2}{2g} + \frac{p}{\varpi} + z = \text{constant}$$

Unsteady irrotational motion
$$\frac{1}{g}\frac{\partial \phi}{\partial t} + \frac{V^2}{2g} + \frac{p}{\varpi} + z = f(t)$$

Steady rotational flow
$$\frac{\partial}{\partial S}\left(\frac{V^2}{2g} + \frac{p}{\varpi} + z\right) = 0$$

Steady flow (rotational or irrotational)
$$\frac{V^2}{gR} = \frac{\partial}{\partial R}\left(\frac{p}{\varpi} + z\right)$$

Unsteady rotational flow
$$\frac{\partial}{\partial S}\left(\frac{V^2}{2g} + \frac{p}{\varpi} + z\right) + \frac{1}{g}\frac{\partial V}{\partial t} = 0$$

All of these formulas have been obtained by exact mathematical integration.

10-3 Generalized Bernoulli Equation

10-3.1 Integration to a Cross Section

10-3.1.1 When the relative curvature of streamlines is small, the Bernoulli equation can be generalized to a cross-section normal to the average streamline direction. To solve this problem first consider the streamline. By continuity, along a streamline: $V\,dA =$ constant, where dA is an infinitesimal area normal to the streamline. Hence, the Bernoulli equation along a streamline can still be written

$$\frac{\partial}{\partial S}\left[\left(\rho\frac{V^2}{2} + p^*\right)V\,dA\right] + \rho\frac{\partial V}{\partial t}V\,dA = 0$$

Integrating along the streamline from position 1 to position 2

$$\left(\rho\frac{V_2^2}{2} + p_2^*\right)V_2\,dA_2 - \left(\rho\frac{V_1^2}{2} + p_1^*\right)V_1\,dA_1 + \int_1^2 \rho\frac{\partial V}{\partial t}V\,dA\,dS = 0$$

This equation is now integrated across a cross section of area A and averaged with respect to time

$$\frac{1}{T}\int_0^T \iint_A \left[\left(\rho_2 \frac{V_2^2}{2} + p_2^*\right)V_2\,dA_2 - \left(\rho\frac{V_1}{2} + p_1^*\right)V_1\,dA_1 + \int_1^2 \rho\frac{\partial V}{\partial t} V\,dA\,dS\right]dt = 0$$

The value of this integration will be computed in the following sections. [Note that the product

$$\left(\rho\frac{V^2}{2} + p^*\right)V\,dA$$

is an energy flux or power. It is seen in elementary hydraulics that the energy flux through a cross section A is given by the expression

$$\text{H.P.} = \frac{1}{550}\left[\frac{V^2}{2g} + \frac{p}{\rho} + z\right]VA$$

in the ft–lb–sec system, where H.P. is in horse power.]

10-3.1.2 Across a section such as BB', the velocity $\mathbf{V}$ can be written as the vector sum (see Fig. 10-6)

$$\mathbf{V} = \mathbf{U} + \mathbf{U}' + \mathbf{V}'$$

$\mathbf{U}$ is in the average streamline direction and is independent of turbulent fluctuations:

$$\mathbf{U} = \frac{1}{T}\int_0^T \frac{1}{A}\iint_A \mathbf{V}\,dA\,dt$$

where A is the cross sectional area and $\mathbf{U}'$ is a measure of the deviation of the time-average direction of $\mathbf{V}$ from the mean value $\mathbf{U}$. ($\mathbf{U}' = -\mathbf{U}$ at the walls.)

$$\mathbf{U}' = \frac{1}{T}\int_0^T (\mathbf{V} - \mathbf{U})\,dt$$

and

$$\iint_A \mathbf{U}'\,dA = 0$$

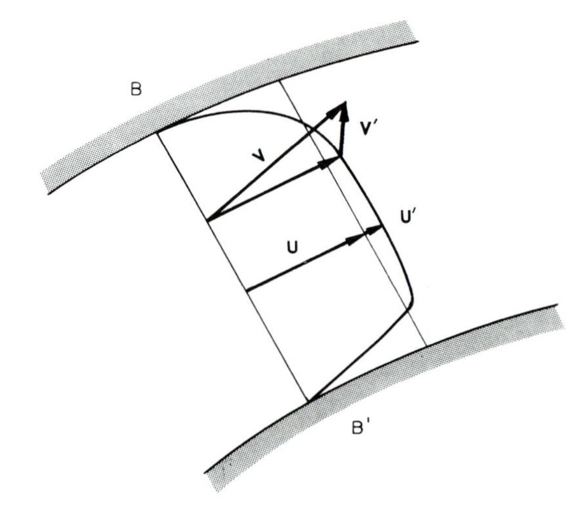

Figure 10-6 *Mean velocity in a cross section.*

Finally, $\mathbf{V}'$ is caused by turbulent fluctuations and is a measure of the deviation of $\mathbf{V}$ from its time average. It varies both in time and direction, and is such that

$$\overline{\mathbf{V}'} = \int_0^T \mathbf{V}'\,dt = 0$$

Across a section such as BB', when the relative curvature of the paths is small, all velocity vectors $\mathbf{V}$ are nearly parallel so that in first approximation $\mathbf{U}'$ and $\mathbf{V}'$ are parallel to $\mathbf{U}$, and one can write $\mathbf{U}' = \sigma\mathbf{U}$, $\mathbf{V}' = \chi\mathbf{U}$; so that $\mathbf{V} \cong \mathbf{U}(1 + \sigma + \chi)$, where σ varies with space only, whereas χ varies both with space and time. By definition, both have an average value of zero. (More rigorously χ should be considered as a tensor.)

10-3.1.3 Across the section the pressure $p^* = p + \rho g z$ will also vary with time and space and one can write similarly

$$p = \bar{p} + p' + p''$$

However, the variations p' and p'' are neglected.

10-3.1.4 Introducing the value of **V** as a function of **U**, the average velocity head term becomes successively

$$\frac{1}{T}\int_0^T \iint_A \rho \frac{V^2}{2} V\, dA\, dt \cong \rho \frac{U^3 A}{2}(1 + 3\overline{\sigma^2} + 3\overline{\chi^2})$$

$$= (1 + \alpha)\, \rho \frac{U^2}{2} Q$$

with $\alpha = 3\overline{\sigma^2} + 3\overline{\chi^2}$. For laminar flow in a circular pipe, there is no turbulence, $V' = 0$, and $3\overline{\chi^2} = 0$. In the case of a paraboloid velocity distributions, as found in elementary hydraulics, it is found that $3\overline{\sigma^2} = 1$. For turbulent flow $3\chi^2 \cong 0.05$ and $3\overline{\sigma^2} \cong 0.05$ to 0.01 (see Section 13-4).

Finally, consider the local inertia term

$$\frac{1}{T}\int_0^T \iint_A \int_S \rho \frac{\partial V}{\partial t} V\, dA\, dt\, dS$$

Simplifying as previously seen, it becomes

$$\rho \int_S A \frac{\partial (U^2/2)(1 + \overline{\sigma^2} + \overline{\chi^2})}{\partial t} dS = \rho A\left(1 + \frac{\alpha}{3}\right) U \frac{\partial U}{\partial t}$$

10-3.2 *Practical Form of the Bernoulli Equation for a Stream Tube*

Taking account of the above correction factors, and dividing by $Q = UA$, the generalized form of the Bernoulli equation for a stream tube is

$$\left(\rho \frac{U_2^2}{2}(1 + \alpha_2) + \bar{p}_2 + \rho g z_2\right) - \left(\rho \frac{U_1^2}{1}(1 + \alpha_1) + \bar{p}_1 + \rho g z_1\right) = \rho \int_S \frac{\partial U}{\partial t}\left(1 + \frac{\alpha}{3}\right) dS$$

If σ^2 and χ^2 and hence α are neglected, dividing by ρg leads to the common form of the Bernoulli equation

$$\left(\frac{U_2^2}{2g} + \frac{\bar{p}_2}{\rho g} + z_2\right) - \left(\frac{U_1^2}{2g} + \frac{\bar{p}_1}{\rho g} + z_1\right) = \frac{1}{g}\int_S \frac{\partial U}{\partial t} dS$$

In the case of a uniform flow in a pipe of length L, the local inertia term becomes

$$\frac{1}{g}\int_S \frac{\partial U}{\partial t} ds = \frac{L}{g}\frac{dU}{dt}$$

This is the formula that must be used to study, for example, surge tanks, locks, etc.

10-4 Limit of Application of the Two Forms of the Bernoulli Equation

10-4.1 *The Two Forms of the Bernoulli Equation*

In the case of steady flow without friction, two forms of the Bernoulli equation are almost identical: the first one is

$$\frac{V^2}{2g} + \frac{p}{\rho g} + z = \text{constant}$$

throughout the fluid, and the other one, valid along a stream tube, is

$$(1 + \alpha)\frac{U^2}{2g} + \frac{p}{\rho g} + z = \text{constant}$$

Strictly speaking, neither of these equations is valid in any real case because the conditions required for establishment of the Bernoulli equation can only be approximated. However, they are essential in many cases. In such cases, it is important to remember the following assumptions:

The first form of the Bernoulli equation is valid for irrotational flow, that is, in convergent flow through short structures. **V** is the local velocity. In case of turbulence, **V** is replaced by the mean local velocity with respect to time: $\bar{\mathbf{V}}$. The streamlines may be curved, but V must never be taken as the mean velocity in a cross section.

The pressure distribution is given as a function of the local value of V and z. The pressure distribution at the walls is not much influenced by the vorticity in thin boundary layers.

The second form of the Bernoulli equation is valid for unidimensional flow where the motion is rotational, but the relative curvature of the paths must be small. U refers to the mean velocity in the cross section, and the kinetic head $U^2/2g$ must be corrected by a factor $(1 + \alpha)$. α is often neglected in practice. It is caused by turbulence and by the variation of the velocity in a cross section. In the case of small curvature, the pressure distribution is hydrostatic in a cross section or more exactly, very slightly smaller in the center of the flow than at the boundary because of the turbulent fluctuations (see Section 8-2.2).

The pressure distribution from one cross section to another varies as a function of U^2, the square mean velocity in these cross sections.

10-4.2 Venturi and Diaphragm as Measuring Devices

In practice it is difficult to know the exact value of the correction factor α. Moreover, a number of assumptions such as small curvature of the paths, negligible head loss, etc., are not always satisfied. It must be understood, therefore, that a venturi used as a device to measure a discharge by a simple application of the Bernoulli equation from pressure variations without correction factors is not an accurate device in itself. It is for this reason that manufacturers must give a calibration curve obtained experimentally by measuring the discharge in a calibrated tank. An overall correction factor, which differs for each kind of venturi, must be given as a function of the Reynolds number at the throat.

A similar statement could be made for a diaphragm where the upstream pressure distribution is preferably given by considering the local value of the velocity near the diaphragm rather than the mean value in a cross section. The downstream pressure distribution is roughly hydrostatic, since the flow is either almost parallel or very slow out of the vena contracta.

10-4.3 Experience of Banki

The application of the Bernoulli equation to a venturi is well known. Despite the approximations that have been indicated previously, it is effectively verified that when V increases in the convergent, p^* decreases as $V^{1/2}$. Conversely, p^* increases as $V^{1/2}$ in the divergent. If the divergent is too rapid, the flow separates and p^* is almost a constant around the jet. This is also the case of a jet arriving in a reservoir.

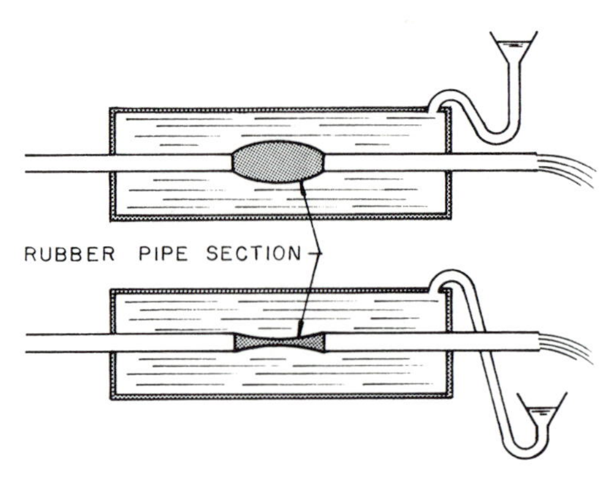

Figure 10-7 *Experience of Banki.*

Now, one can question if any change of p^* also changes V according to the Bernoulli equation. This could be realized by an experiment, initially conducted by Banki, in which the pressure variations are transmitted to the inside flow of a pipe through a membrane (see Fig. 10-7). When the pressure within the tank increases, the pressure, transmitted through the membrane in the pipe, also increases; hence V decreases, and the rubber membrane expands. Also, when the pressure within the tank decreases, V will increase and the rubber membrane will contract. This paradoxical result is in accordance with the Bernoulli equation. However, this experience is difficult to realize because of flow separation. This fact also demonstrates the inherent deficiencies of the Bernoulli equation applied to a stream tube. Finally, the change in the rubber shape changes the flow discharge in the pipe and the motion is unstable.

10-5 Definition of Head Loss

10-5.1 Steady Uniform Flow

Consider the case of steady flow in a pipe. The head loss may be calculated by theory in a number of cases where the flow is laminar and the cross section is of simple shape, such as circular or square. But in the case of turbulent flow, the value of head loss cannot be obtained by theory and must be measured by experiment.

In this case, the Bernoulli equation permits us to define the value of the head loss ΔH between two considered cross sections by the difference between the total heads at these points

$$\left[(1 + \alpha_1)\frac{U_1^2}{g} + \frac{p_1^*}{\rho g}\right] - \left[(1 + \alpha_2)\frac{U_2^2}{g} + \frac{p_2^*}{\rho g}\right] = \Delta H$$

This does not present any difficulty as long as the flow is uniform. In this case

$$\alpha_1 = \alpha_2 \qquad U_1 = U_2 \qquad \Delta H = p_1^*/\rho g - p_2^*/\rho g$$

10-5.2 Head Loss at a Sudden Change in Flow

A sudden change in a uniform pipe, such as caused by a diaphragm or a bend, has an effect on the flow velocity distribution and the head loss at a great distance downstream. The head loss due to this change may be obtained by extrapolating the pressure lines as shown by Fig. 10-8. (A pressure given by a piezometer located near a discontinuity, a bend or an intake has no value in evaluating the head loss because the flow is not uniform. The pressure measurement is locally influenced by a complex flow pattern.) However, in the case of nonuniform flow, it is more difficult because $U_1^2/2g$ is different from $U_2^2/2g$; in particular, α_1 is different from α_2 and they are unknowns. Even if they are considered as known, the definition of such a head loss involves lack of accuracy as evidenced by Fig. 10-9. Moreover, it is impossible to separate the value of head losses caused by a succession of close sudden changes. A linear addition of the various head losses cannot be made because of the interaction between the effects of successive sudden changes.

10-5.3 Head Loss in a Free-Surface Flow

It is interesting to note that any change in a free surface flow gives no extra head loss. Indeed, the initial upstream level and downstream level are always the same provided the considered cross sections are far enough from the discontinuity as is illustrated by Fig. 10-10. The increase of head loss in one place is always compensated by a decrease of head loss in another place. Hence, the head loss definition of a discontinuity or sudden change in a free-surface flow, such as that caused by a grid, must be specified by the relative location of two cross-sectional planes between which the head loss is considered.

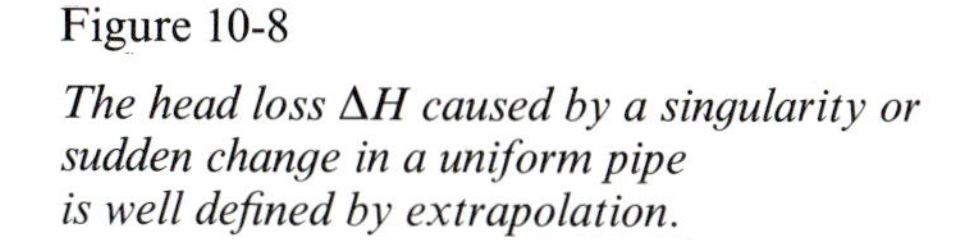

Figure 10-8

The head loss ΔH caused by a singularity or sudden change in a uniform pipe is well defined by extrapolation.

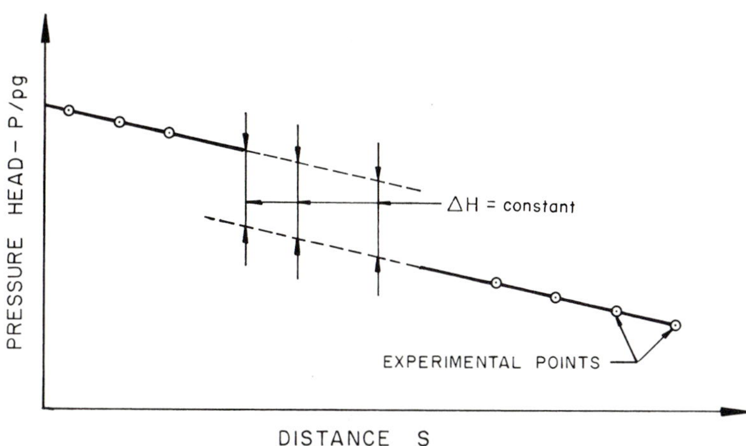

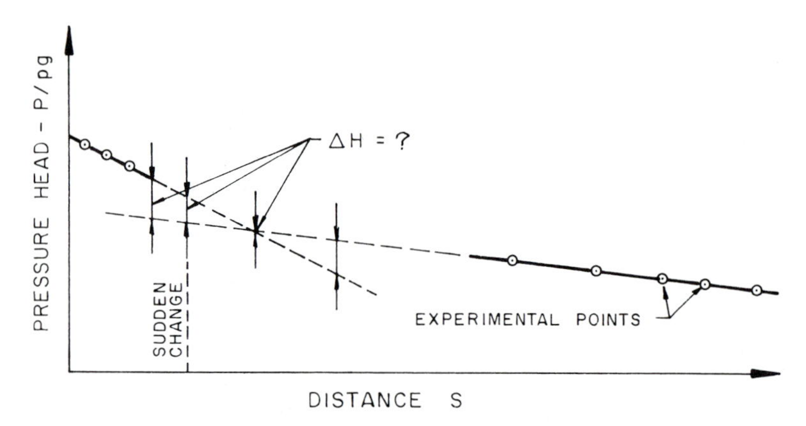

Figure 10-9

The head loss caused by a sudden change in a nonuniform flow is subject to various interpretations.

However, in case of sediment transport, the statement may not hold true as illustrated in Figs. 10-11 and 10-12, as the sediment tends to accumulate upstream, proportionally to the head loss.

10-5.4 Effect of Local Inertia upon Head Loss

By having an effect on the velocity distribution in a pipe, and on the corresponding shear stress, local inertia has an important effect on the value of the head loss (see Fig. 10-13).

Head loss in a given flow at a given time cannot be considered theoretically as equal to the value of the head loss of the steady flow, which would have the same instantaneous value of mean velocity.

The head losses for the same value of the mean velocity are different for steady flow, accelerated flow, and decelerated flow. This may also be noticeable for a discontinuity such as the bottom orifice of a surge tank, where the flow pattern is influenced by an instability phenomenon.

However, owing to lack of experimental data, unsteady flows are often studied with a head loss given by an empirical law experimentally obtained in the case of steady flow. The studies of tide in estuary and storm surge are based on such assumptions. As in the case of steady flow, the Bernoulli equation may be used to define the head loss in an unsteady flow, but even more difficulties are encountered in determining the head loss experimentally in unsteady flow than in steady flow.

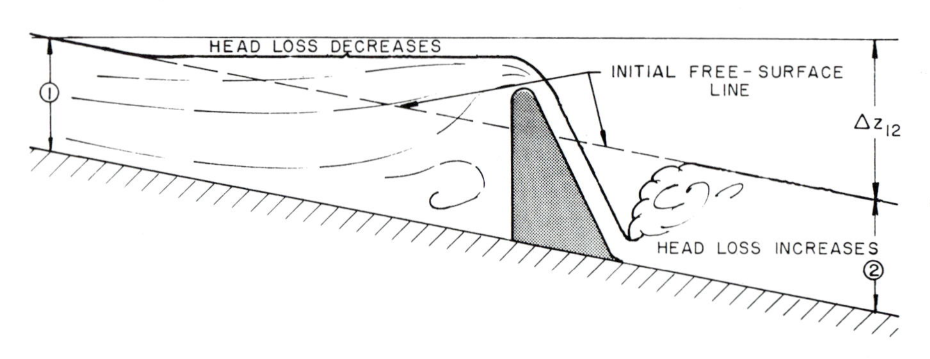

Figure 10-10

A discontinuity in a free surface flow does not change the total value of the head loss Δz_{12}.

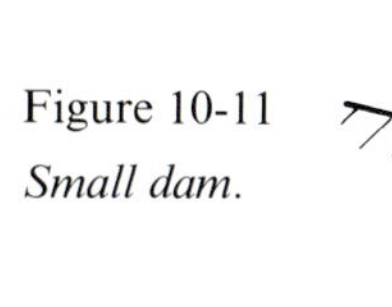

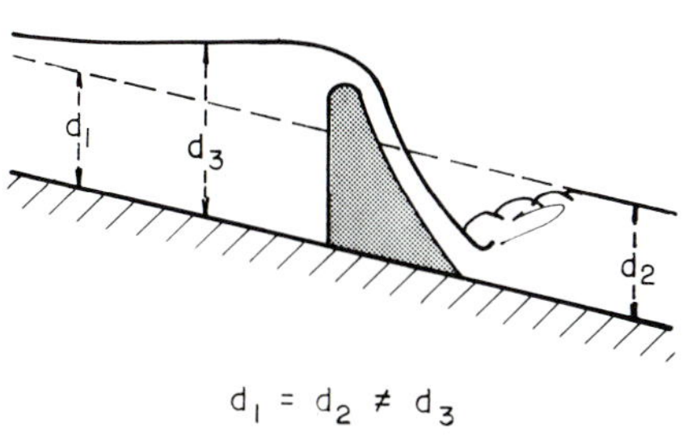

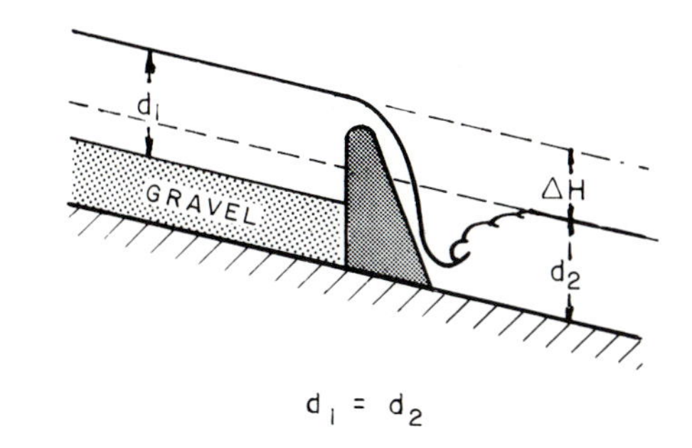

Figure 10-11
Small dam.

10-5.5 An Application to Surge Tank

Because of its importance in engineering practice, an example of the generalized Bernoulli equation for unsteady motion is given. The case of a surge tank in the case where the discharge in the penstoke is zero is analyzed (Fig. 10-14). The application of the Bernoulli equation between points a and b gives

$$\left(\frac{V_a^2}{2g} + \frac{p_a}{\rho g} + z_a\right) - \left(\frac{V_b^2}{2g} + \frac{p_b}{\rho g} + z_b\right) = (\text{head loss})_{ab} + \frac{1}{g}\int_a^b \frac{\partial V}{\partial t}\, dL.$$

$V_a^2/2g$ is negligible as V_a in the reservoir is very small. $V_b^2/2g$ is also very small and is usually neglected. Moreover, $p_a = p_b =$ atmospheric pressure. The head loss term includes the head loss at the entrance of the gallery, the head loss in the gallery $\Delta H = Lf\,V^2/2gD$, and the head loss due to the bottom diaphragm of the surge tank $KV_d^2/2g$. The head loss in the surge tank is usually negligible as is $V_b^2/2g$. (L is the length of the gallery and f the head loss coefficient.)

The local inertia term is usually small enough in the reservoir and in the surge tank to be neglected. It is taken into account in the gallery only. Hence,

$$\frac{1}{g}\int_a^b \frac{\partial V}{\partial t}\, dL = \frac{L}{g}\frac{dV}{dt} = \frac{L}{gf'}\frac{dQ}{dt}$$

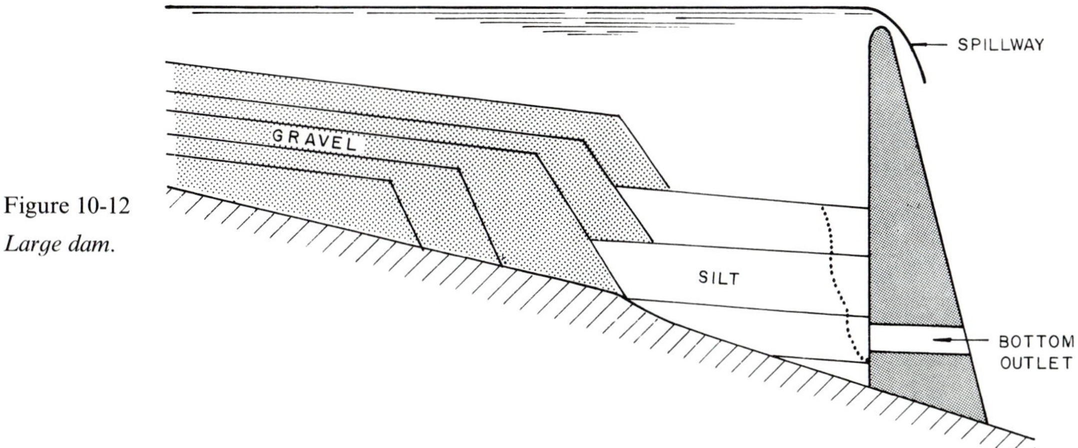

Figure 10-12
Large dam.

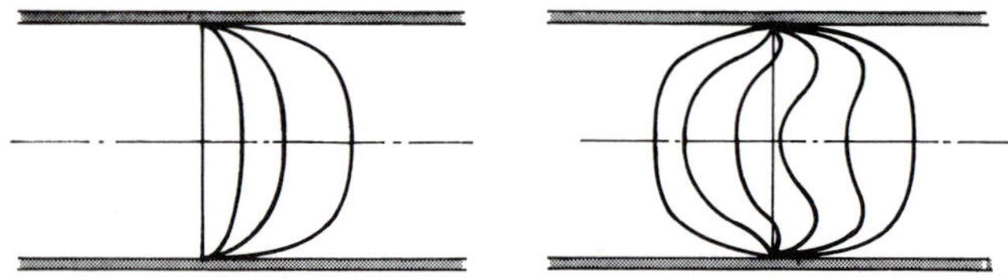

Figure 10-13 *A comparison of velocity distributions within a pipe.*

where Q is the discharge and f' is the cross section of the gallery. In the case of a relatively short gallery, the value for L could be increased by a correction factor to take into account the local inertia of the near-radial flow near the entrance. Finally, with $Z = z_a - z_b$, the basic dynamic equation for studying a surge tank is

$$Z = \left(\frac{fL}{D} + k_{a-b}\right)\frac{V^2}{2g} + \frac{L}{gf}\frac{dQ}{dt}$$

where k_{a-b} is a friction coefficient for singularities between a and b. The continuity equation is $Q\,dt = F\,dZ$, F being the cross section at the free surface of the surge tank.

PROBLEMS

10.1 The velocity around the limit of a circular cylinder is given by the equation $\mathbf{V} = 2\mathbf{U} \sin\theta$ where $\mathbf{U}$ is the velocity at infinity, and where the pressure is p_∞. Determine the pressure distribution around the cylinder.

Figure 10-14 *Surge tank.*

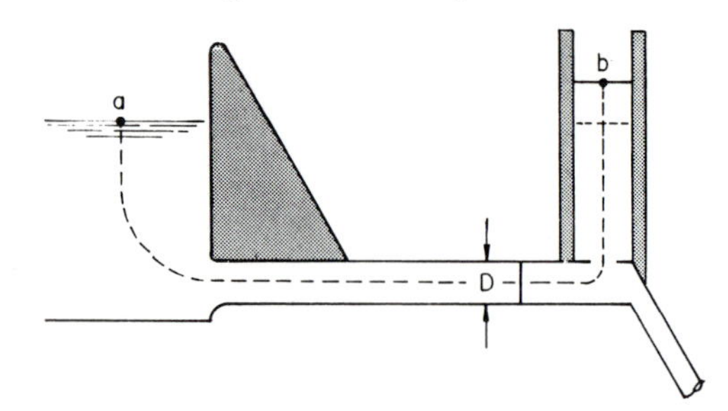

10.2 Demonstrate that the following equality is valid for steady flow:

$$2(w\eta - v\zeta) = -\frac{\partial}{\partial x}\left(\frac{V^2}{2} + \frac{p}{\rho} + gz\right)$$

Two other similar relationships are obtained by circular permutations.

10.3 The velocity potential function for a flow beyond a sphere of radius R is

$$\phi = U\left(\frac{R^3}{2r^2} + r\right)\cos\theta$$

Determine the velocity and pressure distribution around the sphere.

$$v_r = \frac{\partial\phi}{\partial r} \qquad v_\theta = \frac{1}{r}\frac{\partial\phi}{\partial\theta}$$

In view of the results, explain the shape that a drop of rain and an air bubble in water take.

10.4 Demonstrate that the Bernoulli equation is valid along a streamline using the two following fully developed forms of the convective terms of the momentum equation

$$\rho\left(u\frac{\partial u}{\partial x} + v\frac{\partial u}{\partial y} + w\frac{\partial u}{\partial z}\right)$$

and

$$\rho\left[\frac{\partial}{\partial x}\left(\frac{V^2}{2}\right) + 2(\eta w - \zeta v)\right]$$

10.5 Two adjacent tanks have horizontal cross sections S_1 and S_2, respectively. The difference of level between these two tanks at time $t = 0$ is $h_1 - h_2 = h$. An orifice of cross section A is open between the two tanks at time $t = 0$. The discharge coefficient of the orifice is 0.6. Give the expression for the time T after which the level in the two tanks is the same.

10.6 Consider four reservoirs, A, B, C, and D, connected as shown in Fig. 10-15 (not drawn to scale) in which the level

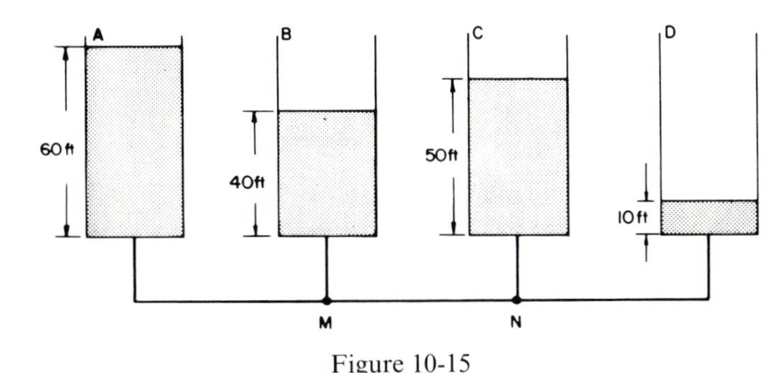

Figure 10-15

is maintained at

$$A: z_A = 60 \text{ ft } (18.28 \text{ m}) \qquad B: z_B = 40 \text{ ft } (12.19 \text{ m})$$
$$C: z_C = 50 \text{ ft } (15.24 \text{ m}) \qquad D: z_C = 10 \text{ ft } (3.04 \text{ m})$$

respectively. The pipe between A and B is 10 in. (25.40 cm) in diameter and 3000 ft (914.4 m) long. The pipe between C and D is 12 in (30.48 cm) in diameter and 6000 ft (1828.8 m) long, and finally, the connecting pipe MN is 5500 ft (1676.4 m) long, M being 1000 ft (304.8 m) from reservoir A, and N being 2000 ft (609.6 m) from reservoir D. The friction coefficients f of these pipes are 0.20 for the 10-in. and the 12-in.-diameter pipes, and 0.224 for the pipe between M and N. The diameter of the pipe MN is such that the discharge through MN is 1.2 ft^3/sec (33,980 cm^3/sec). Determine the discharges between AM, MB, CN, and ND and the diameter of the pipe MN.

10.7 Consider a hydroelectric installation including a large reservoir where the level remains practically constant, a horizontal gallery length L and circular cross section f', a tank of horizontal cross section F and a penstoke as shown in Fig. 10-16. The head loss in the gallery is $P = \pm P_0(W/W_0)^2$, where W is the average water velocity as a function of time, and subscript 0 refers to steady-state conditions.

1. Demonstrate that the governing equation for the elevation z in the surge tank is

$$\frac{LF}{gf'}\frac{d^2z}{dt^2} + z + P = 0$$

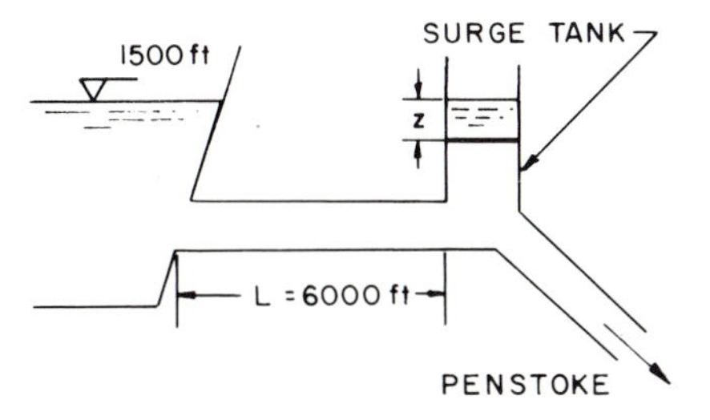

Figure 10-16

2. Give the period of oscillation of the motion in the gallery. (P will be neglected for this calculation.)
3. Give the amplitude of oscillation of z in the surge tank in the case for which the initial discharge $Q_t = fW_a$ is suddenly stopped to a zero value and to a smaller value $Q_t' = fW_a'$. (P will again be neglected.) Explain qualitatively the influence of P.

10.8 Establish the basic equations of motion for unsteady flow in parallel pipes and in series.

10.9 Establish the equation of motion caused by the sudden opening of a gate for a manifold such as that shown in Fig. 10-17.

Figure 10-17

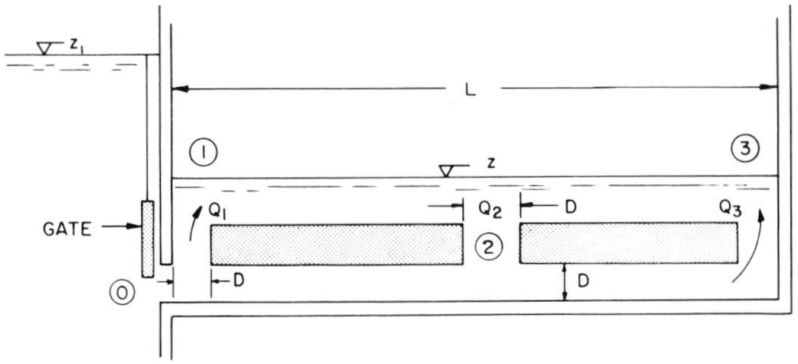

Chapter 11

Flow Pattern, Stream Function, Potential Function

11-1 General Considerations on the Determination of Flow Pattern

The laws that govern the motion of an infinitely small particle of fluid have been established in Part One (Chapters 2 to 9). Some general relationships among velocity, pressure, and gravity, such as that given by the Bernoulli equation, were deduced by general exact integration, independent of the boundary conditions (Chapter 10).

The pressure p (or velocity V) may be determined from these general relationships after insertion of the value V (or p).

Also, the differential equations derived from the continuity principle and momentum equation allow us to theoretically solve directly any particular problem, that is, to determine the velocity V (or pressure p) when the boundary conditions are specified. These boundary conditions are peculiar for each case. An example of this is given in Section 6-2 (laminar flow on a sloped plane).

However, the boundary conditions are usually too complicated in the majority of cases encountered in engineering practice, so that the use of the mathematical theory is limited to oversimplified cases. Nevertheless, a number of practical problems closely approximate some simple cases that can be studied mathematically.

The purpose of this chapter is to study some of these exact mathematical methods. Moreover, a number of approximate methods—graphic, numerical, or experimental—are based on the same mathematical principles as those explained in this chapter. The approximate methods extend the field of application of the exact methods considerably and take into account cases in which the boundary conditions are not so simple. The graphic flow net method is one of these approximate methods.

It is intended that the word "exact" refer to the mathematical process. The physical exactness will depend upon the limit of validity of the basic assumptions necessary to use such methods.

It has already been indicated that the two unknowns to be determined are the velocity V and the pressure p

and that theoretically, both may be found directly from the momentum equation and continuity relationship. However, in many cases the methods under study provide a knowledge of the relative velocity distribution from the velocity field calculated from the continuity principle, and an assumption such as that of irrotationality only.

To calculate the absolute value of the velocity requires a second step. This second step is simple when the absolute value of the velocity at one point or at one boundary is known.

Then, in a third step, the pressure distribution is determined by application of some of the relationships between V and p, which have been established in Chapter 10, such as the Bernoulli equation.

This chapter deals with the problem of the determination of the velocity field by some analytical methods of particular importance. These analytical methods are based on the use of two mathematical tools that allow a concise description of the complete flow pattern. They are the stream function and the velocity potential function.

11-2 Stream Function

11-2.1 Definition

11-2.1.1 The stream function is a mathematical device to describe a flow pattern concisely by its streamlines. The stream function may be used to calculate any kind of flow of incompressible fluid: rotational or irrotational; steady or unsteady; two-dimensional or three-dimensional; laminar or turbulent; slow or nonslow motion. However, in the case of turbulent motion, the stream function is intended to define only the mean motion with respect to time, i.e., the mean velocity vector $\bar{\mathbf{V}}$. It may also be used to define the mean motion with respect to space of a flow through porous medium whatever the value of the Reynolds number, i.e., for turbulent flow as well as for laminar flow.

Although the stream function may theoretically be defined and used for three-dimensional motion, its calculation is complex and its use has been limited. Hence, in practice the stream function is mainly used in two-dimensional flow; only this case is analyzed in this book.

The stream function may be defined by any one of its characteristics and then the other characteristics may be deduced from this chosen definition. As was done for the velocity potential function, the stream function will be defined first by the velocity components.

11-2.1.2 The stream function is a natural outcome of the continuity relationship

$$\frac{\partial u}{\partial x} + \frac{\partial v}{\partial y} = 0.$$

Indeed, consider a function $\psi(x,y,t)$ = constant such that

$$u = \frac{\partial \psi}{\partial y} \qquad v = -\frac{\partial \psi}{\partial x}.$$

From the continuity relationship given above, it follows that

$$\frac{\partial}{\partial x}\frac{\partial \psi}{\partial y} - \frac{\partial}{\partial y}\frac{\partial \psi}{\partial x} \equiv 0$$

which shows that ψ always satisfies the principle of continuity; in other words, the existence of ψ implies that the continuity relationship is satisfied and conversely the continuity equation implies the existence of a stream function ψ.

11-2.2 Stream Function, Streamlines, and Discharge

11-2.2.1 Now it is shown that such a function ψ = constant is not only the equation of one streamline but of any streamline of the considered flow. This is performed by a simple change of the constant value for ψ.

For this purpose, consider the streamline equation $dx/u = dy/v$ (see Section 1-2.2), which may be written $u\,dy - v\,dx = 0$. Introducing the value of u and v as functions of ψ yields the equation of streamlines in terms of stream function

$$\frac{\partial \psi}{\partial x}dx + \frac{\partial \psi}{\partial y}dy = 0$$

It is the total differential $d\psi$ (with respect to distance) of $\psi(x,y,t)$. Hence, the equation of any streamline expressed as a function of ψ is given by the equation $d\psi = 0$, or in the case of steady flow by $\psi(x,y) =$ constant, or in the case of unsteady flow by $\psi(x,y,t_0) =$ constant.

Changing the value of the constant gives different streamlines of the considered flow, but the function $\psi(x,y)$ keeps the same analytical form. It is for this reason that ψ is called a stream function.

Consider the flow pattern as shown by Fig. 11-1. The discharge dQ passing through an element dn perpendicular to the streamlines is

$$dQ = \frac{\partial\psi}{\partial y}\,dy + \frac{\partial\psi}{\partial x}\,dx = d\psi$$

which is also the total differential of $\partial\psi$ with respect to distance.

It is deduced that

$$V = \frac{dQ}{dn} = \frac{d\psi}{dn}$$

Therefore, the total discharge between two streamlines ψ_1 and ψ_2 is given by their difference:

$$\Delta Q = \psi_2 - \psi_1$$

Figure 11-1 *Discharge in terms of stream function—notation.*

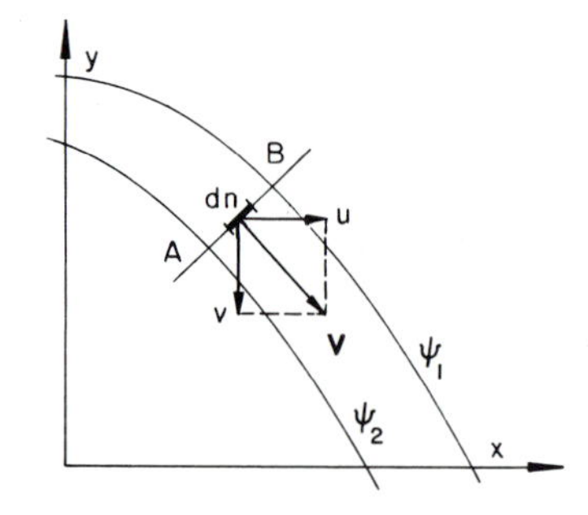

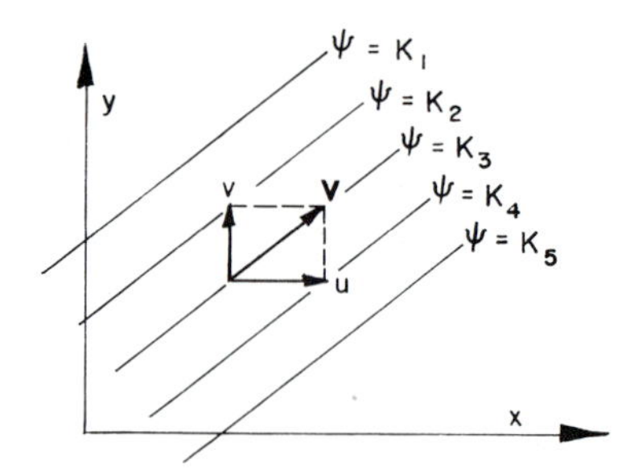

Figure 11-2 *Uniform flow defined by a stream function.*

The average value of V between A and B is

$$V = \frac{\Delta\psi}{\Delta n} = \frac{\Delta\psi}{AB}$$

(see Fig. 11-1).

11-2.2.2 It is verified that the stream function of a uniform flow may take the form $\psi = Ay - Bx$. The velocity components are

$$u = \frac{\partial\psi}{\partial y} = A \qquad v = -\frac{\partial\psi}{\partial x} = B$$

and $V = (A^2 + B^2)^{1/2}$. V does not depend upon x and y; hence, the flow is uniform.

The streamlines are defined by the equation:

$$Ay - Bx = K$$

They are straight lines of slope $y/x = B/A$ and are obtained by giving K various constant values as shown in Fig. 11-2. The discharge between two streamlines is given by the difference between the corresponding values of the constant K.

11-2.3 Stream Function and Rotation

The rate of rotation is $2\zeta = (\partial u/\partial y) - (\partial v/\partial x)$ (see Chapter 2). Expressed as a function of ψ, the rate of rotation becomes successively

$$2\zeta = \frac{\partial}{\partial y}\frac{\partial\psi}{\partial y} - \frac{\partial}{\partial x}\left(-\frac{\partial\psi}{\partial x}\right) = \frac{\partial^2\psi}{\partial y^2} + \frac{\partial^2\psi}{\partial x^2}$$

that is, $2\zeta = \nabla^2\psi$. Hence, an irrotational motion for which $\zeta = 0$ is defined by a stream function ψ, which is a solution of the Laplace equation $\nabla^2\psi = 0$. In other words, $\nabla^2\psi = 0$ defines an irrotational motion which satisfies the continuity principle. It may be easily verified that the example of uniform flow defined by a stream function given in Section 11-2.2.2 is irrotational.

11-2.4 General Remarks on the Use of the Stream Function

The stream function may be used to calculate the flow pattern from the basic equations—continuity relationship and momentum equation—by introducing the values of u and v as a function of ψ. In this case, it must be noted that the problem exists in finding only one unknown (ψ) instead of two (u and v), but from its own definition, the order of the basic differential equation increases by one degree.

For example, consider the equations which are used to study the boundary layer theory as they have been established in Sections 4-5.2.4 and 5-4.2.

$$\text{Continuity:} \quad \frac{\partial u}{\partial x} + \frac{\partial v}{\partial y} = 0$$

$$\text{Momentum:} \quad u\frac{\partial u}{\partial x} + v\frac{\partial u}{\partial y} = \nu\frac{\partial^2 u}{\partial y^2}$$

The first equation allows definition of ψ as

$$u = \frac{\partial\psi}{\partial y} \quad \text{and} \quad v = -\frac{\partial\psi}{\partial x}$$

Then, introducing these values in the momentum equation yields

$$\frac{\partial\psi}{\partial y}\frac{\partial^2\psi}{\partial x\partial y} - \frac{\partial\psi}{\partial x}\frac{\partial^2\psi}{\partial y^2} = \nu\frac{\partial^3\psi}{\partial y^3}$$

This involves the calculation of only one unknown ψ, but the equation is now of the third order rather than second order as it was when the motion was expressed by the two velocity components of u and v. Briefly, the stream function permits the transformation of a system of two equations with two unknowns u and v into one equation of higher order with only one unknown.

When boundary conditions are introduced, this equation gives the theoretical value of ψ after successive integrations from which u and v are afterward obtained by simple differentiation.

11-3 Velocity Potential Function

11-3.1 Use

The velocity potential function has been defined in Section 2-4.2. It should be recalled that the velocity potential function is defined as a function of (x,y,z,t) such that when differentiated with respect to space in any direction, it yields the velocity in that direction. For example, for one direction **S** the velocity in that direction $\mathbf{V}_S$ is such that $V_S = (\partial\phi/\partial S)$.

It is interesting to note the following parallel: The velocity potential function is a natural mathematical outcome from the assumption that the motion is irrotational $(\partial u/\partial y) - (\partial v/\partial x) = 0$ in the same way that the stream function is a natural mathematical outcome from the continuity relationship $(\partial u/\partial x) + (\partial v/\partial y) = 0$.

As with the stream function, the velocity potential function is a mathematical device to describe a flow pattern concisely.

The velocity potential function may be used for any kind of irrotational flow: steady or unsteady; two-dimensional or three-dimensional. It may be used to study turbulent motion, provided the velocity potential function refers to the mean motion with respect to time.

It may also be used to study a flow through porous medium provided it refers to the mean motion with respect to space, and that the Reynolds number is smaller than 1 (see Section 9-3.1). However, with the exception of this last case, it may be used only when friction effects are negligible, and in short convergent structures. When used for the divergent part of a flow, it must be realized that convective inertia forces often cause separation and wakes, and that the velocity potential function has a limit of applicability. If the surface of separation of wakes is known, the flow from stagnant zones may also be defined by a potential function provided the friction effects are negligible, as they are in the convergent part of the flow.

11-3.2 Equipotential Lines and Equipotential Surfaces

By definition an equipotential line in a two-dimensional motion and an equipotential surface in a three-dimensional motion are defined by the fact that ϕ keeps a constant value at any point of this line or of this surface

$$\phi(x,y,z) = \text{constant} = K$$

or

$$\phi(x,y,z,t_0) = \text{constant} = K$$

That is,

$$\frac{\partial \phi}{\partial x} dx + \frac{\partial \phi}{\partial y} dy + \frac{\partial \phi}{\partial z} dz = 0 \quad \text{or} \quad d\phi = 0$$

Changing the value of the constant K gives various equipotential lines or surfaces in the same way that various streamlines were obtained when this operation was performed with the stream function ($\psi = K$).

In contrast, the velocity vector and the streamlines are always perpendicular to the equipotential lines or equipotential surfaces. Consider the equation of an equipotential line given above in the case of a two-dimensional flow:

$$\frac{\partial \phi}{\partial x} dx + \frac{\partial \phi}{\partial y} dy = 0$$

or

$$u\, dx + v\, dy = 0$$

It is deduced that the slope of an equipotential line is $dy/dx = -(u/v)$, which is normal to the slope of a streamline (see Section 11-2.2.1). More generally, this may also be deduced from the fact that $\partial\phi/\partial x$, $\partial\phi/\partial y$, $\partial\phi/\partial z$ are the direction cosines of the perpendicular to the surface defined by $\phi = K$.

11-3.3 Velocity Potential Function and Continuity

11-3.3.1 It is to be recalled that introducing ϕ in the continuity relationship (see Chapter 3-3.4)

$$\frac{\partial u}{\partial x} + \frac{\partial v}{\partial y} + \frac{\partial w}{\partial z} = 0$$

leads to

$$\frac{\partial^2 \phi}{\partial x^2} + \frac{\partial^2 \phi}{\partial y^2} + \frac{\partial^2 \phi}{\partial z^2} = 0 \quad \text{or} \quad \nabla^2 \phi = 0$$

Similarly, introducing ψ in the equation stating that the flow is irrotational leads to $\nabla^2 \psi = 0$.

Hence, a two-dimensional irrotational flow may be found as a solution of either

$$\nabla^2 \phi = 0 \quad \text{or} \quad \nabla^2 \psi = 0$$

The following table summarizes the previous considerations:

Continuity		*Irrotationality*
Definition of ψ	⤫	Definition of ϕ
Expressed as $\nabla^2 \phi = 0$		Expressed as $\nabla^2 \psi = 0$

In a word, both $\nabla^2 \phi = 0$ and $\nabla^2 \psi = 0$ define an irrotational motion which satisfies the continuity principle. But $\nabla^2 \psi \neq 0$ is not compatible with the existence of ϕ.

11-3.3.2 The simplest example of motion in which the velocity potential may be used is the two-dimensional uniform flow for which the velocity potential function is

$$\phi = Ax + By$$

as shown in Fig. 11-3.

The velocity components at any point are

$$u = \frac{\partial \phi}{\partial x} = A \qquad v = \frac{\partial \phi}{\partial y} = B \quad \text{and} \quad V = (A^2 + B^2)^{1/2}$$

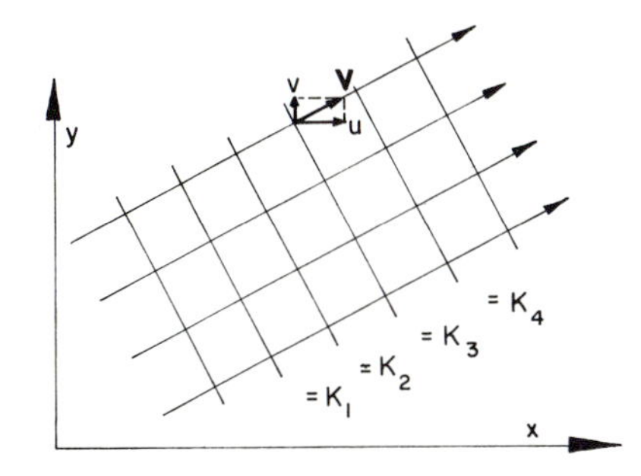

Figure 11-3 *Uniform flow defined by a velocity potential function.*

That is the very same flow as that given by the stream function

$$\psi = Ay - Bx$$

The equipotential lines are given by equating ϕ to a constant value K

$$\phi = Ax + By = K$$

They are straight lines of slope

$$\frac{y}{x} = -\frac{A}{B}$$

It may be noticed that these equipotential lines are perperpendicular to the streamlines (see Section 11-3.2).

11-3.4 General Remarks on the Use of the Velocity Potential Function

Introduction of ϕ rather than u, v, w in the basic momentum equation and continuity relationship reduces the number of unknowns from three (or two in the case of a two-dimensional motion) to one. However, the order of differentiation is increased by one degree. Then the system of equations to be solved has the general form

Continuity: $\nabla^2\phi = 0$

Momentum:

$$\rho\frac{\partial \phi}{\partial t} + \frac{\rho}{2}\left[\left(\frac{\partial \phi}{\partial x}\right)^2 + \left(\frac{\partial \phi}{\partial y}\right)^2 + \left(\frac{\partial \phi}{\partial z}\right)^2\right] + p + \rho g z = f(t)$$

This momentum equation is often introduced as a free-surface condition for which p is constant. But, in that case, another unknown must be introduced: $z = \eta(x,y,t)$, which is the equation of the free surface.

The boundary conditions at a fixed boundary are $\partial\phi/\partial n = 0$. They indicate that the velocity component in a direction perpendicular to the boundary is zero. An irrotational flow under pressure is determined, at least in relative value, from continuity $\nabla^2\phi = 0$ and fixed boundary condition $\partial\phi/\partial n = 0$ only.

11-4 Steady Irrotational Two-Dimensional Motion, Circulation of Velocity

11-4.1 A Review, An Example, Polar Coordinates

11-4.1.1 Table 11-1 establishes a parallel between the stream and potential functions. The conditions summarized in this table are those satisfied by irrotational two-dimensional motion.

These characteristics involve others that have permitted the development of a number of very versatile tools to study steady, irrotational, two-dimensional motions. For this reason, this kind of motion has taken on great importance in hydrodynamics and also in engineering practice since many three-dimensional motions can be analyzed successfully by neglecting the vertical or one horizontal component. For example, the flow in a wide river when the backwater curve effect is small, or the flow toward a well, may often be considered as two-dimensional motion. (The reader is referred to Chapter 2 to distinguish when a flow may be considered as irrotational and when the method described below can be used.)

11-4.1.2 The simple example of uniform flow has already been shown. Another example of irrotational two-dimensional flow is that defined by the stream function: $\psi = xy$. Giving ψ various constant values, it can be seen that the

Table 11-1 *Stream and potential functions*

Stream function	*Potential function*
Continuity	Irrotationality
$\frac{\partial u}{\partial x} + \frac{\partial v}{\partial y} = 0$	$\frac{\partial u}{\partial y} - \frac{\partial v}{\partial x} = 0$
permit definition of	
The stream function ψ	The velocity potential function ϕ
The streamlines are defined by	The equipotential lines are defined by
$d\psi = 0$	$d\phi = 0$
$\psi = K$	$\phi = K$
The velocity components are	
$u = \frac{\partial \psi}{\partial y}$	$u = \frac{\partial \phi}{\partial x}$
$v = -\frac{\partial \psi}{\partial x}$	$v = \frac{\partial \phi}{\partial y}$
$\mathbf{V} = \left(\frac{\partial \psi}{\partial \mathbf{n}}\right)$	$\mathbf{V} = \left(\frac{\partial \phi}{\partial \mathbf{S}}\right)$
dn is the part of an equipotential line defined by	dS is the part of a streamline defined by
$d\phi = 0$	$d\psi = 0$
dn is normal to the streamlines	dS is normal to the equipotential lines
Irrotationality is expressed by	Continuity is expressed by
$\nabla^2 \psi = 0$	$\nabla^2 \phi = 0$

streamlines are represented by a family of rectangular hyperbolas that represent a flow toward a plate perpendicular to the incident motion, as shown in Fig. 11-4. Such a motion is irrotational because $u = x$, $v = -y$, and

$$\frac{\partial u}{\partial y} - \frac{\partial v}{\partial x} = \frac{\partial^2 \psi}{\partial y^2} + \frac{\partial^2 \psi}{\partial x^2} \equiv 0$$

Therefore, a velocity potential function exists.

$$u = \frac{\partial \phi}{\partial x} = \frac{\partial \psi}{\partial y} = x \qquad v = \frac{\partial \phi}{\partial y} = -\frac{\partial \psi}{\partial x} = -y$$

Hence

$$\phi = \int x\,dx = \tfrac{1}{2}x^2 + f(y)$$

$$\phi = \int y\,dy = -\tfrac{1}{2}y^2 + f(x)$$

It is easy to verify that

$$\phi = \tfrac{1}{2}(x^2 - y^2)$$

satisfies these two conditions.

The equipotential lines defined by ϕ = constant form a family of rectangular hyperbolas which are always perpendicular to the streamlines.

11-4.1.3 Before studying some typical flow patterns, it is useful to establish some fundamental formulas in polar coordinates. Referring to Fig. 11-5, it is seen that

$$\mathbf{V} = \mathbf{u} + \mathbf{v} = \mathbf{v}_r + \mathbf{v}_\theta$$

and

$$v_r = u \cos\theta + v \sin\theta$$
$$v_\theta = -u \sin\theta + v \cos\theta$$

Also

$$x = r\cos\theta$$

$$y = r\sin\theta$$

$$\frac{\partial x}{\partial r} = \cos\theta \qquad \frac{1}{r}\frac{\partial x}{\partial \theta} = -\sin\theta$$

$$\frac{\partial y}{\partial r} = \sin\theta \qquad \frac{1}{r}\frac{\partial y}{\partial \theta} = \cos\theta$$

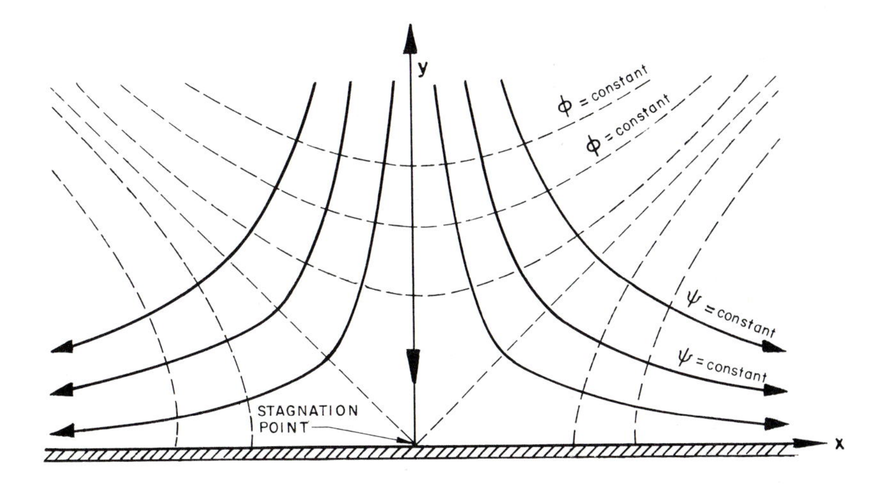

Figure 11-4

Flow toward a plate

Now consider ψ in terms of these coordinates. Introduce ψ into the above equations to obtain

$$v_r = \frac{1}{r}\left[\frac{\partial \psi}{\partial y}(r\cos\theta) + \frac{\partial \psi}{\partial x}(-r\sin\theta)\right]$$

$$= \frac{1}{r}\left[\frac{\partial \psi}{\partial y}\frac{\partial y}{\partial \theta} + \frac{\partial \psi}{\partial x}\frac{\partial x}{\partial \theta}\right] = \frac{1}{r}\frac{\partial \psi}{\partial \theta}$$

and

$$v_\theta = -\frac{\partial \psi}{\partial y}\sin\theta - \frac{\partial \psi}{\partial x}\cos\theta$$

$$= -\frac{\partial \psi}{\partial y}\frac{\partial y}{\partial r} - \frac{\partial \psi}{\partial x}\frac{\partial x}{\partial r} = -\frac{\partial \psi}{\partial r}$$

Figure 11-5 *Polar coordinates notation.*

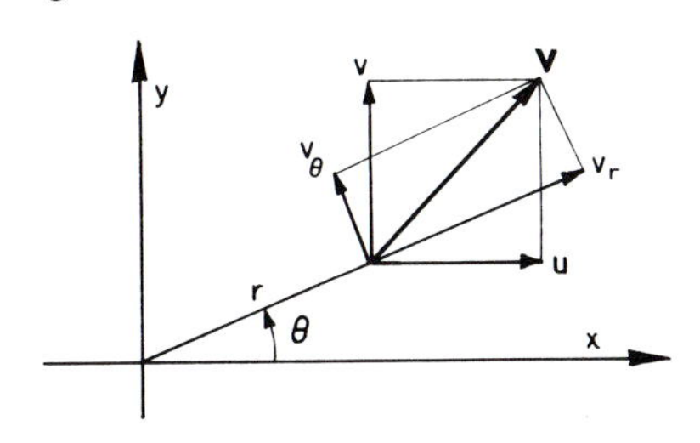

In the same manner, the equations for ϕ become

$$v_r = \frac{\partial \phi}{\partial x}\cdot\frac{\partial x}{\partial r} + \frac{\partial \phi}{\partial y}\cdot\frac{\partial y}{\partial r} = \frac{\partial \phi}{\partial r}$$

and

$$v_\theta = \frac{1}{r}\left[\frac{\partial \phi}{\partial x}(-r\sin\theta) + \frac{\partial \phi}{\partial y}(r\cos\theta)\right]$$

$$= \frac{1}{r}\left[\frac{\partial \phi}{\partial x}\frac{\partial x}{\partial \theta} + \frac{\partial \phi}{\partial y}\frac{\partial y}{\partial \theta}\right] = \frac{1}{r}\frac{\partial \phi}{\partial \theta}.$$

Finally

$$v_r = \frac{1}{r}\frac{\partial \psi}{\partial \theta} = \frac{\partial \phi}{\partial r} \qquad v_\theta = -\frac{\partial \psi}{\partial r} = \frac{1}{r}\frac{\partial \phi}{\partial \theta}$$

By a similar calculation, the condition for irrotationality of a two-dimensional flow

$$2\zeta = \frac{\partial u}{\partial y} - \frac{\partial v}{\partial x} = 0$$

becomes in polar coordinates

$$2\zeta = \frac{1}{r}\frac{\partial(rv_\theta)}{\partial r} - \frac{1}{r}\frac{\partial v_r}{\partial \theta} = 0$$

Note when $v_r = 0$, rv_θ = constant. It is a free vortex.

11-4.2 Elementary Flow Patterns

11-4.2.1 Many cases encountered in engineering practice closely approximate some standard flow patterns. A great number of them are obtained by a combination or a transformation of three elementary flow patterns. These three basic patterns are, as described in Fig. 11-6.

1. Uniform flow, studied as an example in Sections 11-2.2.2 and 11-3.2
2. Radial flow: source or sink
3. Circular flow or vortex flow, which is an irrotational flow with a *circulation of velocity*. If one or more vortices are included, the resulting complex flow pattern is still irrotational. However, the circulation of velocity may not be zero if the area defined by the path of integration includes one vortex.

11-4.2.2 Some examples of elementary combinations of flow patterns without circulation are, as shown in Fig. 11-7,

Figure 11-6 *Basic flow patterns.*

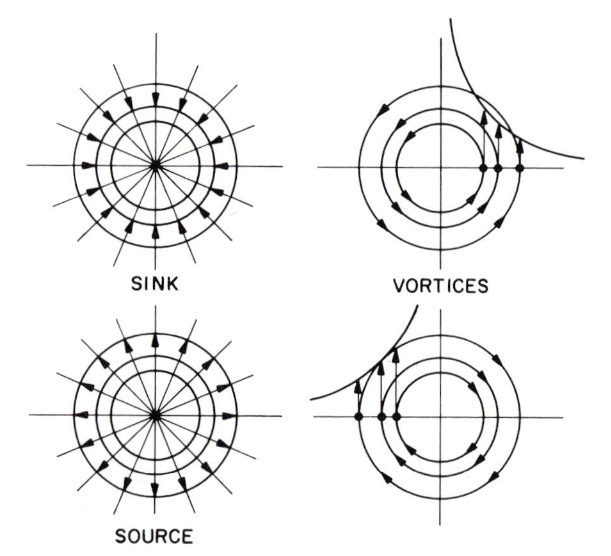

1. One source and one sink
2. A doublet: a source and a sink at the same point
3. Flow past a half-body: a source and uniform flow
4. Flow past a cylinder: a doublet and a uniform flow
5. Flow past a long body: (Rankine body) or a streamlined fixed body: a source, a sink, and a uniform flow, or a source, a series of sinks, and a uniform flow

11-4.2.3 Some elementary combinations of flow patterns with circulation of velocity are as shown in Fig. 11-8

1. Spiral vortex: sink and vortex
2. Flow past a cylinder with circulation of velocity
3. The flow past a cylinder with circulation may be transformed by a conformal mapping operation to the flow around a wing: this is the theory of an airfoil (see Section 11-6.4).

11-4.2.4 A source is a flow that moves radially outward from a point assumed to be infinitely small (Fig. 11-6). A sink is a flow radially inward to a point.

In practice, such a flow is fairly well represented by the flow through a porous medium toward a well of small diameter, insofar as the vertical component is small, i.e., insofar as the curvature of the water table is small.

However, as previously mentioned, its main interest lies in the fact that complex flow patterns usually encountered in engineering practice may be obtained by a combination of sources, sinks, and other elementary kinds of flow.

Let Q be the discharge of the source. The components of velocity at any point are $v_\theta = 0$ (for the purpose of symmetry) and

$$v_r = \frac{Q}{2\pi r} = \frac{\partial \phi}{\partial r} = \frac{1}{r}\frac{\partial \psi}{\partial \theta}$$

Stream function and velocity potential function are given by direct integration. They equal, respectively,

$$\psi = \frac{Q}{2\pi}\theta \qquad \phi = \frac{Q}{2\pi}\ln r$$

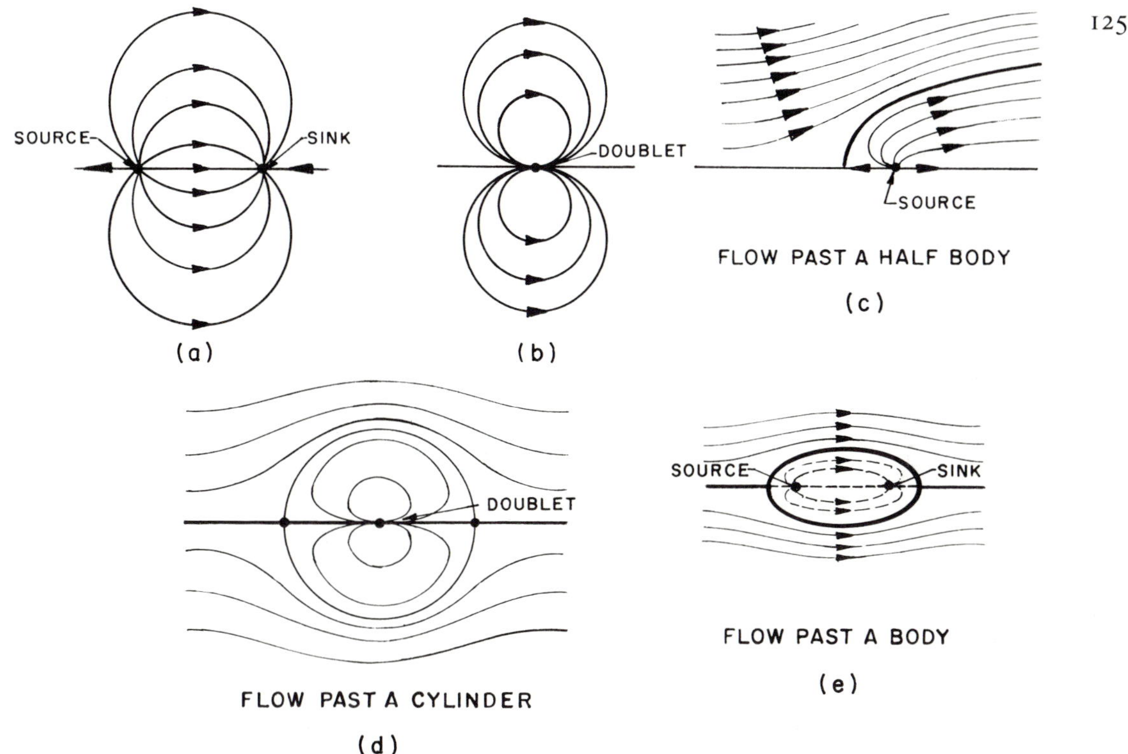

Figure 11-7

Examples of combination of basic flow patterns without circulation.

Equipotential lines, given by ϕ = constant, are circles (r = constant). Streamlines, given by ψ = constant, are straight radial lines (θ = constant). Changing Q to $-Q$ gives the velocity potential function and stream function of a sink.

It is easy to verify that the velocity potential function of a three-dimensional source, where $V = Q/4\pi r^2$ is $\phi = -(Q/4\pi r)$. In this case, the equipotential surfaces ϕ = constant are spheres (r = constant).

11-4.2.5 A vortex is a flow in which the streamlines are concentric circles (Fig. 11-9). In a "forced vortex" water turns as a monolithic mass, the velocity being proportional to the distance from the center (Section 2-4.1).

The flow under study is a "free vortex." In a free vortex the velocity distribution is governed by the law $v_\theta r$ = constant $= K/2\pi$. It may be seen that when r tends to zero, v_θ tends to infinity. Such a motion is irrotational.

As there is no radial flow

$$v_r = \frac{\partial \phi}{\partial r} = \frac{1}{r}\frac{\partial \psi}{\partial \theta} = 0$$

Hence, one obtains

$$v_\theta = V = \frac{K}{2\pi r} = \frac{1}{r}\frac{\partial \phi}{\partial \theta} = -\frac{\partial \psi}{\partial r}$$

which yields

$$\phi = \frac{K}{2\pi}\theta \qquad \psi = -\int \frac{K}{2\pi r}\,dr = -\frac{K}{2\pi}\ln r$$

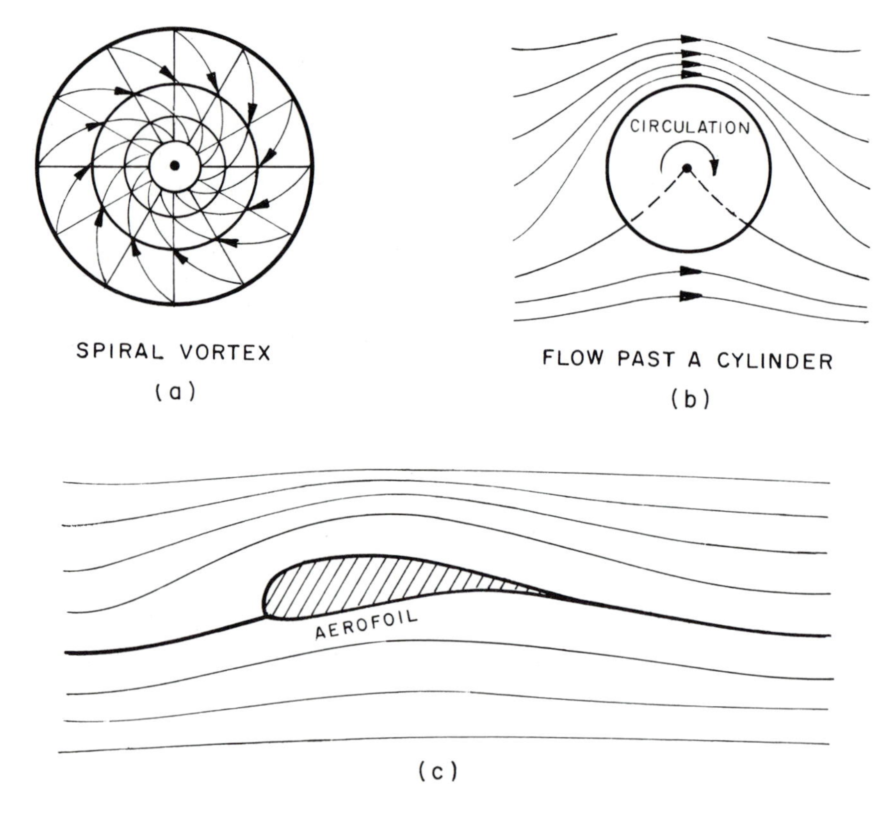

Figure 11-8

Examples of flow combinations with circulation of velocity

The flow pattern is very much the same as that of a source or a sink, but the streamlines and equipotential lines are interchanged. As the flow is irrotational, the Bernoulli equation may be applied throughout the fluid:

$$\frac{V^2}{2g} + \frac{p^*}{\varpi} = \text{constant}$$

This yields

$$\frac{K^2}{8\pi^2 r^2 g} + \frac{p^*}{\varpi} = \text{constant}$$

It is interesting to note that when r tends to zero, p^*/ϖ tends to $-\infty$. Hence, the presence of vortices in a flow is a very important cause of cavitation when air is not admitted into the core from a free surface. Then capillarity forces take on importance when $r \to 0$.

11-4.3 Circulation of Velocity —Definition

Circulation is a mathematical concept on which the theories of wings, airfoils, blades of pumps or turbines, propellers, fans, the Magnus effect (which causes deviation of a tennis ball), some sand motions in a flow, etc. are based.

Circulation is given by the integral along a closed curve S of the tangential velocity component V_S

$$\Gamma = \int_S V_S \, dS$$

It can be demonstrated that Γ is equal to zero in an irrotational flow.

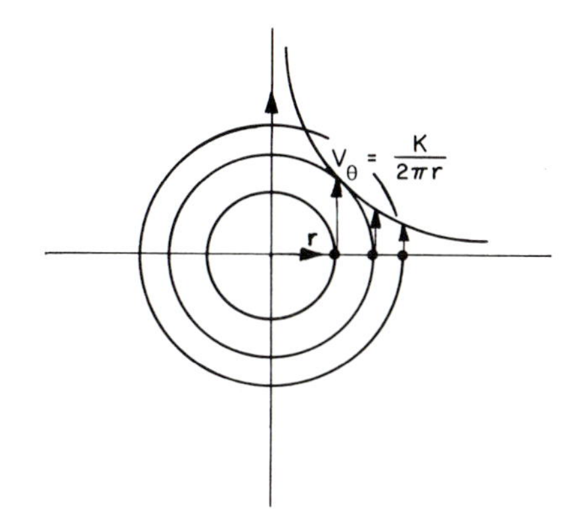

Figure 11-9 *Vortex.*

There is an exception if the closed curve is around a point which is the center of a vortex. Then

$$\Gamma = v_\theta 2\pi r$$

and since

$$v_\theta = \frac{K}{2\pi r} \qquad \Gamma = K$$

Such a flow is called irrotational with circulation. The circulation along a closed curve in a rotational flow is generally different from zero, and it may be demonstrated that when the closed curve is around an elementary area $dx\, dy$, $d\Gamma = \zeta\, dx\, dy$. It can also be demonstrated that the circulation Γ is equal to the flux of the vector rotation of components ζ, η, ξ through the considered area limited by the curve S. Only the definition of the circulation is given here, since it is important to know at least its definition. Its use requires further study beyond the scope of this book.

11-4.4 Combination of Flow Patterns

11-4.4.1 As previously seen, a great number of very complicated flows are obtained by simple addition of the three basic flow patterns studied in the previous paragraphs: (1) uniform flow; (2) radial flow: source or sink; (3) circular flow: vortex.

Examples are first given; then the conditions for adding flow patterns, velocity potential functions, or stream functions are analyzed by consideration of the boundary conditions. Also, some more general considerations of the methods of calculation in hydraulics are given.

11-4.4.2 It has been seen that a uniform flow may be defined by $\phi_1 = Ax$, or $\psi_1 = Ay$. A source may be defined by $\phi_2 = (Q/2\pi) \ln r$, or $\psi_2 = (Q/2\pi)\theta$. Their addition gives the pattern defined by the velocity potential function

$$\phi = \phi_1 + \phi_2 = Ax + \frac{Q}{2\pi} \ln r$$

and the stream function

$$\psi = \psi_1 + \psi_2 = Ay + \frac{Q}{2\pi} \theta$$

This flow pattern is presented in Fig. 11-10. It may be noticed that a central streamline completely separates the source from the outside part of the plane. This streamline may be considered as the round nose body of a pier, for example. In elevation, the upper half of the flow pattern might be regarded as the flow of wind above a hill.

Streamlines and equipotential lines may be obtained graphically from the two basic flow patterns. It is sufficient to add a value $\psi_1 = K_1$ (or $\phi_1 = K'_1$) to a value $\psi_2 = K_2$ (or $\phi_2 = K'_2$) in such a way that $K_1 + K_2$ (or $K'_1 + K'_2$) are always equal to a constant value K.

For example, the intersection of $\psi_1 = 4$ with $\psi_2 = 4$ gives $\psi = 8$. The intersection of $\psi_1 = 3$ with $\psi_2 = 5$ also gives $\psi = 8$. The line joining all the intersections for which $\psi = 8$ is the streamline marked $\psi = 8$. The drawing is very simple when the same interval $\Delta\psi$ (or $\Delta\phi$) is chosen in the two elementary flow patterns. In the case of Fig. 11-10, this interval $\Delta\psi$ is unity.

11-4.4.3 Similarly, it can be demonstrated that one source and one sink of same intensity and located at the same point form a doublet defined by the stream function $\psi_1 = -(K \sin \theta/r)$. The addition of a doublet with a uniform flow $\psi_2 = Ur \sin \theta$ gives a streamline in the shape of a

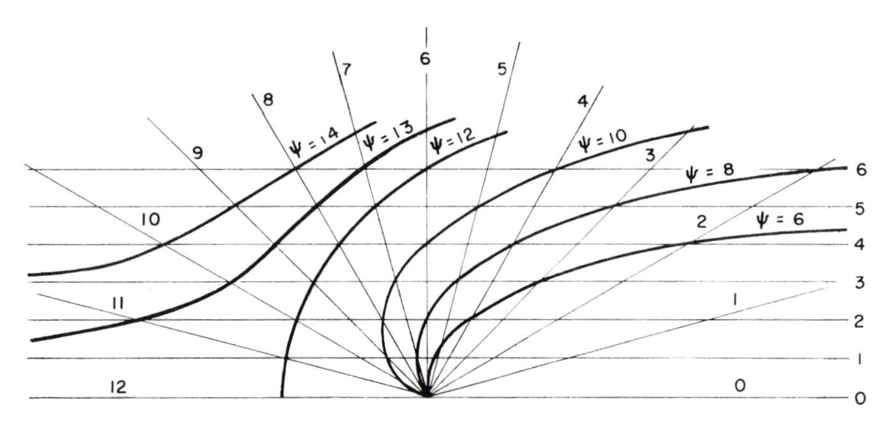

Figure 11-10

Graphic method of addition of flow patterns.

cylinder. Hence the outside flow pattern is considered as the flow of a perfect fluid around a cylinder. The stream function for flow around a cylinder is

$$\psi = -\frac{K \sin \theta}{r} + Ur \sin \theta$$

or

$$\psi = U\left(r - \frac{R^2}{r}\right) \sin \theta$$

where $R = (K/U)^{1/2}$ and U is the velocity at infinity. It can be demonstrated that R is the radius of the cylinder. The potential function is found equal to

$$\phi = -U\left(r + \frac{R^2}{r}\right) \cos \theta$$

The velocity distribution around the cylinder is

$$V = v_\theta|_{r=R} = \frac{1}{r}\frac{\partial \phi}{\partial \theta} = 2U \sin \theta$$

and the pressure distribution is

$$(p - p_\infty) = \tfrac{1}{2}\rho U^2[1 - 4 \sin^2 \theta]$$

where p_∞ is the pressure at infinity. It can be verified that the net pressure force on the cylinder

$$F = 4 \int_0^{\pi/4} p \cos \theta \, R \, d\theta$$

is nil. This result is general. The total force exerted by uniform stream of a perfect fluid on a submerged body, without circulation of velocity, is nil. This is the *paradox of d'Alembert*.

11-4.4.4 If $\phi_1, \phi_2, \ldots, \phi_n, \ldots$ are solutions of $\nabla^2\phi = 0$, any combination $\phi = \phi_1 + \phi_2 + \cdots + \phi_n + \cdots$ is also a solution of $\nabla^2\phi = 0$ and is therefore a possible flow pattern. A similar rule exists for the stream function ψ, and solutions of $\nabla^2\psi = 0$. However, boundary conditions have to be satisfied, and will not necessarily allow the simple addition of elementary solutions to yield the final potential (or stream) function.

11-4.4.5 Consider the flows presented in Fig. 11-11.

In the first case of a flow under pressure, an addition of solutions characterized by velocities V_1 and V_2 at a given point does not change the flow pattern because the pattern does not depend upon the absolute value of the velocity. In the second case of flow with a free surface, the flow pattern is changed because the slope of the free surface changes with V. The solutions cannot be added, as they depend upon the absolute value of the velocity. This stems from the fact that the flow depends upon a non-linear equation: the momentum equation; or more specifically the Bernoulli equation, in which the elevation of the free surface is related to the square of the velocity. The first flow pattern

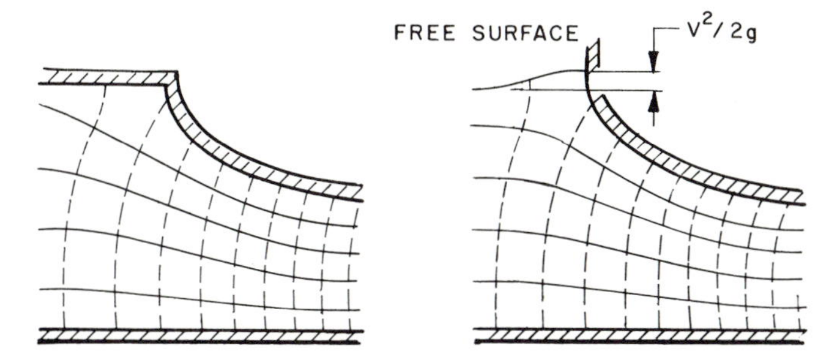

Figure 11-11 *The flow pattern depends upon the shape of the free surface.*

under pressure may be drawn directly from the fixed boundary, which defines two streamlines. This flow pattern depends only on linear relationships:

1. The continuity: $\frac{\partial u}{\partial x} + \frac{\partial v}{\partial y} = 0 \quad \text{or} \quad \nabla^2 \phi = 0$

2. The irrotationality: $\frac{\partial u}{\partial y} - \frac{\partial v}{\partial x} = 0 \quad \text{or} \quad \nabla^2 \psi = 0$

3. The boundary condition: $\frac{\partial \phi}{\partial n} = 0$

This boundary condition involves the continuity only.

This flow pattern does not depend upon the absolute value of velocity, only upon its relative value. In a word, the solution for the flow pattern under pressure within given boundary is unique.

Then the pressure distribution may be calculated in a final, independent step by application of the momentum equation in the form of the Bernoulli equation.

In the second case, the flow has a free surface. This free surface is unknown and must be calculated. Both the nonlinear momentum equation and the continuity equation must be taken into account.

The boundary condition at the free surface p = constant involves a force and must be introduced in the momentum equation to calculate the shape of the free-surface streamlines. In turn, this shape has an effect on the flow pattern.

Hence, the flow pattern and the velocity field on one side, and the pressure distribution and free-surface streamlines on the other side, cannot be calculated independently by successive steps as in the previous case. The flow pattern depends upon the absolute value of the velocity, which may be known only by a combination of linear equations (continuity) with the nonlinear momentum equation. The assumption of irrotationality may be introduced in the momentum equation, but this does not make the free surface condition linear. The considerations given above lead to some more general remarks on the importance of the boundary conditions.

11-5 Reflections on the Importance of Boundary Conditions

11-5.1 New Theoretical Considerations on the Kinds of Flow

From the previous considerations, it is seen that in any kind of flow the method to be used to determine the flow pattern depends upon the kind of boundary conditions and upon the assumption of irrotationality. From this point of view, two major categories of motion may be distinguished that are encountered in all methods in hydrodynamics: analytical, numerical, and graphic methods, or methods based on an analogy. The major categories are on one hand the irrotational motions under pressure and slow motion, and on the other hand, the free-surface flow and flow with friction force.

11-5.2 Irrotational Motion under Pressure and Slow Motion

The first category includes all irrotational motions under pressure, or motions considered as such, and slow motion in which the quadratic terms are negligible.

11-5.2.1 In the case of flow under pressure, the streamlines at the boundary are fully determined since they are coincident with this boundary. The boundary conditions are expressed to satisfy the continuity principle, that is, that the velocity is tangential to the boundary. The flow pattern depends completely upon linear equations only, expressing the continuity and the irrotationality principles, after which the flow pattern is relatively easily known.

The velocity field gives the relative value of the velocity. The absolute value can be known when the velocity is determined at one point, either given by a boundary condition or calculated by application of the momentum equation at this boundary.

Finally, the pressure distribution is determined from the knowledge of the velocity at any given point by application of the momentum equation.

11-5.2.2 In the case of slow motion, the motion is mathematically considered as infinitely small, even with a free surface. Hence, all the quadratic terms may be neglected and the momentum equation becomes linear. The free surface is considered to be known at the beginning and is denoted by a horizontal line. In that case, various solutions of flow patterns may be added.

For example, if ϕ_1 is the velocity potential function of a periodic gravity wave at the first order of approximation, that is when the convective inertia term is neglected, and if ϕ_2 is the potential function of another wave traveling in the opposite direction, ϕ_1 and ϕ_2 are determined by a system of linear equations, as will be seen in Section 16-3.

Hence, $\phi = \phi_1 + \phi_2$ is the potential function of the resultant motion.

11-5.3 Free-Surface Flow and Flow with Friction Forces

The second category of motion includes all motions with free surfaces or the motions for which the friction forces have a nonnegligible effect, causing the motion to be rotational.

11-5.3.1 The free-surface condition involves a force (p = constant). This force can only be inserted in the momentum equation, which is an equality of force.

The flow pattern is now determined from nonlinear boundary conditions as well as from the linear continuity relationship.

Difficulties arise from this nonlinearity and the fact that the free surface is unknown. ϕ_1 and ϕ_2 as defined in Section 11-5.2.2 may also be calculated at a higher order of approximation from the nonlinear free-surface conditions, which take into account the convective inertia term. However the solution ϕ, representing the two nonlinear waves, cannot be obtained by simple addition of the new solutions ϕ_1 and ϕ_2 given at a high order of approximation. It must be calculated from the basic equations. Similar considerations prevail in the case of irregular waves traveling at different velocities in the same direction. There is nonlinear interaction.

In conclusion, in order that the velocity potential functions and the stream functions may be added, they must depend upon a linear and homogeneous equation only. Also, the boundary conditions should be homogeneous.

11-5.3.2 Similarly, a friction force (resulting in a rotational term different from zero) gives a boundary condition $V = 0$. Such a boundary condition must also be introduced in the momentum equation. The nonlinearity of the momentum equation, caused by the convective inertia term, is the major cause of difficulty in studying this kind of flow.

These mathematical difficulties show the importance of the irrotational motion under pressure and of the slow motions in hydrodynamics, even if they only represent very approximately the natural conditions.

11-6 Flow Net

11-6.1 Flow Net Principle

The flow net is a family of equipotential lines and a family of streamlines representing a complete two-dimensional flow pattern, as was shown in Fig. 11-4.

The equalities $V = \partial\phi/\partial S = \partial\psi/\partial n$ for a finite difference, take the form $V = \Delta\phi/\Delta S = \Delta\psi/\Delta n$.

First, $\Delta\psi$ is chosen to be a constant in the complete velocity field, which means that the discharge ΔQ between

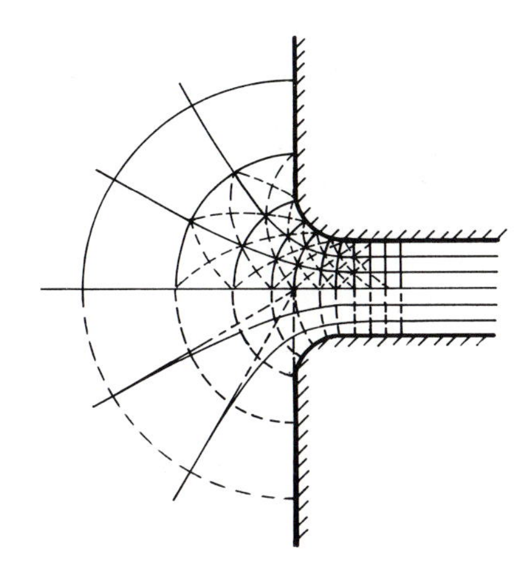

Figure 11-12 *Flow net started in radial and uniform flows.*

two adjacent streamlines is the same ($\Delta\psi = \Delta Q$) (see Section 11-2.3).

Second, the interval $\Delta\phi$ is chosen to be equal to $\Delta\psi$, which leads to $\Delta S = \Delta n$. ΔS is the streamline element, whereas Δn is an equipotential line element with right-angle intersections. Hence, ΔS and Δn are the two sides of a curvilinear square, which tends to be an exact square when ΔS and Δn tend to the infinitessimals dS and dn.

This characteristic of a two-dimensional irrotational flow permits one to draw a complete flow pattern as a mesh of squares, as shown in Fig. 11-12. At any point, the velocity direction is given by the streamline. The magnitude of the velocity given in relative value, is inversely proportional to the square sides.

The graphic procedure for construction of a flow net depends on whether the flow is under pressure or with a free surface.

11-6.2 Flow under Pressure

11-2.6.2 The first case in which the flow net method may be of very great use is for flow between two fixed boundaries, which corresponds to a flow under pressure.

A flow net around a solid body with well-determined boundary conditions at infinity follows the same rules of construction. It should also be noted that many flows at the free surface, such as in a wide and relatively shallow river in which rotation is about a horizontal axis, may be defined by a two-dimensional velocity potential function and studied by the flow net method (see Section 2-6.4). However, wave effects or backwater curves must be neglected. All these types of flow are determined by the same method.

11-6.2.2 The flow pattern is started in the regions in which the velocity distribution is known, such as in a uniform or radial flow (Fig. 11-12). Then a number of streamlines are selected as a function of the desired accuracy, taking into account that this number could easily be increased in a given area if a greater local accuracy is required. Then the equipotential lines are drawn intersecting the streamlines (including the boundaries) perpendicularly, and forming squares with the streamlines.

The simplest method of checking the correctness of the drawing is to draw the diagonal lines of the square mesh. These diagonals should, themselves, form smooth curves that intersect each other perpendicularly. This is *Prasil's method*, as demonstrated in Fig. 11-12 (see also Fig. 11-14). If these diagonals do not intersect, a second drawing is made by superimposition of transparent paper to correct the first mistakes, repeating the process until the desired result is obtained. As a rule, three successive drawings are sufficient to obtain an accurate flow net by trial and error.

11-6.2.3 The limits of validity of the flow net method in studying flow under pressure are the same as those imposed by the assumption of irrotational motion. That is, the flow net method may be applied to study short convergent flow, or flow through porous medium when the Reynolds number is smaller than 1.

Divergent flow causing separation and wakes, long structures for which the friction forces cause the motion to be rotational, and unsteady motion cannot usefully be studied by the flow net method. In the case of a wake, a flow net method may be used if the separation line is determined (see Fig. 11-13). The pressure and velocity

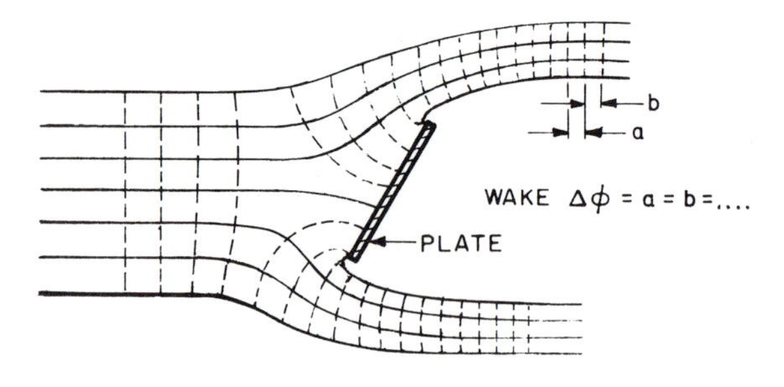

Figure 11-13 *Wake.*

are then considered to be constant along this line. Its determination is relevant to the method for flow with a free surface, which is the subject of the next section.

11-6.3 Flow Net with Free Surface

11-6.3.1 When the free-surface boundaries are known by previous experiment, the same method as that explained to construct a flow net under pressure may be used. Moreover, a free-surface condition is given that determines the distance between equipotential lines.

Three cases may be distinguished.

1. Flow through porous medium: the vertical distances between successive equipotential lines, following the rule $\Delta\phi$ = constant, are constant as has been shown in Section 9-2.2.3 (Fig. 9-4).
2. Horizontal high velocity flow: a flow through an orifice or from a gate, with a contraction and under a high head (Fig. 11-14). In this case, by application of the Bernoulli equation, $V = [2g(H + z)]^{1/2}$, H being the total head, whereas z refers to the exact elevation of any point under consideration above the level downstream of the gate. In many cases, z is always small compared with H, and V is considered to be a constant at the free surface. Hence, the distances between equipotential lines are equal. In a word, such a flow is determined by considering that the gravity force in the downstream part of the gate or the orifice is negligible.
3. Flow with vertical velocity component (over a weir): the velocity at the free surface varies with z. According to the Bernoulli equation, $V = (2gz)^{1/2}$. Hence, the distance between the free surface and the first streamline is given by (z is measured downward herewith)

$$\frac{\Delta\phi}{V} = \frac{\text{constant}}{(2gz)^{1/2}} = \Delta n = \Delta S$$

11-6.3.2 When the free surface is not known from previous experiments, the continuity and the momentum equations, which give conditions such as those presented in (1), (2), and (3) above, should be sufficient to determine the free surface streamline and the complete flow pattern. The solution proceeds as follows: A tentative streamline is drawn intuitively. Then the distances between equipotential lines are calculated as shown in Section 11-6.3.1. The flow net drawn on this basis must be found to be consistent with the given fixed boundary. If it is not consistent, a second trial is made by modification of the free surface, and so the solution proceeds. It is easy to conceive that such a trial and error method is tedious and inaccurate.

Therefore, although such a procedure is theoretically possible, it is unrealistic to attempt to determine a flow

Figure 11-14 *Horizontal high-velocity flow.*

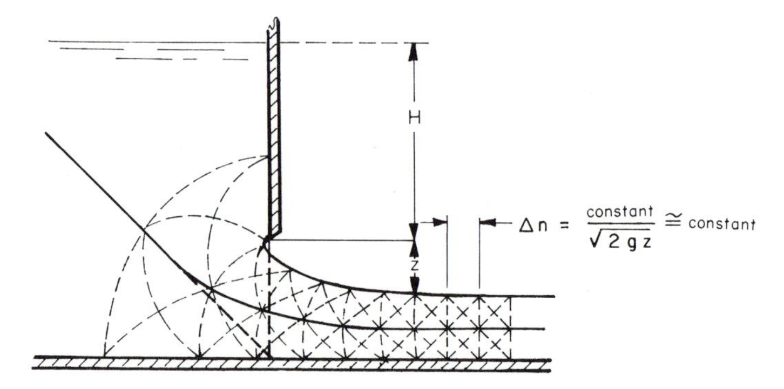

net with a free surface without an experiment. This is even more true when it is a flow between two free surfaces such as a free-falling jet. Most often the necessary experiments for determining the free surface are self-sufficient for practical engineering purposes. Because a model must be built, it can also be used for measuring the pressure distribution; and the determination of the flow net is then a purely academic exercise.

It is more accurate to determine the flow pattern by a numerical scheme and computer. However, the programmed numerical scheme is based on the same trial-and-error method.

11-6.4 Other Methods, Conformal Mapping

11-6.4.1 There are a number of methods for drawing a flow net. All of them are based on the same principles, and a similar difficulty is encountered in the determination of a free surface. The most satisfactory solution is found by using a trial-and-error numerical method with a computer. The relaxation method is also based on numerical calculus.

An analogy with an electric field is very often used. In this case, the analogous equipotential lines between boundaries at different voltages are measured directly. This method is used with liquid resistance mesh, wetted earth, etc.

The relaxation method and electrical method may be easily extended to three-dimensional irrotational flow.

Another analogical method is based on the fact that the mean motion of laminar flow at constant thickness may be considered to be irrotational (see Section 2-6.4).

Finally, because the mean motion of laminar flow through a porous medium is irrotational, it is very easy to use the analogical method to study any two-dimensional or three-dimensional patterns.

11-6.4.2 It is beyond the scope of this book to develop the powerful mathematical tool of conformal mapping for studying two-dimensional irrotational flows with or without circulation of velocity. Only the principle is described below.

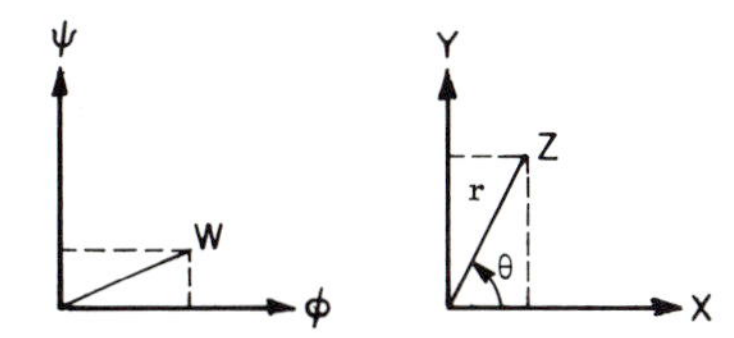

Figure 11-15 *Notation for conformal mapping.*

Conformal mapping (Fig. 11-15) is based on the use of complex numbers and functions of a complex variable. Two complex planes are used. In one, $W = \phi + i\psi$, whereas in the other $Z = x + iy = re^{i\theta}$. A relationship between the two planes is developed such that $W = f(Z)$.

A conformal mapping operation consists of establishing a relationship between each point of a given flow pattern in the x, y plane and a point of another flow pattern in the ϕ, ψ plane. The first is often the real flow under study; the second is often a uniform or a simpler flow pattern. Successive conformal mapping operations may also be done in order to pass step by step from a very complex flow pattern to a uniform flow.

Conformal mapping can also be used for determining free streamlines (*Schwartz–Christoffel transform*). However, its application requires that gravity forces be neglected.

11-6.4.3 For example, consider the transformation

$$W = U\left(Z + \frac{R^2}{Z}\right)$$

where $W = \phi + i\psi$ is the equation for a uniform flow in the W plane, i.e., in the system of ϕ, ψ axes. This flow is parallel to the ϕ axis,. and the streamlines defined by ψ = constant are perpendicular to the ψ axis.

The above relationship is the transformation of a flow around a cylinder of radius R into a uniform flow. This can be seen when the real and imaginary parts are separated.

This is easily done by substituting $Z = re^{i\theta}$ into the transformation and using the relationship $e^{i\theta} = \cos\theta + i\sin\theta$.

$$W = \phi + i\psi$$

$$= U\left(re^{i\theta} + \frac{R^2}{r}e^{-i\theta}\right)$$

$$= U\left(r\cos\theta + \frac{R^2}{r}\cos\theta + ir\sin\theta - i\frac{R^2}{r}\sin\theta\right)$$

$$= U\left(r + \frac{R^2}{r}\right)\cos\theta + iU\left(r - \frac{R^2}{r}\right)\sin\theta$$

Then it is seen that the potential function is

$$\phi = U\left(r + \frac{R^2}{r}\right)\cos\theta$$

and the stream function ψ is

$$\psi = U\left(r - \frac{R^2}{r}\right)\sin\theta$$

These two functions are those of a flow around a cylinder (see Section 11-4.4.3, *vide supra*).

11-6.4.4 The following transformations can be studied by using a similar approach:

Uniform flow: $W = (a + ib)Z$

Source at $Z = A$: $W = \dfrac{Q}{2\pi}\ln(Z - A)$

Vortex as $Z = A$: $W = -\dfrac{iK}{2\pi}\ln(Z - A)$

Spiral vortex at $Z = A$: $W = \dfrac{1}{2\pi}(Q - iK)\ln(Z - A)$

Source at $-A$, sink at $+A$: $W = \dfrac{Q}{2\pi}\ln\dfrac{Z + A}{Z - A}$

Flow through an aperture: $Z = \cosh W$

Flow past a cylinder with circulation of velocity: $W = U\left(Z + \dfrac{R^2}{Z}\right) - \dfrac{iK}{2\pi}\ln Z$

Flow at a wall angle: $W = Z^n$, $\theta = \dfrac{\pi}{n}$

PROBLEMS

11.1 Draw a square mesh in a two-dimensional bend such as shown in Fig. 11-16 and calculate the relative pressure distribution on both boundaries and along the streamline starting from point A at the center of the upstream pipe.

11.2 Give the expression for the Navier–Stokes equations as a function of the stream function $\psi(x,y)$ in the case of two-dimensional motion

11.3 Demonstrate that the velocity potential function for a three-dimensional source is $\phi = -(Q/4\pi r)$.

11.4 Determine the stream function and the potential function for a uniform flow of velocity $\mathbf{V}$ inclined at an angle α with the X axis.

Figure 11-16

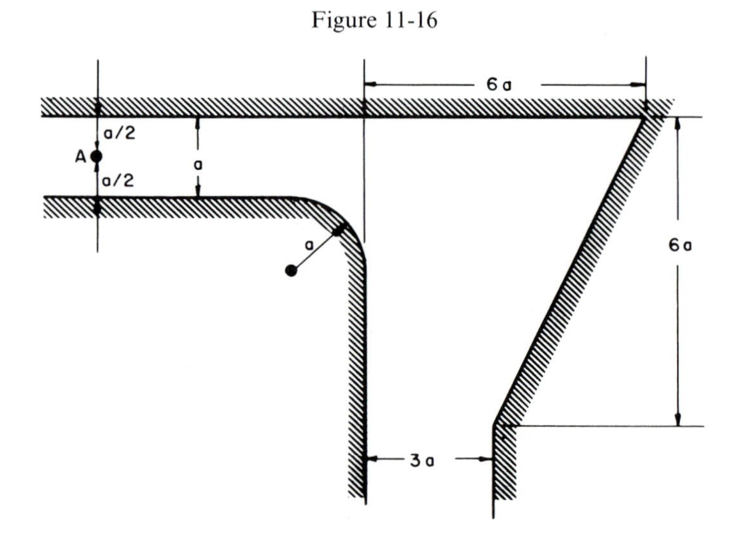

11.5 Sketch the streamlines and equipotential lines for a flow past a cylinder of radius R. Determine the corresponding stream function.

11.6 Study the various characteristics of a flow defined by the stream function $\psi = -x^2$. Determine if such a flow is rotational, and calculate the vorticity. Is the fluid compressible? Plot the streamlines and the equipotential lines.

11.7 Consider a uniform flow in the positive X direction. The velocity varies linearly from $V = 0$ at $y = 0$ to $V = 10$ ft/sec (3.048 m/sec) $y = 10$ ft (3.048). Determine the expression for ψ.

11.8 Draw the flow pattern from a source to a sink by graphical means.

11.9 Consider a flow around a cylinder defined by the potential function $\phi = -U(r - (R^2/r)) \cos\theta$. At which distance is the fluid velocity disturbed by the cylinder by more than 50%, 10%, and 1%? Sketch these three lines of influence around the circle.

11.10 Consider a free surface sink vortex in which the vertical component of velocity will be neglected. Calculate the elevation of the free surface $\eta(r)$.

11.11 Consider the potential function

$$\phi = \frac{Q}{2\pi} \ln r + \frac{K}{2\pi}\theta$$

Calculate the stream function and the general equations for equipotential lines and streamlines. Draw the corresponding flow pattern assuming that $Q = K$ and $Q = \frac{1}{4}K$ successively by means of graphic superposition.

11.12 Demonstrate that the potential function of a doublet is

$$\phi = \frac{K \cos\theta}{r}$$

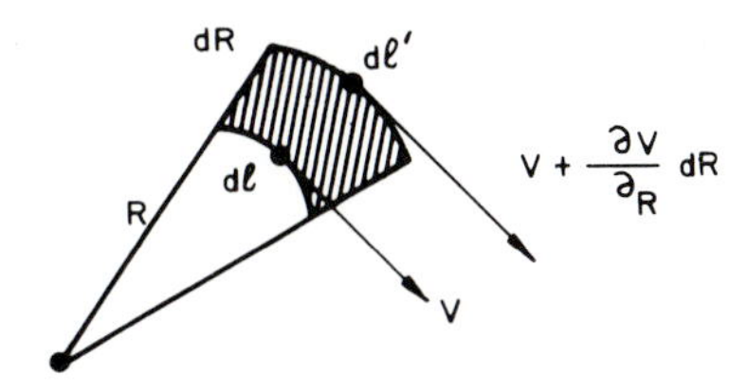

Figure 11-17

and demonstrate that streamlines and equipotential lines are circles.

11.13 Demonstrate that

$$d^2\Gamma = \left(\frac{V}{R} + \frac{\partial V}{\partial R}\right) dR\, dl \quad \text{and} \quad \frac{\partial H}{\partial R} = \frac{d^2\Gamma}{dR\, dl}\frac{V}{g}$$

where

$$H = \frac{V^2}{2g} + \frac{p}{\rho g} + z,$$

V is the particle velocity, Γ the circulation of velocity, dR an element perpendicular to the streamlines, and dl an element of streamlines, as shown in Fig. 11-17.

11.14 The stream function for a flow past a cylinder with circulation of velocity is

$$\psi = U\left(r - \frac{R^2}{r}\right)\sin\theta - \frac{\Gamma}{2\pi}\ln r$$

where Γ is the circulation. Determine the position of the stagnation points on the cylinder as a function of Γ. Demonstrate that the total force exerted by the flow per unit length of the cylinder is $F = \rho U \Gamma$.

11.15 Calculate the potential function for a flow past a "Rankine" body. The stream function is, as shown in Fig. 11-18

$$\psi = \frac{Q}{2\pi}(\theta_1 - \theta_2) + Ur\sin\theta$$

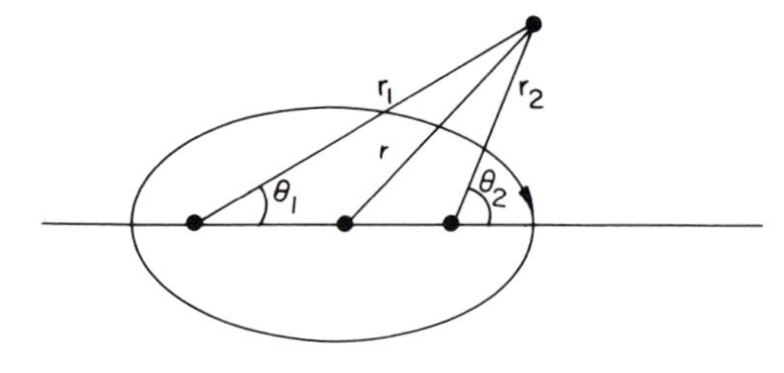

Figure 11-18

Determine the shape of the Rankine body and calculate the pressure around it in terms of the value of the pressure at infinity p_∞.

11.16 Demonstrate by finite differences that, in an irrotational flow, ψ_1 [the value of the stream function at a point (1)] given by

$$\psi_1 = \tfrac{1}{4}(\psi_2 + \psi_3 + \psi_4 + \psi_5)$$

The subscripts 1–5 refer to points 1–5, as shown in Fig. 11-19.

11.17 In the case of a flow past an aperture of length $2C$ and defined by the conformal mapping transformation $Z = C \cosh W$ where $Z = x + iy$ and $W = \phi + i\psi$, demonstrate that the streamlines in the z planes are defined by a family of hyperbolas and that the equipotential lines are defined by a family of ellipses of same foci.

Figure 11-19

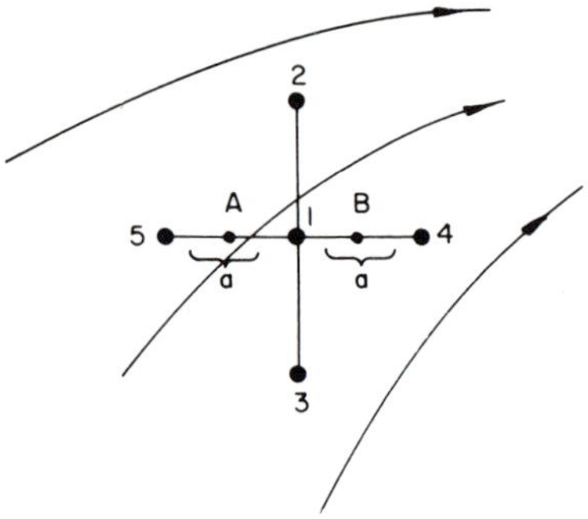

Chapter 12

The Momentum Theorem and Its Applications

12-1 External Forces and Internal Forces

12-1.1 Considerations on Forces

The momentum equation $\mathbf{F} = m(d\mathbf{V}/dt)$ has been expressed in differential form for an elementary fluid particle of unit volume and mass ρ (see Chapter 6). It may be recalled that this momentum equation takes the form of the Navier–Stokes equation, which equates the inertia force of a unit volume with the corresponding applied forces.

The applied forces have been divided into external and internal forces (see Chapter 5). The internal forces are caused by pressure and friction. They are, by definition, vectorially equal to zero and do not contribute to a net torque on the considered particle. This definition is based upon Newton's third law stating that action equals reaction.

The external forces are divided into surface forces caused by pressure and friction, and a body force caused by gravity. These forces have a total sum different from zero, hence impart a motion to the elementary fluid particle.

Consider two adjacent fluid particles, as shown in Fig. 12-1. The external forces acting on the two adjacent sides sum vectorially to zero according to Newton's third law as previously stated. Hence, only the external forces acting on the outer limits of this group of two elementary fluid

Figure 12-1 *External forces at the boundaries of two adjacent fluid particles may be considered as internal forces in order to study their overall motion.*

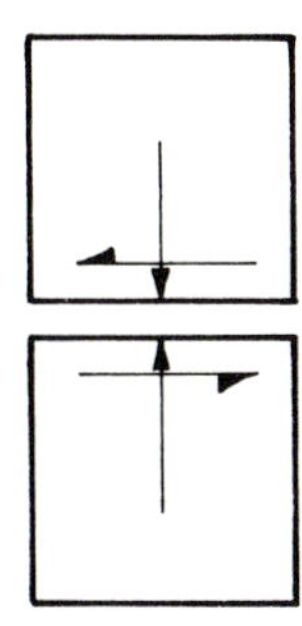

particles affect their overall motion. Therefore, consideration of these external forces permits a theoretical analysis of the combined overall motion of these two particles, but does not permit an analysis of the relative motion of the one particle with respect to the other.

Generalizing for a definite mass of fluid composed of a large number of elementary fluid particles shows that all internal forces sum to zero and will produce no net torque on the definite mass of fluid. The overall motion of this mass of fluid depends only upon the external forces applied to it. Consequently, although this simplification does not permit a study of the internal motion within the mass of fluid nor of the fine structure of the flow pattern, a study of its overall effect may be possible.

12-1.2 Considerations on Energy

Rather than express the momentum equation $\mathbf{F} = m(d\mathbf{V}/dt)$ for an elementary fluid particle as an equality of applied forces and inertia, consider an equality of work and kinetic energy:

$$\mathbf{F} \cdot \mathbf{dS} = m \frac{d\mathbf{V}}{dt} \cdot \mathbf{dS} = m \frac{d\mathbf{V}}{dt} \cdot \mathbf{V}\, dt = d\left(\frac{mV^2}{2}\right)$$

The separation between internal and external forces is always possible. Hence

$$\Sigma(\mathbf{F}_e \cdot \mathbf{dS}) + \Sigma(\mathbf{F}_i \cdot \mathbf{dS}) = d\left(\frac{mV^2}{2}\right)$$

However, in spite of the fact that the total sum of the internal forces is zero by definition ($\Sigma\mathbf{F}_i = 0$), the work of these internal forces does not equal zero, i.e., $\Sigma(\mathbf{F}_i \cdot \mathbf{dS}) \neq 0$.

To illustrate this point, consider a uniform flow in a pipe (see Fig. 12-2). The external forces acting at the wall-boundary have a total sum that is different from zero, thereby tending to move the pipe in the direction of the flow. But the existing internal forces sum to zero. However, these internal forces do work and this work is the cause of the head loss. The head loss expresses the transformation of energy lost by friction into heat.

Thus, insofar as the energy equality is concerned, internal forces may not be neglected.

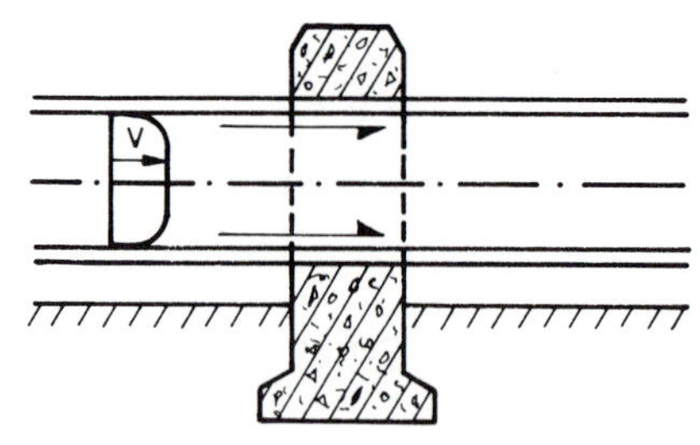

Figure 12-2 *External forces tend to move the pipe downstream, but they do not work. The head loss is caused by the work of internal forces, which have a total sum equal to zero.*

12-1.3 Field of Application

From the previous considerations, it may be deduced that, in practice, the difference in application of the momentum equality and the energy equality lies in the emphasis on the importance of internal forces, insofar as the phenomena being studied are concerned, which is illustrated by a number of examples in this chapter.

A considerable number of hydraulic problems are simplified by the fact that the sum of the internal forces is zero. It is for this reason that the momentum theorem is so often used.

The momentum theorem is used to calculate the overall effects of a mass of fluid, however complex the flow, without dealing with the fine structure of the flow pattern. However, to apply the method which consists of considering only the external forces to calculate the change in momentum requires a perfect knowledge of boundary conditions at the extremities of the mass of fluid under study. This point is illustrated in Section 12-4.

12-1.4 Momentum Theorem and Navier–Stokes Equation

The momentum equation for a finite mass of fluid can be established in several ways as the Bernoulli equation. One could make use of the basic Navier–Stokes equation by integrating all the forces causing motion of an elementary particle of fluid mass ρ to the forces involved in the motion of a definite mass of fluid m. This is evident, as the Navier–Stokes equation is the momentum equation for a mass of fluid of unit volume.

Instead, a direct vectorial demonstration is given for an arbitrary mass of fluid of finite dimensions. Although the momentum theorem is used primarily to solve problems in steady flow, the more general case of unsteady flow is considered here. This method illustrates the difficulties encountered in the application of the momentum theorem to unsteady motion.

12-2 Mathematical Demonstration

12-2.1 Concept

By definition, the product of mass and velocity is momentum. Hence the momentum of an elementary particle fluid of mass ρ is $\rho\mathbf{V}$. Therefore, the total momentum of a definite mass of volume D, in which the velocity vector varies both with time and direction, is $\iiint_D \rho\mathbf{V}\, dD$, where dD is an element of the volume D.

Assuming a mass of fluid is bounded at a given time t by a surface A, the same mass of fluid at time $(t + dt)$ will be bounded by a surface A', quite similar to A. These two surfaces define three domains: D_1, D_2, and D_3. Whereas D_2 has a finite dimension, D_1 and D_3 are, by definition, infinitely small, because the interval of time dt is infinitely small (see Fig. 12-3). Successive values of the total momentum of the fluid in these three domains is calculated.

Figure 12-3 *Notation for momentum of a moving mass of fluid.*

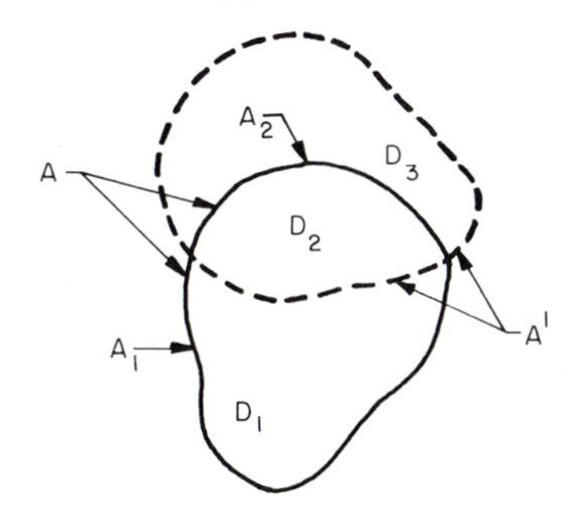

12-2.2 Change of Momentum with Respect to Time

The momentum of fluid enclosed in the common part D_2 at time t is $\iiint_{D_2} \rho\mathbf{V}\, dD$. At time $t + dt$, the velocity becomes $\mathbf{V} + (\partial\mathbf{V}/dt)\, dt$ and the momentum becomes

$$\iiint_{D_2} \rho\left(\mathbf{V} + \frac{\partial\mathbf{V}}{\partial t}\, dt\right) dD$$

Hence, the difference or variation of momentum during the interval of time dt is the difference $\iiint_{D_2}(\partial\rho\mathbf{V}/\partial t)\, dt\, dD$. Note that the integral is the product of a finite number D_2 times an infinitesimal number $(\partial\rho\mathbf{V}/\partial t)\, dt$. Dividing by dt, the variation of momentum per unit time is $\iiint_{D_2}(\partial\rho\mathbf{V}/\partial t)\, dD$.

This term has a zero value in the case of a steady flow because $\partial\mathbf{V}/\partial t = 0$.

12-2.3 Change of Momentum with Respect to Space

The momentum of fluid enclosed in domain D_1 (see Fig. 12-4) at time t is $\iiint_{D_1} \rho\mathbf{V}\, dD$ which is dimensionally the product of a finite number $|\mathbf{V}|$ and an infinitesimal number D_1. Domain D_1 may be considered as containing elementary cylinders of base dA and sides parallel to the velocity vector $\mathbf{V}$.

The volume of an elementary cylinder is $dD = dA\, V_n\, dt$, where V_n is the projected value of $\mathbf{V}$ on a perpendicular to dA. It is deduced that $\iiint_{D_1} dD = \iint_{A_1} dA\, V_n\, dt$ in which A_1 is the part of A, which defines the limit of domain D_1.

Figure 12-4 *Momentum notation.*

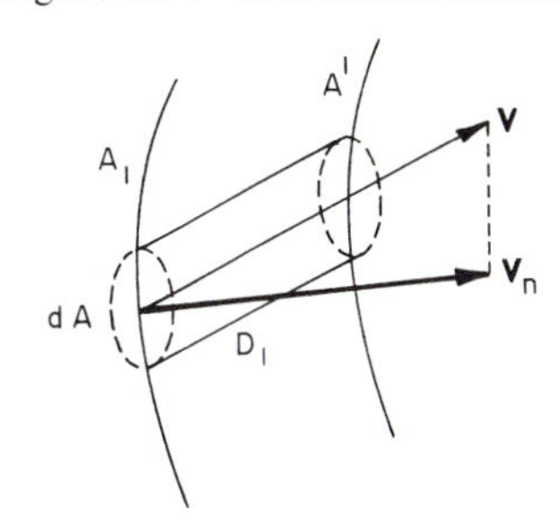

Hence, the momentum of fluid enclosed in D_1 becomes

$$\iiint_{D_1} \rho \mathbf{V} \, dD = dt \iint_{A_1} \rho \mathbf{V} V_n \, dA$$

This is given by a surface integral rather than by a volume integral throughout the volume D_1.

The momentum of fluid enclosed in domain D_3 at time $(t + dt)$ is

$$\iiint_{D_3} \rho \left(\mathbf{V} + \frac{\partial \mathbf{V}}{\partial t} dt \right) dD.$$

The first integral $\iiint_{D_3} \rho \mathbf{V} \, dD$ is the product of a finite number $|\mathbf{V}|$ and an infinitesimal number D_3, whereas the second integral $\iiint_{D_3} \rho (\partial \mathbf{V}/\partial t) \, dt \, dD$ is a product of two infinitesimal numbers $|(\partial \mathbf{V}/\partial t) \, dt|$ and D_3. Hence, this second integral may be neglected.

A process of calculation similar to that just demonstrated above shows that

$$\iiint_{D_3} \rho \mathbf{V} \, dD = -dt \iint_{A_2} \rho \mathbf{V} V_n \, dA$$

where A_2 is the part of A that defines the limit of domain D_3.

Now the difference of momentum between domains D_3 and D_1 at time $(t + dt)$ and time t respectively, caused by the variation of velocity with respect to space is

$$dt \iint_{A_2} \rho \mathbf{V} V_n \, dA + dt \iint_{A_1} \rho \mathbf{V} V_n \, dA$$

because the discharge of momentum entering the domain is affected by a negative sign, whereas the discharge of momentum leaving the domain is affected by a positive sign; considering $A = A_1 + A_2$, and dividing by the interval of time dt, the difference of momentum per unit time or momentum flux is

$$\iint_A \rho \mathbf{V} V_n \, dA$$

Finally, the total change of momentum per unit time with respect to both time and space is equal to the sum of the external forces. Therefore

$$\Sigma \mathbf{F}_e = \iiint_D \frac{\partial \rho \mathbf{V}}{\partial t} \, dD + \iint_A \rho \mathbf{V} V_n \, dA$$

12-3 Practical Application of the Momentum Theorem—Case of a Stream Tube—Specific Force

12-3.1 Application to a Stream Tube

The first step in the application of the momentum theorem consists of choosing the limits of the mass of fluid to which the momentum theorem can be applied. These limits are chosen in a section in which the boundary conditions are well known, i.e., fixed boundaries or cross sections where the motion may be considered as unidimensional.

Because the momentum theorem is a vector equality, it is convenient to choose some axes of reference. Generally, the main flow direction indicates one of the axes to be considered; the vertical axis may be another one.

Finally, the momentum equality is written by projecting all the forces involved on these axes of reference.

In many cases of unidimensional flow, only the equalities of force in the direction of flow are of interest.

For a streamtube flow, as shown in Fig. 12-5, the terms of the momentum equation, presented in Section 12-2.3, applied to a fluid mass within cross sections A_1 and A_2 take the following forms:

$$\Sigma \mathbf{F}_e = \rho \iiint_D \frac{\partial \mathbf{V}}{\partial t} \, dA \, dL + \rho Q(\mathbf{V}_2 - \mathbf{V}_1)$$

Figure 12-5 *Momentum theorem applied to a stream tube—notation*

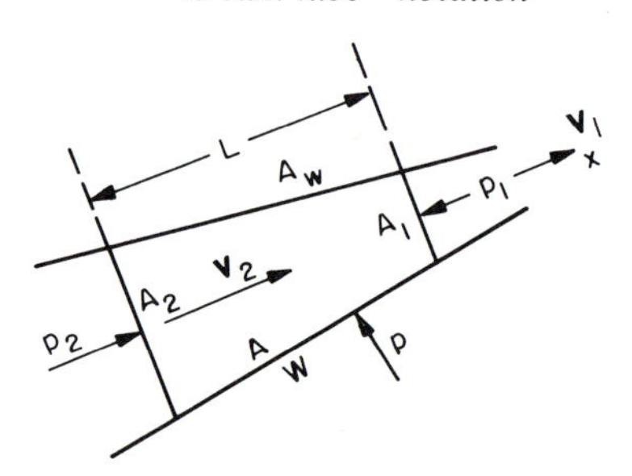

An averaging correction factor $(1 + \alpha/3)$ should be applied to the term $\rho Q\mathbf{V}$. This correction is neglected in the following (see Section 10-3.1).

The external forces are composed of pressure forces, gravity forces, and shearing forces. For reasons discussed below, the latter two forces are often neglected, leaving only the pressure forces to consider.

The pressure forces on sections A_1 and A_2 may be considered as consisting of two components

1. The sum of the forces caused by constant pressures p_2 and p_1, i.e., $p_2 A_2 - p_1 A_1$.
2. Hydrostatic forces applied to the center of gravity of the cross sections (Fig. 12-7). In the case of a free surface two-dimensional flow on a horizontal bottom of depth h, the external forces per unit width are:

$$\Sigma F_e = \rho g\left(\frac{h_2^2}{2} - \frac{h_1^2}{2}\right)$$

The pressure $p(x)$ along the walls limiting the stream tube could be given at any point by the Bernoulli equation or by assumptions based on physical observations. If all the other terms are known, this becomes the only unknown in the equation, and the momentum theorem provides a way of finding the value of its integral in the flow direction. [Note that the pressure force at the limits of the considered stream tube is expressed along the axis in the mean direction of flow only (see Fig. 12-6).]

Figure 12-6 *Pressure forces on the limit of a stream tube.*

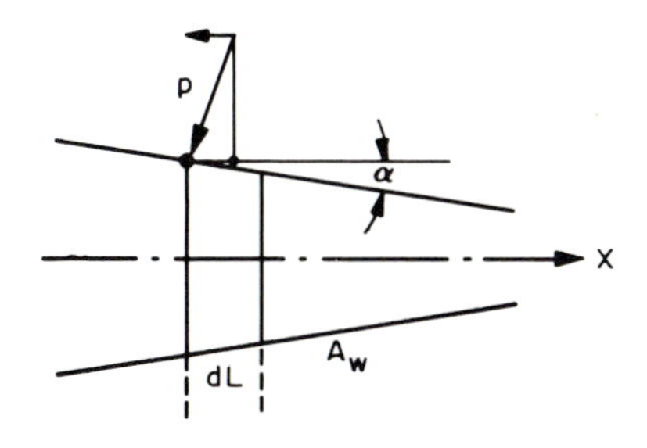

When the momentum theorem is applied along a horizontal axis, the gravity force is zero.

The shearing forces, acting at the limit of the considered volume of fluid, are often neglected in the case of short, rapidly varied flows, such as given by a sudden enlargement or a hydraulic jump (see Section 12-3.3.).

12-3.2 *Specific Force*

The specific force in a cross-section A is the sum of the momentum flux and the pressure force per unit of weight of fluid, i.e.

$$\left(\frac{V^2}{g} + \frac{p}{\rho g}\right)A$$

In order to illustrate previous considerations and to provide a guide for further applications, some examples are given with an emphasis on all the necessary assumptions not usually given in elementary textbooks on hydraulics.

12-3.3 *Hydraulic Jump on a Horizontal Bottom*

From observation, it is common knowledge that the flow pattern in a hydraulic jump is extremely complicated. However, consideration of the external forces only and change of momentum at the boundaries permits the study of this complex phenomenon without having to deal with the complicated fine structure of the flow.

First, the flow limits are chosen in a plane at which the flow pattern is well known, i.e., far enough from the front of the hydraulic jump for the mean flow to be parallel to the bottom (Fig. 12-7).

Figure 12-7 *Hydraulic jump—notation.*

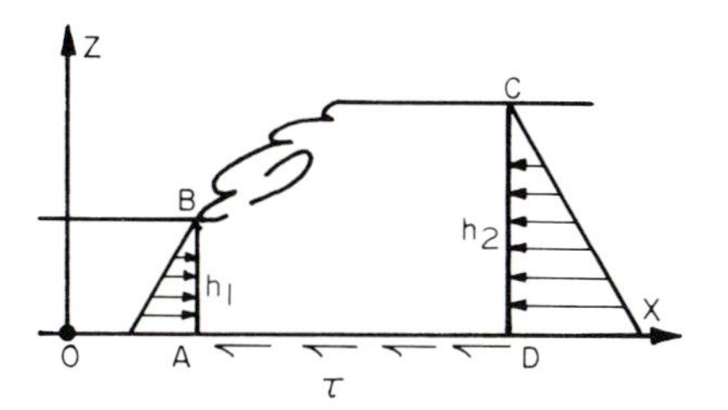

Second, two reference axes are chosen. One axis will obviously be chosen in the direction of flow OX.

The external forces to be considered in the OX direction are

1. The pressure forces at the boundaries, i.e., on the vertical planes AB and CD, having a total sum in the OX direction different from zero. The pressure distribution is hydrostatic.
2. The shearing stresses caused by friction on the boundaries, including the free surface, and on the planes BC and AD in a direction opposite to OX. In such a short structure, however, these shearing stresses are negligible.

Hence, the external forces acting in the OX direction are per unit width

$$\Sigma \mathbf{F}_e = \rho \mathbf{g}\left(\frac{h_1^2}{2} - \frac{h_2^2}{2}\right)$$

The difference in momentum with respect to time is

$$\frac{d(m\mathbf{V})}{dt} = \rho q \,\Delta \mathbf{V} = \rho Q(\mathbf{V}_2 - \mathbf{V}_1)$$

Equating the change of momentum with the external forces leads to

$$\rho q(\mathbf{V}_2 - \mathbf{V}_1) = \rho \mathbf{g}\left(\frac{h_1^2}{2} - \frac{h_2^2}{2}\right)$$

This finally becomes, after some elementary transformations

$$\frac{q^2}{g h_1 h_2} = \frac{h_1 + h_2}{2}$$

This is the hydraulic jump equation. Choosing a second axis in the vertical direction will give an equality between the atmospheric pressure force acting on the free surface, the gravity force, which is equal to the total weight of water in volume $ABCD$, and the external force acting vertically upward on the bottom of the hydraulic jump. There is no change of momentum in this direction.

12-3.4 *Hydraulic Jump in a Tunnel*

Consider the case of a partially open gate in a tunnel submitted to a high upstream pressure, as shown in Fig. 12-8. An air vent is often necessary to avoid cavitation effects. When conditions for a hydraulic jump are satisfied, the water flow acts as an ejector and a quantity of air is sucked into the tunnel. Because of the head loss in the air vent, the pressure at the free surface is smaller than atmospheric pressure. Therefore, the external forces must take into account this difference in pressure. The simplest solution is obtained using the absolute value $P = p + p_a$ of the pressure which gives

$$\Sigma F_e = \left(P_1 A_1 + \rho g \frac{A_1 h_1}{2}\right) - \left(P_2 A_2 + \rho g \frac{A_2 h_2}{2}\right)$$

P_2 is greater than the atmospheric pressure because of the head loss in the downstream part of the tunnel. The momentum flux in cross section (1) is $\rho Q^2/A_1$. In cross section (2) it is $\rho Q(Q + Q_a)/A_2$ because of air entrainment. Q_a is the discharge of air at pressure P_2; ρQ is the mass per unit time; and $(Q + Q_a)/A$ is the velocity. The mass of air is neglected, but the air discharge has an influence on the velocity of the water.

Figure 12-8 *Hydraulic jump in a tunnel.*

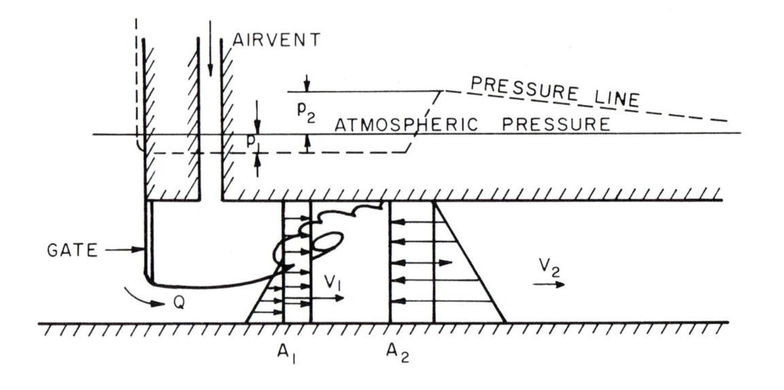

12-3.5 *Paradox of Bergeron*

Consider a tank on wheels as in Fig. 12-9. On one side the pressure distribution is hydrostatic, whereas on the other side the pressure head is transformed into velocity head. The difference of total forces acting on both sides of the tank could be obtained by calculating, successively, a flow net, the pressure distribution on the two sides, and the forces. However, the momentum theorem gives the total value of the force directly as

$$F = \rho Q V = \rho C_c A (2gh)^{1/2}$$

where C_c is the coefficient of contraction.

Because of this force and the fact that other forces (such as gravity and atmospheric pressure) have a horizontal component equal to zero, the tank has a tendency to move in the opposite direction of the jet. This is the principle of jet propulsion.

Now suppose that water is present outside the tank and that the tank is heavier than the buoyant force (Fig. 12-10). In this case, the tank does not move. The force caused by the jet is equal to the force caused by the very complicated motion inside the tank. This may be considered as another application of Newton's third law—action is equal and opposite to reaction—and the momentum theorem must be applied to the total mass of water. This is what is known as the Paradox of Bergeron.

Figure 12-9 *Jet reaction principle.*

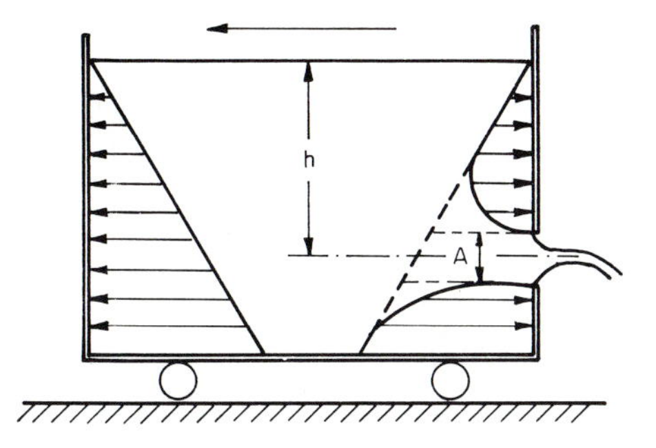

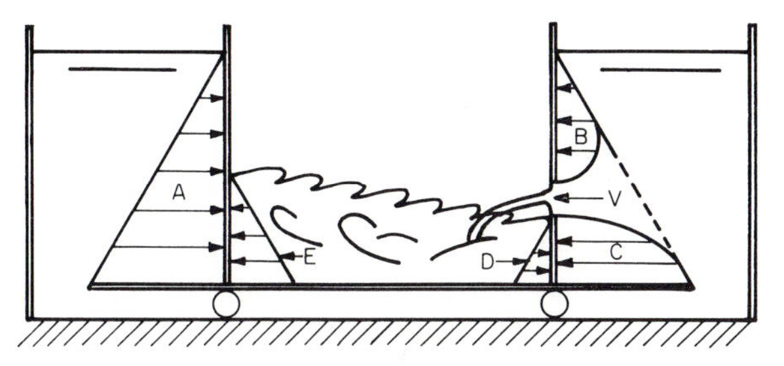

Figure 12-10 *Paradox of Bergeron:* $A - (B + C) = E - D$.

The same result is obtained in the case shown by Fig. 12-11. The jet acts on the wall of the downstream tank. The tank does not move. Finally, consider the two tanks as shown in Fig. 12-12 in which it is assumed that the holes have the same cross section. One of the holes is closed by means of a plane held in place by the jet from the left tank. The area of pressure $ABCD$ equals the area $A'B'C'B'$ or $ABE + FDC = E'B'C'F'$. Considering the forces on the plate we obtain $F_L = F_R$ when

$$\rho A C_d (2gz_1) = \rho g A z_2$$

i.e.

$$2C_d z_1 = z_2$$

With $C_d = 0.60$ for an orifice, it is possible for z_2 to be $1.2\ z_1$ by the simple insertion of a plate. Shaping the hole

Figure 12-11 *Paradox of Bergeron:* $A - (B + C) = B - E$.

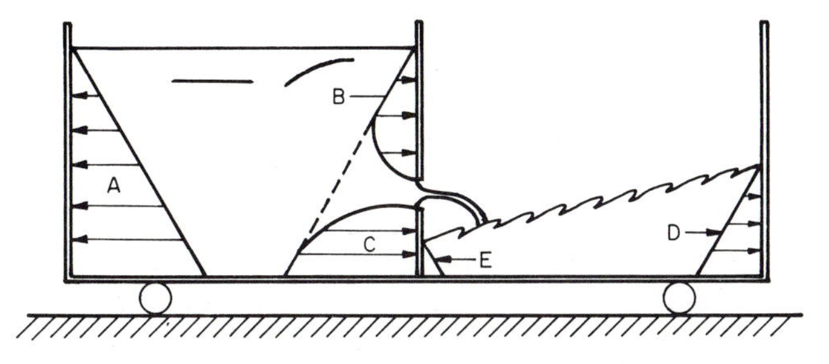

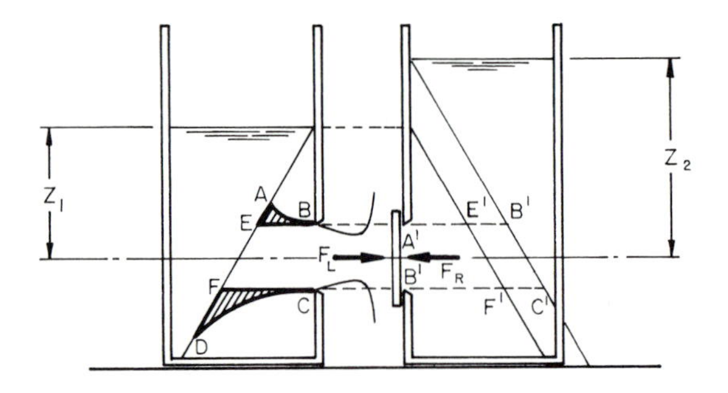

Figure 12-12 *Another paradox:* *$AEB + FCD = E'B'C'F'$.*

in the left tank so that $C_d \to 1.0$, z_2 can be made almost equal to $2z_1$. If the jet returns through 180°, z_2 could be equal to $4z_1$.

All of this may be physically explained by the consideration of external forces (AEB and DCF) transformed into momentum.

12-3.6 Jet and Intake

Let us consider the case of a curved pipe surrounded by water as shown in Fig. 12-13. When the water is expelled from the pipe, the total force exerted on the pipe is $\mathbf{F} = \rho Q \mathbf{V}$. The pipe tends to move in a direction opposite to that of the jet. For example, if a pool is filled with a flexible hose, the hose will continuously wiggle. When the water is now sucked by the pipe, the total force on the pipe is obtained by calculating the difference of pressure on both sides (inside and outside) of the bend. The projected cross section is A. The force is $F_e = p_e A$ outside. Inside the bend, the force projected on A is $F_i = p_i A + \rho Q V$. This is caused by the change of momentum flux direction in the bend. However, application of the Bernoulli equation gives

$$p_i = p_e - \rho \frac{V^2}{2} - \rho K \frac{V^2}{2}$$

where K is a head loss coefficient at the entrance approximately equal to unity, as seen in elementary hydraulics. Therefore

$$F_i = (p_e A - \rho V^2 A) + \rho Q V = p_e A = F_e$$

The total force on the pipe is zero. The pipe does not move, as can be verified by siphoning a pool with a hose. Therefore, one can now imagine a boat equipped with vertical pipes bent horizontally, as shown in Fig. 12-14. The pressure variation at the bottom end of the pipe that results from heaving will cause the water to move alternately up and down. (This phenomenon can be enhanced by resonance.) Even though the previous theory applied to steady flow and local inertia effects are neglected, it is seen that when the water goes up inside the tube, the boat does not move, whereas when the water is expelled, the boat will tend to move forward. The boat is moved by wave energy.

Figure 12-13 *Jet reaction compared to intake.*

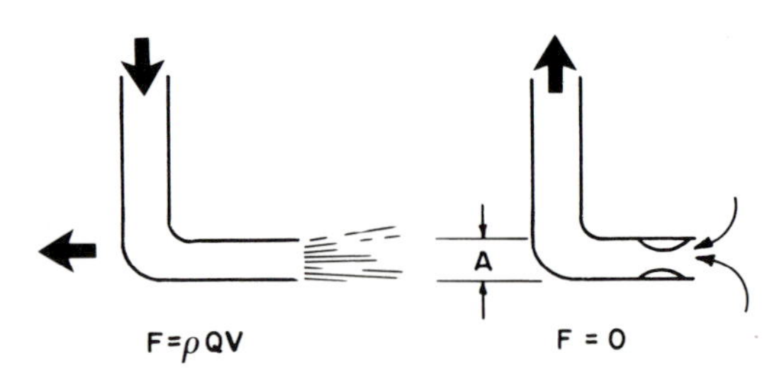

Figure 12-14 *When the water is expelled, a reaction is exerted.*

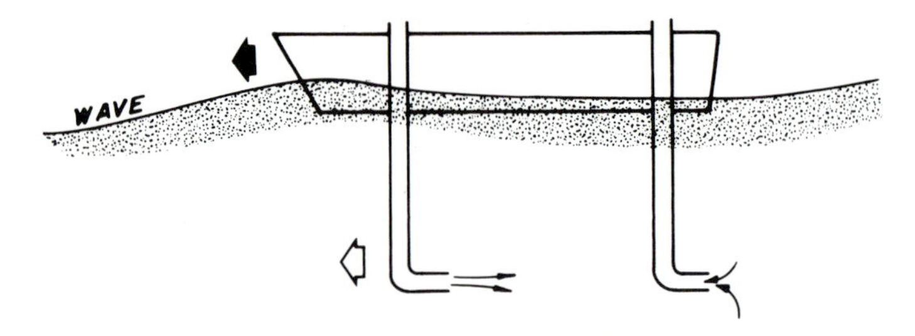

12-4 Difficulties in the Application of the Momentum Theorem

12-4.1 Difficulties in the Case of Unsteady Flow

In the case of steady flow, the integral with respect to volume $D(\iiint_D (\partial \rho \mathbf{V}/\partial t)\, dD = 0)$ and the momentum theorem becomes simply

$$\Sigma \mathbf{F}_e = \iint_A \rho \mathbf{V} V_n \, dA$$

As for external forces, only the boundary conditions for A appear in this equation because the momentum flux is given by a surface integral. Hence its application does not require a knowledge of the fine structure of flow within the domain D but only on the surface A.

In the case of unsteady flow, the volume integral $\iiint_D (\partial \rho \mathbf{V}/\partial t)\, dD$ is different from zero and requires a knowledge of the flow patterns within D as a function of time and space. Thus, the momentum theorem is difficult to apply to unsteady flow and as such, is less frequently used except when $\mathbf{V}(t)$ can easily be determined.

In the case of steady motion, the momentum theorem permits the analysis of complex motions. This leads one to think about the precautions that have to be taken in the application of this theorem.

One difficulty in the application of the momentum theorem is in the choice of the boundary and the boundary conditions. It is impossible to calculate the external forces without having a rough idea of the flow pattern. This often requires experimentation or knowledge of similar previous experiments. Some examples to illustrate these considerations are given below.

12-4.2 Sudden Enlargement

The external forces involved are the pressure forces on parts (1) and (2) in Fig. 12-15. It is generally assumed that the pressure p^* exerted by wall S is the same as the pressure p_1^* at the end of the smaller pipe. If the flow is laminar, the streamlines have such a curvature that this assumption is incorrect (see Fig. 12-16). However, it is known by experiment that when the Reynolds number is greater than a critical value, the flow enters the wider pipe in the form of a jet. This jet, often unstable, generates by friction some secondary currents in the corners.

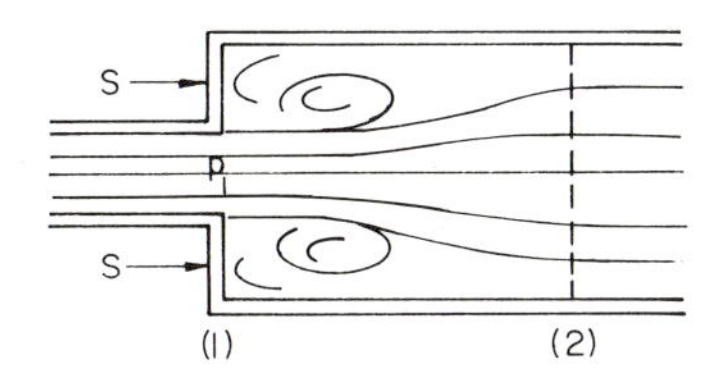

Figure 12-15 *Sudden enlargement.*

If the velocity is small enough for the convective inertia to be negligible, it is true that the pressure distribution at the cross section (1) is hydrostatic and this usual assumption is valid. In fact, the eddies caused by friction induce a centrifugal force that causes the pressure force to be slightly greater than the force calculated with the previous assumption. However, the assumption is quite valid for practical purposes.

12-4.3 Hydraulic Jump Caused by a Sudden Deepening

This example illustrates a case where it is impossible to calculate external force without experiments.

Suppose that a channel has a sudden deepening in order to fix the position of a hydraulic jump (Fig. 12-17). The external force exerted by the vertical wall changes as shown in the diagram, depending upon the exact location of the

Figure 12-16 *Sudden enlargement. Laminar flow.*

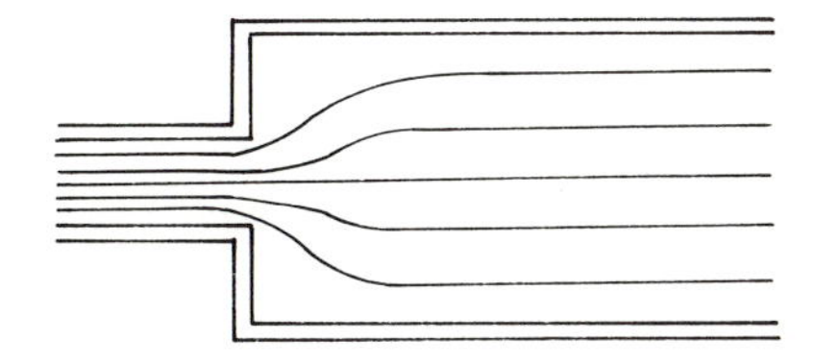

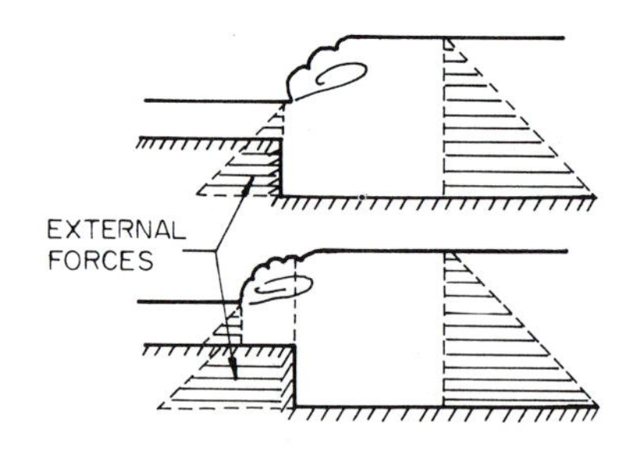

Figure 12-17 *Hydraulic jump on a sudden deepening.*

jump. Only systematic experimentation gives the factors that influence this phenomenon, and the corresponding theory is dictated by the results of this experiment.

12-4.4 *Hydraulic Jump on a Slope*

The study of a hydraulic jump on a slope involves a body force—the force of gravity—which definitely influences the flow conditions (Fig. 12-18). The gravity force to be considered is the component of the total gravity force in the direction of the main flow due to the weight of water included between the two usual limits where the flow is parallel to the bottom. Hence this force $\rho g A l \sin \alpha$ is a function of the length of the hydraulic jump l. This length could be roughly estimated by experimentation, but it is evident that it cannot be determined with great accuracy. However, the length of the hydraulic jump must satisfy the equation

$$\rho Q(V_2 - V_1) = \rho g\left(\frac{h_1^2}{2} - \frac{h_2^2}{2}\right) + \rho g A l \sin \alpha$$

Figure 12-18 *Hydraulic jump on a slope (l = length of jump).*

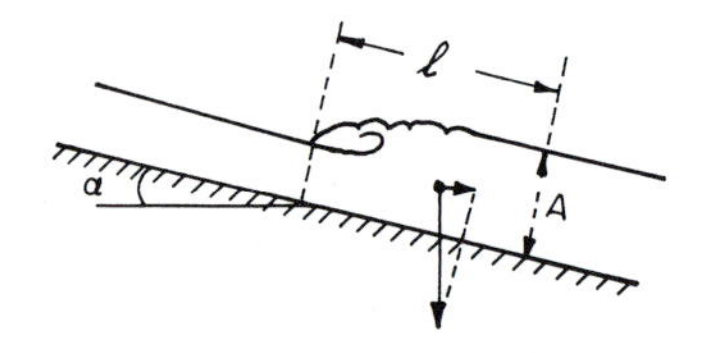

12-4.5 *Intake*

Consider a free-surface flow as shown in Fig. 12-19. The rise of the water level in part $ABCD$ may theoretically be calculated by applying the momentum equation in the OX direction to the mass of water enclosed in $EFBC$. Thus, if one assumes that the flow is perpendicular to the limit GD:

$$\rho Q V = \rho g\left[\frac{(d + \Delta d)^2}{2} - \frac{d^2}{2}\right]$$

which gives $\Delta d = V^2/g$ and not $V^2/2g$, as may be expected following a superficial analysis.

However, the practical result is often closer to $V^2/2g$ than V^2/g. This is not because the momentum theory is wrong, but because the boundary conditions are wrong. The velocity at section GD is not perpendicular to the cross section and the momentum theorem should be applied to the mass $BCDJHGEF$ in order to include the difference in external forces applied at DJ and HG. Unfortunately, these external forces cannot be estimated by theory.

12-4.6 *Unsteady Flow—Translatory Wave*

As an example of unsteady flow where the momentum theorem may be used, the case of a translatory wave is analyzed (Fig. 12-20). The wave is traveling at a velocity **W**.

Figure 12-19 *External forces due to the intake cannot be determined.*

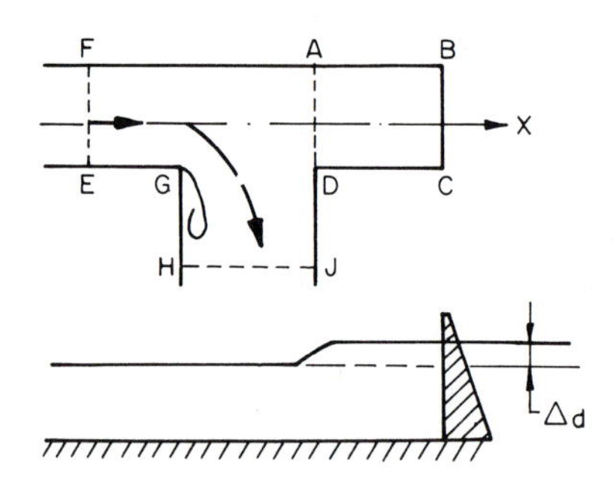

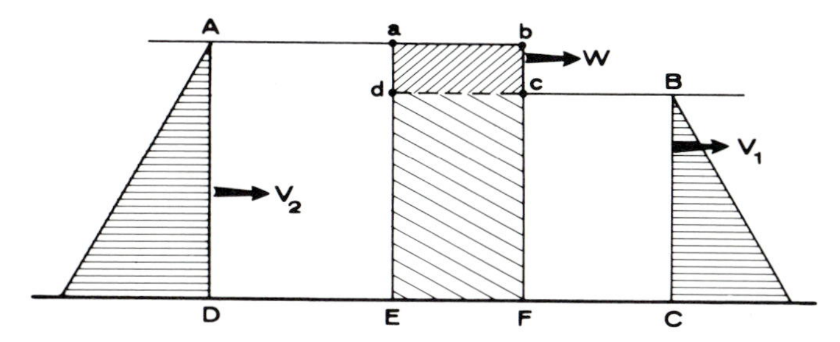

Figure 12-20 *Translatory wave.*

It is assumed that the velocities $\mathbf{V}_1$ in front of the wave and $\mathbf{V}_2$ behind it are constant along a vertical plane. Referring to Section 12-2, applying the momentum theorem to the fluid within the *ABCD* control sections, and letting h be the height represented by line segment *BC* and $h + \Delta h$ the height represented by *AD*, one finds successively: *External forces*:

$$\Sigma F_e = \rho g[\tfrac{1}{2}(h + \Delta h)^2 - \tfrac{1}{2}h^2]$$

Change of momentum with respect to space or variation of momentum flux:

$$\rho \iint_A V V_n \, dA = \rho V_2^2(h + \Delta h) - \rho V_1^2 h$$

Change of momentum with respect to time: The momentum of volume *AaED* and *BcFC* remains unchanged. The change of momentum $\iiint_D [\partial(\rho \mathbf{V})/\partial t] \, dD$ contains two terms: one due to the variation of mass, the other due to the variation of velocity. The variation of mass is due to the addition of fluid *abcd*, which is $\rho \Delta h W \, dt$, or $\rho W \Delta h$ per unit of time. This mass takes a velocity V_2. Therefore, the corresponding variation of momentum is:

$$V \frac{dM}{dt} = \rho \Delta h W V_2$$

The quantity of fluid within *dcFE* is $Wh \, dt$ and this fluid is subjected to a change of velocity from V_1 to V_2. Therefore, the variation of momentum of *dcFE* is:

$$M \frac{dV}{dt} = \rho h W(V_2 - V_1)$$

Finally, the momentum balance is

$$\tfrac{1}{2}\rho g[(h + \Delta h)^2 - h^2] = \rho V_2^2(h + \Delta h) - \rho V_1^2 h + \rho \Delta h W V_2 + \rho h W(V_2 - V_1)$$

which, when combined with the continuity equation

$$V_2(h + \Delta h) - V_1 h = W \Delta h$$

yields

$$W = V_1 + (gh)^{1/2}\left[\frac{1}{2}\frac{h + \Delta h}{h}\left(\frac{h + \Delta h}{h} + 1\right)\right]^{1/2}$$

12-5 Momentum Versus Energy

In this section, the field of application of the momentum theorem is analyzed and compared with the field of application of the principle of energy conservation.

As has been seen in Section 12-1.1, the main difference in application of these two methods lies in knowing the connection of the internal forces with the phenomena to be studied.

The momentum theorem is used to study an overall effect, whatever the complexity of the flow. The principle of conservation of energy is used to study phenomena linked to the internal motion and to the very fine structure of the flow pattern.

It is easy to conceive that the second method will be more quickly limited in its scope when hydraulic problems are analyzed. It has been seen that a number of assumptions, such as irrotationality, are necessary before this second method can be used. These limitations are linked with the difficulties of integration of the basic Navier–Stokes equations.

A number of examples will now be used to illustrate the above points.

12-5.1 Irrotational Flow without Friction

The total thrust of a jet on a fixed or a movable plane, the force on the bucket of a spillway, or the total horizontal force exerted on a partially open gate in a tunnel, etc., may be calculated by the momentum theorem (Fig. 12-21).

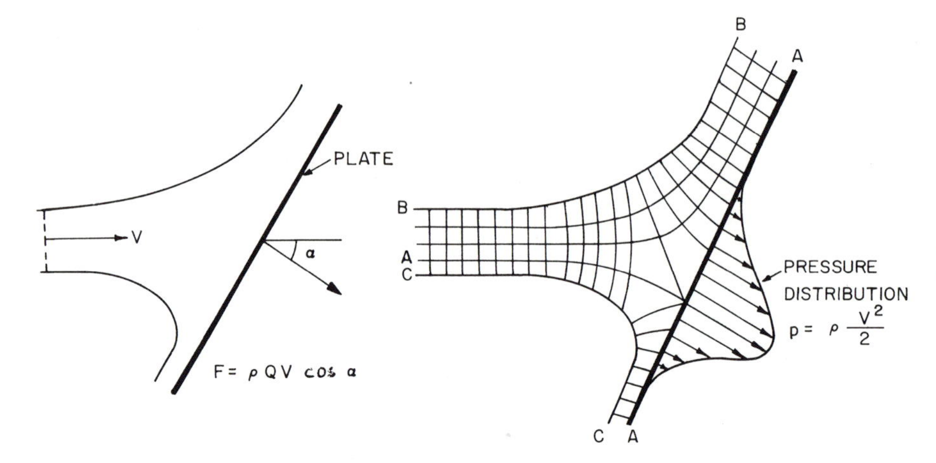

Figure 12-21

Total thrust is given by the momentum theorem. The pressure distribution is given by the Bernoulli equation and by a flow net.

In contrast, the precise pressure distribution caused by the above conditions of flow may be calculated by the following method:

The flow is assumed to be irrotational and flow net may be drawn, which gives the velocity distribution. The pressure distribution is given by the application of the Bernoulli equation expressing the conservation of energy. It is evident that the total thrust may also be deduced by this process of calculation, by an integration of pressure forces $\int p \, dA$, and this could be compared to the result given by the momentum theorem.

However, the result given by the momentum theorem, which is obtained without any assumptions, is more exact provided that the boundary conditions are well known.

12-5.2 Unidimensional Rotational Flow

The momentum theorem may be used to analyze a number of phenomena, such as the head loss at a sudden enlargement or a hydraulic jump, for example, whatever the complexity of the flow.

Combining the force–momentum equation with the energy–work equation gives the value of the head loss by calculating the difference in total heads: $(V^2/2g) + (p/\varpi) + z$. Both the force–momentum equation and the energy–work equation are valid to study a diverging flow where the head loss may be neglected, but there cannot be any application of the energy–work equation to a sudden enlargement without the introduction of another term expressing the head loss, despite the identity of equations in their differential form.

12-5.3 Mechanics of Manifold Flow

The mechanics of manifold flow is of particular interest as an illustration of the previous considerations on the fields of application of the Bernoulli and momentum equations. Consider a flow as shown by Fig. 12-22. If the motion is two-dimensional, a first method of analysis assumes the motion to be irrotational. In such a manner it is possible to calculate the flow pattern by conformal mapping, although this method is far from being valid because of the friction forces.

It is also possible to apply the energy equation to the two stream tubes which separate into the main conduit flow and the flow in the lateral, if a term is included to express energy losses for both stream tubes. Unfortunately, it is impossible by theory to establish the value of this term. One can assume that head loss for the lateral pipe stream tube is that of a bend, whereas for the main flow it is that of

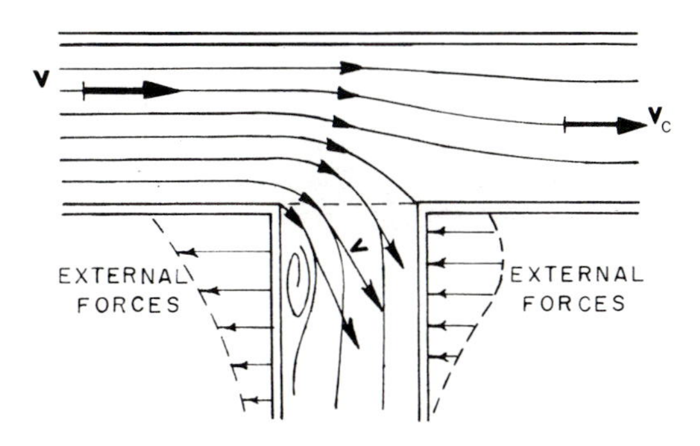

Figure 12-22 *Manifold flow.*

a sudden enlargement $(V - V_c)^2/2g$. But systematic experimental results do not verify this assumption.

A similar simplified approach consists of writing the momentum equation for the flow at the junction, provided that a term is included for the momentum of the flow at an angle in the lateral or the corresponding unbalanced external force component acting on the walls of the lateral.

Lack of knowledge of one significant term, makes direct application of these methods impossible without recourse to experiment.

12-5.4 Conclusion

For any problem dealing with steady flow motion for which only the overall effect is of interest, the momentum theorem can be applied because of its great simplicity. This is because the sum of the internal forces is considered to be zero. However, when using the momentum theorem, one must be very careful in estimating the external forces and boundary conditions. Often an experiment may be necessary to establish these unknowns.

When more details about the flow characteristics are required, the system of differential equations giving the fine structure of the flow directly must be solved completely. But the validity of this solution is quickly limited because of the number of assumptions which must be introduced in order to simplify the system of equations to be solved.

PROBLEMS

12.1 Derive the momentum equation by integrating the Eulerian equation to a finite volume. Determine the correcting terms due to turbulence by integrating the Reynolds equation to a finite volume.

12.2 Consider the flow through a pipe of radius R_0 ended by two circular disks of radius R and separated by a small distance h, as shown in Fig. 12-23. Calculate the total force exerted by the flow on the lower disk by assuming that the flow between the two disks is radial and that the total discharge is Q. Take $R_0 = h$ first, and then discuss the case when $R_0 \neq h$.

12.3 Consider a two-dimensional flow such as shown on Fig. 12-24. Draw the corresponding flow net and determine the pressure distribution from A to B (assuming no separation at C and D), and calculate the total force on AB by integrating the pressure distribution as a function of the fluid discharge Q. Calculate the same force by application of the momentum theorem. Explain the discrepancy between these two results.

Figure 12-23

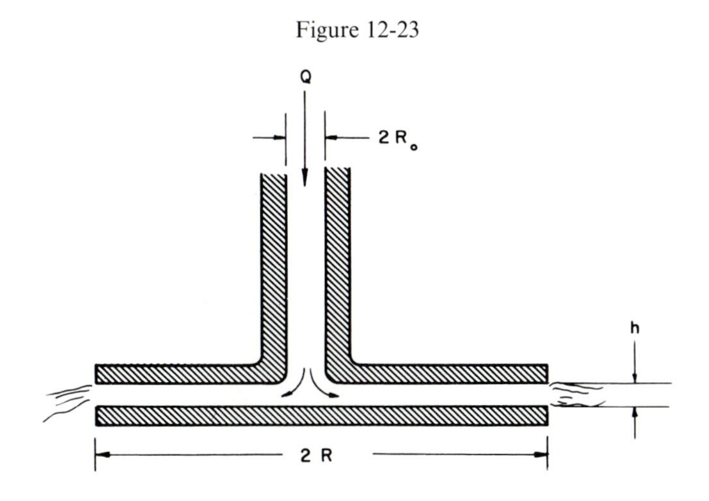

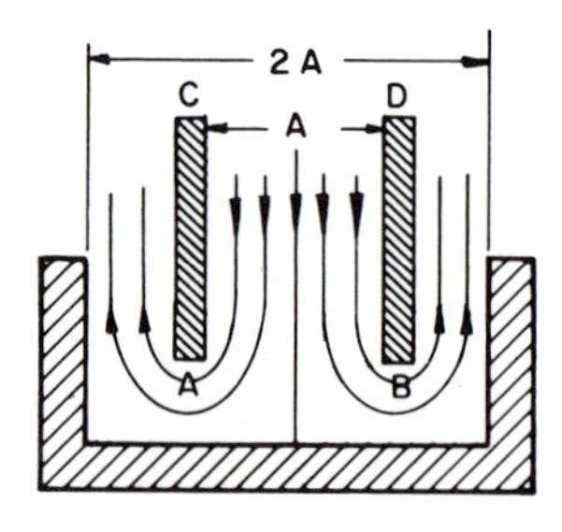

Figure 12-24

12.4 Demonstrate the following relationship for a hydraulic jump in a rectangular horizontal channel

$$\frac{y_2}{y_1} = \tfrac{1}{2}((1 + 8F_1^2)^{1/2} - 1)$$

where y_1 and y_2 are the upstream and downstream water depths, respectively, $F_1 = V_1/(gy_1)^{1/2}$ is the Froude number of the upstream flow, where V_1 is the average flow velocity.

12.5 Consider the flow as shown in Fig. 12-25

1. Draw two flow nets at two different scales to analyze the pressure distribution at the entrance of the gallery and against the gate.
2. Calculate the integral of the horizontal and vertical components of the pressure forces acting against the gate. Compare the result of this total horizontal sum with that obtained by applying the momentum theorem.
3. Is there any risk of cavitation? [Pressure head $\gtrsim -32$ ft (10 m).]
4. Give the values of u and v and $\partial u/\partial x$, $\partial u/\partial y$, $\partial v/\partial x$, $\partial v/\partial y$ along OY for $x = 0$ at the entrance of the gallery from A to B.

12.6 Consider the three following two-dimensional flows as illustrated by Fig. 12-26. The first flow (1) is a sudden enlargement under high pressure; the second flow (2) is a gently diverging flow; and the third (3) is a hydraulic

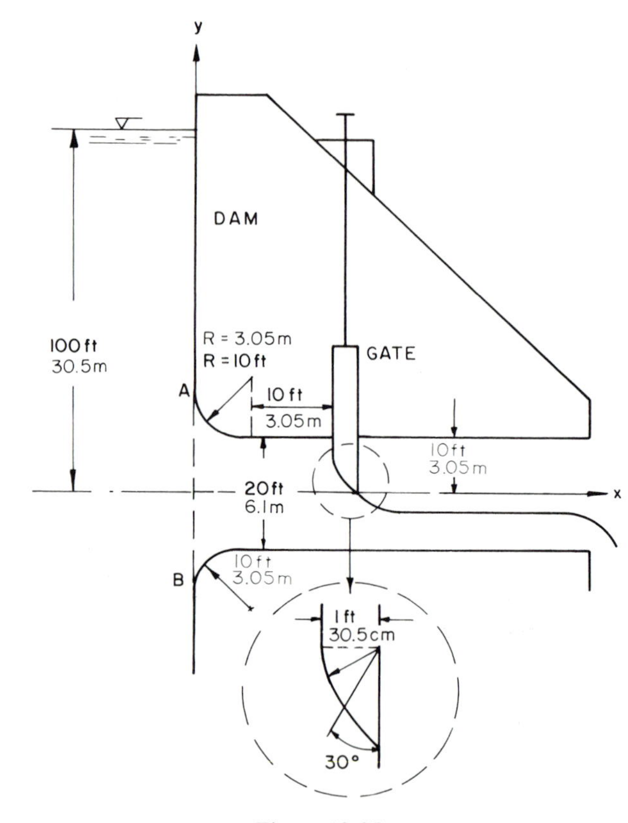

Figure 12-25

jump. In these three cases it is assumed that the two end water depths h_1 and h_2 are each constant and that the fluid discharge per unit of width q is the same.

By application of the momentum theorem and the Bernoulli equation to these three cases, determine the value of the external forces and the head losses (neglect the shearing stress at the wall).

12.7 Find the value of the contraction coefficient in the case of the circular orifice (called the Borda mouthpiece) as shown in Fig. 12-27. The contraction coefficient C_c is defined by the ratio of the smallest cross section of the jet to the cross section of the orifice.

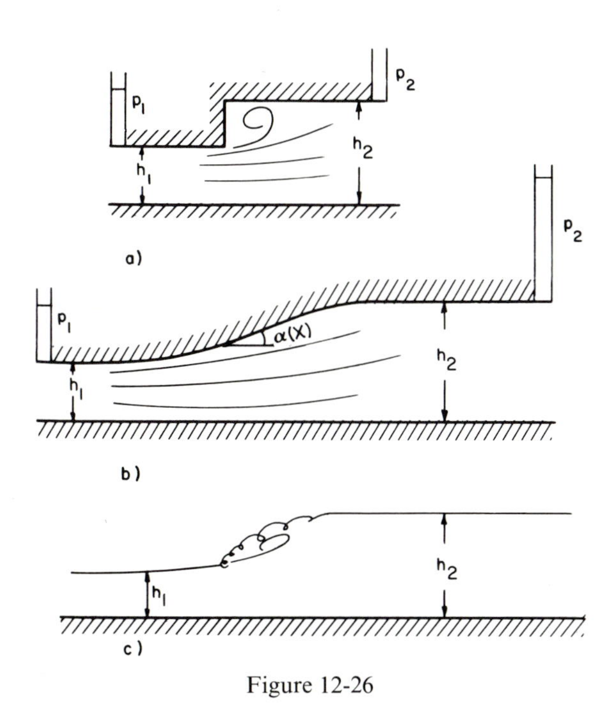

Figure 12-26

12.8 Consider a weir such as shown in Fig. 12-28. Determine the expression for z as a function of the head above the weir edge H. It will be assumed that the weir is aerated, i.e., the atmospheric pressure is applied on the free surface. The discharge per unit length is $q = 0.5H(2gH)^{1/2}$ and V will be taken equal to $0.1(2gh)^{1/2}$. What error is made when we neglect angle α of the falling water with the vertical?

12.9 A jet hits a plane perpendicularly. The discharge of the jet is $Q = 2\ \text{ft}^3/\text{sec}$ ($56630\ \text{cm}^3/\text{sec}$) and the particle velocity is $V = 20$ ft/sec (609 cm/sec). The plane is moving at a velocity U ($U < V$) in the direction of the jet. Calculate (1) the total force exerted by the jet on the plane as a function of the velocity U; (2) the power of the jet in horsepower; (3) the power transmitted by the jet to the plane as a function of the velocity U; (4) the efficiency defined as a ratio of these two powers when U varies from O to V.

Do the same calculations assuming that the plane is replaced by a bucket as shown on Fig. 12-29.

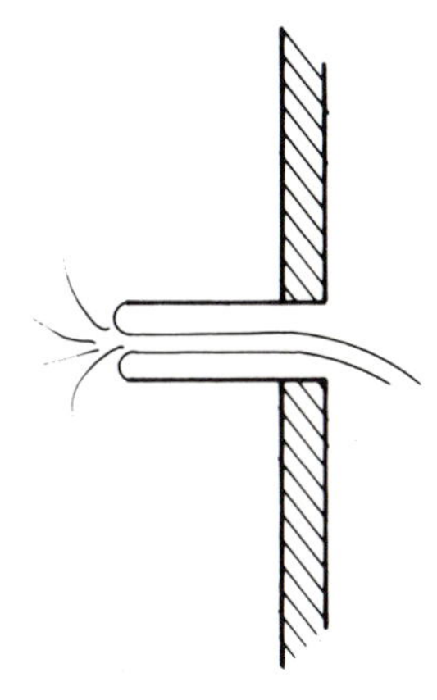

Figure 12-27

12.10 Consider the case of a hydraulic jump created by an abrupt drop h of the bottom of a channel. Demonstrate the two following relationships between the upstream water depth y_1 and the downstream water depth y_2.

$$\frac{V_1^2}{gy_1} = \frac{1}{2}\frac{y_2/y_1}{1 - y_2/y_1}\left[1 - \left(\frac{y_2}{y_1} - \frac{h}{y_1}\right)^2\right]$$

or

$$\frac{V_1^2}{gy_1} = \frac{1}{2}\frac{y_2/y_1}{1 - y_2/y_1}\left[\left(\frac{h}{y_1} + 1\right)^2 - \left(\frac{y_2}{y_1}\right)^2\right]$$

depending upon the assumption for the value of the pressure distribution on the vertical wall forming the abrupt drop.

Figure 12-28

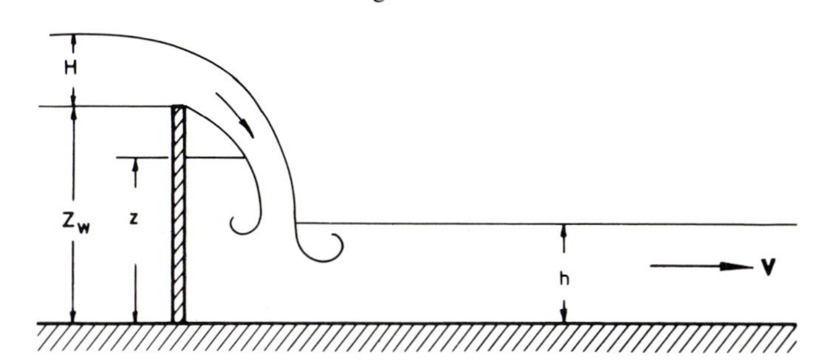

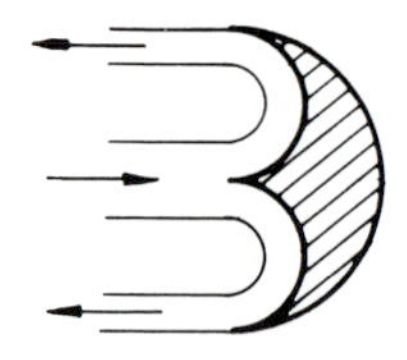

Figure 12-29

12.11 Consider a horizontal convergent between two cross sections $A_1 = 2$ ft^2 (1858 cm^2) and $A_3 = 1$ ft^2 (929 cm^2). At section A_1, the pressure $p_1 = 12$ psi (0.84 kg/cm^2) and $V_1 = 6$ ft/sec (182.8 cm/sec). The shearing force exerted by water is $\tau = \rho f V^2$, where $f = 0.05$ and V is the average velocity as a function of the area of the cross section. Determine the head loss and the total force exerted by the convergent on its anchor as a function of the length of the convergent.

Determine the total force exerted on the anchor as a function of the length of convergent in the case where the convergent is bent by 90°.

Now, neglecting friction force and taking a length of convergent of 10 ft (304.8 cm) determine the total force on the anchor in the straight and curved convergent in the case in which V_1 is time dependent such that V_1 ft/sec $= 6 \sin (2\pi/T)t$ and $T = 20$ sec [or V cm/sec $= 182.8 \sin (2\pi/T)t$].

12.12 Consider the spillway defined by Fig. 12-30. The coefficient of discharge C, defined by $A = Ch(2gh)^{1/2}$, is a function of h/h_n such as

h/h_n	0.2	0.4	0.8	1
C	0.394	0.425	0.470	0.490
h/h_n	1.2	1.4	1.6	2
C	0.504	0.518	0.532	0.552

It will be assumed that $h_n = 8$ ft (243 cm) and $h = 12$ ft (365 cm).

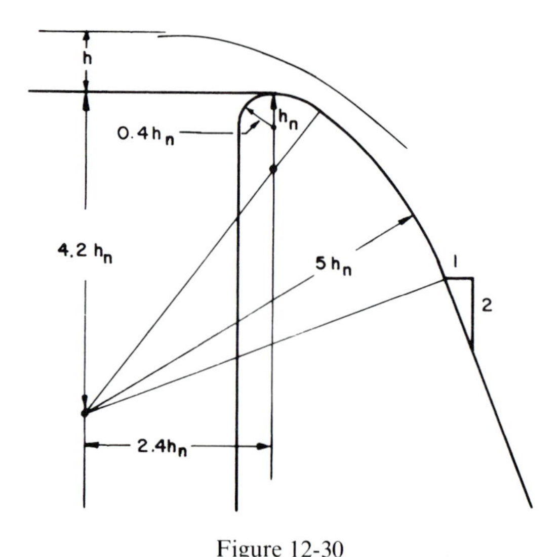

Figure 12-30

1. Calculate the discharge per linear foot of spillway (or meter of spillway).
2. Draw the flow net by successive approximation.
3. Determine the pressure distribution from the velocity field and establish if there is any risk of cavitation.
4. This spillway is ended by a skijump (bucket) of 70 ft (21.34 m). radius as shown on Fig. 12-31. Calculate the pressure distribution on the bucket (without drawing any flow net) and integrate it in order to obtain the total force on the bucket.
5. Determine the total force on the bucket by application of the momentum theorem. Compare the results of 4 and 5.
6. Calculate the distance D between the foot of the dam and the location of impact of the jet.

12.13 Establish, by choosing a number of simplifying assumptions and by making use of the momentum theorem and the Bernoulli equations, the set of equations giving the distribution of discharges through manifolds with two holes, three holes, four holes, . . . , n holes of same cross section and subjected to the same pressure. The head

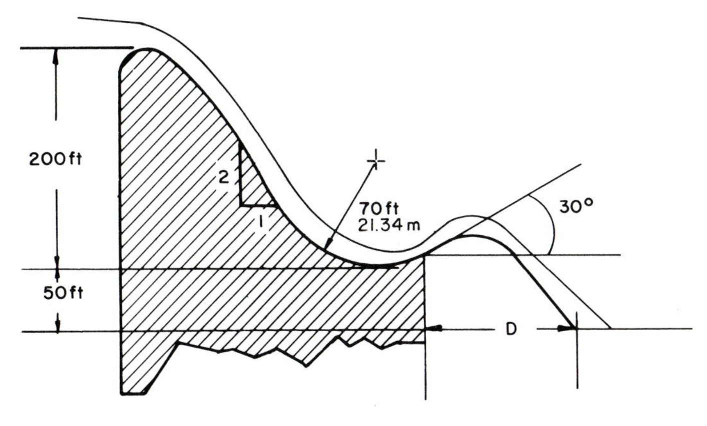

Figure 12-31

losses through the normal section of the main pipe will be neglected.

12.14 A boat is equipped with a windmill. The axis of the windmill is parallel to the axis of the boat. This windmill generates the power used by the propeller. Can the boat go upwind? If it can, establish the condition for this to be done. Assume that efficiency of the system windmill-propeller is unity, and that the drag of the boat is proportional to the square of its velocity. Can the boat go faster than the wind when she travels in the wind direction?

Chapter 13

The Boundary Layer, Flow in Pipes, Drag, and Added Mass

13-1 General Concept of Boundary Layer

13-1.1 Definition

13-1.1.1 As a viscous flow passes a solid boundary such as a flatplate or a streamlined body, the influence of viscosity on the flow field is usually confined within a thin layer near the boundary. Outside this layer, the effect of the viscosity is vanishingly small, the fluid behaves like a perfect fluid. This physical picture suggests that the entire flow field can be divided into two domains, each of which can be treated separately for the purpose of simplifying the mathematical analysis, as shown in Fig. 13-1.

The first domain is called the boundary layer, which is a thin layer right in the neighborhood of the boundary. In this domain, the flow velocity is zero at the wall and increases rapidly to the velocity corresponding to the free stream velocity. In the boundary layer, the velocity distribution is strongly influenced by the friction forces.

In the second domain, the influence of viscosity is small. The friction forces can be neglected in comparison to the inertia forces. Hence, the viscous terms in the Navier–Stokes equations may be neglected. The fluid can be assumed to be nonviscous and can be considered as irrotational.

The pressure in the boundary layer, as will be shown later, is approximately equal to the pressure at the limit of the free stream.

Figure 13-1 *Two flow domains.*

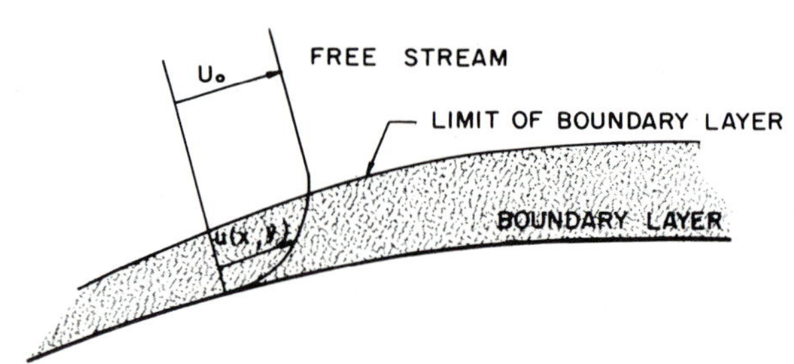

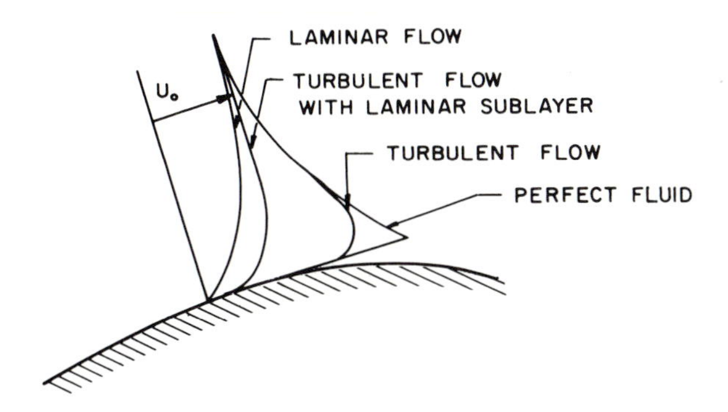

Figure 13-2 *Relative velocity distributions as a function of the flow regime.*

13-1.1.2 The larger the value of the Reynolds number, the thinner the boundary layer.

Consequently, at very high Reynolds numbers, the average flow motion with respect to time is very close to that of a perfect fluid. This point has been explained qualitatively in Section 8-1.1, and is further illustrated in Fig. 13-2.

13-1.2 Thickness of Boundary Layer

13-1.2.1 The definition of thickness of the boundary layer is to a certain extent arbitrary because the transition of velocity from zero to the ambient velocity takes place asymptotically. Because the velocity increases very rapidly from the wall to the free system velocity, it is possible to specify the thickness of the boundary layer beyond which the effects of the wall friction are rather small. For example, the distance from the wall at which the velocity differs from the free stream velocity by 1% is often used. Other quantities used to describe the extent of the boundary layer are the displacement thickness δ^*, and the momentum thickness θ. The significance of these quantities is defined in the following.

13-1.2.2 Because of the existence of the friction forces, a certain amount of fluid is slowed down within the boundary layer. The corresponding decrease in flow rate is

$$\int_{y=0}^{\infty} (U_0 - u)\, dy$$

The displacement thickness δ^* is the value by which the wall will have to be shifted in order to give the same total discharge if the fluid were frictionless. Consequently, δ^* is defined by the equality, as shown in Fig. 13-3,

$$\delta^* U_0 = \int_{y=0}^{\infty} (U_0 - u)\, dy$$

i.e.,

$$\delta^* = \frac{1}{U_0} \int_{y=0}^{\infty} (U_0 - u)\, dy \simeq \frac{1}{U_0} \int_{0}^{\delta} (U_0 - u)\, dy$$

13-1.2.3 The momentum flux is reduced in the boundary layer. As a measure of reduction of momentum flux, the momentum thickness is then defined in a manner similar to that for the displacement thickness. The reduction of the momentum flux in the boundary layer is

$$\int_{y=0}^{\infty} \rho u (U_0 - u)\, dy$$

The momentum thickness θ is then defined by

$$\rho U_0^2 \theta = \rho \int_{0}^{\infty} u (U_0 - u)\, dy$$

i.e.,

$$\theta = \frac{1}{U_0^2} \int_{y=0}^{\infty} u (U_0 - u)\, dy \cong \frac{1}{U_0^2} \int_{0}^{\delta} u (U_0 - u)\, dy$$

Figure 13-3 *Displacement thickness (area ABC equals area CDE).*

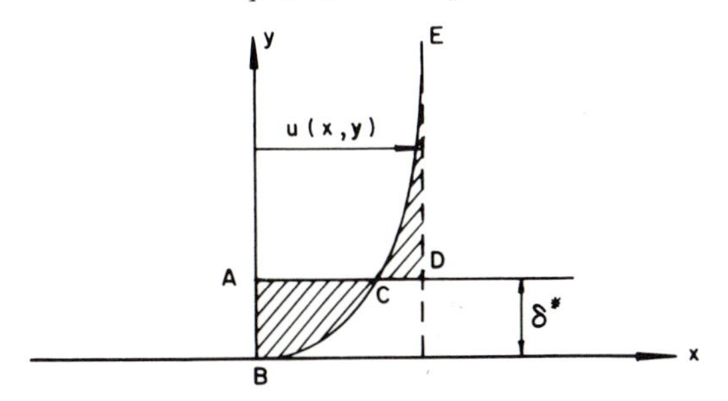

13-2 Laminar Boundary Layer

13-2.1 Steady Uniform Flow over a Flat Plate

13-2.1.1 The Navier–Stokes equations in the case of a two-dimensional steady motion are

$$u\frac{\partial u}{\partial x} + v\frac{\partial u}{\partial y} = -\frac{1}{\rho}\frac{\partial p^*}{\partial x} + \nu\left(\frac{\partial^2 u}{\partial x^2} + \frac{\partial^2 u}{\partial y^2}\right)$$

$$u\frac{\partial v}{\partial x} + v\frac{\partial v}{\partial y} = -\frac{1}{\rho}\frac{\partial p^*}{\partial y} + \nu\left(\frac{\partial^2 v}{\partial x^2} + \frac{\partial v}{\partial y^2}\right)$$

and the continuity equation is

$$\frac{\partial u}{\partial x} + \frac{\partial v}{\partial y} = 0$$

As mentioned in the previous section, the velocity varies rapidly along the y axis. It is zero at the wall and reaches the free stream velocity at a distance of the order of the thickness of the boundary layer δ. In contrast, the velocity varies very slowly along the plate (see Fig. 13-4). Therefore, all derivatives in the y direction must be much larger than the derivatives in the x direction (see Sections 4-5.2.4 and 5-4.2). Consequently, in the first equation

$$\nu\frac{\partial^2 u}{\partial x^2} \ll \nu\frac{\partial^2 u}{\partial y^2}$$

and the term $\nu(\partial^2 u/\partial x^2)$ may be neglected. Furthermore, the velocity v across the boundary layer is of much smaller value than the velocity u along the boundary layer. As a result, the terms that contain the velocity v in the second equation are of much smaller value than are the terms in the first equation, and they may therefore be neglected. Finally, the second equation becomes

$$\frac{\partial p^*}{\partial y} = 0.$$

This states that the pressure is hydrostatic along a perpendicular to the plate and that p^* depends only on x, and can be determined from the nature of the flow in the free stream. Because p^* is a function of x only, one has the equality

$$\frac{\partial p^*}{\partial x} = \frac{dp^*}{dx}$$

and the equations of motion become

$$u\frac{\partial u}{\partial x} + v\frac{\partial u}{\partial y} = -\frac{dp^*}{dx} + \nu\frac{\partial^2 u}{\partial y^2}$$

which is often called the boundary-layer equation. In addition to the above equation, the boundary conditions

$$u = v = 0 \quad \text{at} \quad y = 0$$

$$u = U_0 \quad \text{when} \quad y \to \infty$$

and the continuity equation determine the flow field near the flat plate.

In the particular case in which U_0 is constant, i.e., in the case of a steady uniform flow over a flat plate with zero

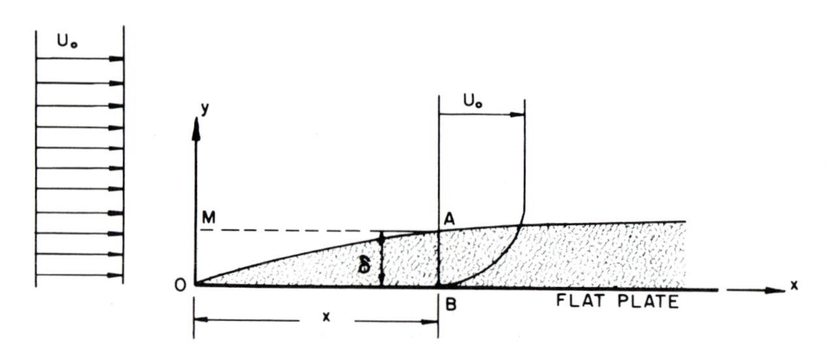

Figure 13-4

Schematic drawing of the boundary layer.

incidences, as a consequence of the Bernoulli equation applied in the second domain, one has $(\partial p^*/\partial x) \equiv 0$.

13-2.1.2 As shown in Fig. 13-4, the momentum flux per unit width through OM is equal to $\rho U_0^2 \delta$. The momentum flux through AB is smaller. However, because $u < U_0$ in the boundary layer, the definition of θ shows that $\theta \lesssim \delta$. Hence, the difference of momentum is also of order $\rho U_0^2 \delta$. In contrast, the total friction force per unit width between between the sections OM and AB is

$$\int_0^B \mu \frac{\partial u}{\partial y} dx.$$

By definition of boundary-layer thickness, $\partial u/\partial y \sim U_0/\delta$.

Since the difference of momentum flux between sections OM and AB is caused by the friction force along OB, one may equate the difference of momentum flux with the friction force. Consequently, one has

$$\int \mu \frac{U_0}{\delta} dx \sim \rho U_0^2 \delta$$

or

$$\delta \sim \left(\frac{\mu x}{\rho U_0}\right)^{1/2} = \left(\frac{\nu x}{U_0}\right)^{1/2}$$

This equation is valid only in the case of uniform pressure distribution along the plate. If it is assumed that the velocity profiles at all distances x from the leading edge are similar, the velocity profiles $u(y)$ for varying distances x can be made identical by choosing the proper scale factors for $u(y)$ and y. The free stream velocity U_0 and the boundary layer thickness δ are these scale factors. Hence, one has the similarity relationship

$$\frac{u}{U_0} = \Phi(\eta)$$

defining $\Phi(\eta)$, where $\eta = y/\delta$, i.e.,

$$\eta = y\left(\frac{U_0}{\nu x}\right)^{1/2}$$

13-2.1.3 Let us now consider the case of steady flow over a flat plate where $\partial p^*/\partial x = 0$, and let us introduce the stream function $\psi(x,y)$ such that $u = \partial\psi/\partial y$, $v = -(\partial\psi/\partial x)$ whereby the boundary-layer equation becomes

$$\frac{\partial\psi}{\partial y}\frac{\partial^2\psi}{\partial x\partial y} - \frac{\partial\psi}{\partial x}\frac{\partial^2\psi}{\partial y^2} = \nu\frac{\partial^3\psi}{\partial y^3}$$

which is a third-order nonlinear differential equation. From the above definition it is seem that ψ is equal to $\int_0^y u\, dy$. If $u = U_0\Phi(\eta)$, and $y = \eta(\nu x/U_0)^{1/2}$ are substituted in this equality the result is

$$\psi = f(\eta)(U_0 \nu x)^{1/2}$$

Note that y is considered as a function of η.

The boundary-layer equation presented above can now be transformed into an ordinary differential equation as follows:

$$u = \frac{\partial\psi}{\partial y} = \frac{\partial\psi}{\partial\eta}\frac{\partial\eta}{\partial y} = f'(\eta)\frac{1}{\delta}(U_0\nu x)^{1/2} = U_0 f'(\eta)$$

$$v = -\frac{\partial\psi}{\partial x} = \frac{1}{2}\left(\frac{\nu U_0}{x}\right)^{1/2}(\eta f' - f)$$

$$\frac{\partial u}{\partial y} = \frac{\partial^2\psi}{\partial y^2} = U_0\left(\frac{U_0}{\nu x}\right)^{1/2} f''(\eta)$$

$$\frac{\partial u}{\partial x} = \frac{\partial^2\psi}{\partial x\partial y} = -\frac{U_0\eta}{2x}f''(\eta)$$

$$\frac{\partial^3\psi}{\partial y^3} = U_0\left(\frac{U_0}{\nu x}\right)f'''(\eta)$$

Substituting these into the boundary-layer equation, one obtains the Blasius equation $2f''' + ff'' = 0$, and the boundary conditions $f = f' = 0$ at $\eta = 0$, and $f' = 1$ at $\eta = \infty$.

13-2.1.4 The general solution of the Blasius equation cannot be given in a closed form. However, the solution can be obtained through power-series expansion. The power-series expansion near $\eta = 0$ is assumed to be of the form of

$$f(\eta) = A_0 + A_1\eta + A_2\frac{\eta^2}{2!} + A_3\frac{\eta^3}{3!} + \cdots$$

where A_n are constants. From the boundary conditions $f = f' = 0$ at $\eta = 0$, one obtains $A_0 = A_1 = 0$.

Substituting the power series with $A_0 = A_1 = 0$ into the Blasius equation, one obtains

$$2A_3 + 2A_4\eta + (A_2^2 + 2A_5)\frac{\eta^2}{2!} + (4A_2A_3 + 2A_6)\frac{\eta^3}{3!} + \cdots = 0$$

This must be equal to zero for any value of η, which can be verified only if all the coefficients of each term are equal to zero. Consequently, one has

$$A_3 = A_4 = A_6 = A_7 = A_9 = \cdots = 0$$
$$A_5 = -\tfrac{1}{2}A_2^2;\ A_8 = -\tfrac{11}{2}A_2A_5 = \tfrac{11}{4}A_2^3;\ \cdots$$

Hence, all the coefficients can be expressed as a function of A_2. The constant A_2 can be determined from the second boundary condition: $y \to \infty$, $u = U_0$; or $\eta \to \infty$, $f' = 1$. Once A_2 is obtained, $f(\eta)$ can be calculated. The solution $f(\eta)$ together with $f'(\eta)$ and $f''(\eta)$ are plotted in Fig. 13-5. This gives the Blasius solution of the laminar boundary-layer equations.

13-2.1.5 The boundary-layer thickness for the steady uniform flow over a flat plate as defined in Section 13-1.2.1 (that is, the distance from the wall at which $u = 0.99\ U_0$) can be obtained from Fig. 13-5, where $\eta \approx 5.0$, i.e.,

$$\eta = \frac{\delta}{(\nu x/U_0)^{1/2}} \approx 5.0$$

Hence, the boundary-layer thickness δ becomes

$$\delta \approx 5.0\frac{x}{(xU_0/\nu)^{1/2}}$$

13-2.1.6 From the numerical calculation in Section 13-2.1.4, or the graph presented in Fig. 13-4, one obtains $f''(0) = 0.332$. Therefore, the shear stress at the wall is

$$\tau_0(x) = \mu\left(\frac{\partial u}{\partial y}\right)_{y=0} = \mu U_0\left(\frac{U_0}{\nu x}\right)^{1/2} f''(0) = \frac{0.332}{(R_x)^{1/2}}\rho U_0^2$$

where $R_x = U_0 x/\nu$ is the Reynolds number based on the distance x from the leading edge of the plate.

The resistance force on one side of the plate over length l per unit width is equal to

$$D = \int_0^l \tau_0\, dx = \frac{0.664\ \rho U_0^2 l}{(U_0 l/\nu)^{1/2}}$$

and the resistance coefficient C_f is

$$C_f = \frac{D}{\frac{1}{2}\rho l U_0^2} = \frac{1.328}{(R_l)^{1/2}}$$

where $R_l = U_0 l/\nu$ is the Reynolds number based on the length of the plate.

Figure 13-5 *Solution of the Blasius equation.*

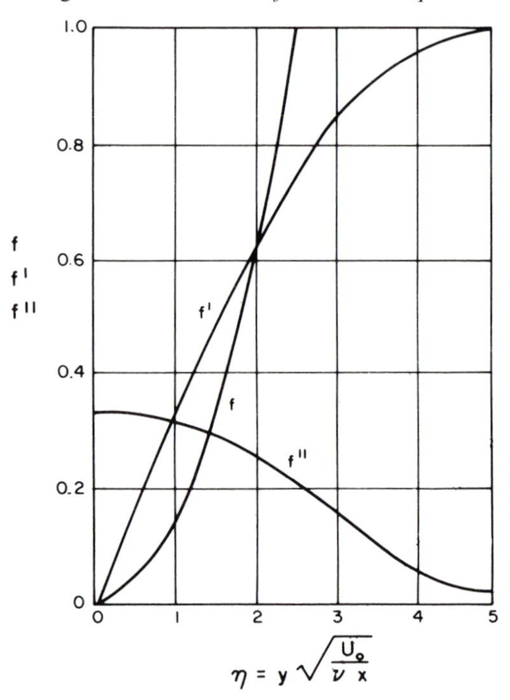

13-2.2 *Momentum Integral Equation for Boundary Layer*

13-2.2.1 As demonstrated in the previous simple example of a laminar boundary layer on a semiinfinite flat plate at zero incidence, the calculations are cumbersome and time consuming. It is desirable for practical use to find some approximate method to evaluate the necessary quantities. In particular, the case in which $(\partial p^*/\partial x)$ can no longer be considered as zero, such as in flow past a wedge, can be analyzed by the momentum integral method developed by von Kármán.

Consider an element $ABCD$ of a two-dimensional steady flow as shown in Fig. 13-6. The momentum integral method consists of applying the momentum theorem to this element, i.e., equating the variation of momentum flux among the boundaries AD, BC, and CD, to the applied forces. The applied forces consist of the pressure force acting on the boundaries and the shear stress on the wall. Each of these will be considered separately. The equality will give the momentum integral equation. Of course, in the case in which the flow is unsteady, the additional term resulting from unsteady flow also has to be considered in the equation.

13-2.2.2 The discharge through AD is

$$\int_0^\delta u\,dy\bigg|_{x=x_1}$$

and the discharge through CB is

$$\int_0^\delta u\,dy\bigg|_{x=x_1+dx} = \int_0^\delta u\,dy\bigg|_{x=x_1} + \frac{d}{dx}\left(\int_0^\delta u\,dy\right)dx$$

The net outflow over the vertical control surface is equal to the difference of these discharges:

$$\frac{d}{dx}\left(\int_0^\delta u\,dy\right)dx$$

This amount of flow must be supplied through the top boundary for the sake of continuity.

13-2.2.3 Similarly the net x momentum outflow through a vertical control surface is equal to

$$\rho\frac{d}{dx}\left(\int_0^\delta u^2\,dy\right)dx$$

On the other hand, because the velocity at the limit of the boundary layer is U_0, the x momentum inflow over the top is equal to the mass flow ρU_0 times the fluid discharge

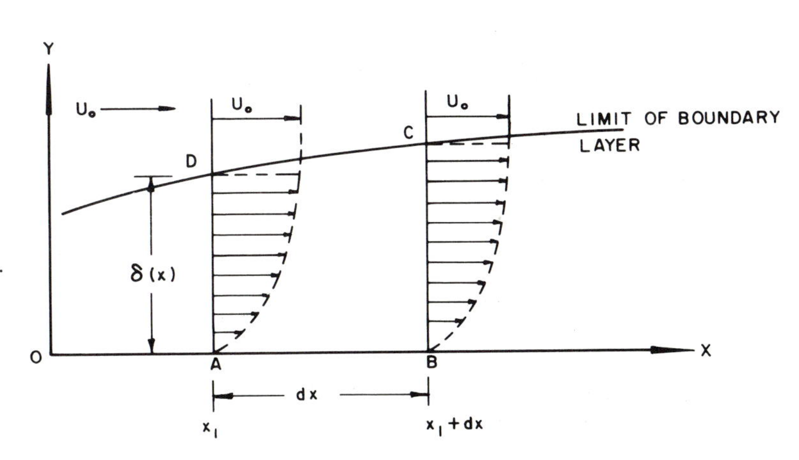

Figure 13-6
Diagram for the momentum integral method.

through the control surface. This fluid discharge was shown in the previous section to be

$$\frac{d}{dx}\left(\int_0^\delta u\,dy\right)dx$$

Therefore, the x momentum inflow over the top boundary DC is

$$\rho U_0\frac{d}{dx}\left(\int_0^\delta u\,dy\right)dx$$

and the total variation of x momentum flux through $ABCD$ is

$$\rho U_0\frac{d}{dx}\left(\int_0^\delta u\,dy\right)dx-\rho\frac{d}{dx}\left(\int_0^\delta u^2\,dy\right)dx$$

13-2.2.4 The pressure forces acting on the limit of the element $ABCD$ in the OX direction are now considered.

The pressure forces on AD are $p\delta$, and on CB they are

$$\left(p+\frac{dp}{dx}dx\right)\left(\delta+\frac{d\delta}{dx}dx\right)$$

Since the variation of boundary-layer thickness with distance is small, $(d\delta/dx)\,dx$ may be neglected, but $(dp/dx)\,dx$ may not be small.

The pressure force on DC acting in the OX direction is also neglected because $d\delta/dx$ is small.

Finally, the net pressure force remains: $-\delta(dp/dx)\,dx$, This last term may also be expressed in terms of the velocity U_0 as follows:

From the Bernoulli equation applied to the irrotational flow outside the boundary layer, one has

$$\frac{dp}{dx}=-\rho U_0\frac{dU_0}{dx}$$

Hence, the net pressure force is

$$-\delta\frac{dp}{dx}dx=\rho U_0\delta\frac{dU_0}{dx}dx$$

13-2.2.5 Now that all the terms have been established, it is possible to write the momentum integral equation for the element $ABCD$. One obtains the momentum integral by equating the momentum flux (in the x direction) and the net pressure force to the shear force on the boundary, that is,

$$\tau_0\,dx=U_0\frac{d}{dx}\left(\int_0^\delta \rho u\,dy\right)dx-\frac{d}{dx}\left(\int_0^\delta \rho u^2\,dy\right)dx$$
$$+\rho U_0\delta\frac{dU_0}{dx}dx$$

By dividing the above equation by dx and rearranging the terms, one obtains

$$\tau_0=\frac{d}{dx}\left[\int_0^\delta \rho u(U_0-u)\,dy\right]+\frac{dU_0}{dx}\int_0^\delta \rho(U_0-u)\,dy$$

Introducing the displacement thickness, δ^* and the momentum thickness θ, the momentum integral can be written

$$\frac{\tau_0}{\rho}=\frac{d}{dx}(U_0^2\theta)+\delta^*U_0\frac{dU_0}{dx}$$

Because no assumption is being made as to the nature of the flow, this method is applicable to laminar as well as turbulent flows. However, in turbulent flow, the velocity should be considered to be the mean value with respect to time.

13-2.2.6 The momentum integral method is now applied to the simple case, already investigated, of a steady uniform laminar flow over a flat plate.

In this case, U_0 is constant. Therefore, the momentum integral becomes

$$\frac{\tau_0}{\rho U_0^2}=\frac{d\theta}{dx}$$

First, one assumes a velocity profile such as

$$u(y)=a_0+a_1y+a_2y^2$$

where a_0, a_1, a_2 are constants that can be determined from boundary conditions. For example

$$u = 0 \quad \text{at} \quad y = 0 \quad \text{implies} \quad a_0 = 0$$

$$u = U_0 \quad \text{at} \quad y = \delta \quad \text{implies} \quad a_1\delta + a_2\delta^2 = U_0$$

$$\frac{du}{dy} \cong 0 \quad \text{at} \quad y = \delta \quad \text{implies} \quad a_1 + 2a_2\delta = 0$$

Therefore

$$a_1 = 2\frac{U_0}{\delta}$$

$$a_2 = -\frac{U_0}{\delta^2}$$

and the velocity profile is

$$\frac{u}{U_0} = 2\left(\frac{y}{\delta}\right) - \left(\frac{y}{\delta}\right)^2$$

One can find the momentum thickness θ to be

$$\theta = \int_0^\delta \frac{u}{U_0}\left(1 - \frac{u}{U_0}\right) dy = \tfrac{2}{15}\delta$$

and

$$\frac{\tau_0}{\rho} = \nu \frac{\partial u}{\partial y}\bigg|_{y=0} = \frac{2\nu U_0}{\delta}$$

Substituting these equalities into momentum integral equation, one has

$$\frac{15\nu}{U_0} = \delta\frac{d\delta}{dx}$$

Integrating and using the boundary condition $x = 0$, $\delta = 0$, one obtains

$$\delta = 5.5\left(\frac{\nu x}{U_0}\right)^{1/2} = \frac{5.5x}{(\mathrm{R}_x)^{1/2}}$$

and the normalized shear stress is equal to

$$\frac{\tau_0}{\rho U_0^2} = \frac{\nu}{U_0^2}\frac{\partial u}{\partial y}\bigg|_{y=0} = \frac{\nu}{U_0^2}\cdot\frac{2U_0}{\delta} = \frac{0.366}{(\mathrm{R}_x)^{1/2}}$$

Both the boundary-layer thickness and normalized shear stress are close to the exact value obtained in Sections 13-2.1.5 and 13-2.1.6.

13-2.2.7 For flows with pressure gradient, Pohlhausen suggested using a fourth-order polynomial

$$\frac{u}{U} = a_0 + a_1\eta + a_2\eta^2 + a_3\eta^3 + a_4\eta^4$$

and adding to the boundary conditions used above, the conditions

$$\nu\frac{\partial^2 u}{\partial y^2} = \frac{1}{\rho}\frac{dp}{dx} = -U_0\frac{dU_0}{dx} \quad \text{at} \quad y = 0$$

and

$$\frac{\partial^2 u}{\partial y^2} = 0 \quad \text{at} \quad y = \delta$$

13-2.3 Uniform Unsteady Flow over an Infinite Flat Plate

13-2.3.1 Because the plate is of infinite length, the derivatives with respect to x should be zero. That is, $\partial u/\partial x = 0$. From the continuity equation

$$\frac{\partial u}{\partial x} + \frac{\partial v}{\partial y} = 0$$

it follows that $\partial v/\partial y = 0$. Hence, v is identical to zero because it is zero at the boundary. Furthermore, the pressure p^* is constant everywhere, because of the infinite fluid field. Finally, the Navier–Stokes equation becomes

$$\frac{\partial u}{\partial t} = \nu\frac{\partial^2 u}{\partial y^2}$$

This equation is linear. A number of exact solutions can be found as functions of the boundary conditions. If the fluid is moving at a speed $U_0(t)$ and the plate is fixed, the boundary conditions are

$$u = 0 \quad \text{at} \quad y = 0$$

$$u = U_0(t) \quad \text{when} \quad y \to \infty$$

If the fluid at infinity is fixed and the plate is moving at velocity $U_0(t)$, the boundary conditions are

$$u = U_0(t) \quad \text{at} \quad y = 0$$
$$u = 0 \quad \text{when} \quad y \to \infty$$

13-2.3.2 The case of an impulsive motion of an infinite flat plate is given as an example. In that case, $u = 0$ for all y when $t \leq 0$, and $u = U_0$ at $y = 0$ when $t > 0$, and $u = 0$ when $y \to \infty$.

The partial differential equation can be transformed into an ordinary differential equation by introducing the dimensionless variable

$$\eta = \frac{y}{(2\nu t)^{1/2}}$$

After performing the required differentiations and substituting into the equation, one obtains

$$\frac{d^2u}{d\eta^2} + \eta \frac{du}{d\eta} = 0$$

Integrating with respect to η, one gets

$$u(y,t) = C_1 \int_0^{y/(2\nu t)^{1/2}} e^{-\eta^2/2}\, d\eta + C_2$$

where C_2 can be determined by the boundary condition $u = U_0$, for $\eta = 0$. This gives $C_2 = U_0$. The constant C_1 is determined from the initial condition $u = 0$ at $t = 0$ ($\eta = y/(2\nu t)^{1/2} = \infty$). Substituting into the above equation, one has

$$C_1 \int_0^{\infty} e^{-\eta^2/2}\, d\eta + U_0 = 0$$

The above integral has a value of $(\pi/2)^{1/2}$; hence $C_1 = -(2/\pi)^{1/2} U_0$.

Substituting C_1, C_2 into the equation, one obtains the velocity distribution

$$u = U_0 \left[1 - (2/\pi)^{1/2} \int_0^{y/(2\nu t)^{1/2}} e^{-\eta^2}\, d\eta \right]$$

or

$$u = U_0 \operatorname{erfc} \frac{y}{(2\nu t)^{1/2}}$$

The velocity distribution is presented in Fig. 13-7. It is clear that the velocity profile for different times are similar. They can be reduced to one curve by using the dimensionless variables u/U_0 and $\eta = y/(2\nu t)^{1/2}$

13-2.3.3 As an infinite flat plate oscillates parallel to itself, the governing equation of motion is the same as the impulsive motion of an infinite flat plate, that is,

$$\frac{\partial u}{\partial t} = \nu \frac{\partial^2 u}{\partial y^2}$$

The boundary conditions may be given by

$$u(0, t) = U_0 \cos kt$$
$$u(\infty, t) = 0$$

as the plate oscillates periodically.

Figure 13-7 *Velocity distribution near the infinite flat plate under impulsive motion.*

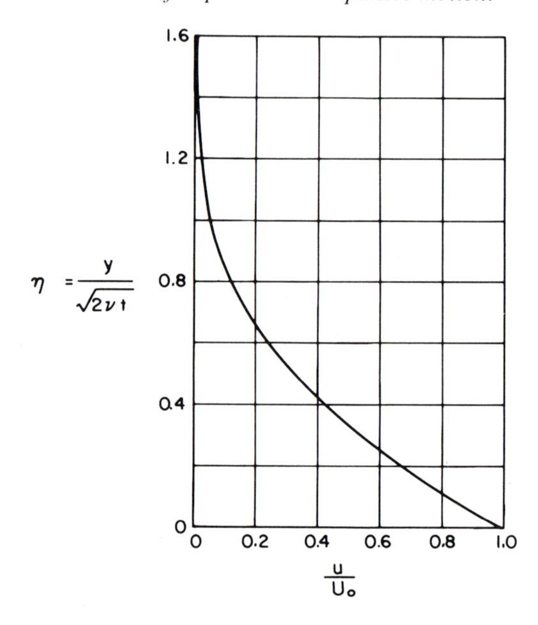

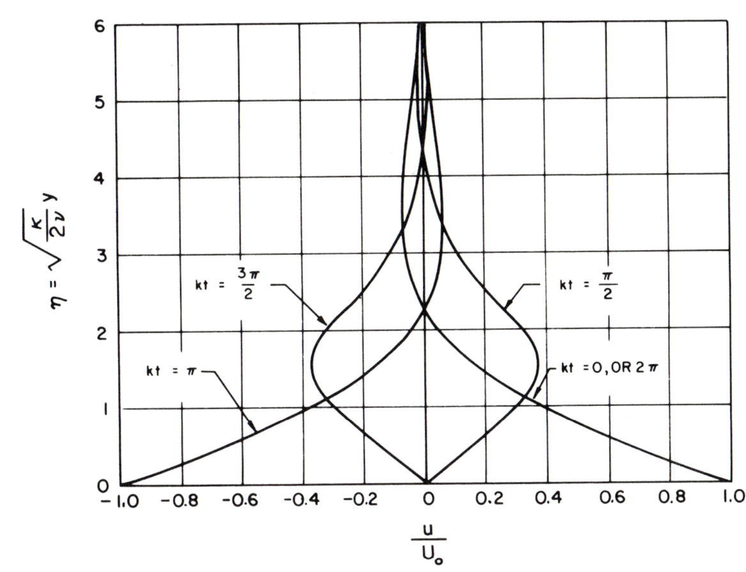

Figure 13-8

Velocity distribution near an oscillating infinite flat plate.

The solution of this equation is

$$u(y,t) = U_0 \exp\left[-\left(\frac{k}{2\nu}\right)^{1/2} y\right] \cos\left[kt - \left(\frac{k}{2\nu}\right)^{1/2} y\right]$$

The velocity profile, $u(y,t)$ has the form of damped harmonic oscillation, with an amplitude of $U_0 \exp[-(k/2\nu)^{1/2} y]$. The amplitude decreases exponentially from the plate. The velocity profiles for several instants of time are plotted in Fig. 13-8.

A similar solution applies in the case where the fluid is moving at a velocity $u = U_0 \cos kt$, and the plate is fixed. Such solution is of particular interest for studying the motion in the boundary layer of a periodic gravity wave and the wave damping by bottom friction.

13-3 Turbulent Boundary Layer

13-3.1 General Description

13-3.1.1 For a laminar boundary layer, it has been seen that its thickness increases with the distance x from the edge of the plate. As this boundary-layer thickness increases, the flow has a tendency to become turbulent. The criteria of the transition from laminar to turbulent are usually based on the Reynolds number $U_0 x/\nu$, in which U_0 is the free-stream velocity, x is the distance from the edge of the plate, and ν is the kinematic viscosity.

The value of Reynolds number (or the location x) at which the boundary layer becomes turbulent depends somewhat on the turbulence level of the free stream; it ranges from 10^5 to 10^6. The shear stress acting on the boundary is much larger in the turbulent boundary layer than in the laminar boundary layer; therefore, the determination of the location of this transition is not only of theoretical interest but has practical uses.

After transition, the main part of the flow in the boundary layer is turbulent. However, immediately adjacent to the wall, the turbulent fluctuations are suppressed by the presence of the wall. The flow field in this region can be divided into three domains—the laminar sublayer, the turbulent boundary layer, and the free stream (see Fig. 13-9). If the boundary is rough, the laminar sublayer may be destroyed by the presence of the roughness elements. A detailed discussion of this is given in Section 13-4.3.

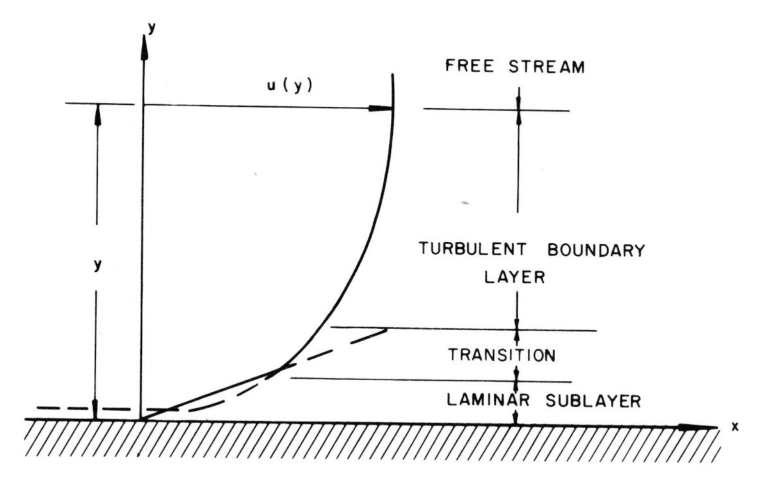

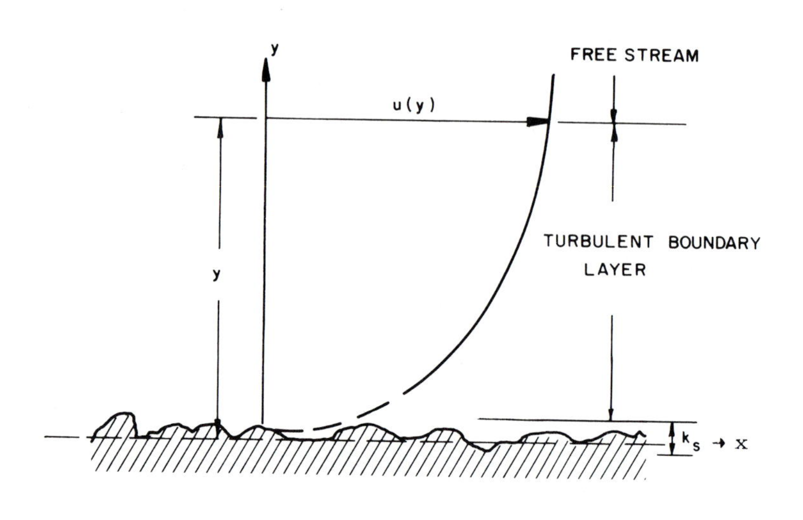

Figure 13-9

Turbulent velocity distributions near the wall.

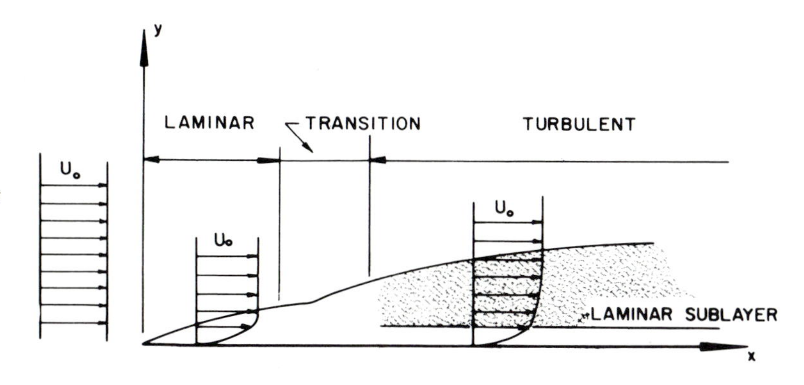

Figure 13-10

Schematic representation of the laminar and turbulent boundary layer.

13-3.1.2 The velocity distribution of a *laminar sublayer* is determined by the viscous force, that is, $\tau = \rho\nu(du/dy)$. Because this layer is very thin, it is reasonable to assume that τ is constant within this layer and equal to the shear stress at the wall τ_0. After integration, the equation becomes

$$u = \frac{\tau_0}{\rho}\frac{y}{\nu}$$

If, by definition,

$$u_* = \left(\frac{\tau_0}{\rho}\right)^{1/2}$$

then one finally obtains

$$\frac{u}{u_*} = \frac{u_* y}{\nu}$$

u_* is called the *shear velocity*.

The name laminar sublayer does not mean that the flow in this region is entirely laminar. Strong eddies generated in the turbulent flow often break through this thin layer and form turbulent spots in the sublayer. Therefore, to avoid confusion, the name *viscous sublayer* is sometimes used.

13-3.1.3 In the case of a *turbulent boundary layer*, the effect of turbulent fluctuation creates a large turbulent shear stress, whereas the effect of viscous shear is very small. Therefore, the velocity distribution is determined by the effect of turbulent shear stress which results in a logarithmic velocity distribution, as discussed in Section 8-3.5.

13-3.1.4 The effect of the boundary shear stress in the case of *free stream flow* is small. Therefore, the flow field can be determined by considering that the flow is non-viscous.

However, there are no sharp boundaries between each region, and the concept of each domain is to some extent qualitative. A schematic drawing of the three domains of flow and the flow pattern before the formation of the turbulent boundary layer is given in Fig. 13-10.

13-3.2 Resistance and Boundary-Layer Growth on a Flat Plate

Owing to the complicated flow conditions in the turbulent boundary layer, the exact solution of the equation of motion is not possible. One mathematical method available at present consists of determining the characteristics of the turbulent boundary layers by application of the momentum integral method described in Section 13-2.2.

The momentum integral method is used in the turbulent boundary layer to evaluate the variations with distance of the thickness of this boundary layer and the boundary shear stress. The use of this integral method involves the assumptions of a velocity profile at one location and similar profiles along the boundary. In the turbulent boundary layer, it is rather difficult to assume a velocity profile with sufficient accuracy because of the complicated flow field. Therefore, some experimental results have to be used as a base in order to get a good prediction of the variations of thickness of the boundary layer and the resistance with distance. Based on the experimental results the velocity profile can be represented by

$$\frac{u}{u_*} = 8.74\left(\frac{u_* y}{\nu}\right)^{1/7}$$

where $u_* = (\tau_0/\rho)^{1/2}$ is the shear velocity. Because the velocity at $y = \delta$ is equal to U_0, then

$$\frac{U_0}{u_*} = 8.74\left(\frac{\delta u_*}{\nu}\right)^{1/7}$$

or $\tau_0 = 0.0225\,\rho U_0^2(\nu/U_0\delta)^{1/4}$. From the first two equations, one can immediately derive

$$\frac{u}{U_0} = \left(\frac{y}{\delta}\right)^{1/7}$$

From the definition of the momentum thickness, one has successively

$$\theta = \int_0^\delta \frac{u}{U_0}\left(1 - \frac{u}{U_0}\right) dy$$

$$= \delta \int_0^1 \left(\frac{y}{\delta}\right)^{1/7}\left[1 - \left(\frac{y}{\delta}\right)^{1/7}\right] d\left(\frac{y}{\delta}\right) = \frac{7}{72}\delta$$

On substituting momentum thickness θ and the wall shear stress τ_0 into the momentum integral equation,

$$\frac{\tau_0}{\rho U_0^2} = \frac{d\theta}{dx} \qquad U_0 = \text{constant}$$

one obtains

$$0.0225\left(\frac{\nu}{U_0\delta}\right)^{1/4} = \frac{7}{72}\frac{d\delta}{dx}$$

This is the differential equation for δ. This equation is integrated from the initial value $\delta = \delta_0$ at $x = x_0$, where the boundary layer starts to become turbulent, to a given point x measured from the leading edge of the plate. The result is

$$\delta^{5/4} - \delta_0^{5/4} = 0.29\left(\frac{\nu}{U_0}\right)^{1/4}(x - x_0)$$

If one assumes that the boundary layer becomes turbulent at the edge of the plate, the initial value above can be replaced by $\delta = 0$ at $x = 0$. The above equation gives directly

$$\delta = 0.37\left(\frac{U_0}{\nu}\right)^{-1/5} x^{4/5}$$

This indicates that the boundary-layer thickness increases with the power $x^{4/5}$, whereas in the laminar boundary layer, the thickness increases with $x^{1/2}$.

The resistance force per unit width of length l is

$$R = \int_0^l \tau_0\, dx = \rho U_0^2 \theta(l)$$

$$= \tfrac{7}{72}\rho U_0^2 \delta = 0.036\,\rho U_0^2\left(\frac{U_0}{\nu}\right)^{-1/5} l^{4/5}$$

and the resistance coefficient C_f is

$$C_f = \frac{R}{\frac{1}{2}\rho U_0^2 l} = 0.072\left(\frac{U_0 l}{\nu}\right)^{-1/5}$$

In the range of Reynolds number $5 \times 10^5 < R_e < 10^7$, the last equation gives very good agreement with experimental results.

The value of the resistance coefficient for the turbulent boundary layer as well as the laminar boundary layer are plotted in Fig. 13-11.

13-4 Flow in Pipes

13-4.1 Steady Laminar Flow in Pipes

Laminar flow in pipes rarely occurs in practice. However, a rather full discussion is given because the simple and rational analysis is of some help in the understanding of

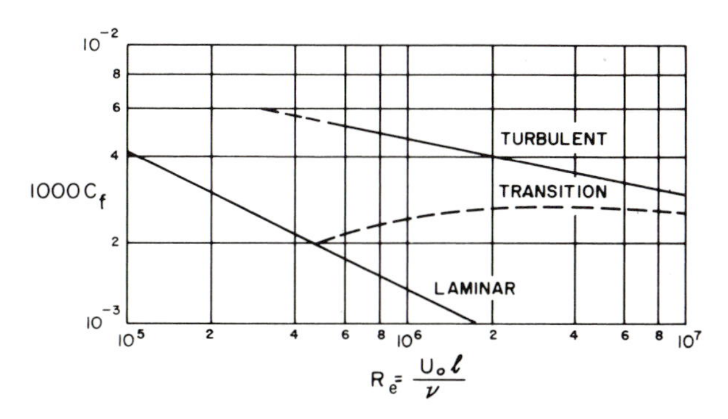

Figure 13-11 *Resistance on a smooth flat plate.*

the turbulent flow where conditions are so complicated that a complete theoretical treatment is not possible.

The flow conditions can be determined directly from the application of the Navier–Stokes equation. However, it is much simpler to derive the equation of motion directly from a consideration of force equilibrium. Consider a cylindrical element of length l and radius r (Fig. 13-12). The equilibrium of the shear stress τ and the pressure drop Δp is

$$\Delta p \pi r^2 = 2\pi r \tau l$$

It is assumed that Δp is a constant across a pipe section, a result which can be derived from the Navier–Stokes equation. For the laminar flow, the shear stress is simply, $\tau = \mu(du/dr)$. Substituting into the above equation, one obtains

$$du = \frac{\Delta p}{2l\mu} r\, dr$$

Figure 13-12 *Force equilibrium for a cylindrical element.*

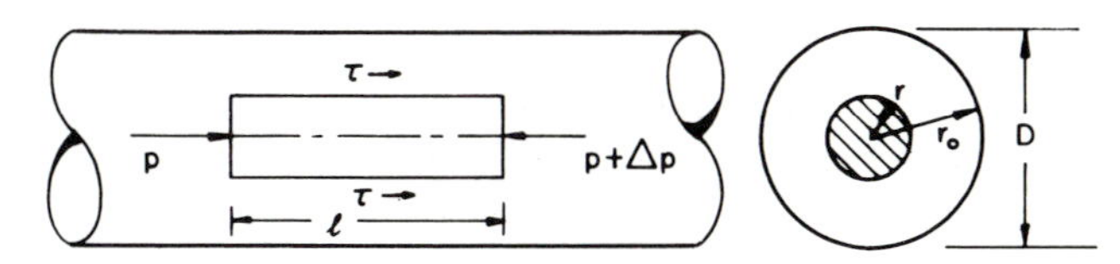

After integration, the velocity profile becomes

$$u = \frac{\Delta p}{2l\mu}\frac{r^2}{2} + c$$

The constant c can be determined from the boundary condition $u = 0$ at $r = r_0$. Therefore, the velocity distribution in the pipe is

$$u = \frac{\Delta p}{4l\mu}(r_0^2 - r^2)$$

which has the form of a symmetrical paraboloid. By integration, the average velocity V

$$V = \frac{Q}{\pi r_0^2} = \frac{1}{\pi r_0^2}\int_0^{r_0} u 2\pi r\, dr = \frac{\Delta p r_0^2}{8l\mu}$$

Rearranging the above equation, one obtains

$$\frac{\Delta p}{\rho g} = 64 \frac{\mu l}{V\rho D^2}\frac{V^2}{2g}$$

where $D = 2r_0$.

This equation can be compared with the Darcy–Weisbach equation for head loss, obtained by dimensional analysis, that is

$$\Delta H = \frac{\Delta p}{\rho g} = f\frac{l}{D}\frac{V^2}{2g}$$

where ΔH is the head loss and f is the friction factor. One obtains immediately the friction factor f for the laminar flow in a circular pipe

$$f = \frac{64\mu}{VD\rho} = \frac{64}{\mathrm{R_e}}$$

13-4.2 Turbulent Velocity Distributions and Resistance Law for Smooth Pipes

13-4.2.1 As the flow in the pipe becomes turbulent, the analytical determination of the velocity distribution is not possible. As in the case of turbulent boundary layer, the velocity profile must be determined based on logical assumptions and experimental verifications. One of the best-known assumptions in regard to the velocity distribution near the wall is based on the assumption that the velocity u at a distance y from the wall depends on the tangential stress τ_0 and on the viscosity μ and density ρ. Therefore, one may write in the most general case that $F(\tau_0,u,y,\mu,\rho) = 0$ (It is understood that u means the average velocity with respect to time and should actually be written $\bar{u}$. The bar will be omitted in the following sections for the sake of simplicity.) Based on the dimensional analysis, one obtains the dimensionless form

$$\frac{u}{(\tau_0/\rho)^{1/2}} = f\left[\frac{(\tau_0 y/\rho)^{1/2}}{\nu}\right]$$

which is similar to the case of flow over a flat plate (see Section 13-3.2). The functional relationship in the laminar sublayer has been derived in Section 13-3.1 to be

$$\frac{u}{u_*} = \frac{u_* y}{\nu}$$

where $u_* = (\tau_0/\rho)^{1/2}$. Outside the laminar sublayer and a transitional layer the turbulent stress dominates and the velocity profile follows logarithmic law, which has been derived in Section 8-3.5 as

$$\frac{u}{u_*} = \frac{1}{k}\ln y + C_1$$

In writing the dimensionless form, the velocity distribution reads

$$\frac{u}{u_*} = \frac{1}{k}\ln\frac{yu_*}{\nu} + C_2$$

where

$$C_2 = -\frac{1}{k}\ln\frac{u_*}{\nu} + C_1$$

The value of C_2 and the range of validity of these two equations which describes the velocity distribution in laminar sublayer and the turbulent flow have to be determined experimentally.

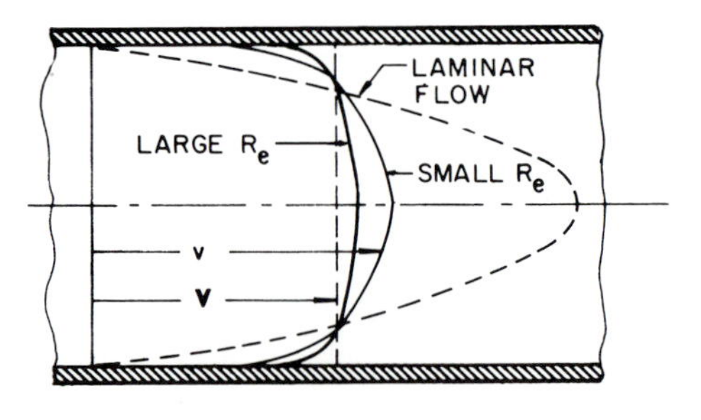

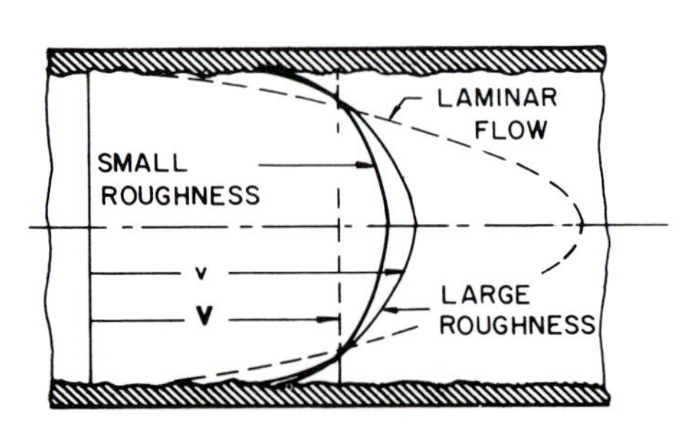

Figure 13-13 *Typical velocity distribution in pipes.*

Many experiments have been performed for measuring the velocity distribution in circular pipes. Typical velocity profiles are shown on Fig. 13-13a. The results are also presented in terms of the dimensionless variables u/u_* and $u_* y/\nu$ in Fig. 13-14. It indicates that at low values of $u_* y/\nu$, ($u_* y/\nu < 5$) the velocity follows the linear relationship

$$\frac{u}{u_*} = \frac{u_* y}{\nu}$$

For values of $u_* y/\nu > 30$, the experimental curve follows the logarithmic law which can be approximated by the equation

$$\frac{u}{u_*} = 5.75\log_{10}\frac{yu_*}{\nu} + 5.5$$

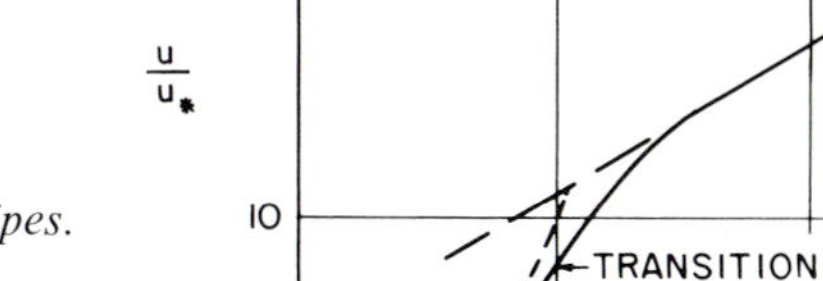

Figure 13-14

Universal velocity distribution law for smooth pipes.

However, in the range $u_* y/\nu$ between 5 to 30, where both turbulent and viscous effects are of equal importance, the velocity profile deviates from both of the above equations. Mathematical analysis fails to give correct prediction. This region is usually called a buffer region or a transition region.

13-4.2.2 Usually the thickness of the laminar sublayer and the layer of transition is very small in comparison with the size of the pipe. Therefore, in computing the average velocity, one could just use the logarithmic velocity distribution without introducing any significant error. The average velocity may then be obtained by substituting

$$\frac{u}{u_*} = 5.75 \log_{10} \frac{yu_*}{\nu} + 5.5$$

into the following equation:

$$\begin{aligned} U &= \frac{1}{r_0^2 \pi} \int_0^{r_0} 2\pi r u \, dr \\ &= \frac{2u_*}{r_0^2} \int_0^{r_0} r\left(5.75 \log_{10} \frac{u_*(r_0 - r)}{\nu} + 5.5\right) dr \end{aligned}$$

After integration, one obtains

$$\frac{U}{u_*} = 5.75 \log_{10} \frac{u_* r_0}{\nu} + 1.75$$

On the other hand, it has been seen that the Darcy–Weisbach equation gives the value of the head loss ΔH as function of the friction coefficient f and the average velocity U as follows:

$$\Delta H = f \frac{l}{D} \frac{U^2}{2g}$$

where l and D are the length and diameter of the pipe, respectively. Also, $\rho g \, \Delta H(\pi D^2/4)$ is the difference of pressure forces acting on two cross sections separated by a distance l. This force is balanced by the shearing force $\tau_0 \pi D l$. Equating these two forces yields

$$\Delta H = \frac{4}{g} \frac{\tau_0}{\rho} \frac{l}{D}$$

Inserting $u_* = (\tau_0/\rho)^{1/2}$ and eliminating ΔH by considering this above equality with the Darcy–Weisbach equation yields

$$\frac{U}{u_*} = \left(\frac{8}{f}\right)^{1/2} = 5.75 \log_{10} \frac{u_* r_0}{\nu} + 1.75$$

Since

$$\frac{u_* r_0}{\nu} = \frac{1}{2}\left(\frac{f}{8}\right)^{1/2} \frac{UD}{\nu}$$

The above equation can be further written as

$$\frac{1}{f^{1/2}} = 2.04 \log_{10} \mathrm{R_e} f^{1/2} - 0.91$$

where $\mathrm{R_e} = UD/\nu$. By modifying the constant slightly to agree with the results obtained from experiments, that is to change the constants in the above equations 2.04 and 0.91 to 2.0 and 0.8, respectively, one has

$$\frac{1}{f^{1/2}} = 2.0 \log_{10} \mathrm{R_e} f^{1/2} - 0.8$$

which is Prandtl's universal law of friction for smooth pipes.

13-4.3 Effect of Roughness

13-4.3.1 The effect of the roughness element on the flow depends on the thickness of the laminar sublayer. If the laminar sublayer is so thick that it covers the roughness, the roughness has no effect. The surface then can be considered to be hydrodynamically smooth. If the size of the roughness elements is large compared with the laminar sublayer, the effect of viscosity becomes small and no longer enters explicitly into the picture. The surface is then considered to be completely rough. In this case the shear stress depends only on the roughness, the specific density ρ, and the velocity u at some distance y from the wall. Some typical velocity profiles obtained in rough pipes are presented in Fig. 13-13b. Following the same procedure for the flow in smooth pipes, one could establish a dimensionless functional relationship for the closely packed uniform sand roughness elements in completely rough regime:

$$\frac{u}{u_*} = f\left(\frac{y}{k_s}\right)$$

where k_s is the sand size. (If the sand is not closely packed or nonuniform, one should also take into account the concentration distribution and shape of the roughness elements.) In the case in which the wall is not completely rough, then an additional dimensionless parameter $k_s u_*/\nu$ should also be included. Therefore, the general function should read

$$\frac{u}{u_*} = f\left(\frac{y}{k_s}, \frac{k_s u_*}{\nu}\right)$$

This general functional relationship has been determined by experiments and can be approximated by the equation

$$\frac{u}{u_*} = 5.75 \log_{10} \frac{y}{k_s} + B$$

where B depends on the "shear Reynolds number," $k_s u_*/\nu$. The value of B obtained by experiments is shown in Fig. 13-15. As indicated in this figure, the value of B characterizes three regimes:

1. Hydraulically smooth regime previously considered in Section 13-4-2.1, where

$$0 \leq \frac{k_s \mu_*}{\nu} \leq 5$$

Figure 13-15 *Roughness function B in terms of shear Reynolds number $k_s u_*/\nu$.*

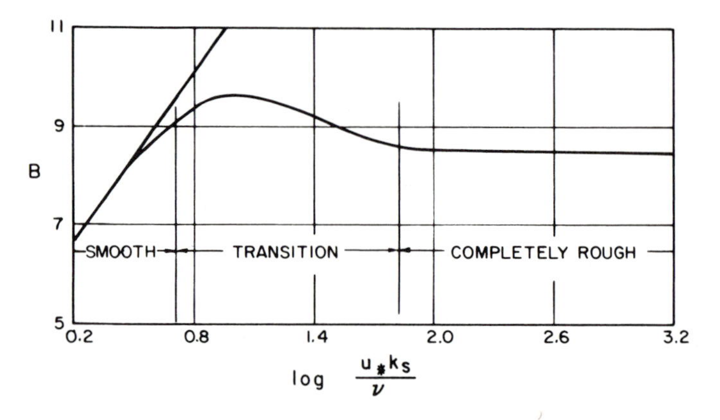

In this regime, the size of roughness is so small it is covered by the laminar sublayer.

2. Transition regime:

$$5 \leq \frac{k_s \mu_*}{\nu} \leq 70$$

Some of the roughness elements extend outside the laminar sublayer and contribute some resistance through form drag.

3. Completely rough regime:

$$\frac{k_s \mu_*}{\nu} > 70$$

All the roughness elements are exposed outside the laminar sublayer or one may say that the laminar sublayer has been destroyed completely by the roughness elements. The turbulent action extends all the way to the rough wall. Further increase of shear and the Reynolds number does not bring any change of flow patterns. Therefore, B remains independent of shear Reynolds number.

13-4.3.2 The resistance coefficient in the completely rough regime can be evaluated the same way as in the case of smooth pipe. The final form of the resistance equation reads

$$\frac{1}{f^{1/2}} = 2.0 \log_{10} \frac{r_0}{k_s} + 1.74$$

Experiments were performed first by Nikuradse, who used closely packed sand grain roughness elements and obtained the resistance diagram shown in Fig. 13-16.

The velocity distribution and resistance formula discussed are based on the closely packed sand grain roughness used by Nikuradse. In this case, k_s is the actual sand size. However, if a different type of sand is used or sand particles are not packed closely, the resistance offered to the flow will be different. Therefore, sand size alone is not enough to describe the velocity distribution and resistance.

Figure 13-16 *Resistance formula for pipes roughed with Nikuradse's sand roughness.*

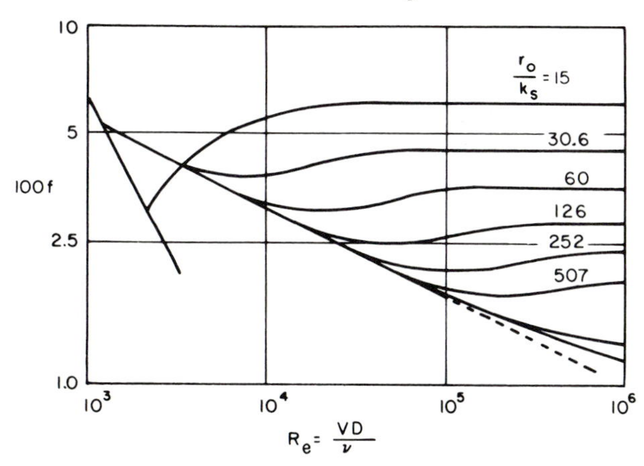

13-5 Drag on Immersed Bodies

13-5.1 Drag on a Body in Steady Flow

13-5.1.1 It has been seen that the total force exerted by a current on a cylinder is zero (see Section 11-4.4.3). This result is general, for the case of a perfect fluid, i.e., the total force exerted on a body by a perfect fluid without circulation of velocity is always zero. This is called the paradox of D'Alembert.

For the case in which a circulation is introduced to the fluid (see Sections 11-4.2 and 11-4.3) a force perpendicular to the incident velocity is exerted on the body. It can be demonstrated that this force is proportional to the velocity V of the fluid and the strength of the circulation. It is this force that causes the lift of an aerofoil.

The problem now under study is that of a real fluid in which case a boundary layer develops along the body and induces a drag. This drag is caused by the shearing force acting on the body and to the wake. This leads us to discuss the problem of boundary-layer separation.

13-5.1.2 The flow field near a flat plate in parallel flow and at zero incidence is quite simple because the pressure in the entire flow field remains constant. In the case of flow

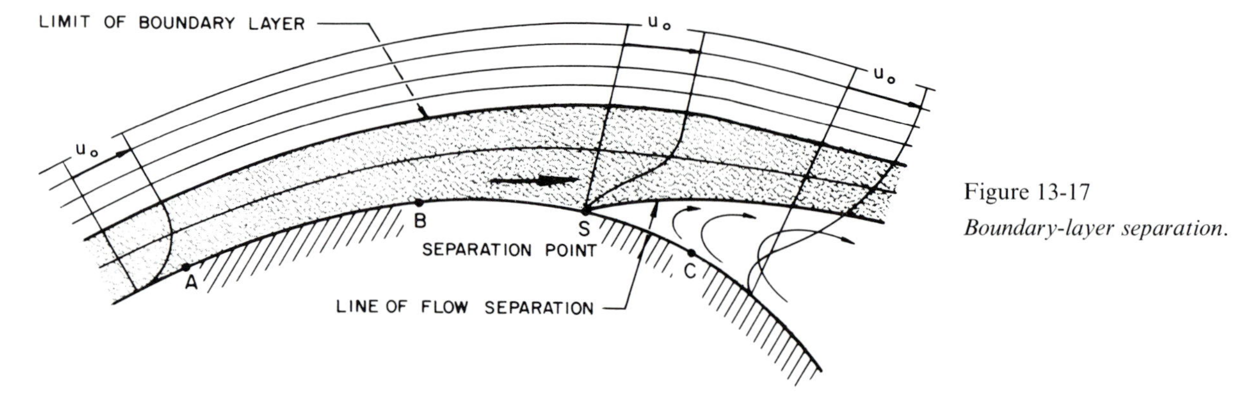

Figure 13-17
Boundary-layer separation.

about a blunt body such as a cylinder, the pressure along the surface of the body, which is impressed on the boundary from the external flow, is not constant. As shown in Fig. 13-17 the fluid particles are accelerated from A to B and decelerated from B to C. Hence, the pressure decreases from A to B and then increases from B to C. This can be seen by application of the Bernoulli equation. Since the fluid is viscous, a certain amount of kinetic energy is lost by the friction within the thin boundary layer as the fluid particles move along the boundary. The remaining energy may be too small to overcome the increasing pressure toward the point C. As a result of this, the fluid particles being influenced by the external pressure may move in the reverse direction and cause flow separation behind the body at point S. The flow field behind the separation is very irregular and is characterized by large turbulent eddies. This region is usually called *turbulent wake*, although the wake may also be laminar when the Reynolds number is small (<40 in the case of a cylinder).

Because of the existence of the wake, the flow field changes radically as compared with that in frictionless flow. The main flow which separates from either side of the boundary does not meet right behind the body. It leaves a "dead zone" in which the pressure remains close to its value at the separation point, which is always less than the pressure at the forward stagnation point. Therefore, a large net force will act on the body resulting from the pressure difference. This force is called *form drag*.

13-5.1.3 In principle, the total drag exerted on a sphere moving with constant velocity in the infinite flow field is the sum of the friction drag (or shear drag) and the form drag. If the velocity is low enough ($R_e < 1$), the inertia terms in the Navier–Stokes equations may be neglected. The drag can be obtained analytically and is given by Stokes' law

$$\mathbf{F} = 3\pi\rho\nu\mathbf{V}D$$

where $\mathbf{V}$ is the relative velocity of the body with respect to the water and D the diameter of the sphere. The drag coefficient C_D, which is defined from the equation

$$F = \frac{C_D A\rho V^2}{2}$$

is then equal to $24/R_e$, where A is the cross-sectional area of the sphere, R_e is the Reynolds number VD/ν.

As the Reynolds number R_e increases, the flow separates from the surface of the sphere, beginning at the rear stagnation point, where the adverse pressure gradient is greatest. As the flow separates from the boundary, the form drag, which is a function of the area of separation and the square of velocity, becomes important. The drag coefficient C_D will deviate from the line $C_D = 24/R_e$, and start to level off. Figure 13-18 indicates the variation of the drag coefficient C_D with Reynolds number R_e.

As the Reynolds number reaches 2×10^3, the drag coefficient becomes almost constant. However, in the

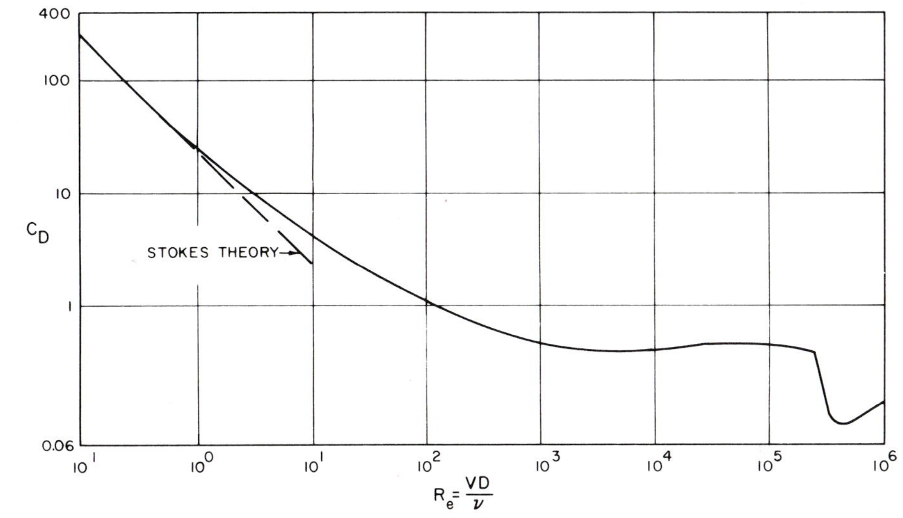

Figure 13-18
Drag coefficient for sphere.

range of Reynolds number 2×10^5 to 3×10^5, the drag coefficient is suddenly reduced. The reason for this lies in the transition of boundary layer from laminar to turbulent. This transition brings a violent mixing in the boundary layer. As a result, the fluid particles near the boundary gain additional kinetic energy, which enables them to withstand the adverse pressure gradient better and to move the separation point somewhat downstream, as illustrated in Fig. 13-19. This results in a sudden decreasing of the drag coefficient near the Reynolds number 3×10^5 as shown in Fig. 13-18. As the transition that occurs depends on the roughness of the sphere, and also slightly on the turbulence level in the free stream, the drag coefficient near this critical region is not a unique function of Reynolds number.

13-5.1.4 As shown in Fig. 13-20, the relationship between drag coefficient and the Reynolds number for a circular cylinder with axis normal to the direction of motion is, in general, similar to that for a sphere. However, rather peculiar phenomena which are not ordinarily found in the flow around a sphere can be observed in flow around a cylinder. In the range of Reynolds number between 40 and 5000, one can see a regular pattern of vortices which move alternately clockwise and counterclockwise downstream, as shown in Fig. 13-21. This is known as the Kármán vortex street. The vortex street moves with a velocity V_e, which is somewhat smaller than the free stream velocity U_0. Von Kármán found that the vortex street is unstable except at the spacing $h/l = 0.281$, and that the drag experienced by the cylinder depends on the width of the vortices h and on the velocity ratio V_e/U_0:

$$F = \rho U_0^2 h \left[2.83 \frac{V_e}{U_0} - 1.12 \left(\frac{V_e}{U_0} \right)^2 \right]$$

Because the vortex developed behind the cylinder is unsymmetrical, a time-dependent circulation of velocity is induced around the cylinder. The cylinder will experience a side push that continually reverses its direction. Therefore the cylinder may tend to oscillate from one side to the other, particularly if its natural frequency of oscillation is in resonance with the frequency of the vortex shedding.

The shedding frequencies k in the Kármán vortex street behind a circular cylinder have been measured. Those measurements indicate that the dimensionless frequency

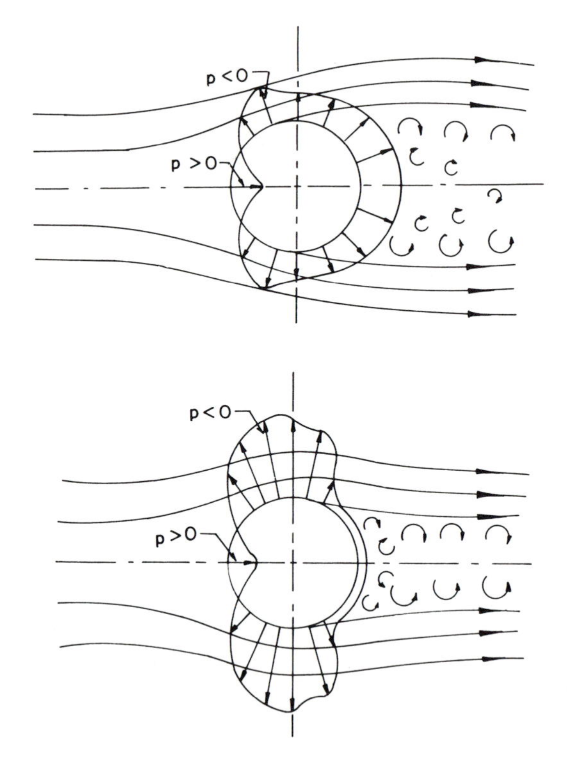

Figure 13-19 *Changing location of separation and pressure distribution as a result of boundary-layer transition.*

known at the Strouhal number $S = kD/U_0$ depends uniquely on the Reynolds number. The Strouhal number is an indication of the relative importance of local inertia as compared to convective inertia.

An experimental curve which can be used to determine the frequency of the vortex shedding is given in Fig. 13-22. From this curve, one could determine the shedding frequency which is useful information in practical design.

13-5.2 Drag Due to Unsteady Motion: The Added Mass Concept

13-5.2.1 The concept of added mass, or virtual mass, or induced mass, is of particular importance in the study of the forces acting on a body accelerating or decelerating in still water, or on a fixed body subjected to an unsteady current.

It is recalled that under steady state conditions the total force acting on a fixed body by a current without circulation is nil in the case of a perfect fluid. It is the paradox of D'Alembert. In the case of a real fluid, it has been seen that the force is a complex function of the Reynolds number.

Under unsteady conditions, another force has to be added, whether the fluid be perfect or real. The value of this force is analyzed below.

13-5.2.2 When a body of mass M moves in still water at a speed U, it has a kinetic energy $\frac{1}{2}MU^2$. This body automatically induces a fluid motion around it which tends to zero when the distance from the body tends to infinity. The exact law of decay depends upon the shape of the body. However, far from the body it can be said that the fluid particle velocity $V(x,y,z,t)$ decreases as $1/R^3$ in the case of a three-dimensional flow and $1/R^2$ in the case of a two-dimensional flow; R being the distance of the considered fluid particle from the center of the body.

The total kinetic energy of the fluid surrounding the body is then

$$\iiint_{\lim}^{\infty} \tfrac{1}{2}\rho V^2(x,y,z,t)\, d\sigma$$

where lim is the limit of the body and $d\sigma$ is an elementary volume (or area in the case of a two-dimensional motion).

The total kinetic energy of the system can then be written

$$W = \tfrac{1}{2}U^2\left[M + \rho \iiint_{\lim}^{\infty} \left(\frac{V}{U}\right)^2 d\sigma\right]$$

The quantity

$$M' = \rho \iiint_{\lim}^{\infty} \left(\frac{V}{U}\right)^2 d\sigma$$

is the added mass. It is the mass of fluid which, moving at speed **U**, will have the same kinetic energy as the total mass of fluid. W is the work that is required to give the body speed U, or it is the work that would be required to

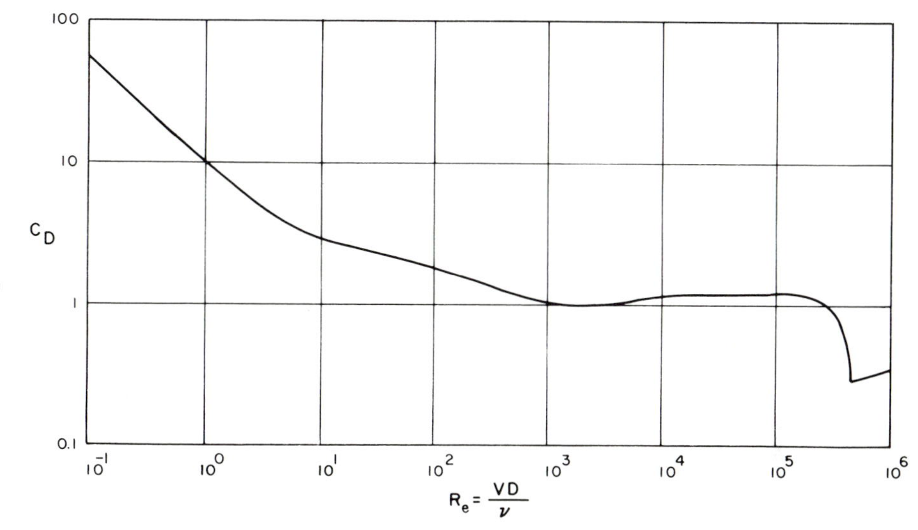

Figure 13-20
Drag coefficient for circular cylinder

stop it. It is seen that this work also includes the work required to move the fluid around it: $\frac{1}{2}M'U^2$. Once this work is produced, the body will continue to travel in a perfect fluid at a constant velocity U.

13-5.2.3 It is pointed out that because V decreases with R^{-3} (or R^{-2} in the two-dimensional case), $(V/U)^2$ varies with distance as R^{-6} (or R^{-4}), whereas the integral of $d\sigma$ varies as R^3 (or R^2). Consequently, the integral for M' has a finite value.

It is seen also that in the general case M' is a function of the absolute value of U and consequently of the Reynolds number UD/ν and other empirical parameters characterizing the flow (such as UT/D for periodic motion where D is a typical dimension of the body). Consequently, M' will also be in general a function of time. However, in the case of a perfect fluid, $V(x,y,z,t)/U$ is independent of U, but depends upon the flow pattern only. If one refers V to a coordinate system moving at velocity U this ratio is also independent of time. Hence, the integral of the coefficient $V(x,y,z,t)/U$ is independent of the value of U and the time as well. In a word, M' is a constant associated with the body and the specific mass of the fluid only.

13-5.2.4 Let us now consider the total force to move the body. It is equal to the sum of the inertia of this body

Figure 13-21 *Von Kármán vortex street.*

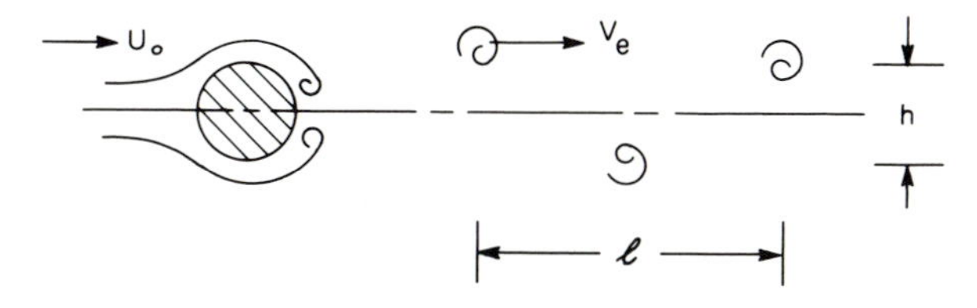

Figure 13-22 *Relationship between the Strouhal number and the Reynolds number.*

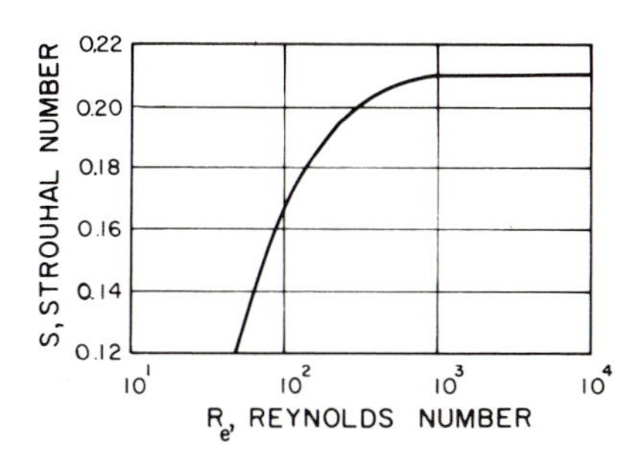

itself and the inertia of the fluid surrounding it, i.e.,

$$\mathbf{F} = M\frac{d\mathbf{U}}{dt} + \rho \iiint_{\lim}^{\infty} \frac{d\mathbf{V}}{dt}\, d\sigma$$

which can still be written as

$$\mathbf{F} = (M + M')\frac{d\mathbf{U}}{dt}$$

where

$$M' = \frac{\rho(d/dt)\iiint_{\lim}^{\infty} V\, d\sigma}{dU/dt}$$

It is not evident *a priori* that the two definitions for M' are identical. As a matter of fact the integral

$$\iiint_{\lim}^{\infty} V\, d\sigma$$

may diverge as the distance from the body tends to infinity. Hence, in the case of moving body the force $\mathbf{F}' = M'\, d\mathbf{U}/dt$ should be determined from the force exerted by the fluid on the body or vice versa, i.e.

$$F' = \iint_s p \cos\theta\, ds$$

where p is the pressure around the body, $\cos\theta$ is the angle of the perpendicular to ds with the main direction of the motion, and s the area of the body. (Fig. 13-23). Given V (or ϕ), p can be determined by application of the Bernoulli equation. In general, the integral of $\rho\ V^2/2$ being zero (paradox of D'Alembert), the integral of $\rho\ \partial\phi/\partial t$ only is significant, so that finally, replacing p by $-\rho\partial\phi/\partial t$

$$M' = \frac{-\iint \rho(\partial\phi/\partial t)\cos\theta\, ds}{dU/dt}$$

Of course, the equality force–momentum can also be obtained by differentiating the equality work–energy as follows

$$\frac{d}{dt}(W) = \frac{d}{dt}\left[\tfrac{1}{2}(M + M')U^2\right]$$

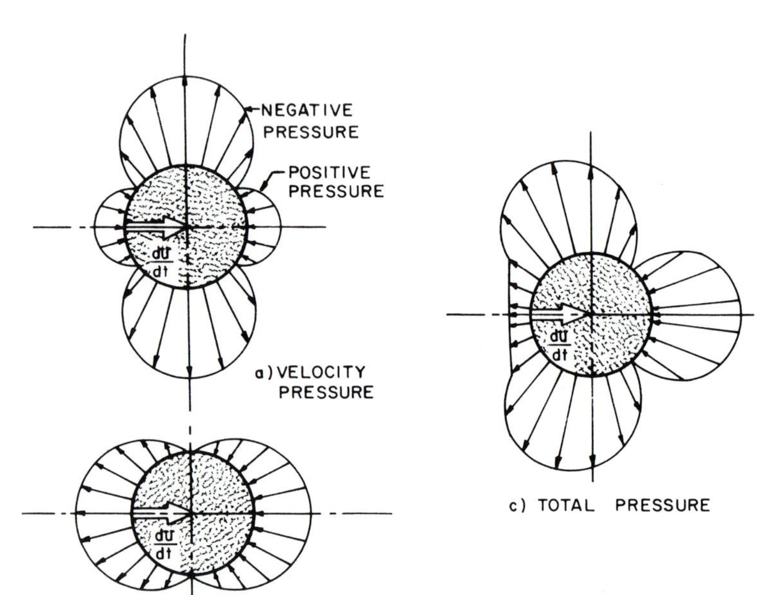

Figure 13-23

An example of distribution of pressure force on an accelerating body in still water.

This gives

$$F\frac{dL}{dt} = U(M + M')\frac{dU}{dt}$$

Since $dL/dt = U$, the equality force–momentum is obtained. Still, this operation is done by assuming M' constant and $dF/dt = 0$. Actually, a more rigorous demonstration will not require this limitation.

In conclusion, for all practical purposes, the added mass is determined by calculating the integral

$$M' = \rho \iiint \left(\frac{V}{U}\right)^2 d\sigma$$

or in the case of an irrotational flow

$$M' = -\frac{\iint \rho(\partial\phi/\partial t)\cos\theta\, ds}{dU/dt}$$

Then this value of M' will be used for determining the force $M'(dU/dt)$. The shortcomings of these simple demonstrations are not discussed in this book. It is only pointed out that one of the limits of the integral, namely the body limit, could be time dependent.

13-5.2.5 The case of a moving circular cylinder of radius R is given here as an example. The velocity potential for a cylinder moving through a fluid at rest is given by superimposing upon the steady state pattern of a flow around a cylinder, a uniform velocity U (see Section 11-4.4.3) i.e.,

$$\phi = -U\left(r + \frac{R^2}{r}\right)\cos\theta + Ur\cos\theta$$

It is seen that this operation nullifies the uniform flow component and the potential function is that of a doublet:

$$\phi = -U\frac{R^2}{r}\cos\theta$$

The fluid velocity at any point has a magnitude given by

$$V^2 = \left[\left(\frac{1}{r}\frac{\partial\phi}{\partial\theta}\right)^2 + \left(\frac{\partial\phi}{\partial r}\right)^2\right]$$

This gives

$$V(r,\theta,t) = \frac{R^2}{r^2}U(t)$$

where $U(t)$ is the velocity of the body. The total kinetic energy of the fluid per unit length of the cylinder is then successively

$$T = \int_0^{2\pi}\int_{r=R}^{\infty} \rho\frac{1}{2}\frac{R^4U^2}{r^4}\, r\, dr\, d\theta$$

$$= \rho\pi R^4U^2\int_{r=R}^{\infty}\frac{dr}{r^3}$$

$$= \tfrac{1}{2}\rho\pi R^2U^2$$

It is seen that the added mass is $M' = \rho\pi R^2$, i.e., the mass of a cylinder of radius R having the same density as the fluid. It is also seen that the total force to move the body is

$$\mathbf{F} = (M + M')\frac{d\mathbf{U}}{dt}$$

i.e.,

$$\mathbf{F} = (\rho_b + \rho)\pi R^2\frac{d\mathbf{U}}{dt}$$

where ρ_b is the density of the body. It can be verified that $\mathbf{F}' = M'(d\mathbf{U}/dt)$ is the total force exerted by the fluid on the body as the sum of all the pressure forces in the direction of the motion:

$$F' = \int_0^{2\pi} p\cos\theta\, R\, d\theta$$

The pressure distribution around a moving cylinder in the case of an unsteady motion is given by

$$-\frac{p}{\rho} = \frac{\partial\phi}{\partial t} + \frac{1}{2}V^2 + f(t)$$

i.e.,

$$\frac{p}{\rho} = R\frac{dU}{dt}\cos\theta + \tfrac{1}{2}U^2[1 - 4\sin^2\theta]$$

Since the integral of the quadratic term is zero, the total force acting on the cylinder is

$$F' = \int_0^{2\pi} \rho R^2 \frac{dU}{dt} \cos^2 \theta \, d\theta = \rho\pi R^2 \frac{dU}{dt}$$

i.e.,

$$M' = \rho\pi R^2$$

13-5.2.6 Let us now consider the case of a fixed body subjected to an unsteady current. The total force exerted by water on the body is still

$$F = \iint_s p \cos \theta \, ds$$

which, in the case of an irrotational motion without circulation of velocity is identical to

$$-\iint \rho \frac{\partial \phi}{\partial t} \cos \theta \, ds$$

This integral is twice the value of the same integral in the case of a moving body in a still fluid, and consequently

$$\mathbf{F} = 2M' \frac{d\mathbf{U}}{dT}$$

It is interesting to note that in the first case of a moving body in still water, the same force is found provided $M = M'$, i.e., the body has the same average density as the fluid. This is true of floating systems and submarines.

13-5.2.7 The case of a fixed circular cylinder subjected to an unsteady fluid flow is given here as an example.

The potential function for the motion is then

$$\phi = -U(t)\left(r + \frac{R^2}{r}\right) \cos \theta$$

and

$$\rho \frac{\partial \phi}{\partial t}\bigg|_{r=R} = -\rho 2R \frac{dU}{dt} \cos \theta$$

It is seen that the pressure component due to local inertia is in this case twice the value of the pressure component in the case of a moving cylinder.

Inserting this value in the previous integral yields

$$\mathbf{F} = 2\rho\pi R^2 \frac{d\mathbf{U}}{dT} = 2M' \frac{d\mathbf{U}}{dT}$$

13-5.2.8 In the case of a real fluid, this inertial force still exists, but because of viscosity, separation, and wake, there is also a superimposed quadratic force. The following empirical formula is often proposed

$$F = \rho C_D A \frac{U^2}{2} + \rho C_M \text{ vol} \frac{dU}{dt}$$

where A is the cross section of the body perpendicular to the flow ($A = 2R$ in the case of a cylinder), "vol" is the volume of the body, C_D is the drag coefficient, and C_M the inertial coefficient. It is seen by comparison with the previous result that $C_M = 2$ in the case of a cylinder. As a matter of fact, both C_M and C_D are not constants but complex functions of the reduced frequency $D^2/\nu T$ (Valensi number), the Reynolds number UD/ν, and time; $(D^2/\nu T)/(UD/\nu)$ is a Strouhal number, which is also called the *Iversen modulus.* The added mass of a floating body is also a function of frequency, even in the case of a perfect fluid, as the waves generated by the body is a function of frequency.

13-5.2.9 The previous formula is used by engineers to calculate the wave force on small vertical piles. The horizontal velocity component $u(x,z,t)$ then replaces $\mathbf{U}$, and the formula is called the *Morison formula.* Assuming that the pile disturbance on the velocity field is small, u is obtained by a wave theory. It is not entirely clear if the experiment in unidirectional acceleration of a cylinder through a fluid is strictly applicable to the wave force problem. Also the mass transport due to wind stress near the free surface increases considerably the particle velocity as compared to the particle velocities given by classical wave theories (see Section 16-1.5).

13-5.2.10 In the case in which both a body moves with a time dependent velocity $\mathbf{V}(t)$, in a time dependent fluid flow, which has a velocity $\mathbf{U}(t)$ at a distance from the body, the total force to be exerted to move a body of mass M is

$$\mathbf{F} = M \frac{d\mathbf{V}}{dt} + M' \frac{d}{dt}(\mathbf{V} - \mathbf{U})$$

13-5.3 The Case of Time-Dependent Mass

13-5.3.1 In the case of (for example) an underwater missile popping up vertically from the tube of a submarine, and subjected to a steady current U, an inertial force acting in the direction of the current is added to the drag force. This added force is due to the variation of mass of the body subjected to the cross-flow, i.e., the momentum equation is now written

$$\mathbf{F} = \frac{dm\mathbf{U}}{dt} = \mathbf{U} \frac{dm}{dt}$$

where dm/dt is proportional to the rate of volume of missile exposed to the fluid flow, i.e., the product

$$\int A \, dV \cong \frac{\pi D^2}{4} V$$

where V is the exit velocity of the missile and A is the cross section of the missile of diameter D.

Therefore,

$$F \cong \rho \frac{\pi D^2}{4} U V C_M$$

where C_M is a dimensionless coefficient.

13-5.3.2 In the case of fixed vertical piles subjected to wave action, the fluid velocity is time dependent and, owing to the variation of elevation of the free surface, the volume of pile into the water also time dependent. In this case:

$$\mathbf{F} = \frac{dm\mathbf{V}}{dt} = m \frac{d\mathbf{V}}{dt} + \mathbf{V} \frac{dm}{dt}$$

The first term results from the added mass, as defined in Section 13-5.2 (*vide supra*). The second term is the result of the free surface variation and is equal to

$$\rho C_M u_s \frac{\pi D^2}{4} w_s$$

where C_M is a dimensionless coefficient, u_s and w_s are the free surface horizontal and vertical particle velocity components at the pile location ($w_s = \partial\eta/\partial t$, $\eta(t)$ being the free surface elevation), and D is the pile diameter. This force is often neglected. Actually, this neglect introduces a large error on the calculation of wave force and torque, in the case of piles subjected to large-amplitude shallow water waves.

PROBLEMS

13.1 The dissipation energy thickness δ^{**} of a boundary layer is defined by the equation:

$$\rho U_0^3 \delta^{**} = \rho \int_0^\infty u(U_0^2 - u^2)\, dy$$

where the right-hand term is the flux of dissipated energy by friction. Calculate the value of δ^*, θ, and δ^{**} as functions of δ in the cases where

1. $u = U_0 \dfrac{y}{\delta}$ for $y < \delta$ and $u = U_0$ for $y \geq \delta$
2. $u = U_0\left[2\dfrac{y}{\delta} - \left(\dfrac{y}{\delta}\right)^2\right]$ for $y < \delta$ and

   $u = U_0$ for $y \geq \delta$

13.2 Obtain the transverse velocity component v for the laminar boundary layer along a plate.

13.3 Determine the coefficients A_n of the Blasius theory up to A_{11} as a function of A_2. Demonstrate that only

A_{3n+2} are different from zero and establish a reference formula for A_{3n+2} as a function of A_2. Present the expression of $f(\eta)$ as a power series as a function of A_2 and determine the value of A_2 (it is found that $A_2 = 0.332$). Determine the value of $f(\eta)$, $f'(\eta)$, $f''(\eta)$ at $y = 0$.

13.4 The thickness of the laminar boundary layer on a semiflat plate can be evaluated through the von Kármán momentum integral formula by assuming a proper velocity profile. If the velocity profile is assumed to be a polynomial

$$u = a_0 + a_1 y + a_2 y^2 + a_3 y^3 + a_4 y^4$$

1. Give the proper boundary conditions and use these boundary conditions to determine the five constants.
2. Calculate the shear stress along the plate.
3. Obtain the thickness of boundary layer by use of von Kármán's momentum integral formula.

13.5 Consider the steady boundary layer on a convergent channel with flat walls (Fig. 13-24). The boundary-layer equations along the wall parallel to the x axis are

$$u\frac{\partial u}{\partial x} + v\frac{\partial u}{\partial y} = -\frac{1}{\rho}\frac{\partial p}{\partial x} + \nu\frac{\partial^2 u}{\partial y^2}$$

$$\frac{\partial p}{\partial y} = 0 \qquad \frac{\partial u}{\partial x} + \frac{\partial v}{\partial y} = 0$$

The free-stream velocity is given in the form $u(x) = -(u_0/x)$. Introducing the similarity transformation

$$\eta = \frac{y}{x}\left(\frac{u_0}{\nu}\right)^{1/2}$$

as well as the stream function

$$\psi(x,y) = -(\nu u_0)^{1/2} f(\eta)$$

1. Find the ordinary differential equation for the stream function and the boundary conditions.
2. Obtain the velocity distribution.

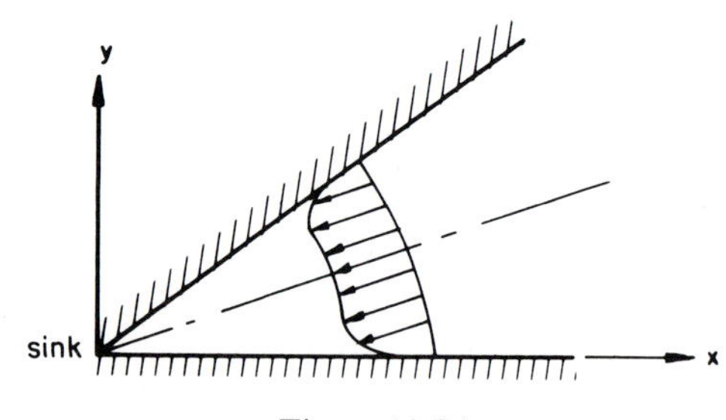

Figure 13-24

13.6 Establish the momentum integral equation for unsteady boundary layer.

13.7 Determine the frictional force on an oscillating plane covered by a layer of fluid of thickness h. The frequency of oscillation is k and the fluid has kinematic viscosity ν.

13.8 A fixed amount of discharge Q flows uniformly down a semiinfinite plate started at $x = 0$. The fluid has viscosity μ and density ρ and it accelerates with the gravitational acceleration g as shown in Fig. 13-25. The free surface is assumed to be of constant pressure. The fluid in contact with the plate forms a boundary layer. The thickness of the boundary layer increases until it reaches the free surface as shown in Fig. 13-25. The flow will continue to accelerate and the fluid layer will become thinner until it reaches an asymptotic value.

1. Derive the momentum integral equation before the boundary layer reaches the free surface.
2. Assume a parabolic velocity profile

$$\frac{u}{u_0} = a_0 + a_1\left(\frac{y}{\delta}\right) + a_2\left(\frac{y}{\delta}\right)^2$$

   Determine the constants a_0, a_1, a_2 by use of appropriate boundary conditions. Substitute the determined velocity profile into the momentum integral equation and derive a differential equation for $\delta(x)$.
3. Try a solution of the form $\delta = \beta x^n$. Determine the values of β and n from the integral equation.

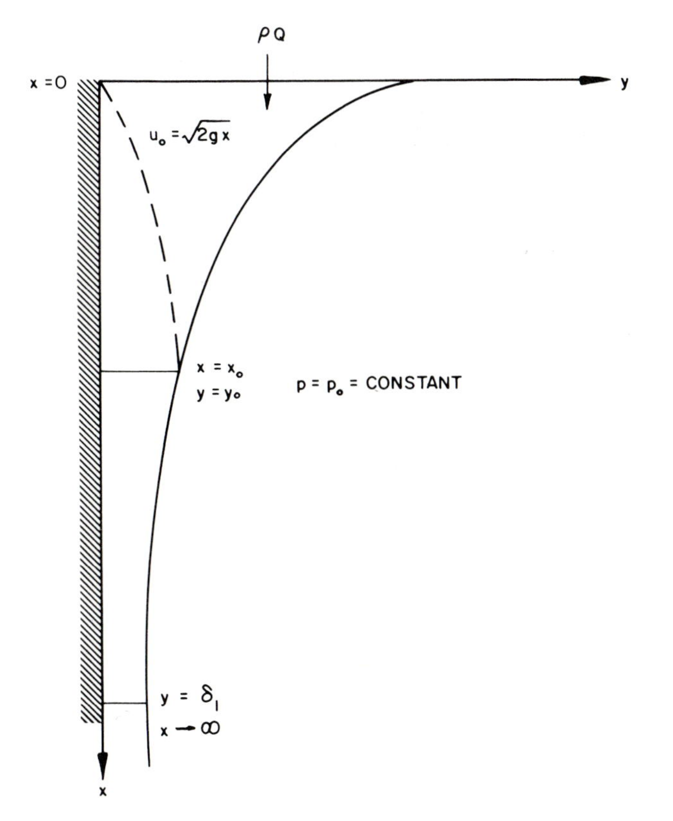

Figure 13-25

4. Determine the distance x_0 at which the boundary layer reaches the free stream.
5. Determine the thickness of the boundary layer δ_0 at the location where the boundary layer reaches the free stream.
6. Derive the momentum integral equation for the flow regime $x > x_0$.
7. Using the velocity profile derived in question (2), derive the differential equation of $\delta(x)$ for $x > x_0$.
8. Obtain the relationship between the distance x and the layer thickness.
9. Obtain the layer thickness δ_3 when x approaches infinity.

13.9 Derive the resistance equation for turbulent flow in rough pipes

$$\frac{1}{f^{1/2}} = 2 \log_{10} \frac{r_0}{k_s} + 1.74$$

13.10 The potential function for a two-dimensional flow around a cylinder of radius R is

$$\phi = U\left(r + \frac{R^2}{r}\right) \cos \theta$$

where U is the velocity at infinity. Give the pressure distribution around the cylinder in the case where

1. $U = \text{constant} = U_0$
2. $U = U_0 \sin kt$

Determine the total force acting on a cylinder by integration of the pressure.

13.11 Calculate the added mass for a sphere, taking into account the fact that the velocity potential for a sphere of radius R moving at velocity U in a spherical coordinate system (r,θ,Φ) in a still fluid is

$$\phi = \frac{UR^3}{2r^2} \cos \theta$$

13.12 The horizontal velocity component due to a linear periodic gravity wave in deep water is

$$u = \frac{H\pi}{T} e^{2\pi z/L} \cos (kt - mx)$$

Calculate the maximum total force exerted on a vertical cylinder of 5 ft (152 cm) diameter by a wave $H = 20$ ft (609 cm), $T = 10$ sec. The drag coefficient $C_D = 1$, and the inertial coefficient $C_M = 2$, $k = 2\pi/T$, $m = 2\pi/L$ and $L = gT^2/2\pi$.

Chapter 14

Open-Channel Hydraulics

14-1 Uniform Flow, Normal Depth, and Critical Depth

14.1.1 Hydraulic Radius and the Chezy Formula

14-1.1.1 Three kinds of approaches are used to obtain the equation used in open-channel hydraulics. However, all of these mathematical forms originate from the Newtonian equation, and a criterion for choosing one rather than another does not really exist.

Most of the basic equations of this chapter can be obtained from (1) a simplified form of the Eulerian equations in which an empirical friction term is included, (2) the generalized Bernoulli equation in which the condition at the free surface $p = p_a$ (atmospheric pressure) is inserted, or (3) by a direct application of the balance of forces in which the significant terms only are included.

14-1.1.2 Consider a uniform flow parallel to the axis OX (see Fig. 14-1). Since the motion is steady, $\partial(u,v,w)/\partial t = 0$, and since the motion is uniform, v and $w = 0$, and $\partial u/\partial x = 0$. Hence, it is easily verified that all the inertial terms are zero. Moreover, the pressure forces acting on each side of cross

Figure 14-1 *Uniform flow.*

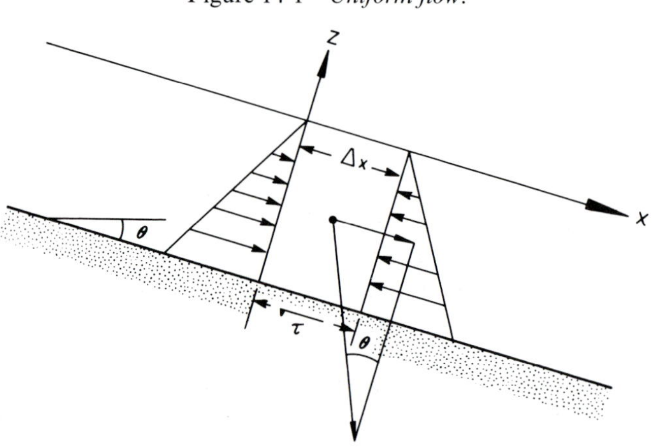

section A of an element of fluid $A\,\Delta x$ are in equilibrium. The OZ components of the pressure force and gravity also balance independently of the flow velocity. Thus, the only significant forces which remain are the gravity component in the OX direction and the shearing forces:

$$\rho g A\,\Delta x \sin\theta = \Delta x \int_0^P \tau\, dP$$

where P is the "wetted perimeter" (i.e., the length of the perimeter of the cross section A which is underwater) and τ is the shearing stress per unit area.

14-1.1.3 In general, τ is not constant, except around a circular pipe due to symmetry. However, owing to secondary currents, the variation of τ can often be considered as negligible (see Section 8-2.3).

The previous equation is then written

$$\rho g R_H \sin\theta = \tau \qquad \text{where} \qquad R_H = \frac{A}{P}$$

Note that the "hydraulic radius" R_H has the dimension of a length. It is easily verified that in the case of a rectangular section of depth h and width l

$$R_H = \frac{lh}{l + 2h}$$

and R_H tends to h when $l \to \infty$; i.e., in the case of a large river. In the case of a circular pipe of radius R

$$R_H = \frac{\pi R^2}{2\pi R} = \frac{R}{2}$$

14-1.1.4 In the case of a river or a channel, the Reynolds number is generally large, and the flow is fully turbulent. The shearing stress can then be assumed to be related to the average velocity **V** by a quadratic function such as

$$\tau = \rho f V^2$$

where f is a dimensionless friction factor. When this is combined with the equation $\rho g R_H \sin\theta = \tau$ the following equation for V is obtained

$$V = \left(\frac{g}{f}\right)^{1/2} (R_H \sin\theta)^{1/2}$$

The Chezy coefficient of dimension $(LT^{-2})^{1/2}$ is $C_h = (g/f)^{1/2}$; thus $f = g/C_h^2$. The slope S is generally small, so that $S = \tan\theta \cong \sin\theta$ and $V = C_h(R_H S)^{1/2}$. This is the Chezy formula.

The discharge $Q_n = VA$ is then

$$Q_n = AC_h(R_H S)^{1/2} = KS^{1/2}$$

where $K = AC_h(R_H)^{1/2}$ is the *conveyance* of the channel and depends upon the geometry of the cross section of the channel and the water depth only. Q_n is the *normal discharge*, which is defined as a function of water depth for a given channel.

14-1.15 The Chezy coefficient and f can only be determined by experiment. It is found that in the case of turbulent flow over a rough bottom

$$C_h = \frac{1.486}{n} R_H^{1/6}$$

where R_H is in feet and n is the *Manning coefficient*; n is given as a function of relative roughness and in practice varies between 0.01 and 0.05. Inserting this expression in the Chezy formula gives the *Manning formula*

$$V = \frac{1.486}{n} R_H^{2/3} S^{1/2}$$

and the conveyance

$$K = \frac{1.486}{n} A R_H^{2/3}$$

14-1.2 Normal Depth and Transitional Depth

14-1.2.1 The normal depth h_n is defined as the distance between the lowest part of the channel and the free surface of a uniform flow. It is determined by the equality

$$\frac{Q_n}{S^{1/2}} = K(h_n)$$

where $K(h_n)$ is the function that characterizes the conveyance of the channel. In the case of a wide rectangular channel, since $R_H = h_n$, the normal depth h_n is

$$h_n = \left(\frac{q_n}{C_h S^{1/2}}\right)^{2/3}$$

where q_n is the normal discharge per unit width. It is seen that, in the case of a uniform flow and for a given channel, there is a unique relationship between normal depth and normal discharge.

14-1.2.2 In the case of a nonuniform flow, the term $d/dx(V^2/2)$ is no longer zero; hence gravity force and friction force do not balance exactly. The water depth h is different from the normal depth; h is then a "transitional depth," which varies with distance and can be larger or smaller than h_n. For example, the transitional depth upstream of a dam is larger than the normal depth.

It is assumed that the values or formulas for the friction coefficients f, C_h, and n obtained in the case of a steady uniform flow with normal depth are still valid in the case of a nonuniform flow with a transitional water depth. In general, the lack of accurate information on the value of the friction coefficient makes this approximation compatible with the inherent error in the original formula. Consequently, the conveyance K is a general function of h which can also be used for nonuniform flow.

14-2 Specific Energy, Specific Force, and Critical Depth

14-2.1 Definition of the Specific Energy and Specific Force

The quantity

$$E = \frac{(1 + \alpha)V^2}{2g} + h = \frac{(1 + \alpha)Q^2}{2gA^2} + h$$

where V is the average velocity, h the maximum water depth, Q the discharge, and A the cross section, is called the specific energy; α is a positive coefficient which appears in this formula because V is average velocity. For the sake of simplicity, α will be neglected in the following discussion. It is seen that the specific energy is the sum of the kinetic energy ($V^2/2g$) and the potential energy per unit of weight with respect to the bottom of the channel h (but not with respect to a horizontal datum) (see Fig. 14-2).

The specific force is

$$I = \iint_A \left(\frac{V^2}{g} + \frac{p}{\rho g}\right) dA = \frac{Q^2}{gA} + aA$$

where a is the distance between the center of gravity of the cross section A and the free surface. It is recalled that the specific force is the sum of the momentum per unit time and per unit weight and the integral of pressure force to a cross section of the channel per unit weight.

In the case of a two-dimensional channel (rectangular cross section), $A = hl$, $a = h/2$, and $q = Q/l$; thus E and I per unit width become

$$E = \frac{q^2}{2gh^2} + h$$

$$I = \frac{q^2}{gh} + \frac{h^2}{2}$$

respectively. Writing

$$h^* = \frac{h}{E}, \qquad q^* = \frac{q}{E(2gE)^{1/2}}$$

one obtains the equation

$$q^* = h^*(1 - h^*)^{1/2}$$

which is universal.

Figure 14-2 *Notation for an open channel.*

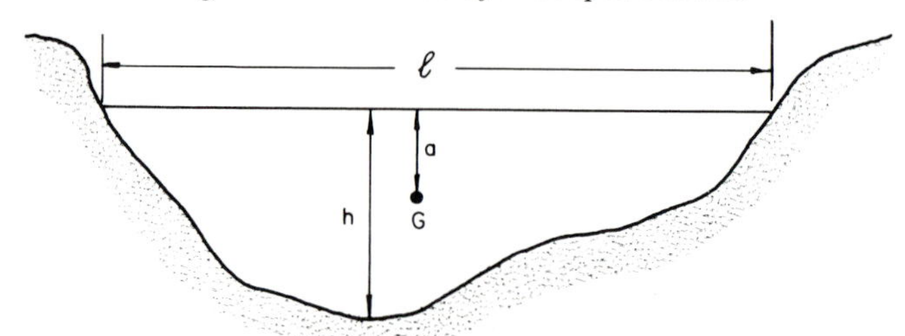

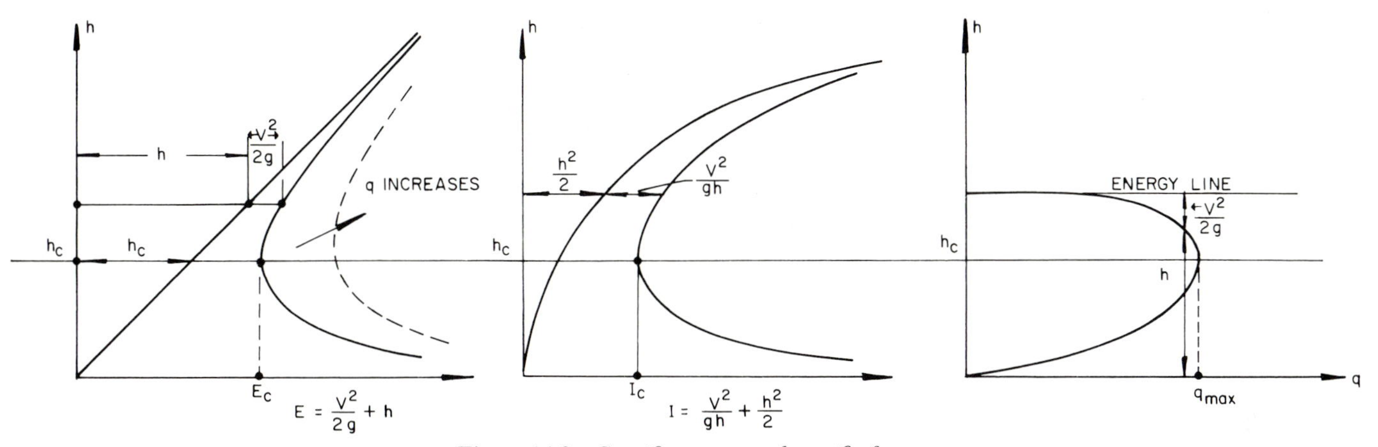

Figure 14-3 *Specific energy and specific force.*

14-2.2 Definition of the Critical Depth

14-2.2.1 Let us consider the two functions $E(q,h)$ and $I(q,h)$ in the case of a rectangular channel. It is seen that for a given value q, $E(h)$, and $I(h)$ vary as shown on Fig. 14-3a,b.

In particular, the minimum values for E and I given by $\partial E/\partial h = 0$ and $\partial I/\partial h = 0$ are obtained for the same value of $h = h_c$, when

$$\frac{q^2}{gh_c^3} = 1$$

i.e.,

$$h_c = \left(\frac{q^2}{g}\right)^{1/3} \qquad \text{or} \qquad V = (gh_c)^{1/2}$$

in which case

$$E = E_c = \tfrac{3}{2}h_c$$

$$I = I_c = \tfrac{3}{2}h_c^2$$

14-2.2.2 Similarly, the function $q(h)$ for a given value $E =$ constant varies as shown in Fig. 14-3c. It is easily verified that the maximum value for q given by $\partial q/\partial h = 0$, is also found to be for $h = h_c$.

14-2.2.3 In the case of a complex cross section $A(h)$ the critical depth is defined by

$$\frac{\partial E}{\partial h} = 1 - \frac{2}{A^3}\frac{Q^2}{2g}\frac{\partial A}{\partial h} = 0$$

Since $\partial A/\partial h = l$, $l(h)$ being the width of the river at the free surface, one has

$$\frac{Q^2}{g} = \frac{A^3}{l}$$

Because A and l are functions of h only, it is then possible to establish the curve h_c as a function of the critical discharge Q_c for any kind of cross section (see Fig. 14-4). It is easily verified that in the case of a rectangular cross section, the previous value h_c is found.

Figure 14-4 *Variation of the critical depth with discharge.*

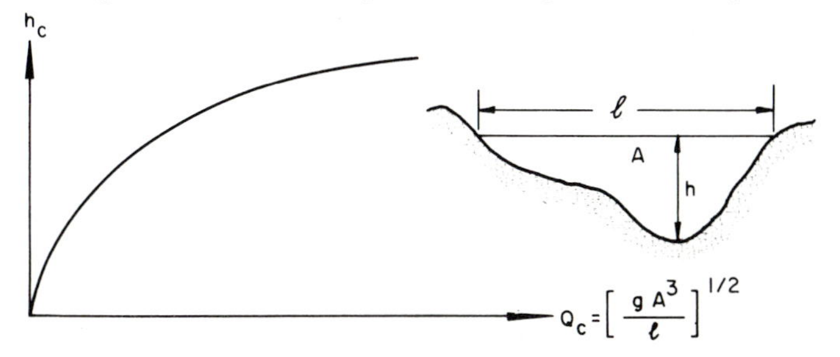

14-3 Tranquil Flow and Rapid Flow

14-3.1 Definition

There exist two possible values, h_1 and h_2, for the same values of E and q, defined by

$$E = h_1 + \frac{V_1^2}{2g} = h_2 + \frac{V_2^2}{2g}$$

and

$$q = h_1 V_1 = h_2 V_2$$

such that

$$\frac{q^2}{2g}\frac{h_1^2 - h_2^2}{h_1^2 h_2^2} = h_1 - h_2$$

By inserting the value for $h_c = (q^2/g)^{1/3}$, one obtains

$$\frac{2h_1^2 h_2^2}{h_2 + h_1} = h_c^3$$

The larger depth, say h_1, corresponds to a "subcritical" or "tranquil flow" in which case $V_1 < (gh)^{1/2}$. The smaller one, say h_2, corresponds to a "supercritical" or "rapid flow" in which case $V_2 > (gh)^{1/2}$.

14-3.2 Critical Slope

It has been seen that the normal discharge is a function of the conveyance and the bottom slope: $Q_n = K(S)^{1/2}$, whereas the critical discharge is a function of the cross section and the width of the channel only $Q_c = (gA^3/l)^{1/2}$.

There is a slope, the "critical slope," for which the normal discharge, i.e., the discharge of a uniform flow, is equal to the critical discharge, and consequently the normal depth h_n is equal to the critical depth h_c.

Consequently, the critical slope is defined by

$$AC_h(R_H S_c)^{1/2} = \left(\frac{gA^3}{l}\right)^{1/2}$$

i.e.,

$$S_c = \frac{g}{C_h^2}\frac{A}{R_H l}$$

and in the case of a large rectangular channel $S_c = g/C_h^2 = f$. When $S > S_c$, one has $h_n < h_c$ corresponding to a value of the Froude number $V/(gh)^{1/2} < 1$: the flow is subcritical or tranquil. When $S < S_c$, one has $h_n > h_c$ corresponding to a value of the Froude number $V/(gh)^{1/2} > 1$, the flow is supercritical or rapid.

14-3.3 Free Surface Disturbance and Flow Control

Since any disturbance travels at an approximate speed $(gh)^{1/2}$, the occurrence of $V > (gh)^{1/2}$ means no perturbation can travel in an upstream direction. Also, when the flow characteristics change from a tranquil upstream flow to a rapid downstream flow such as at the top section of a weir, the discharge is controlled at this critical cross section.

14-4 Gradually Varied Flow

14-4.1 Basic Equations

14-4.1.1 The generalized Bernoulli equation with friction can be applied to a gently sloped free surface flow as follows:

$$\frac{d}{dx}\left(\frac{V^2}{2g} + \frac{p}{\rho g} + z\right) = -\frac{\tau}{\rho g R_H}$$

i.e.,

$$\frac{d}{dx}\left(\frac{Q^2}{2gA^2} + \frac{p}{\rho g} + z\right) = -\frac{Q^2}{C_h^2 R_H A^2}$$

where

$$\frac{Q^2}{C_h^2 R_H A^2} = \frac{Q^2}{K^2} = S_f$$

S_f is the slope of the energy line or head loss. Then, insert $p = p_a$ (atmospheric pressure) as a constant and call z the elevation of the free surface with respect to a horizontal datum. Since

$$\frac{dz}{dx} = -S + \frac{dh}{dx}$$

(see Fig. 14-5) the result is

$$\frac{d}{dx}\left(\frac{Q^2}{2gA^2} + h\right) = S - \frac{Q^2}{K^2}$$

The quantity in parentheses is recognized to be the specific energy.

14-4.1.2 Since

$$\frac{d}{dx}\left[\frac{Q^2}{2gA^2}\right] = -\frac{Q^2}{g}\frac{l}{A^3}\frac{dh}{dx}$$

one has finally,

$$\frac{dh}{dx} = S\,\frac{1 - (Q^2/K^2S)}{1 - (Q^2l/gA^3)}$$

which is the fundamental equation for gradually varied flow. Considering that

$$\frac{Q^2}{K^2S} = \frac{S_f}{S} \quad \text{and} \quad \frac{Q^2l}{gA^3} = \frac{C_h^2R_HS_fl}{gA} = \frac{S_f}{S_c}$$

this equation can be written

$$\frac{dh}{dx} = S\,\frac{1 - (S_f/S)}{1 - (S_f/S_c)}$$

In the case of a large rectangular channel ($R_H \cong h$) one obtains, after some elementary operations,

$$\frac{dh}{dx} = S\,\frac{h_n^3 - h^3}{h_c^3 - h^3}$$

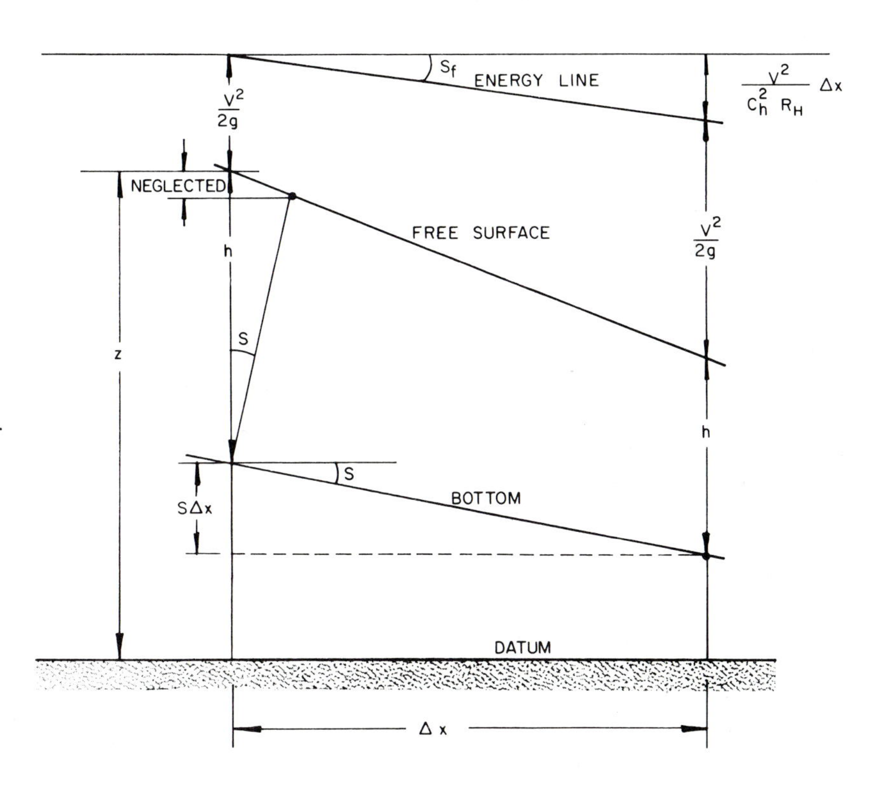

Figure 14-5
Notation for gradually varied flow.

14-4.2 Backwater Curves

14-4.2.1 In general, the free-surface variation depends upon (1) the sign of S; (2) the sign of the numerator, depending upon the water depth h by comparison with the normal depth h_n; and (3) the sign of the denominator, depending upon the water depth h by comparison with the critical depth h_c.

It is to be noted that $h > h_n$ or $h < h_n$ may involve $h > h_c$ or $h < h_c$, depending whether $h_n > h_c$ or $h_n < h_c$. In practice, the first calculation will consist of comparing the value of h_n and h_c.

For example, consider the case of a flow in a channel with a positive slope: $S > 0$, in which one finds $h_n > h_c$. In the case in which $h > h_n$, both the numerator and denominator are negative, so dh/dx is positive. The water depth continuously increases with distance. It is the case of a backwater curve due to a dam on a river with a gentle slope (see Fig. 14-6, case M_1).

Table 14-1 and Fig. 14-6 illustrate all the cases that may be encountered.

Area A corresponds to the case in which the transitional water depth is larger than both the normal depth h_n and the critical depth h_c. Area B corresponds to the case where the transitional water depth is between h_n and h_c. Area C corresponds to the case where the transitional water depth is smaller than both h_n and h_c. In depth variation B (backwater) means that the water depth tends to increase in the direction of the flow. D (draw-down) means that the water depth tends to decrease in the direction of the flow. U (uniform) means that the water depth is constant.

All other cases can be found systematically by a simple analysis of the basic equation. Similarly, the backwater curves on channels of variable slope, continuous or discontinuous, can be analyzed and calculated.

14-4.2.2 In engineering practice, the calculation of backwater curves can be done either by an exact integration in some simple cases such as a rectangular channel (the Bresse formula), or most often numerically by application of a finite difference method over a succession of intervals Δx. Also, a number of approximate methods and graphical methods exist. A number of tables and graphs can be found in the technical literature for a dimensionless "channel unity." As usual, the most direct approach based on a finite difference method is the most practical and can be easily programmed for high-speed computers.

Table 14-1 *Backwater curves*

	Areas				*Water depth variations*
Bottom slopes	*A*	*B*	*C*	*Water depths*	
Horizontal	—			$h > h_n > h_c$	—
$(S = 0)$		H_2		$h_n > h > h_c$	D
			H_3	$h_n > h_c > h$	B
Adverse	—			$h > \lvert h_n \rvert > h_c$	—
$(S < 0)$		A_2		$\lvert h_n \rvert > h > hc$	D
			A_3	$\lvert h_n \rvert > h_c > h$	B
Gentle	M_1			$h > h_n > h_c$	B
$(0 < S < S_c)$		M_2		$h_n > h > h_c$	D
			M_3	$h_n > h_c > h$	B
Critical	C_1			$h > h_c = h_n$	B
$(S = S_c)$		C_2		$h_c = h = h_n$	U
			C_3	$h_c = h_n > h$	B
Steep	S_1			$h > h_c > h_n$	B
$(S > S_c > 0)$		S_2		$h_c > h > h_n$	D
			S_3	$h_c > h_n > h$	B

Notation as shown in Fig. 14-6.

14-5 Rapidly Varied Flow

14-5.1 Hydraulic Jump

14-5.1.1 When the flow is rapid upstream and tranquil downstream, the transition is done through a hydraulic jump. Let us consider the simple case of a rectangular

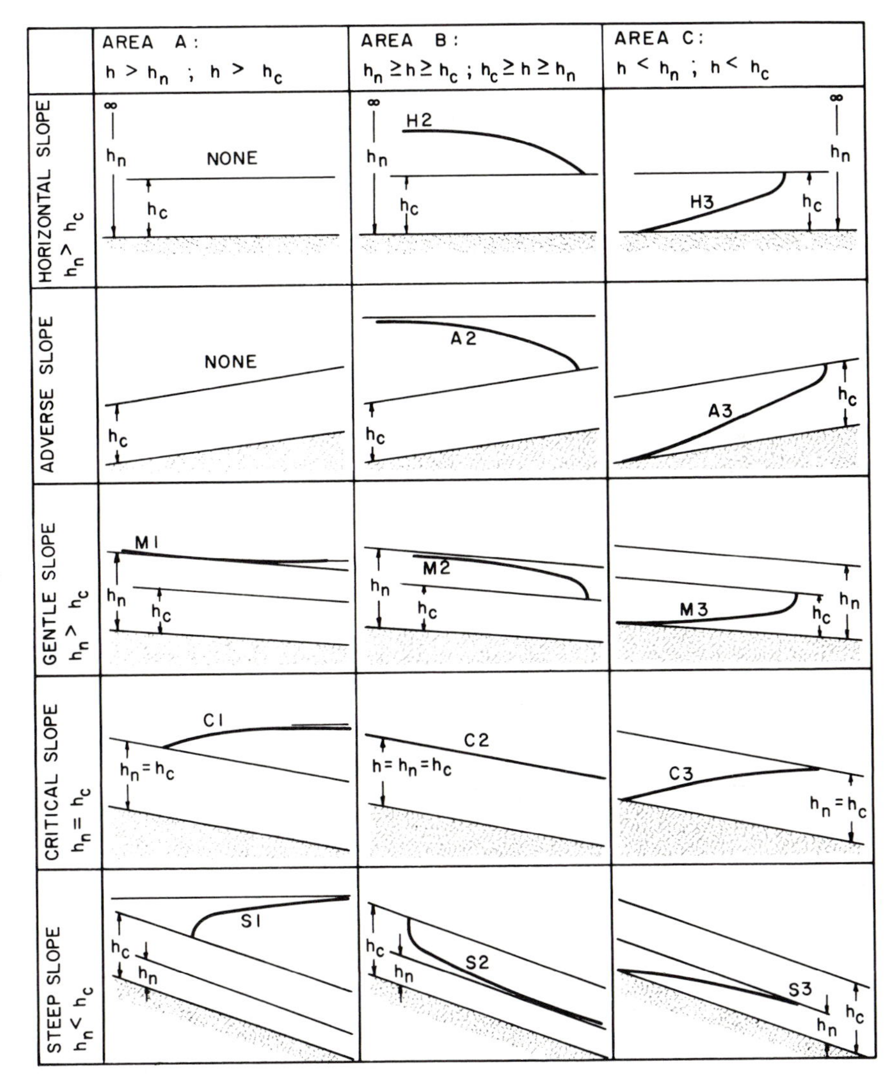

Figure 14-6

Different kinds of backwater curves.

channel of width unity. The discharge is q. The momentum theorem applied to the hydraulic jump is then (see Section 12-3.3):

$$\rho \frac{q^2}{h_1} + \rho g \frac{h_1^2}{2} = \rho \frac{q^2}{h_2} + \rho g \frac{h_2^2}{2}$$

where h_1 and h_2 are the "conjugate" depths. For every upstream value $h_1 < h_c$, there corresponds a value $h_2 > h_c$.

The backwater curve $h_1'(x)$ for rapid flow is calculated by application of the Bernoulli equation from the upstream conditions. The backwater curve for tranquil flow $h_2'(x)$ is also determined by application of the Bernoulli equation

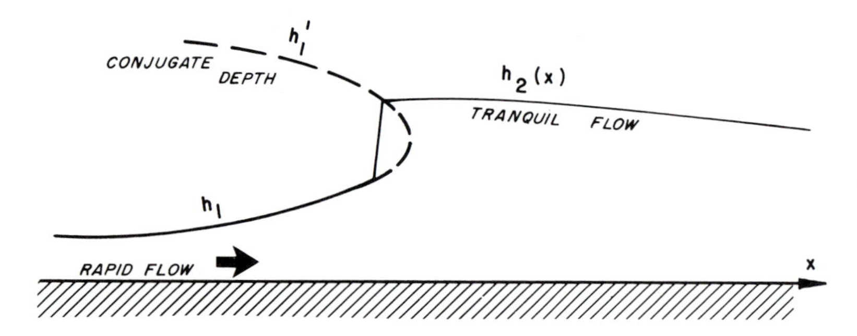

Figure 14-7

Determination of the location of an hydraulic jump.

from the downstream conditions. The hydraulic jump takes place between these two curves at a location where the two water depths are *conjugated*. Therefore, it is sufficient to calculate for each value $h_1 < h_c$ the corresponding conjugate depth $h_1' > h_c$ (Fig. 14-7). The hydraulic jump takes place at the location where the backwater curve $h_2(x)$ for tranquil flow, crosses the curve $h_1'(x)$, conjugate of $h_1(x)$.

The location of such a hydraulic jump is somehow unstable. It could be stabilized by an obstacle, or a sudden deepening. Such a system is used in water power installations as a stilling basin for dissipating energy downstream spillways or bottom outlets.

14-5.1.2 The head loss caused by a hydraulic jump is determined by the difference

$$\Delta H = \left(\frac{q^2}{2gh_1^2} + h_1\right) - \left(\frac{q^2}{2gh_2^2} + h_2\right)$$

which gives

$$\Delta H = (h_2 - h_1)\left[\frac{q^2(h_2 + h_1)}{2gh_1^2h_2^2} - 1\right]$$

Inserting the hydraulic jump equation (see Section 12-3.3)

$$\frac{q^2}{gh_1h_2} = \frac{(h_1 + h_2)}{2}$$

one obtains

$$\Delta H = \frac{(h_2 - h_1)^3}{4h_1h_2}$$

and the energy loss per unit time is

$$\frac{dE}{dt} = \rho g q \, \Delta H$$

14-5.2 Effect of Flow Curvature

The pressure distribution is no longer hydrostatic in the case of rapidly varied flow. The path curvature has a nonnegligible influence on the flow behavior. Furthermore, a uniform flow near critical depth is very unstable. The flow motion may then enter the category of flow motion with path curvature.

Call R the radius of curvature of the free surface, such that

$$\frac{1}{R} = \frac{d^2h/dx^2}{[1 + (dh/dx)^2]^{3/2}} \simeq \frac{d^2h}{dx^2}$$

Assume that the curvature is linearly distributed from the bottom ($z = -h$) to the free surface, such that

$$\frac{1}{R(z)} = \frac{h + z}{h}\frac{d^2h}{dx^2}$$

The centrifugal acceleration is V^2/R where V is the average velocity and the dynamic equation along a vertical becomes

$$\frac{\partial p}{\partial z} = -\rho g + \rho V^2\left[\frac{h + z}{h}\frac{d^2h}{dx^2}\right]$$

which, when integrated, is

$$p(z) = -\rho g z + \rho \frac{V^2}{h} \frac{d^2 h}{dx^2} \left[hz + \frac{z^2}{2} \right]$$

Boussinesq has made an extensive use of this equation for determining the condition when a flow motion is stable or may become undulated, or when it results in an undulated hydraulic jump.

Also, the study of flow over a rapidly variable slope such as over a weir has been investigated by taking flow curvature into account. This subject will not be developed further in this book.

PROBLEMS

14.1 Determine the value of the hydraulic radius R_H for a circular conduit as a function of the maximum water depth.

14.2 A trapezoidal channel has a width at the bottom of 40 ft (12.19 m) and a bank slope of 45°. The bottom slope $S = 0.002$, and the Manning coefficient $n = 0.03$. Calculate the curve for a normal depth y_n as a function of the discharge Q_n up to $Q = 1000$ ft^3/sec (28.32 m^3/sec). Calculate the curve giving the critical discharge Q_c as a function of the critical depth y_c and determine the value of the discharge for which critical depth is equal to normal depth.

14.3 It is possible to define a channel that has a constant hydraulic radius R_H whatever the water depth h?

14.4 Demonstrate that the free surface slope of a steady gradually varied flow in an open channel is equal to the sum of the energy slope and the slope due to the velocity change $d(V^2/2g)/dx$.

14.5 Demonstrate that

$$\frac{dh}{dx} = S \frac{1 - (Q/Q_n)^2}{1 - (Q/Q_c)^2}$$

where S is the bottom slope, Q the discharge, Q_n the normal discharge, and Q_c the critical discharge. Also demonstrate that, in the case where the Manning formula is used for large rectangular channels,

$$\frac{dh}{dx} = S \frac{1 - (h_n/h)^{10/3}}{1 - (h_c/h)^3}$$

Demonstrate also that in a rectangular channel of variable width $l(x)$

$$\frac{dh}{dx} = \frac{S - S_f + (Q^2/gA^3) dl/dx}{1 - (Q^2 l/gA^3)}$$

14.6 From the equation

$$\frac{dh}{dx} = S \frac{h^3 - h_n^3}{h^3 - h_c^3}$$

and depending upon the sign of S and the relative value of h, h_n, and h_c, determine all the possible kinds of backwater curves that may be encountered in a rectangular channel.

14.7 Determine all the kinds of water surface profiles that are encountered in the two cases in which the normal depth is larger than the critical depth and at the opposite, where the critical depth is larger than the normal depth in Fig. 14-8, next page. The opening of the gate O will be considered as a variable in such a way that O can be smaller or larger or both than h_c and h_n.

14.8 Establish under which conditions there are two normal depths and two critical depths in a circular closed conduit. Determine all the possible backwater curves that may exist in such a conduit.

14.9 Establish the required condition for establishing a hydraulic jump on a sudden bottom drop corresponding to various assumptions on the exact location of the jump with respect to the sudden bottom drop.

14.10 Demonstrate that the ratio of gain of potential energy defined by the difference $h_2 - h_1$ to the loss of kinetic

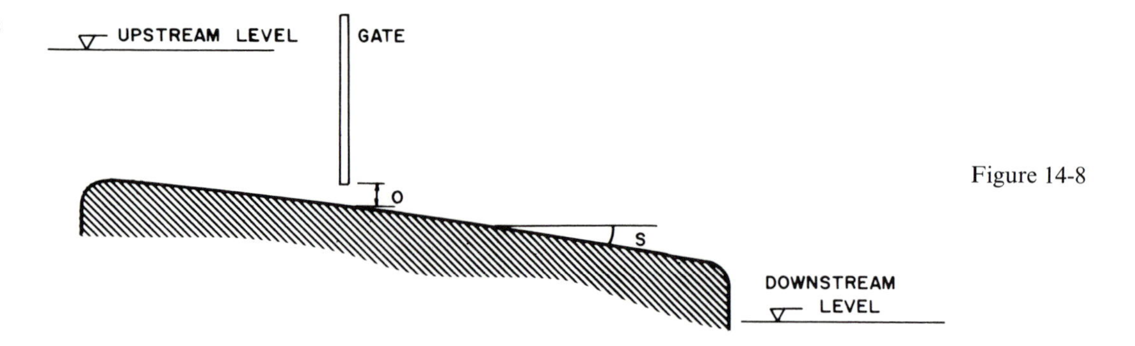

Figure 14-8

energy defined by the difference

$$\frac{V_1^2 - V_2^2}{2g}$$

across a hydraulic jump in a rectangular channel is

$$\frac{4h_1h_2}{(h_1 + h_2)^2}$$

14.11 Demonstrate that the efficiency of a hydraulic jump, defined as the ratio of the specific energy after and before the jump, is

$$\frac{E_2}{E_1} = \frac{(8F_1^2 + 1)^{3/2} - 4F_1^2 + 1}{8F_1^2(2 + F_1^2)}$$

where

$$F = \frac{V_1}{(gh_1)^{1/2}}$$

Subscripts 1 and 2 refer to upstream and downstream values, respectively.

14.12 Consider a vertical jet hitting a horizontal large circular plate at its center. The plate is ended by a weir providing a quasiconstant water depth h. Determine the location of the circular hydraulic jump as a function of the discharge Q of the jet and h. The bottom friction forces will be neglected.

REFERENCES FOR PART TWO

Bakhmeteff, B. A., *Hydraulics of Open Channels.* McGraw-Hill, New York, 1932.

Caratheodory, C., *Conformal Representation.* Cambridge University Press, London, 1941.

Chow, V. T., *Open Channel Hydraulics.* McGraw-Hill, New York, 1959.

Hinze, J. O., *Turbulence.* McGraw-Hill, New York, 1959.

Kellogg, O. D., *Foundations of Potential Theory.* Murrey Printing Co., New York, 1929.

Kober, H., *Dictionary of Conformal Representations.* Dover Publications, Inc., New York, 1952.

Lamb, H., *Hydrodynamics*, 6th ed., Dover Publications, New York, 1945.

Landau, L. D., and Lifshitz, L. D., *Fluid Mechanics.* Pergamon Press, London, 1959.

Lin, C. C., *The Theory of Hydrodynamics Stability.* Cambridge University Press, London, 1955.

McNown, J. S., Hsu, E. Y., and Yih, C. S., Applications of the relaxation technique in fluid mechanics. *Trans., ASCE, 120*: 650–686, 1955.

Milne-Thomson, L. M., *Theoretical Hydrodynamics*, 2nd ed., Macmillan, New York, 1950.

Prandtl, L., *Essentials of Fluid Mechanics*, Hafner, New York, 1952.

Rouse, H., editor, *Advanced Mechanics of Fluids.* Wiley & Sons, New York, 1959.

Rouse, H., editor, *Engineering Hydraulics*, Proceedings of the 4th Hydraulic Conference, Iowa, Wiley & Sons, New York, 1949.

Sabersky, R. H., and Acosta, S. A., *Fluid Flow—A First Course in Fluid Mechanics.* Macmillan, New York, 1964.

Schlichting, H., *Boundary Layer Theory.* McGraw-Hill, New York, 1960.

Streeter, V. L., editor, *Handbook of Fluid Mechanics*, McGraw-Hill, New York, 1961.

Taylor, G. I., Statistical theory of turbulence. *Proc. Roy. Soc. Lond. A*, *151*, 1935.

Townsend, A. A., *The Structure of Turbulent Shear Flow.* Cambridge University Press, London, 1956.

Von Kármán, T., *Collected Works.* Butterworth Scientific Publications, London, 1956.

Von Kármán, T., and Howarth, L., On the statistical theory of isotropic turbulence. *Proc. Roy. Soc. Lond.*, *164*, 1937.

PART THREE

Water Wave Theories

Chapter 15

An Introduction to Water Waves

15-1 A Physical Classification of Water Waves and Definitions

15-1.1 On the Complexity of Water Waves

15-1.1.1 The aim of this chapter is to present the theories for unsteady free surface flow subjected to gravitational forces. Such motions are called water waves, although pressure waves (such as acoustic waves) in water are also water waves. They are also called gravity waves, although atmospheric motions are also waves subjected to gravity.

From the physical viewpoint, there exist a great variety of water waves. Water wave motions range from storm waves generated by wind in the oceans to flood waves in rivers, from seiche or long period oscillations in harbor basins to tidal bores or moving hydraulic jumps in estuaries, from waves generated by a moving ship in a channel to tsunami waves generated by earthquakes or to waves generated by underwater nuclear explosions.

From the mathematical viewpoint, it is evident that a general solution does not exist. Even in the simpler cases, approximations must be made. One of the important aspects of water wave theories is the establishment of the limits of validity of the various solutions due to the simplifying assumptions. The mathematical approaches for the study of wave motion are as varied as their physical aspects. As a matter of fact, the mathematical treatments of the water wave motions embrace all the resources of mathematical physics dealing with linear and nonlinear problems as well. The main difficulty in the study of water wave motion is that one of the boundaries, namely the free surface, is one of the unknowns.

Water wave motions are so varied and complex that any attempt at classification may be misleading. Any definition corresponds to idealized situations which never occur rigorously but are only approximated. For example, a pure two-dimensional motion never exists. It is a convenient mathematical concept which is physically best approached in a tank with parallel walls. Boundary layer effects and transverse components still exist although they are difficult to detect.

15-1.1.2 It has to be expected that, due to this inherent complexity, a simple introduction to the problem of water waves is a difficult, if not impossible, task. Hence, this chapter should rather be regarded as a guide for the following chapters and for continuing further study beyond the scope of this book.

The full assimilation of the subject leading to a clear-cut understanding of this chapter can only come after a comprehensive study of each existing theory within or beyond the scope of the present book. With this in mind, the following classification is proposed. A physical classification is given first, then the different mathematical approaches and their limits of validity are introduced. Finally, the traditional two great families of water waves are presented.

15-1.2 Oscillatory Waves

From the physical viewpoint, there exist essentially two kinds of water waves. They are the oscillatory waves and the translatory waves. In an oscillatory wave, the average transport of fluid; i.e., the discharge or mass transport, is nil. The wave motion is then analogous to the transverse oscillation of a rope (see Fig. 15-1). A translatory wave involves by definition a transport of fluid in the direction in which the wave travels. For example, a moving hydraulic jump, so called tidal bore or simply bore, is a translatory wave.

15-1.2.1 An oscillatory wave can be progressive or standing. Consider a disturbance $\eta(x,t)$ such as a free surface elevation traveling along the OX axis at a velocity C. The characteristics of a progressive wave remain identical for an observer traveling at the same speed and in the same direction as the wave (Fig. 15-2). In the case where η can be expressed as a function of $(x - Ct)$ instead of (x,t), a "steady state" profile is obtained. $\eta(x - Ct)$ is the general expression for a steady progressive wave traveling in the positive OX direction at a constant velocity C. In the case where the progressive wave is moving in the opposite direction, its mathematical form is expressed as a function of $(x + Ct)$. It is pointed out that the definition of wave velocity C for a nonsteady profile is not strictly valid, since each "wave element" travels at its own speed, resulting in wave deformation.

15-1.2.2 The simplest case of a progressive wave is the wave which is defined by a sine or cosine curve such as

$$\eta = \frac{H}{2} \left\{ \begin{matrix} \sin \\ \cos \end{matrix} \right\} m(x - Ct)$$

Such a wave is called a harmonic wave where $H/2$ is the amplitude and H the wave height.

The distance between the wave crests is the wavelength L, and $L = CT$ where T is the wave period. The wave-number (or number of cycles per unit distance) $m = 2\pi/L$ is the number of wavelengths per cycle. The frequency is $k = 2\pi/T$. Hence the previous equation can be written

$$\eta = \frac{H}{2} \left\{ \begin{matrix} \sin \\ \cos \end{matrix} \right\} 2\pi\left(\frac{x}{L} - \frac{t}{T}\right)$$

H/L is the *wave steepness.*

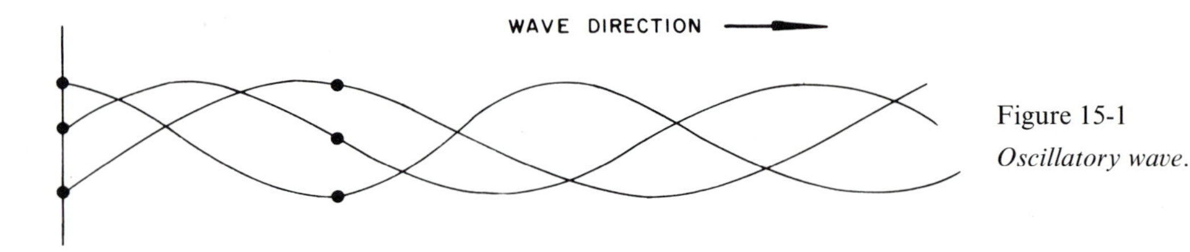

Figure 15-1
Oscillatory wave.

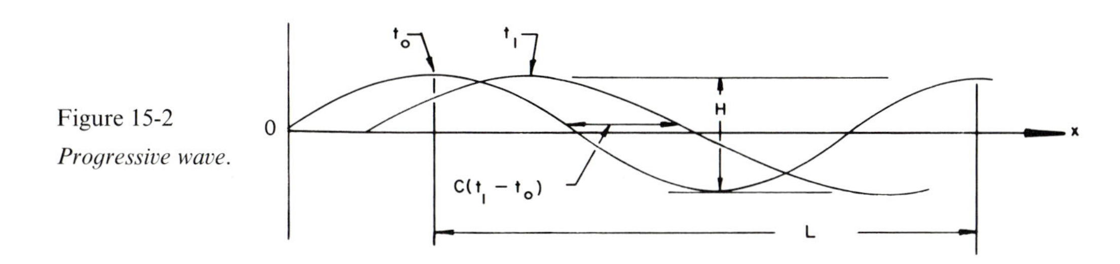

Figure 15-2
Progressive wave.

15-1.2.3 A standing or stationary wave is characterized by the fact that it can be mathematically described by a product of two independent functions of time and distance, such as

$$\phi = H \sin \frac{2\pi x}{L} \sin \frac{2\pi t}{T}$$

or more generally:

$$\phi = \phi_1(x) \cdot \phi_2(t)$$

A standing wave can be considered as the superposition of two waves of the same amplitude and same period traveling in opposite directions. In the case where the convective inertia terms are negligible, the standing wave motion is obtained by a mere linear addition of the solutions for two progressive waves (see Fig. 15-3). The following identity holds:

$$\frac{H}{2} \sin \frac{2\pi}{L}(x - Ct) + \frac{H}{2} \sin \frac{2\pi}{L}(x + Ct)$$

$$= H \sin \frac{2\pi}{L} x \cos \frac{2\pi}{T} t$$

A standing wave generated by an incident wind wave is called *clapotis*. In relatively shallow water ($d/L < 0.05$) it is called a *seiche*. A seiche is a standing oscillation of long period encountered in lakes and harbor basins. The amplitude at the node is zero, and at the antinode it is H.

15-1.2.4 Two waves of same period but different amplitudes traveling in opposite directions form a "partial clapotis" and can be defined linearly by the sum of $A \sin (x - Ct) + B \sin (x + Ct)$. A partial clapotis can also be considered as the superposition of a progressive wave with a standing wave. A partial clapotis is encountered in front of an obstacle which causes partial reflection.

15-1.2.5 It will be seen that the wave velocity is in general a function of the water depth (see Section 16-3.2). Hence, if the depth remains constant, wave crests remain parallel. However, when the depth varies in a direction not parallel to the wave crest, different parts of the crest travel at different speeds, and the direction of propagation of the crest changes. This is the phenomenon of refraction which is encountered when a wave travels from one water depth to another water depth at an angle with the bottom contours.

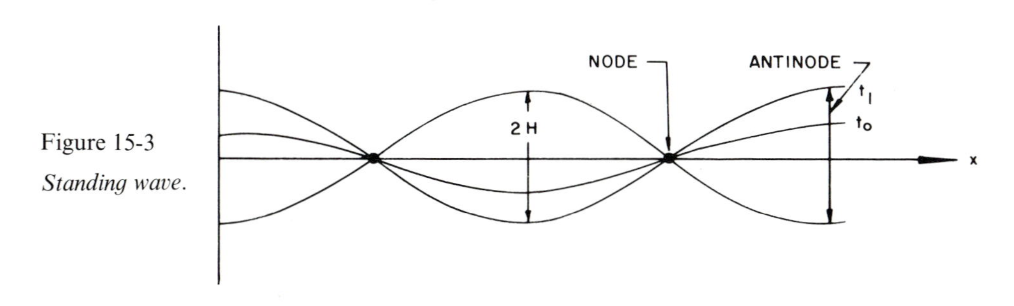

Figure 15-3
Standing wave.

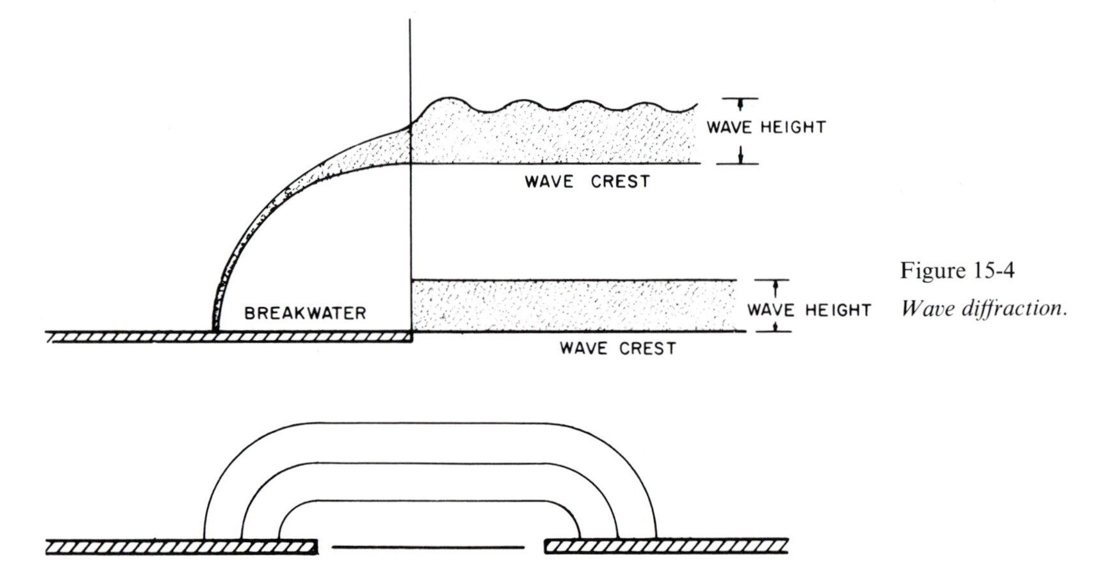

Figure 15-4
Wave diffraction.

15-1.2.6 The phenomenon of diffraction is encountered at the end of an obstacle (Fig. 15-4). It can be considered as a process of transmission of energy in a direction parallel to the wave crest although this definition oversimplifies a more complex phenomenon. The phenomenon of wave diffraction is actually the result of phase differences between the waves radiating from a line source, which is interrupted by the obstacle. It is an end effect.

15-1.2.7 The breaking phenomenon is encountered at sea under wind action (white caps), on beaches (surf), and in tidal estuaries (tidal bores) (see Fig. 15-5). It is a shock wave phenomenon which is also encountered in gas dynamics. The breaking phenomenon is characterized by a high rate of free turbulence and air entrainment associated with a high rate of energy dissipation. Bores generated by wind waves breaking on beaches or by tides of high amplitude in estuaries should be regarded as translatory waves rather than oscillatory waves.

A number of equivalent definitions can be given for the breaking criterion: Breaking occurs (1) when the particle velocity at the crest becomes larger than the wave velocity, (2) when the pressure at the free surface given by Bernoulli equation is incompatible with the atmospheric pressure, (3) when the particle acceleration at the crest tends to separate the particles from the bulk of the water surface, or (4) when the free surface becomes vertical. Accordingly, the following theoretical formulas are generally used:

1. In deep water (Michell limit): $H/L < 0.142$.
2. In intermediate water depth (Miche formula): $H/L < 0.14 \tanh (2\pi\, d/L)$.
3. In the case of a solitary wave, the maximum relative wave height: $H/d < 0.78$, where d is the water depth.

15-1.3 Translatory Waves

In a translatory wave, there is a transport of water in the direction of the wave travel. Some examples of such phenomena are tidal bores or moving hydraulic jumps; waves generated by the breaking of a dam; surges on a dry bed; undulated moving hydraulic jumps; solitary waves; and flood waves in rivers. It is pointed out that oscillatory and translatory waves may sometimes look very much alike and be treated mathematically by the same method.

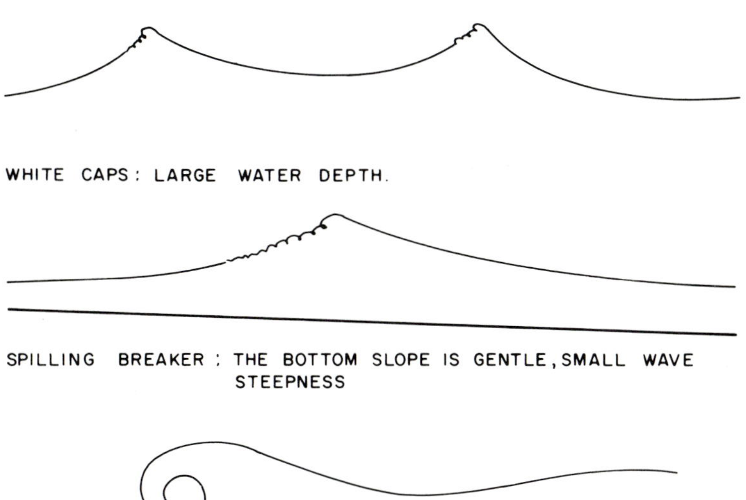

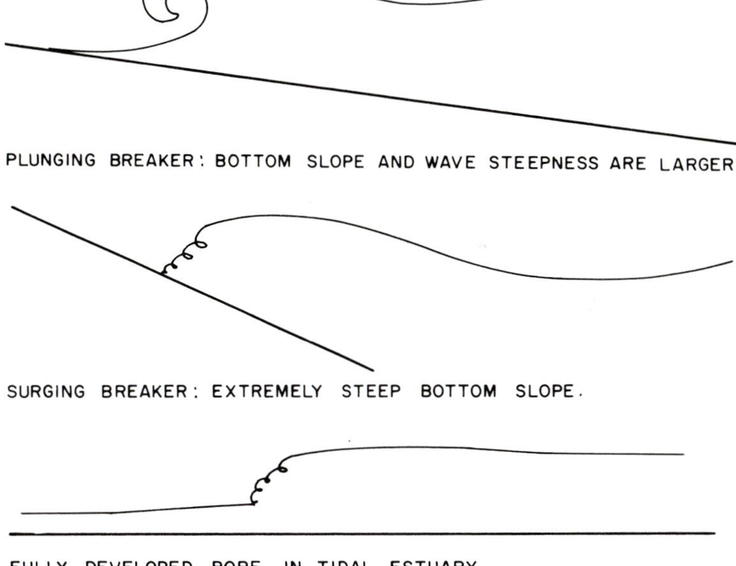

Figure 15-5
Different kinds of wave breakers.

For example, a solitary wave, which is a translatory wave, is characterized by a unique wave crest accompanied by a sudden jump ahead by the water particles under the wave crest. Cnoidal waves (discussed in far greater detail below), which are oscillatory waves, present very similar characteristics. However, in the case of a cnoidal wave, there is a gentle slow return under a long flat trough between wave crests. A solitary wave motion always involves an important net mass transport. A cnoidal wave has a very small mass transport because of this return flow. From the mathematical viewpoint, these two kinds of motion are of the same family, i.e., they are subjected to the same simplifying

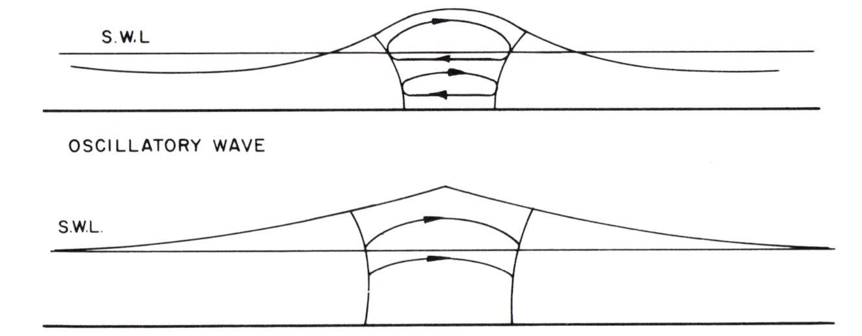

Figure 15-6

Difference between an oscillatory wave and a translatory wave.

assumptions and they obey the same basic equations. The solitary wave is a limited case of the cnoidal wave when the wave period tends to infinity (see Fig. 15-6).

15-2 Mathematical Classification

15-2.1 The Significant Wave Parameters

15-2.1.1 In an Eulerian system of coordinates a surface wave problem generally involves three unknowns: the free surface elevation (or total water depth), the pressure (generally known at the free surface), and the particle velocity. Since a general method of solution is impossible, a number of simplifying assumptions have been made which apply to a succession of particular cases with varying accuracy. In general, the method of solution which is used depends upon nonlinear effects, i.e., the relative importance of the convective inertia terms with respect to the local inertia.

15-2.1.2 However, instead of dealing with these inertial terms directly, it is more convenient to relate them to more accessible parameters. Three characteristic parameters are used. They are:

1. A typical value of the free surface elevation such as the wave height H
2. A typical horizontal length such as the wavelength L
3. The water depth d

Although the relationships between the inertial terms and these three parameters are not simple, their relative values are of considerable help in classifying the water wave theories from a mathematical viewpoint.

For example, it is easily conceived that when the free surface elevation decreases, the particle velocity decreases also. Consequently, when the wave height H tends to zero, the convective inertia term, which is related to the square of the particle velocity, is an infinitesimal of higher order than the local inertia term, which is related linearly to the velocity. In such a case, the convective inertia can be neglected and the equations can be linearized.

Three characteristic ratios can be obtained from H, L, and d. These are H/L, H/d, and L/d. The relative importance of the convective inertia term increases as the value of these three ratios increases.

In deep water (small H/d, and small L/d), the most significant parameter is H/L which is called the wave steepness. In shallow water the most significant parameter is H/d which is called the relative height. In intermediate water depth, it will be seen that a significant parameter which also covers the three cases is $(H/L)(L/d)^3$.

15-2.2 The Methods of Solution

Depending upon the problem under consideration and the range of values of the parameters H/L, H/d, and L/d, three mathematical approaches are used. They are (1) linearization; (2) power series; and (3) numerical methods. Statistical methods are also used to describe the complexity of sea states or waves generated by wind action.

15-2.2.1 The simplest cases of water wave theories are, of course, the linear wave theories, in which the convective inertia terms are completely neglected. These theories are valid when H/L, H/d, and L/d are small, i.e., for waves of small amplitude and small wave length in deep water. For the first reason they are called the "small amplitude wave theories." This is the infinitesimal wave approximation.

The linearized equations are so amenable to mathematical solutions that the linear wave theories are used for an extreme variety of water wave motions. For example, some phenomena which are studied by this method are wave diffraction, waves generated by a moving ship, and waves generated by explosions, even though they may have large amplitudes.

15-2.2.2 Solutions can also be found as a power series in terms of a parameter small compared to others. This small quantity is H/L in deep water or H/d in shallow water. In the first case (development in terms of H/L, or Stokes waves) the first term of the power series is a solution of the linearized equations. In the second case, the first term of the series is already a solution of nonlinear equations; these are the cnoidal waves.

The calculation of the successive terms of the series is rather cumbersome so that these methods are used in a very small number of cases. The most typical case is the two-dimensional progressive periodic wave. In this case, the solution is assumed to be *a priori* that of a steady-state profile, i.e., a function such as $F = f(x - Ct)$ where C is a constant equal to the wave velocity. C is also called the phase velocity.

The simplification introduced by such an assumption is due to the fact that

$$\frac{\partial F}{\partial x} = \frac{\partial F}{\partial (x - Ct)}$$

and

$$\frac{\partial F}{\partial t} = C \frac{\partial F}{\partial (x - Ct)}$$

so that

$$\frac{\partial F}{\partial t} = -C \frac{\partial F}{\partial x}$$

As a result, the time derivatives can be eliminated and replaced by space derivatives. Power series solutions are also found in the case of periodic standing waves and irregular waves.

15-2.2.3 It may happen that a steady-state profile does not exist as a solution, in which case a numerical solution where the differentials are replaced by finite differences, is often used. This occurs for large values of H/d and L/d, when the nonlinear terms such as $\rho u(\partial u/\partial x)$ are relatively large by comparison with the local inertia such as $\rho(\partial u/\partial t)$. This is the case of long waves in very shallow water.

Of course, a numerical method of calculation can be used for solving a linearized system of equations. For example, the relaxation method is used for studying small wave agitation in a basin. Also, an analytical solution of a nonlinear system of equations can be found in some particular cases. Hence it must be borne in mind that these three methods and the range of application which has been given indicate more of a trend than a general rule.

15-2.2.4 Aside from the three previous methods which aim at a fully deterministic solution of the water wave problem, the description of sea state generally involves the use of random functions. The mathematical operations which are used in this case (such as harmonic analysis) generally imply that the water waves obey linear laws, which are the necessary requirements for assuming that the principle of superposition is valid. Consequently, such an approach loses its validity for describing the sea state in very shallow water (large values of H/d and L/d) and in the surf zone.

15-2.3 An Introduction to the Ursell Parameter

15-2.3.1 The potential function for a Stokes wave or irrotational periodic gravity wave traveling over a constant finite depth at a second order of approximation is found to be Equation 15-1.

The series is convergent in relatively deep water, and the term in H is the solution obtained by taking into account the local inertia only, while the term in H^2 is the most significant correction due to convective inertia. Therefore, the relative importance of the convective inertia term can be described by the ratio of the amplitudes of these two terms. Even though this solution loses its validity in very shallow water, it is seen after some simple calculations that the ratio of the amplitude of the second-order term to the amplitude of the first-order term is:

$$\frac{3}{16}\frac{1}{(2\pi)^2}\frac{H}{L}\left(\frac{L}{d}\right)^3 \quad \text{when} \quad A = md = \frac{2\pi d}{L} \to 0$$

since $\cosh A \to 1$ and $\sinh A \to A$. When $(H/L)(L/d)^3$ is very small, the small amplitude wave theory is valid.

If, instead of H, one uses the maximum elevation η_0 above the still water level (η_0 is equal to $H/2$ in the linear theory), the so-called Ursell parameter initially introduced by Korteweg and de Vries is obtained:

$$U_R = \frac{\eta_0}{L}\left(\frac{L}{d}\right)^3$$

When $(\eta_0/L)(L/d)^3 \ll 1$, the linear small amplitude wave theory applies. In principle more and more terms of the power series would be required in order to keep the same relative accuracy as the Ursell parameter increases.

15-2.3.2 Also, in the case of very long waves in shallow water such as flood waves, bore, and nearshore tsunami waves, the Ursell parameter is difficult to use since the interpretation of L is not clear. The relative amplitude H/d is then a more significant parameter to assess the importance of the nonlinear terms. In this case, the vertical component of inertia force is negligible and the only significant term for convective inertia is $\rho u(\partial u/\partial x)$. Then it is possible to calculate the ratio of amplitude of convective inertia to the amplitude of local inertia, $\rho u(\partial u/\partial x)/\rho(\partial u/\partial t)$, directly. Since in very shallow water d/L is very small, one has simply, using only the linear term,

$$u = -\frac{\partial \phi}{\partial x} = -\frac{H}{2}\frac{k}{md}\cos(mx - kt)$$

and it is found that

$$\frac{\rho u \left.\dfrac{\partial u}{\partial x}\right|_{\max}}{\rho \left.\dfrac{\partial u}{\partial t}\right|_{\max}} = \frac{H}{2d}$$

which demonstrates the relative importance of the ratio H/d. Despite these difficulties of interpretation, the Ursell parameter is a useful simple guide, but is not necessarily sufficient for judging the relative importance of the nonlinear effects.

$$\phi = -\overbrace{\frac{H}{2}\frac{k}{m}\frac{\cosh m(d+z)}{\sinh md}\sin(kt - mx)}^{\substack{\text{Linear, first-order term} \\ \text{proportional to } H}} - \overbrace{\frac{3}{8}\left(\frac{H}{2}\right)^2 k \frac{\cosh 2m(d+z)}{\sinh^4 md}\sin 2(kt - mx)}^{\substack{\text{Second-order term} \\ \text{proportional to } H^2}} \qquad (15\text{-}1)$$

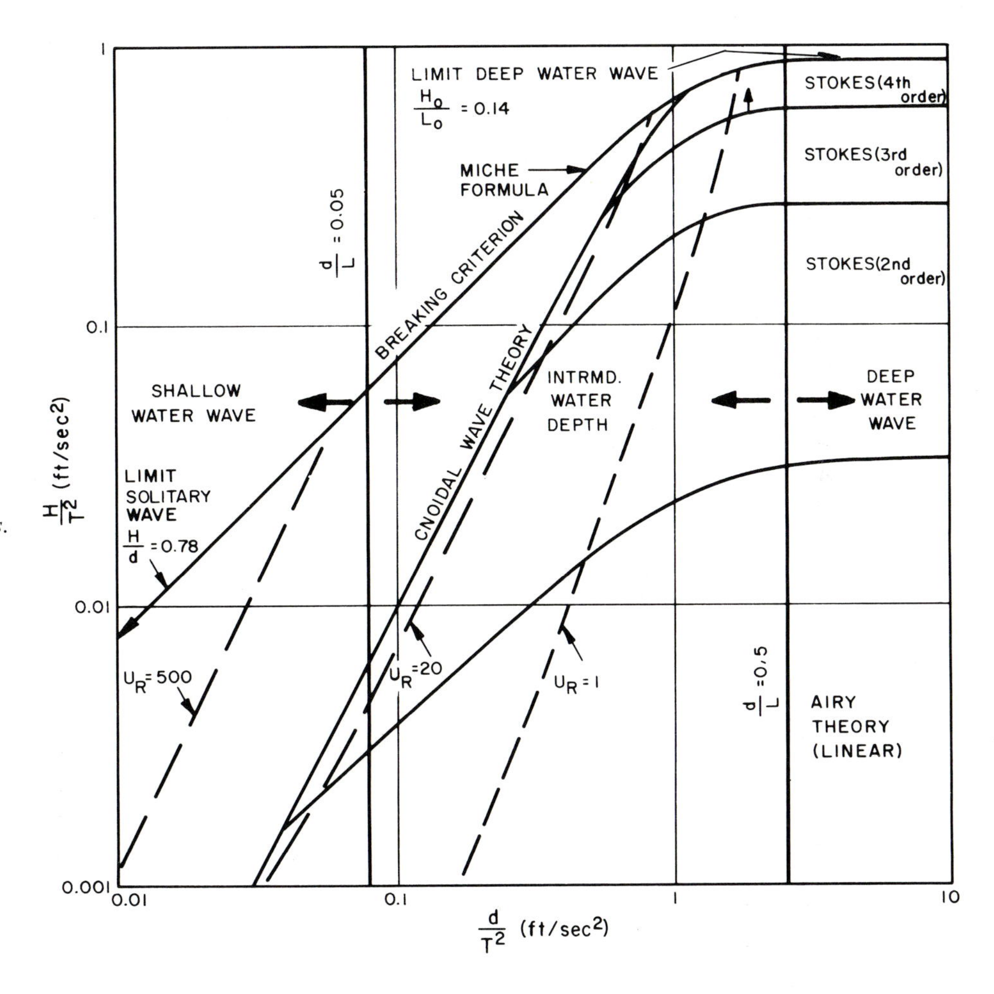

Figure 15-7

Limits of validity for various wave theories.

15-2.3.3 Figure 15-7 indicates approximately the range of validity of the various theories. This graph has been established for two-dimensional periodic waves such as illustrated on Fig. 15-8, but it gives an indication for any kind of water waves. Three corresponding values of the Ursell parameter have been shown. The graph is limited by a breaking criteria which implies that there is a maximum value for the wave steepness which is a function of the relative depth (see Section 15-1.2.7). A comprehensive quantitative investigation of the error which is made by using various theories in various domains has not been done so far; hence, such a graph is somewhat arbitrary and merely qualitative.

15-2.4 The Two Great Families of Water Waves

In hydrodynamics the water wave theories are generally classified in two great families. They are the "small amplitude wave theories" and the "long wave theories." The small amplitude wave theories embrace the linearized

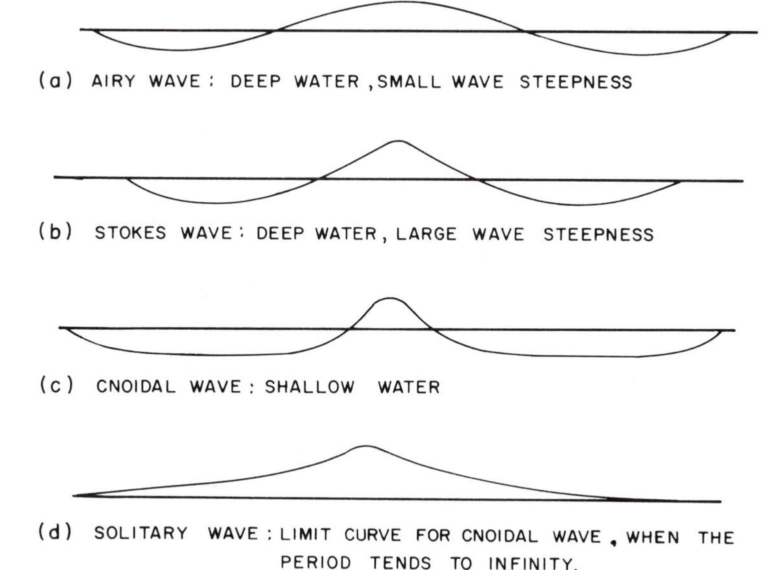

Figure 15-8

A physical illustration of various wave profiles.

theory for infinitesimal amplitude waves and the first categories of power series, i.e., the power series in terms of H/L for finite amplitude waves.

The long wave theories embrace the numerical method of solution mostly used for the nonlinear long wave equations.

These two great families include a number of variations and some intermediate cases presenting some of the characteristics of both families. For example, the cnoidal wave, the solitary wave, and the monoclinal wave are considered as being particular cases (steady-state profile) of the long wave theories, because they are nonlinear shallow water waves.

It can be considered that there exists some arbitrariness in such classification. This arbitrariness is the heritage of tradition, since the wave theories, as any theory, have been developed in a haphazard manner. But it is most important to understand the relative position of these theories with respect to each other, and their limits of validity. The small amplitude wave theories and the long wave theories are now considered separately.

15-3 The Small Amplitude Wave Theory

15-3.1 The Basic Assumption of the Small Amplitude Wave Theory

It has been mentioned in the previous section that the small amplitude wave theory is essentially a linear theory, i.e., the nonlinear convective inertia terms are considered small. It is called the small amplitude wave theory because the equations are theoretically exact when the motion tends to zero even if the convective inertia terms are taken into account. Indeed, in that case the nonlinear terms are infinitesimals of higher order than the linear terms.

This assumption is extremely convenient because the free surface elevation can *a priori* be neglected; i.e., the motion takes place within known boundaries. This assumption is used in order to determine the zero wave motion and such solution is assumed to be valid even if the wave motion is different from zero.

Aside from this assumption, the motion is also most often considered as irrotational. This assumption is compatible with the neglect of the quadratic convective term $\rho\mathbf{V} \times \mathbf{curl\ V}$ (see Section 4-4.3). Then the solution of the problem consists of determining the velocity potential function $\phi(x,y,z,t)$ satisfying the boundary conditions at the free surface and at the limit of the container.

This approach has been proven to be extremely successful even for wave motion of significant magnitude and in shallow water. Moreover, the assumption of linearity permits the determination of a complex motion by superposition of elementary wave motions.

15-3.2 The Various Kinds of Linear Small Amplitude Waves

15-3.2.1 Progressive periodic two-dimensional linear wave motion is the basic motion which leads to the understanding of many other more complex motions. Such a solution is found by assuming that the motion is of the form $A \sin (2\pi/L)(x - Ct)$ where C is a constant. A general solution is obtained as a function of water depth. In deep water, an asymptotic solution is obtained as a limiting case. It is valid when the relative water depth: d/L is larger than 1/2.

The deep water wave theory can also be obtained independently. In very shallow water ($d/L < 1/20$) the solution is the linear shallow water wave theory, which is a special case (a limiting case) of the nonlinear long wave theory. As for the nonlinear long wave theory, the vertical acceleration is negligible and the pressure distribution tends to become hydrostatic, i.e., the pressure is equal to the product of the specific weight of the fluid ρg by the distance from the free surface. Also, the velocity distribution is uniform along a vertical.

Two periodic progressive waves of slightly different period traveling in the same direction form a succession of wave trains that give rise to a beat phenomenon. It has been seen that two progressive periodic waves of the same period and amplitude traveling in opposite directions form a standing wave or clapotis, and in the case where $d/L < 1/20$, a seiche (Fig. 15-9). A periodic wave reflected by a vertical wall at an angle forms a system of "short-crested" waves which appear as a grid of peaks of water moving parallel to the wall.

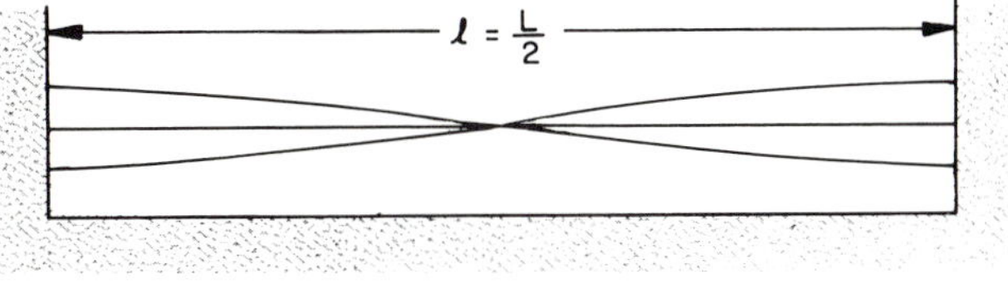

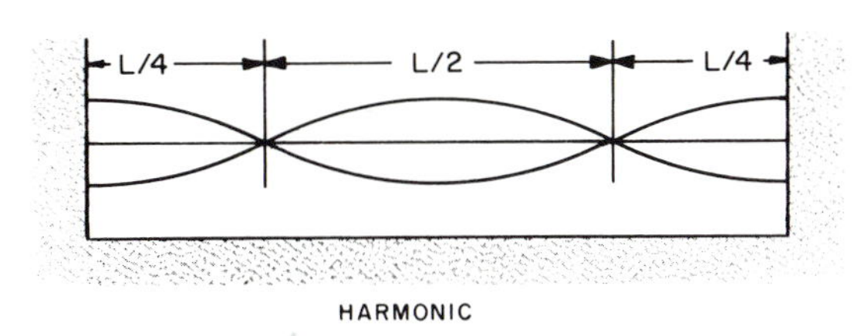

Figure 15-9 *Seiche motion in a two-dimensional basin.*

A great number of three-dimensional periodic motions within complex boundaries can be determined by the small amplitude wave theory. They are the three-dimensional wave motions within tanks of various shapes (rectangular, circular, etc.) with constant or varying depth. The process of wave diffraction by a vertical wall or through a breach is also thoroughly analyzed by the small amplitude wave theory.

Finally, this theory for progressive waves is essentially destined to be the foundation of the study of wind waves, although this phenomenon is random and not periodic. Hence the study of wind waves will require further analysis as described in Appendix A on wave spectra.

15-3.2.2 The small amplitude wave theory is also particularly successful in determining the wave motion created by a sudden disturbance or impulse at the free surface, or at the bottom (see Fig. 15-10). For example, tsunami waves generated by earthquakes can be treated in deep water by mathematical methods of the small amplitude wave theory. Likewise, waves generated by an underwater explosion or by the drop of a stone on the free surface of a body of water receive a similar theoretical treatment. In

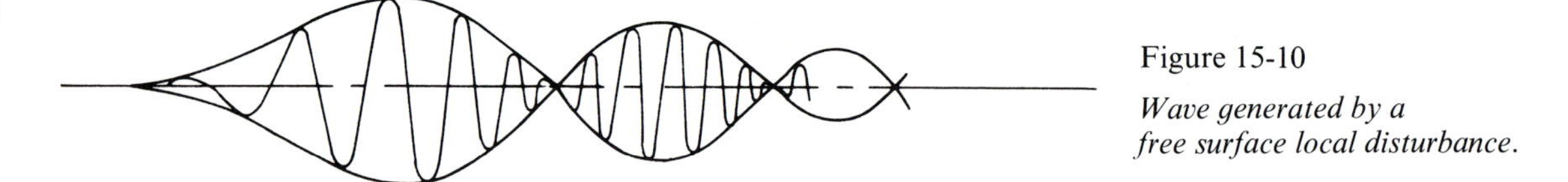

Figure 15-10

Wave generated by a free surface local disturbance.

general, these waves have a cylindrical symmetry but are nonperiodic. One may consider that they have a "pseudo" wave period defined by the time which elapses between two wave crests. This period decreases with time at a given location. Also, in general, this "pseudo" wave period tends to increase with the distance from the disturbance. Waves generated by disturbances often appear as a succession of wave trains, the number of waves within each wave train increasing as the distance from the disturbance increases.

The average wave height also tends to decrease with distance due to the double effects of the increase of wave length with distance and radial dispersion.

15-3.2.3 Finally, the wave motion created by a moving disturbance (ship or atmospheric disturbances) can also be analyzed by application of the small amplitude wave theory. The theoretical wave patterns created by a moving ship are presented in Fig. 15-11. These are called Kelvin waves.

15-3.3 The Finite Amplitude Wave Theories

15-3.3.1 The solution for a progressive harmonic linear wave over a horizontal bottom is a sine function of $(x - Ct)$, so the free surface is perfectly defined by a sine curve. In shallow water, the crest has a tendency to become steeper and the trough flatter, as shown in Fig. 15-8c. In this case, the linear small amplitude wave theory is no longer valid.

15-3.3.2 In the simple case of periodic waves, either progressive or standing, the small amplitude wave theory can be refined by taking into account the convective inertia forces to some extent. It has been indicated in a previous

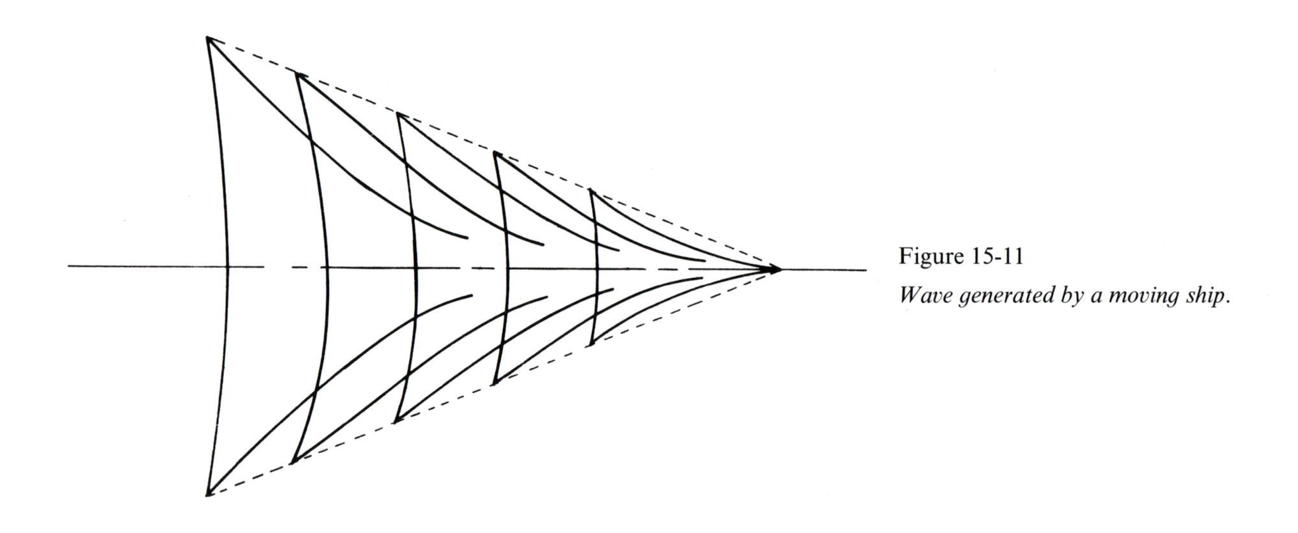

Figure 15-11

Wave generated by a moving ship.

section (Section 15-2.2.2) that this is done by assuming the solution for the motion is given by a power series in terms of a quantity which is small compared to the other dimensions. For example, in the simple case of a periodic progressive or standing two-dimensional wave, it is assumed that the solution for the motion is given as a power series in terms of the wave height H (or of the wave steepness H/L, defined as the ratio of the wave height to the wavelength L). For example, the potential function $\phi(x,z,t)$ will be written:

$$\phi(x,z,t) = H\phi_1 + H^2\phi_2 + H^3\phi_3 + H^4\phi_4$$

The first-order term $H\phi_1$ is found from the linear small amplitude theory by neglecting the nonlinear terms. The other terms are correction terms due to the nonlinear convective inertia. These terms of the series are obtained successively by recurrence formulas. A third-order wave theory, known as a third-order approximation, is a theory in which the calculation has been performed up to the third power of the small quantity. In this case, a third-order approximation is

$$\phi = H\phi_1 + H^2\phi_2 + H^3\phi_3$$

In the case of a harmonic wave, ϕ_2 and ϕ_3 are sinusoidal functions of $n(x - Ct)$ where n is an integer equal to the order of the considered term. The ϕ_n are functions of the relative depth d/L. In practice the complexity of the terms $\phi_2, \phi_3, \ldots$ increases so much as the order of approximation increases that calculation can rarely be performed at a high order of approximation. The formulas for the high order of approximation are so complicated that for their application a set of tables obtained from a high-speed computer is required.

In engineering practice, the first-order wave theory is most often sufficient. However, higher-order wave theory indicates some interesting trends for waves of large steepness (large H/L) in deep water. In very shallow water the convective inertia terms are relatively high and the convergence of the series becomes very slow. The series are not even necessarily uniformly convergent and the function of relative depth d/L loses its meaning.

It has been mentioned that in shallow water the important parameter becomes H/d instead of H/L for deep water. A power series in terms of H/d is most convenient and, in principle, would require fewer terms for a better accuracy. Such power series appear in the cnoidal and solitary wave theories which will be discussed in the following paragraphs (Section 15-4.2.1).

15-3.3.3 Once all the equations of motion and the boundary conditions at the free surface and at the bottom have been specified, an infinite number of solutions may be found. These equations are not sufficient for determining the wave motion.

Two other conditions are required. One is on rotationality and mass transport and is considered in Chapter 17. It is also necessary to specify whether the wave motion should be a progressive wave, a standing wave, or a wave train. For example, in the first case, a solution for steady-state profile has to be found such that the solution appears as a function of $(x - Ct)$ where C is the constant wave velocity. In the second case, a mere addition of two periodic gravity waves traveling in opposite directions can be used in the linear case and for the first term of the power series. However, higher-order terms must be found independently by recurrence formulas, established for the specific type of motion (progressive or standing).

15-4 The Long Wave Theory

15-4.1 The Basic Assumptions of the Long Wave Theory: The Long Wave Paradox

15-4.1.1 The long wave theory applies when the relative depth is very small. Thus, the vertical acceleration can be neglected and the path curvature is small. Consequently, the vertical component of the motion does not influence the pressure distribution, which is assumed to be hydrostatic. However, contrary to the small amplitude wave theory, the free surface is now unknown even during the first step of the calculations. Also, the velocity distribution along a vertical is assumed to be uniform, or the particle velocity is averaged over a vertical. (As in the case of the generalized Bernoulli equation, a correction coefficient close to unity should be included where quadratic terms appear. This refinement is neglected.) Because the equations are nonlinear, the number of analytical solutions is limited to a few particular cases.

While the small amplitude wave theory consists of finding potential function by analytical means, the long wave theory is most often treated by numerical methods or graphical methods and by the use of a high-speed computer.

15-4.1.2 An error, inherent in the simplifying assumptions, is encountered systematically in the treatment of the nonlinear long wave theory. The velocity of the "wave element" is an increasing function of the water depth such as $(gh)^{1/2}$. Therefore, according to the theory, the wave elements carrying the most energy should have a tendency to catch up with the first wave elements ahead of the wave (see Fig. 15-12). A vertical wall of water should soon result, forming a tidal bore. However, although this phenomenon may actually occur physically, when it does occur, it happens much later than predicted by the long wave theory. In particular, in the case of a wave which contains high space derivatives for $\partial\eta/\partial x$ and $\partial u/\partial x$, the long wave theory may no longer be valid. Similarly, the breaking of a long wave on a beach will be predicted sooner than if it were due to the change of bottom depth only.

Figure 15-12 *A physical illustration of the long wave paradox.*

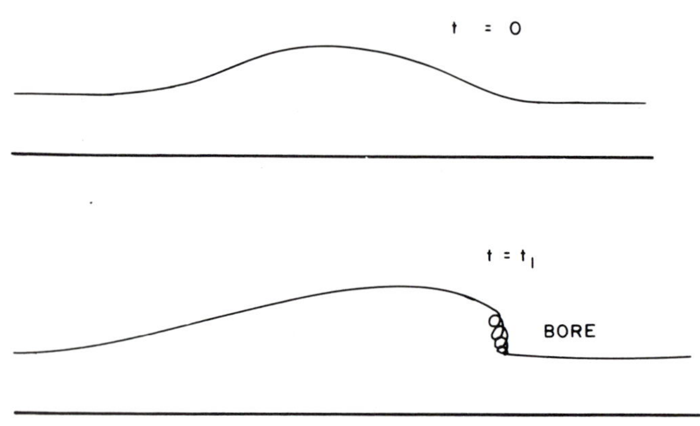

Finally, it is realized that the long wave theory and the steady-state profile are two concepts theoretically incompatible, although steady-state profiles have been observed. This inherent deficiency in the long wave theory is the long wave paradox and is also encountered in gas dynamics and nonlinear acoustics.

The two stabilizing factors which explain the existence of steady-state profile are the vertical acceleration and the bottom friction. They are now considered successively.

15-4.2 Steady State Solutions

15-4.2.1 If one takes vertical acceleration into account, the pressure distribution is no longer hydrostatic. In particular, due to the centrifugal force of water particles under a wave crest, the pressure at the bottom decreases significantly (see Section 14-5.2).

Although nonnegligible, the vertical acceleration can still be linearized by assuming $dw/dt \cong \partial w/\partial t$. Since the vertical component of the motion is small, the convective terms $w(\partial w/\partial x)$ and $u(\partial w/\partial x)$ remain small. If this correction effect is taken into account in the long wave theory, the motion becomes nonlinear horizontally and linear vertically. Even if the nonlinear vertical components are taken into account in a solution obtained as a power series in terms of H/d, they have to be introduced in high-order terms only for the sake of consistency in the approximations. If it is assumed that the solution of the equation of long waves with a correction term for flow curvature is that of steady state, this solution has to be defined as a function of $(x - Ct)$. Such solutions do exist. They are the solutions for solitary waves and cnoidal waves (although in the latter case the pressure is found to be hydrostatic at a first order of approximation). In these cases, the value of the Ursell parameter does not need to be small (see Section 15-2.3.1).

In the case of very long waves where the vertical acceleration and path of curvature are effectively negligible, the choice of the significant length is quite arbitrary. This kind of motion corresponds, however, to a value of the Ursell parameter much larger than unity.

15-4.2.2 A quadratic, or more generally a nonlinear, friction from the bottom slope has a stabilizing effect which may balance the horizontal components to slow down most of the wave elements having the highest particle velocity, i.e., carrying the most energy.

Under certain conditions inherent to the characteristics of the bottom slope, the friction factor, and the water depths ahead and after the transient wave, it may happen that a steady-state translatory wave also exists. It is the monoclinal flood wave which is an exact solution of the long wave equation with bottom friction.

15-5 A Synthesis of Water Wave Theories

15-5.1 A Flow Chart for Water Waves

The flow chart just preceding this page summarizes the previous considerations. It describes the main characteristics of water waves. The two categories of motion are the linear and nonlinear motions, depending upon whether the convective inertia is taken into account or not. Each of these motions may also be subdivided into motion where the pressure is assumed to be hydrostatic or motion with nonnegligible flow curvature. Finally, the motion may or may not be irrotational and the bottom friction may or may not be taken into account.

Only the main theories have been indicated. Variations of these main theories exist in the literature. The details of the mathematics involved in these differences are beyond the scope of this book. An aspect of these is presented in Chapter 17. Due to the limitation of mathematical methods, the most complex cases cannot be analyzed. For example, a theory for nonlinear rotational waves in shallow water with nonnegligible vertical acceleration and quadratic bottom friction does not exist yet.

The flow chart can be used as a guide throughout the following chapters. However, all the theories given in the flow chart will not be studied in detail in this book.

15-5.2 A Plan for the Study of Water Waves

The plan in this book for the study of water waves follows tradition, i.e., the two families of water waves are separated under the titles "small amplitude wave theory" and "long wave theory."

Chapter 16 is entirely devoted to the linear small amplitude wave theory. The case of irrotational frictionless harmonic motion only is considered. Two-dimensional and three-dimensional motions are studied.

Chapter 17 deals with finite amplitude waves and Chapter 18 is devoted to the long wave theory. The theories of solitary waves and monoclinal waves are also presented.

Appendix A is devoted to the concept of wave spectrum as an application of linear theory to phenomena of random characteristics. Appendix B is on similitude and scale model technology, with particular emphasis on engineering problems associated with water waves.

Chapter 16

Linear Small Amplitude Wave Theories

16-1 Basic Equations and Formulation of a Surface Wave Problem

16-1.1 *Notation and Continuity*

The motion is defined with respect to the three axes in a Cartesian coordinates system. OX, OY, and OZ are the mutually perpendicular axes. The OZ axis is taken to be vertical and positive upwards. Any point is defined by the coordinates x, y, and z. The depth is defined by $z = -d$, and is assumed to be constant (see Fig. 16-1). Viscosity forces are neglected. The motion is assumed to be irrotational and the fluid is incompressible.

$$\textbf{curl V} = 0 \quad \text{or} \quad \zeta = \eta = \xi = 0$$

(Note that η in this chapter will be used for the free surface elevation and not for vorticity.) Also,

$$\text{div } \mathbf{V} = 0 \quad \text{or} \quad \frac{\partial u}{\partial x} + \frac{\partial v}{\partial y} + \frac{\partial w}{\partial z} = 0$$

These assumptions result in a number of simplifications.

curl V $= 0$ ensures the existence of a single-valued velocity potential function $\phi(x,y,z,t)$ from which the velocity field can be derived. Thus, the potential function can arbitrarily be defined as $\mathbf{V} = \textbf{grad}\ \phi$ or $\mathbf{V} = -\textbf{grad}\ \phi$. The latter definition is used in this chapter, i.e., $u = -\partial\phi/\partial x$, $v = -\partial\phi/\partial y$, $w = -\partial\phi/\partial z$. The velocity potential function

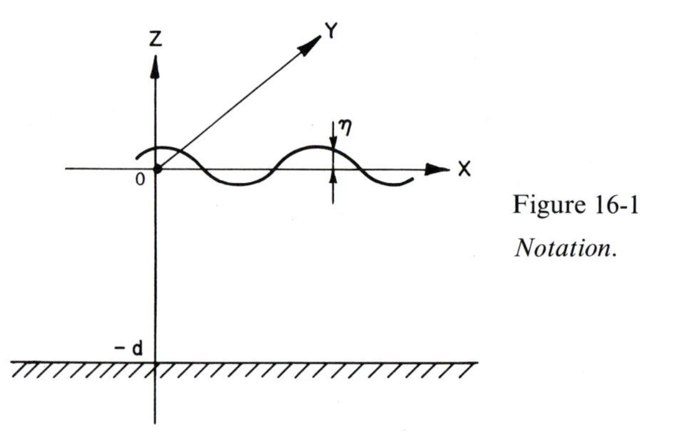

Figure 16-1
Notation.

has to be found from the continuity equation, the momentum equation and the boundary conditions.

The continuity equation div $\mathbf{V} = 0$ is expressed in terms of ϕ by the equation $\nabla^2\phi = 0$. In Cartesian coordinates, it is written as

$$\frac{\partial^2 \phi}{\partial x^2} + \frac{\partial^2 \phi}{\partial y^2} + \frac{\partial^2 \phi}{\partial z^2} = 0$$

16-1.2 The Momentum Equation

The momentum equation for an irrotational flow is given by the following form of the Bernoulli equation (see Section 10-1.2). The minus sign in Equation 16-1 is due to the new definition of ϕ.

$$\underset{\substack{\text{Local}\\\text{inertia}\\\text{term}}}{-\frac{\partial \phi}{\partial t}} + \underset{\substack{\text{Convective}\\\text{inertia}\\\text{term}}}{\frac{1}{2}V^2} + \underset{\substack{\text{Pressure}\\\text{term}}}{\frac{p}{\rho}} + \underset{\substack{\text{Gravity}\\\text{term}}}{gz} = f(t) \quad (16\text{-}1)$$

In this equation, $f(t)$ may depend on t but not on the space variables. The fact that the flow is assumed to be irrotational means that the Bernoulli law is valid throughout the fluid and not only along streamlines.

This equation is nonlinear because of the convective inertia term. This term may be expressed as a function of the potential function ϕ so that

$$\frac{1}{2}V^2 = \frac{1}{2}\left[\left(\frac{\partial \phi}{\partial x}\right)^2 + \left(\frac{\partial \phi}{\partial y}\right)^2 + \left(\frac{\partial \phi}{\partial z}\right)^2\right]$$

The nonlinearity of the motion can be seen clearly. In the case of very slow motion, the convective term is neglected and the Bernoulli equation is written as

$$-\frac{\partial \phi}{\partial t} + \frac{p}{\rho} + gz = f(t)$$

Periodic gravity wave theories often satisfy the condition for slow motion with a fairly good degree of accuracy. The corresponding solutions are mathematically exact when the motion tends to be infinitely small.

16-1.3 Boundary Conditions

16-1.3.1 At a fixed boundary, the fluid velocity is tangential to the boundary, that is, the normal component V_n is zero. In terms of velocity potential ϕ, this condition is written $\partial\phi/\partial n = 0$. In particular on a horizontal bottom

$$w\Big|_{z=-d} = -\frac{\partial \phi}{\partial z}\Big|_{z=-d} = 0$$

16-1.3.2 One of the difficulties encountered in determining the nature of wave motion is due to the fact that one of the boundaries—the free surface—is unknown, except in the case of infinitely small motion in which the free surface is, at the beginning, assumed to be a horizontal line. Hence, another unknown $z = \eta$ appears in wave problems. If one assumes that the free surface, in the most general case of a three dimensional motion, is given by the equation $z = \eta(x,y,t)$, then the variation of z with respect to time t is

$$\frac{dz}{dt} = \frac{\partial \eta}{\partial t} + \frac{\partial \eta}{\partial x}\frac{dx}{dt} + \frac{\partial \eta}{\partial y}\frac{dy}{dt}$$

Introducing the values

$$\frac{dx}{dt} = u = -\frac{\partial \phi}{\partial x} \qquad \frac{dy}{dt} = v = -\frac{\partial \phi}{\partial y}$$

$$\frac{dz}{dt} = w\Big|_{z=\eta} = -\frac{\partial \phi}{\partial z}\Big|_{z=\eta}$$

the free surface equation becomes

$$\frac{\partial \phi}{\partial z}\Big|_{z=\eta} = -\frac{\partial \eta}{\partial t} + \frac{\partial \phi}{\partial x}\Big|_{z=\eta}\frac{\partial \eta}{\partial x} + \frac{\partial \phi}{\partial y}\Big|_{z=\eta}\frac{\partial \eta}{\partial y}$$

This equation is nonlinear and is the *kinematic condition* at the free surface.

Another equation—the dynamic equation—is given by the Bernoulli equation in which the pressure p is considered as constant (and equal to atmospheric pressure). Hence the free surface dynamic condition becomes

$$-\frac{\partial\phi}{\partial t}+\frac{1}{2}\left[\left(\frac{\partial\phi}{\partial x}\right)^2+\left(\frac{\partial\phi}{\partial y}\right)^2+\left(\frac{\partial\phi}{\partial z}\right)^2\right]+g\eta=f(t)$$

Thus, generally ϕ and η appear to be given by the solution of $\nabla^2\phi=0$ with two simultaneous nonlinear boundary conditions at the free surface and a linear boundary condition at the bottom

$$\left.\frac{\partial\phi}{\partial z}\right|_{z=-d}=0$$

16-1.4 The Free Surface Condition in the Case of Very Slow Motion

In the case of slow motion the Bernoulli equation

$$-\frac{\partial\phi}{\partial t}+\frac{p}{\rho}+gz=f(t)$$

becomes

$$-\left.\frac{\partial\phi}{\partial t}\right|_{z=\eta}+g\eta=0$$

at the free surface. The result is:

$$\eta=\left.\frac{1}{g}\frac{\partial\phi}{\partial t}\right|_{z=\eta}$$

provided the function $f(t)$ and any additive constant can be included in the value of $\partial\phi/\partial t$.

Since the motion is assumed to be infinitely small, η may be written

$$\left.\frac{1}{g}\frac{\partial\phi}{\partial t}\right|_{z=0}$$

This approximation leads to an error of the order of those already done in neglecting the convective inertia term.

Consider the kinematic condition: $\partial\eta/\partial x$ and $\partial\eta/\partial y$ are the components of the slope of the free surface and are small as in the case of slow motion (see Fig. 16-2).

The nonlinear terms $(\partial\phi/\partial x)(\partial\eta/\partial x)$ and $(\partial\phi/\partial y)(\partial\eta/\partial y)$ may be neglected. The normal component of the fluid velocity at the free surface is now equal to the normal velocity of the surface itself. This gives with sufficient approximation

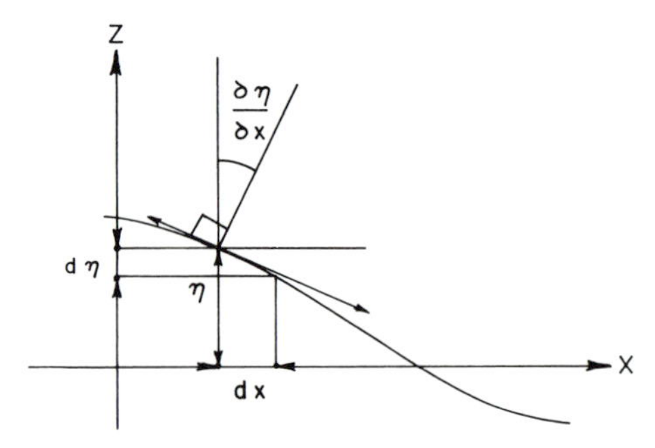

Figure 16-2 *Notation.*

$$\frac{\partial\eta}{\partial t}=-\left.\frac{\partial\phi}{\partial z}\right|_{z=0}$$

η may now be easily eliminated from the dynamic and kinematic conditions. The derivative of η with respect to t in the dynamic condition gives

$$\frac{\partial\eta}{\partial t}=\left.\frac{1}{g}\cdot\frac{\partial^2\phi}{\partial t^2}\right|_{z=0}$$

$\partial\eta/\partial t$ can be eliminated by equating the two above equations. This yields:

$$\left[\frac{\partial\phi}{\partial z}+\frac{1}{g}\cdot\frac{\partial^2\phi}{\partial t^2}\right]_{z=0}=0$$

which is called the *Cauchy–Poisson condition* at the free surface.

16-1.5 Formulation of a Surface Wave Problem

16-1.5.1 Thus ϕ and η appear to be solutions of the following system:

1. Continuity

$$\nabla^2\phi=0\qquad\begin{cases}-d\le z\le\eta(x,y,t)\\ -\infty<x<\infty\\ -\infty<y<\infty\end{cases}\ \text{no boundary}$$

2. Fixed boundary $\partial\phi/\partial n = 0$. In particular, at the bottom

$$\left.\frac{\partial \phi}{\partial z}\right|_{z=-d} = 0$$

3. Free surface $z = \eta(x,y,t)$
   a. Kinematic condition:

$$\left.\frac{\partial \phi}{\partial z}\right|_{z=\eta} = -\frac{\partial \eta}{\partial t} + \left.\frac{\partial \phi}{\partial x}\right|_{z=\eta}\frac{\partial \eta}{\partial x} + \left.\frac{\partial \phi}{\partial y}\right|_{z=\eta}\frac{\partial \eta}{\partial y}$$

   b. Dynamic condition:

$$-\frac{\partial \phi}{\partial t} + \frac{1}{2}\left[\left(\frac{\partial \phi}{\partial x}\right)^2 + \left(\frac{\partial \phi}{\partial y}\right)^2 + \left(\frac{\partial \phi}{\partial z}\right)^2\right] + g\eta = 0$$

   where $f(t)$ is now included in $\partial\phi/\partial t$. However, even in this case, this last equation may be different from zero, and equal to a given function $f(x,y,t)$ in the case of a disturbance created at the free surface.

So formulated, the solution of the system of equations presented in this section is still difficult to determine. First, the equations are nonlinear, and second, the free surface is unknown and is time-dependent.

16-1.5.2 In the case of slow motion, η may be eliminated from the two free surface conditions resulting in the simple Cauchy–Poisson condition

$$\left[\frac{\partial^2 \phi}{\partial t^2} + g\frac{\partial \phi}{\partial z}\right]_{z=0} = 0$$

This leaves only one unknown, ϕ, to be determined from

$$\nabla^2 \phi = 0 \qquad \begin{cases} -d \le z \le \eta = 0 \\ -\infty < x < \infty \\ -\infty < y < \infty \end{cases}$$

$$\left.\frac{\partial \phi}{\partial z}\right|_{z=-d} = 0 \quad \text{and} \quad \left[\frac{\partial^2 \phi}{\partial t^2} + g\frac{\partial \phi}{\partial z}\right]_{z=0} = 0$$

16-2 Method of Solutions

16-2.1 *General Approach*

16-2.1.1 When all the equations are homogeneous and linear, the principle of superposition states that any number of individual solutions may be superimposed to form new functions which constitute solutions themselves. In a linear equation ϕ and its derivatives occur only in the first degree in every term. For example, if ϕ_1 and ϕ_2 are two separate solutions, $a\phi_1 + b\phi_2$ is also a solution, a and b being two arbitrary constants. This basic principle is very important and will be used in the following sections.

16-2.1.2 Most of the solutions with which we are concerned in this chapter are harmonic. This stems from the fact that harmonic functions are quite natural solutions of the basic equations. The solutions characterizing periodic motions may be considered as superposition of harmonic components.

The solution of $\phi(x,y,z,t)$ is usually of the form

$$\phi = f(x,y,z)\cos(kt + \varepsilon),$$

where $k = 2\pi/T$ and T is the wave period. Another form of the solution is

$$\phi = \text{Re}\, f(x,y,z)e^{i(kt+\varepsilon)}$$

Recall that

$$e^{i(kt+\varepsilon)} = \cos(kt + \varepsilon) + i\sin(kt + \varepsilon)$$

"Re" means the real part of the function and ε is the phase of ϕ with respect to the origin of time, $t = 0$. In the following, "Re" will be omitted and it is to be understood that only the real parts of the mathematical expressions are considered.

Introducing this form of ϕ in the free surface condition

$$\frac{1}{g}\cdot\frac{\partial^2 \phi}{\partial t^2} + \frac{\partial \phi}{\partial z} = 0$$

gives

$$\left(\frac{k^2}{g} - \frac{\partial}{\partial z}\right)\phi = 0$$

16-2.1.3 If it is assumed that ϕ is given by a product of functions of each variable alone, then the basic equation $\nabla^2\phi = 0$ may be solved by the separation of variables method. From physical considerations, it may be expected that the solution ϕ will be given by the product of the functions of the horizontal components $U(x,y)$, the vertical component $P(z)$, and the time $f(t)$. Hence

$$\phi = U(x,y) \cdot P(z) \cdot f(t).$$

This value of $\phi(x,y,z,t)$ can be used in the continuity equation, $\nabla^2\phi = 0$. Algebraic manipulation of the result will give:

$$-\frac{\partial^2 U/\partial x^2 + \partial^2 U/\partial y^2}{U(x,y)} = \frac{d^2P/dz^2}{P(z)}$$

This may be written as

$$-\frac{\nabla^2 U}{U} = \frac{P''}{P}$$

Notice that the functions of x and y are on one side of the equal sign, while the functions of z are on the other. The variables have been separated.

It must be said that it was not certain at the beginning that it would have been possible to separate the variables as has been done. However, it will be shown later that this process may be performed for solutions of $\nabla^2\phi = 0$. The right-hand side of the above equation is a function of z. The left-hand side is a function of x and y. Since x and y can vary independently of z and vice versa, the only way in which the function of x and y and the function of z can always be equal (as stated by the above equation) is if the left-hand side and the right-hand side are both equal to the same constant m^2 where m may be real or imaginary.

It will be easily seen that if m is imaginary there is no physical meaning to the solutions in the case of wave motion. Thus, m is chosen to be real and m^2 is always positive.

The equations

$$\frac{P''}{P} = -\frac{\nabla^2 U}{U} = m^2$$

are now reduced to

$$\frac{d^2P}{dz^2} - m^2 P(z) = 0$$

and

$$\frac{\partial^2 U}{\partial x^2} + \frac{\partial^2 U}{\partial y^2} + m^2 U(x,y) = 0$$

These equations will often be written in the shorter form:

$$\begin{cases} \left(\dfrac{d^2}{dz^2} - m^2\right)P = 0 \\ (\nabla^2 + m^2)U = 0 \end{cases}$$

The last equation is the well-known *Helmholtz equation* (also called the wave equation) of mathematical physics.

16-2.2 Wave Motion along a Vertical

The equation

$$\left(\frac{d^2}{dz^2} - m^2\right)P = 0$$

may easily be integrated, giving the general solution

$$P = Ae^{mz} + Be^{-mz}$$

where A and B are constants.

The boundary condition at the bottom,

$$\left.\frac{\partial\phi}{\partial z}\right|_{z=-d} = 0$$

gives for any fixed value of x, y, and t:

$$\left.\frac{dP}{dz}\right|_{z=-d} = 0$$

When this boundary condition is applied to the solution for P the result is

$$mAe^{-md} - mBe^{+md} = 0$$

Hence

$$Ae^{-md} = Be^{+md}$$

Consider the original solution,

$$P = Ae^{mz} + Be^{-mz}$$

Multiply each term on the right by $e^{-md}e^{+md}$. Let

$$Ae^{-md} = Be^{+md} = \tfrac{1}{2}D$$

and substitute this in the equation for P. The result is

$$P = \frac{D}{2}\left(e^{m(z+d)} + e^{-m(z+d)}\right)$$

That is $P = D \cosh m(z + d)$.

Now, substituting P in the expression of ϕ gives

$$\phi = D \cosh m(d + z)U(x,y)f(t)$$

16-2.3 Introduction of the Free Surface Condition: General Solution

The solution for $f(t)$ is given by the Cauchy–Poisson condition at the free surface:

$$\left(\frac{\partial^2\phi}{\partial t^2} + g\frac{\partial\phi}{\partial z}\right)_{z=0} = 0$$

Substitute the value of ϕ obtained in the previous section in this equation. Only the case when $z = 0$ needs to be considered here, since this is the free surface condition. When the resulting equation is divided by the value of ϕ, the following is obtained:

$$f''/f = -gm \tanh md$$

If we let $k^2 = gm \tanh md$, the solution for f is given by the equation $f'' + k^2 f = 0$. The characteristic equation $r^2 + k^2 = 0$ gives $r = \pm ik$. Hence

$$f = \alpha e^{ikt} + \beta e^{-ikt}$$

where α, β are constant coefficients which depend upon the boundary conditions, k is the frequency $2\pi/T$, and T is the wave period. When $\beta = 0$ and the coefficient α is included in the coefficient D, it is found

$$\phi = D \cosh m(d + z)U(x,y)e^{ikt}$$

Since there exist an infinite (but discrete) number of values for k_n and m_n which satisfy the equation

$$k_n^2 = m_n g \tanh m_n d$$

a general solution for ϕ can be written as

$$\phi = \sum_{n=0}^{\infty} D_n \cosh m_n(d + z)U_n(x,y) \exp(ik_n t + \varepsilon_n)$$

where ε_n is a phase constant.

Consider the case of a monochromatic wave. It is convenient to express D as a function of the wave height $2a$. From the free surface dynamic equation for slow motion,

$$\eta = \frac{1}{g}\frac{\partial\phi}{\partial t}\bigg|_{z=0}$$

one obtains

$$\eta = \frac{ikD}{g}\cosh md\,U(x,y)e^{ikt}$$

Considering only the real part, this can be written as (recall $i^2 = -1$):

$$\eta = -\frac{kD}{g}\cosh md\,U(x,y)\sin kt$$

The expressions for ϕ and η become more convenient if we write (aU) for the amplitude of η. Then

$$a = -\frac{kD}{g}\cosh md$$

Hence

$$D = -\frac{ag}{k}\frac{1}{\cosh md}$$

and

$$\phi = -\frac{ag}{k}\frac{\cosh m(d + z)}{\cosh md}U(x,y)e^{ikt}$$

Substituting the relationship $k^2 = mg \tanh md$ leads to

$$\phi = -a \frac{k}{m} \frac{\cosh m(d+z)}{\sinh md} U(x,y) e^{ikt}$$

With the value of D that has been found, the expression for P becomes

$$P(z) = -\frac{ak}{m} \frac{\cosh m(d+z)}{\sinh md} = -\frac{ag}{k} \frac{\cosh m(d+z)}{\cosh md}$$

Under these conditions the wave height at any point is $2aU(x,y)$. $U(x,y)$ is the relative value of the wave height with respect to a plane or a point where it is simply $2a$.

16-3 Two-Dimensional Wave Motion

16-3.1 Integration of the Wave Equation

The differential equation to be solved is $(\nabla^2 + m^2)U = 0$. A general solution of this equation does not exist, but a number of solutions may be found, corresponding to particular boundary conditions. In the case of a two-dimensional wave such as motion encountered in a wave flume.

$$\frac{\partial \phi}{\partial y} = 0 \qquad \frac{\partial U}{\partial y} = 0$$

This reduces the wave equation to

$$\left(\frac{\partial^2}{\partial x^2} + m^2\right) U = 0$$

Solving this, one finds that the solutions for U are given by any linear combination of e^{-imx} and e^{imx} such as,

$$U = A'e^{imx} + B'e^{-imx}$$

In particular if $U = e^{-imx}$, then

$$\phi = -a \frac{k}{m} \frac{\cosh m(d+z)}{\sinh md} e^{i(kt - mx)}$$

or

$$\phi = -a \frac{k}{m} \frac{\cosh m(d+z)}{\sinh md} \cos (kt - mx)$$

This is the velocity potential function of a progressive wave traveling in the OX direction.

If $U = e^{imx}$, the velocity potential function of a wave traveling in the opposite direction is obtained.

If the solution for U is:

$$U = \frac{1}{2} (e^{imx} + e^{-imx}) = \cos mx$$

or

$$U = \frac{1}{2i} (e^{imx} - e^{-imx}) = \sin mx$$

then

$$\phi = -a \frac{k}{m} \frac{\cosh m(d+z)}{\sinh md} \begin{Bmatrix} \cos \\ \sin \end{Bmatrix} mx \cos kt$$

This is the velocity potential function of a standing wave. If A' is different from B', a partial standing wave is obtained. In practice, the values for A' and B' are given by vertical boundary conditions (wave reflection, etc.).

In the most general case of a two-dimensional irregular wave, as may be observed at sea, the velocity potential function ϕ is:

$$\phi = \sum_{n=0}^{\infty} -a_n \frac{k_n}{m_n} \frac{\cosh m_n(d+z)}{\sinh m_n d} \exp [i(k_n t - m_n x + \varepsilon_n)]$$

where ε_n is a phase constant.

When there are two waves only, traveling in the same direction, the velocity potential function describing the "beating" phenomena may be obtained easily.

16-3.2 Physical Meaning: Wavelength

It is easy to see the physical meaning of the coefficient m. Since ϕ and consequently η is periodic with respect to space, $m = 2\pi/L$ and L is the wavelength.

The wavelength is given by

$$k^2 = mg \tanh md$$

and then

$$\left(\frac{2\pi}{T}\right)^2 = \frac{2\pi}{L} g \tanh \frac{2\pi}{L} d$$

that is,

$$L = \frac{gT^2}{2\pi} \tanh \frac{2\pi d}{L}$$

and the wave celerity:

$$C = \frac{L}{T} = \frac{gT}{2\pi} \tanh \frac{2\pi d}{L}$$

In particular when d/L is small (shallow water)

$$\tanh \frac{2\pi d}{L} \cong \frac{2\pi d}{L} \qquad L = T(gd)^{1/2} \qquad C = (gd)^{1/2}$$

and

$$L = T(gd)^{1/2} \qquad C = (gd)^{1/2}$$

When d/L is large (deep water), $\tanh 2\pi d/L = 1$, and $L = gT^2/2\pi$, $C = gT/2\pi$. The values of L and C are given as functions of the depth d and the wave period T on the following nomographs (Figs. 16-3, 16-4, and 16-5).

16-3.3 Flow Patterns

The velocity components are $u = -\partial\phi/\partial x$, $w = -\partial\phi/\partial z$ and the particle orbits are:

$$x = -\int_0^t \frac{\partial \phi}{\partial x} dt \qquad z = -\int_0^t \frac{\partial \phi}{\partial z} dt$$

In the case of a progressive wave:

$$u = ka \frac{\cosh m(d + z)}{\sinh md} \sin (kt - mx)$$

$$w = ka \frac{\sinh m(d + z)}{\sinh md} \cos (kt - mx)$$

The particle orbits are determined by assuming that the motion around a fixed point x_0, z_0 is small, so that one can consider x and z constant in the integration.

$$x = x_0 - a \frac{\cosh m(d + z_0)}{\sinh md} \cos (kt - mx_0)$$

and

$$z = z_0 + a \frac{\sinh m(d + z_0)}{\sinh md} \sin (kt - mx_0)$$

Squaring and adding these two last equations to eliminate t, the equation of an ellipse is obtained

$$\frac{(x - x_0)^2}{A^2} + \frac{(z - z_0)^2}{B^2} = 1$$

It is now seen that x_0 and z_0 are at the center of the ellipse, i.e., can be considered as the position of the particle at rest with the horizontal semimajor axis

$$A = a \frac{\cosh m(d + z_0)}{\sinh md}$$

and the vertical semiminor axis:

$$B = a \frac{\sinh m(d + z_0)}{\sinh md}$$

$B = a$ at the free surface, and $B = 0$ at the bottom (Fig. 16-6). The free surface equation is

$$\eta = \frac{1}{g} \frac{\partial \phi}{\partial t} = a \sin (kt - mx)$$

When $d \to \infty$, $(A/B) \to 1$ and the orbits are circles of radius $R = a \exp (4\pi^2 z/gT^2)$.

In the case of a standing wave, it would be easily found that the paths of particles are straight lines given by

$$\frac{z - z_0}{x - x_0} = -\tanh m(d + z_0) \cot mx_0$$

or

$$\frac{z - z_0}{x - x_0} = \tanh m(d + z_0) \tan mx_0$$

(see Fig. 16-7). They are parabolas at a second order of approximation.

16-3.4 Partial Standing Wave

A partial standing wave is caused by the superimposition of two waves of the same period but travelling in opposite directions and with different amplitudes (Fig. 16-8).

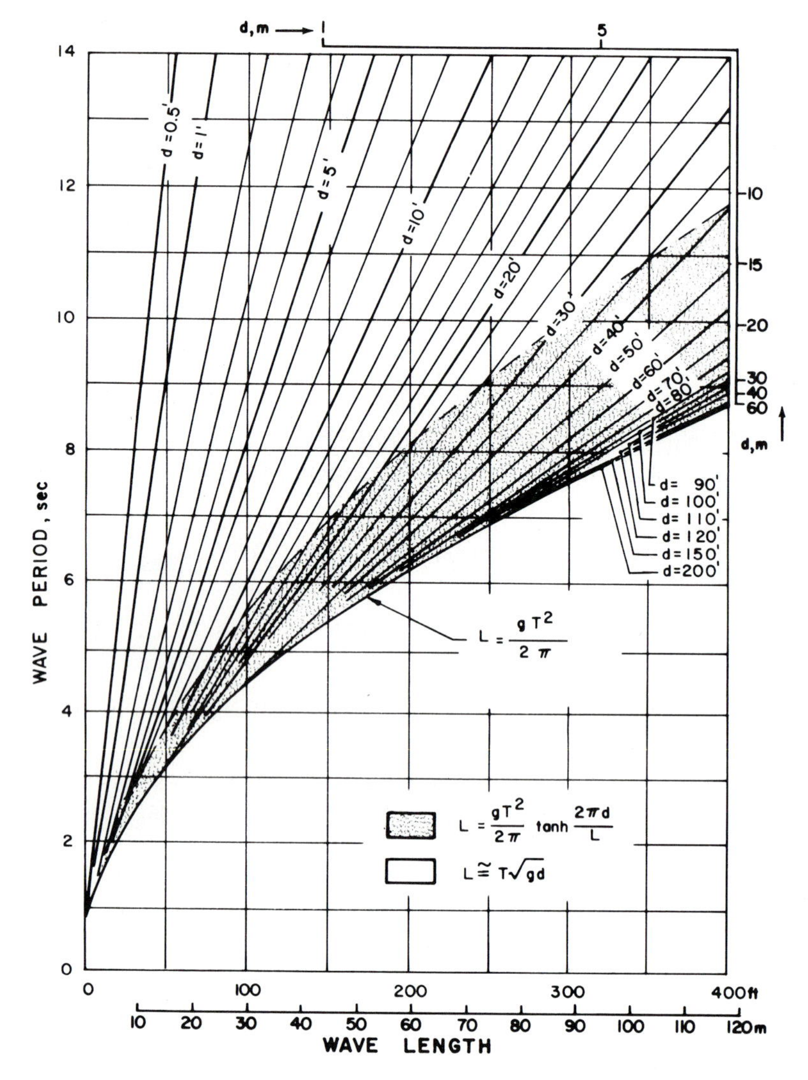

Figure 16-3
Wavelength vs depth and period.

The first-order potential function is

$$\phi = -\frac{k}{m}\frac{\cosh m(d+z)}{\sinh md} \times [\alpha_1 \sin(kt + mx) + \alpha_2 \sin(kt - mx)]$$

The amplitude at the antinode is $(\alpha_1 + \alpha_2)$ and at the node is $(\alpha_1 - \alpha_2)$. It is possible to determine the individual wave height of the two progressive waves by measuring the amplitude at the antinode (maximum) A and at the node B (minimum). Then,

$$\alpha_1 = \tfrac{1}{2}(A + B)$$

$$\alpha_2 = \tfrac{1}{2}(A - B)$$

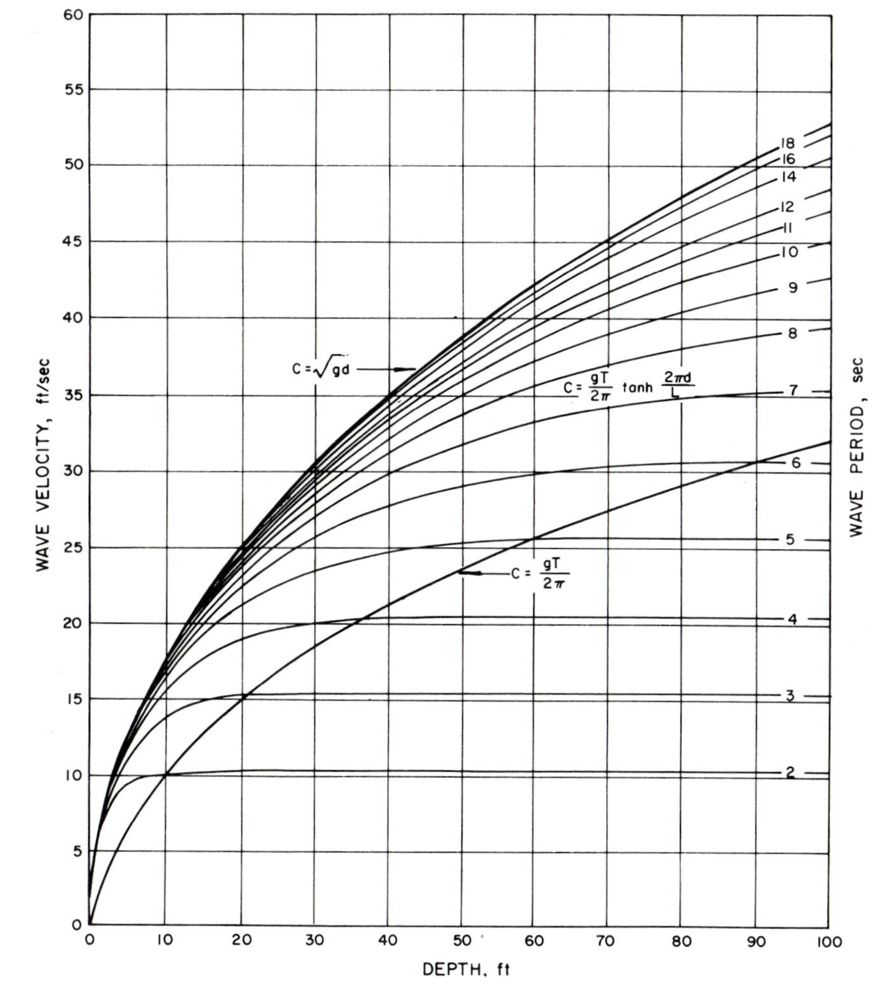

Figure 16-4
Wave velocity vs depth and period.

The reflection coefficient of the obstacle causing the partial standing wave is:

$$R = \frac{A - B}{A + B}$$

The envelopes of the crest and the trough are two sinusoids of amplitude, $(A - B)/2$.

16-3.5 The Use of Complex Number Notation

To show the great simplicity introduced by the operation $W = \phi + i\psi$, W is calculated to a first order of approximation in the case of a two-dimensional monochromatic progressive wave motion. First of all, it is convenient to make a change in the origin of the vertical axis, and to take the horizontal axis on the bottom, such as: $z' = d + z$.

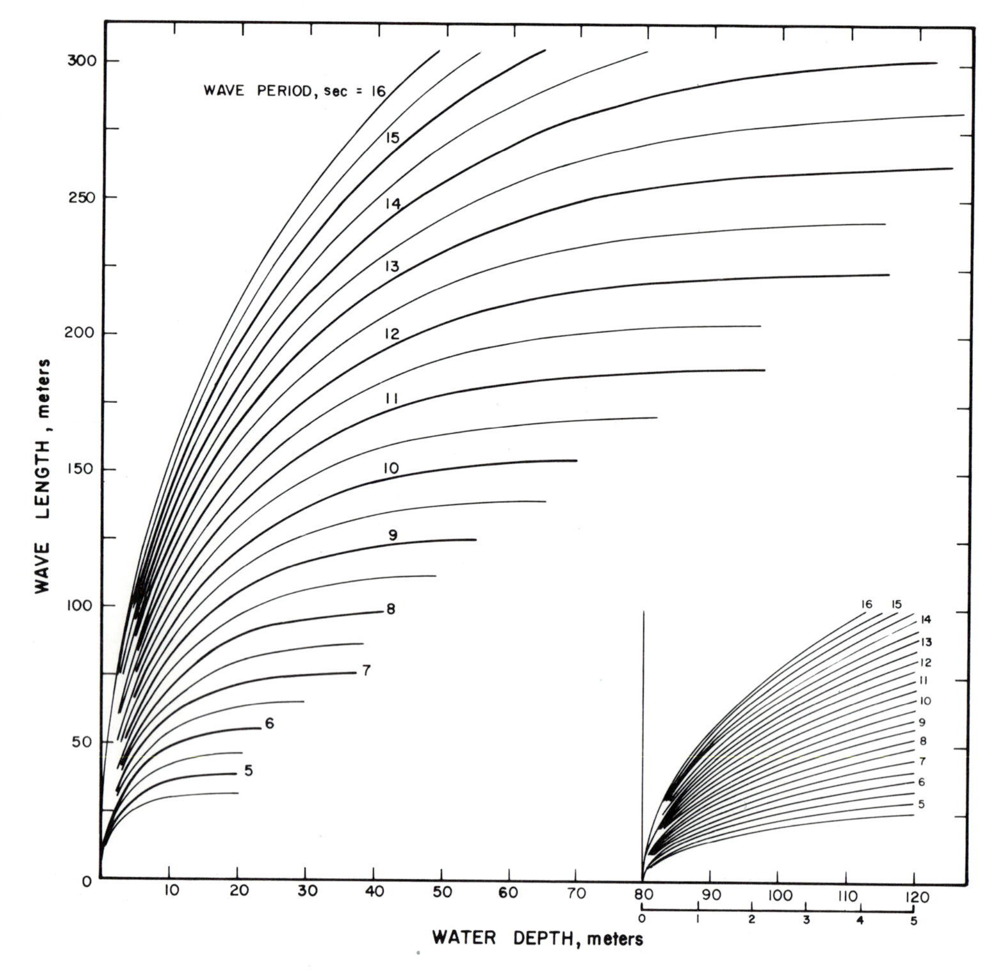

Figure 16-5
Wavelength vs period and depth.

With this new condition the velocity potential function could be written

$$\phi = -a\frac{k}{m}\frac{\cosh mz'}{\sinh md}\cos(mx - kt)$$

If

$$A = -a\frac{k}{m}\frac{1}{\sinh md}$$

then

$$\phi = A\cosh mz'\cos(mx - kt).$$

The stream function ψ is given by one of the following operations:

$$u = -\frac{\partial\phi}{\partial x} = \frac{\partial\psi}{\partial z'} \quad \text{or} \quad w = -\frac{\partial\phi}{\partial z'} = -\frac{\partial\psi}{\partial x}$$

This gives

$$\psi = A\sinh mz'\sin(mx - kt)$$

Now

$$W = \phi + i\psi$$
$$= A[\cosh mz'\cos(mx - kt) + i\sinh mz'\sin(mx - kt)]$$

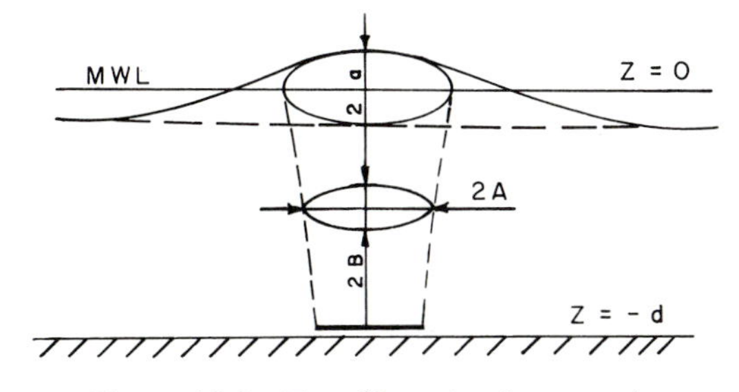

Figure 16-6 *Two-dimensional progressive wave motion.*

The following relationships hold:

$$\cosh mz' = \cos imz'$$
$$\sinh mz' = -i \sin imz'$$

Using this, one obtains

$$W = A[\cos imz' \cos (mx - kt) + \sin imz' \sin (mx - kt)]$$

This can be rewritten as

$$W = A \cos (mx - imz' - kt)$$

Introducing the complex number $Z = x - iz'$, one obtains the very simple relationship

$$W = \phi + i\psi = A \cos (mZ - kt)$$

The velocity potential function ϕ is given by the real part of W, while the stream function ψ is given by the imaginary part.

Similarly, the value of W for a clapotis is

$$W = 2A \sin mZ \cos kt \qquad \text{or} \qquad 2A \cos mZ \cos kt$$

16-4 Three-Dimensional Wave Motion

16-4.1 *Three-Dimensional Wave Motion in a Rectangular Tank*

The separation of variables method is used to solve the equation governing three-dimensional motion,

$$\left(\frac{\partial^2}{\partial x^2} + \frac{\partial^2}{\partial y^2} + m^2\right)U = 0$$

It is found that

$$\frac{\partial^2 U}{\partial x^2} + p^2 U = 0$$

and

$$\frac{\partial^2 U}{\partial y^2} + q^2 U = 0$$

in which p and q are constants, real or complex, such that $m^2 = p^2 + q^2$.

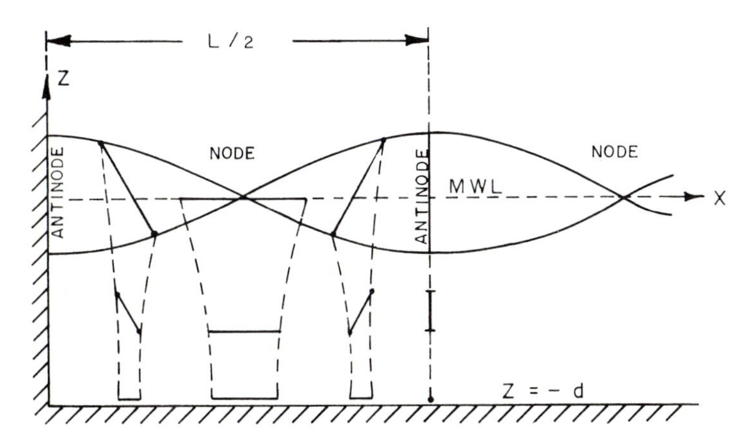

Figure 16-7 *Standing wave.*

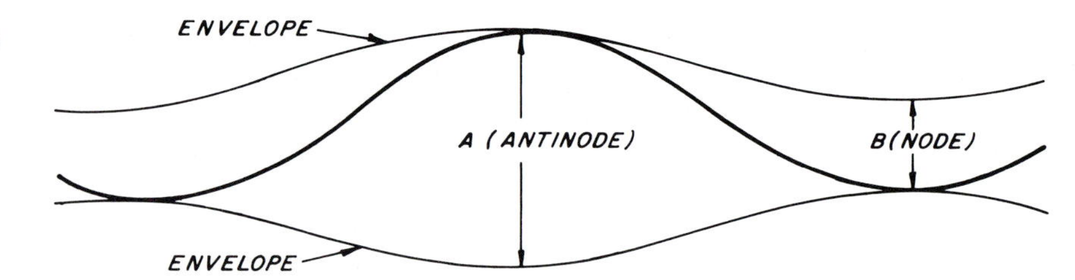

Figure 16-8 *Variation in amplitude in a partial standing wave.*

As an application of three-dimensional motion, the case of a standing wave in a rectangular tank is analyzed. The boundaries are shown in Fig. 16-9.

$$\frac{\partial \phi}{\partial n} = 0 \quad \text{or} \quad \frac{\partial U}{\partial n} = 0 \quad \text{for} \quad \begin{cases} x = 0, x = a \\ y = 0, y = b \end{cases}$$

It is easily verified that the solution of $(\nabla^2 + m^2)U = 0$ is $U = \cos(r\pi x/a)\cos(s\pi y/b)$ with $p = r\pi/a$, $q = s\pi/b$, where r and s are integers. The solution for $\nabla^2\phi = 0$ is

$$\phi = -a\frac{k}{m}\frac{\cosh m(d+z)}{\sinh md}\cos\frac{r\pi x}{a}\cos\frac{s\pi y}{b}$$

The general solution is

$$\phi = \sum_{i=0}^{\infty}\sum_{j=0}^{\infty}\sum_{n=0}^{\infty} -a_n\frac{k_n}{m_n}\frac{\cosh m_n(d+z)}{\sinh m_n d}\cos\frac{r_i\pi x}{a}\cos\frac{s_j\pi y}{b}$$

where

$$k_n^2 = m_n g \tanh m_n d \quad \text{and} \quad p_i^2 + q_j^2 = m_n^2$$

Figure 16-9 *Rectangular basin—notation.*

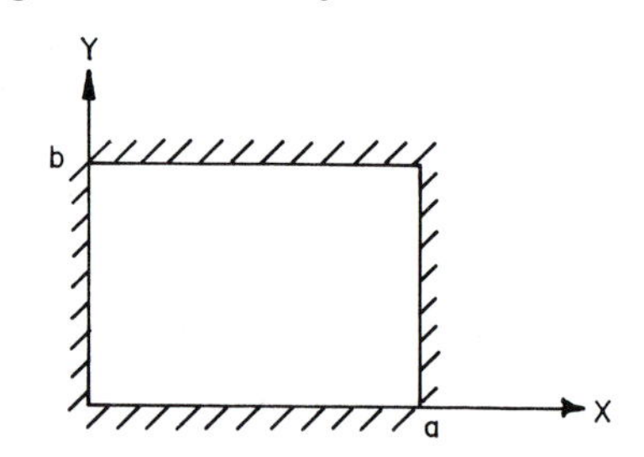

It is interesting to note that the equation of a two-dimensional motion,

$$\left(\frac{\partial^2}{\partial x^2} + m^2\right)U = 0$$

shows that the curvature of the free surface is proportional to the free surface elevation η since $\eta = aU$. Similarly, in the case of three-dimensional motion, one has

$$\left(\frac{\partial^2}{\partial x^2} + \frac{\partial^2}{\partial y^2} + m^2\right)U = 0$$

Then it is the sum of the curvatures in two directions (OX and OY) which is proportional to the free surface elevation. This is the case, for example, of the "short-crested wave."

16-4.2 Cylindrical Wave Motion

The continuity equation expressed in terms of cylindrical coordinates is

$$\frac{\partial^2\phi}{\partial r^2} + \frac{1}{r}\frac{\partial\phi}{\partial r} + \frac{1}{r^2}\frac{\partial^2\phi}{\partial\theta^2} + \frac{\partial^2\phi}{\partial z^2} = 0$$

In the case of a motion with an axis of symmetry so that $\partial^2\phi/\partial\theta^2 = 0$, the solution is

$$\phi = U(r)P(z)e^{ikt}$$

The resulting wave equation $(\nabla^2 + m^2)U = 0$ for cylindrical waves becomes

$$\left(\frac{\partial^2}{\partial r^2} + \frac{1}{r}\frac{\partial}{\partial r} + m^2\right)U = 0$$

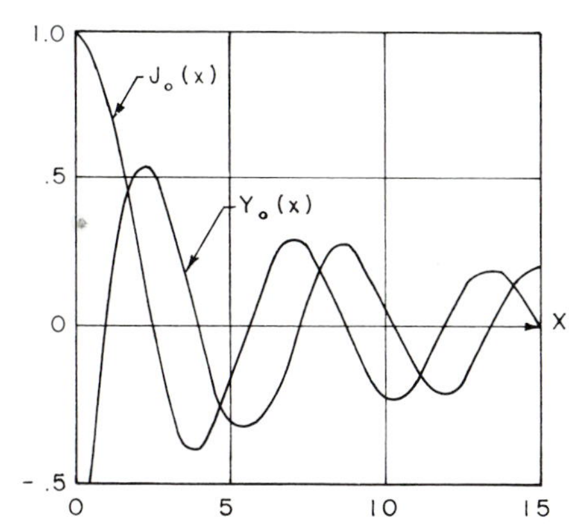

Figure 16-10 *Bessel functions of first order.*

This is the Bessel equation of order zero. Its standard solutions are denoted by $J_0(mr)$ and $Y_0(mr)$, the first being regular at the origin, the second singular at the origin. The variation of J_0 and Y_0 as functions of $x = mr$ are given by Fig. 16-10. Any linear combination of these standard solutions is itself obviously a solution of the equation. Solutions of peculiar interest are given by the Hankel functions $H_0^{(2)}(mr)$ and $H_0^{(1)}(mr)$, which when combined with $e^{\mp ikt}$ represent progressive (diverging or converging) waves. Table 16-1 gives the relationships between J_0, Y_0, and the Hankel functions $H_0^{(2)}(mr)$ and $H_0^{(2)}(mr)$ as well as their physical meanings and their analogy with sinusoidal functions. The appropriate solution and its physical significance is obtained by combining the solution with e^{-ikt} or e^{+ikt} and examining its asymptotic form at large distance from the origin. The asymptotic expressions enable us to identify outgoing waves from incoming waves. When mr becomes very large, it is found that

$$H_0^{(1)}(mr) \approx Ae^{iB} \qquad H_0^{(2)}(mr) \approx Ae^{-iB}$$

$$J_0(mr) \approx A \cos B \qquad Y_0(mr) \approx A \sin B$$

where

$$A = \frac{2}{\pi m r^{1/2}} \quad \text{and} \quad B = \left[mr - \frac{\pi}{4} \right]$$

In a two-dimensional motion, A is replaced by a and B is replaced by mx. It is seen that the wave height, which is

Table 16-1 *Analogy between Bessel functions and sinusoidal functions*

Two-dimensional waves (sinusoidal functions) $\left(\frac{\partial^2}{\partial x^2} + m^2\right)U = 0$		*Circular waves (Bessel functions)* $\left(\frac{\partial^2}{\partial r^2} + \frac{1}{r}\frac{\partial}{\partial r} + m^2\right)U = 0$	
$e^{-imx} = \cos mx - i \sin mx$	Progressive wave in the positive OX direction (wave height = constant)	$H_0^{(2)}(mr) = J_0 - iY_0$	Converging wave (sink) (wave height increases when r decreases) when combined with e^{+ikt}
$e^{+imx} = \cos mx + i \sin mx$	Progressive wave in the negative OX direction (wave height = constant)	$H_0^{(1)}(mr) = J_0 + iY_0$	Diverging wave (source) (wave height decreases when r increases) when combined with e^{+ikt}
$\cos mx = \frac{e^{imx} + e^{-imx}}{2}$	Standing wave (horizontal velocity = 0 when $x = 0$)	$J_0(mr) = \frac{H_0^{(1)} + H_0^{(2)}}{2}$	Standing circular wave (horizontal velocity = 0 when $r = 0$)
$\sin mx = \frac{e^{imx} - e^{-imx}}{2i}$	Standing wave (horizontal velocity maximum for $x = 0$)	$Y_0(mr) = \frac{H_0^{(1)} - H_0^{(2)}}{2i}$	Standing circular wave (horizontal velocity infinite for $r = 0$)

given by the amplitude of U, is proportional to A and decreases as $r^{1/2}$. This may also be demonstrated by conservation of the transmitted wave energy (see Section 16-5).

16-4.3 *Wave Agitation in a Circular Tank*

The boundary condition for a circular tank is

$$\left.\frac{\partial\phi}{\partial r}\right|_{r=R} = 0$$

or

$$v_r|_{r=R} = 0,$$

where R is the radius of the tank. The solution for U is $U = J_n(mr)\cos n\theta$, where J_n is the Bessel function of order n. When $n = 0$, $U = J_0(mr)$ and the motion is a stationary circular wave.

16-5 Energy Flux and Group Velocity

16-5.1 *Power and Energy Flux*

It is seen in Section 10-3.1.1 that the power transmitted by a fluid through a penstoke of cross section A, is given by

$$P = \left\{\frac{V^2}{2g} + \frac{p}{\rho g} + z\right\}AVK$$

Note that $K = \frac{1}{550}$ if P is in horsepower and A is in square feet. The same formula applies to water waves, but due to time dependency the wave power is obtained by averaging with respect to time. It is then called energy flux. The average energy flux per unit of wave crest through a fixed vertical plane parallel to the wave crest is

$$F_{\text{av}} = \frac{1}{T}\int_t^{t+T}\int_{-d}^{\eta}\left(\rho\frac{V^2}{2} + p + \rho g z\right)u\,dz\,dt$$

i.e., by application of the Bernoulli equation where $f(t)$ is assumed to be taken into account in $\partial\phi/\partial t$:

$$F_{\text{av}} = -\rho\frac{1}{T}\int_t^{t+T}\int_{-d}^{\eta}\frac{\partial\phi}{\partial t}\frac{\partial\phi}{\partial x}\,dz\,dt$$

This formula is general and can be applied by inserting the value of the potential function ϕ for any kind of irrotational wave, linear or nonlinear. In the case of a linear periodic progressive wave (see Section 16-3.1)

$$\phi = -a\frac{k}{m}\frac{\cosh m(d+z)}{\sinh md}\cos(kt - mx)$$

Inserting this value in the above expression leads to

$$F_{\text{av}} = \rho\frac{1}{T}\int_t^{t+T}\int_{-d}^{\eta} a^2\frac{k^3}{m}\frac{\cosh^2 m(d+z)}{\sinh^2 md}$$
$$\times \sin^2(kt - mx)\,dz\,dt$$

After integration and the neglection of some high-order terms, this becomes

$$F_{\text{av}} = \tfrac{1}{4}\rho g a^2 C\left(1 + \frac{2md}{\sinh 2md}\right)$$

This tends toward

$$F_{\text{av}}|_{d\to\infty} = \tfrac{1}{4}\rho g a^2\frac{gT}{2\pi}$$

in deep water, and

$$F_{\text{av}}|_{d\to 0} = \tfrac{1}{2}\rho g a^2 (gh)^{1/2}$$

in shallow water.

16-5.2 *Energy per Wavelength and Rate of Energy Propagation*

The stored energy per wavelength per unit length of crest is the sum of the kinetic energy and the potential energy:

$$E = \rho\int_x^{x+L}\int_{-d}^{\eta}\left(\tfrac{1}{2}V^2 + gz\right)dz\,dx$$

By inserting the value of $V^2 = [(\partial\phi/\partial x)^2 + (\partial\phi/\partial z)^2]$ and neglecting some high-order terms, it is found that in the case of a simple linear periodic progressive wave that the potential energy E_p equals the kinetic energy E_K and the total energy per wavelength per unit length of crest is

$$E = E_p + E_K = \tfrac{1}{2}\rho g a^2 L$$

Per unit horizontal area, this is

$$E_{av} = \tfrac{1}{2}\rho g a^2$$

Dividing the energy flux by the stored energy gives the rate of propagation of energy:

$$U_E = \frac{F_{av}}{E_{av}} = \frac{C}{2}\left(1 + \frac{2md}{\sinh 2md}\right)$$

i.e., $U_E = \frac{1}{2}(gT/2\pi)$ in deep water and $U_E = (gd)^{1/2}$ in shallow water, which means that the rate of transmitted energy is half of the wave velocity in deep water and equal to the wave velocity in shallow water.

16-5.3 Group Velocity

Consider the linear superimposition of two progressive waves of the same amplitude and slightly different periods such as

$$\eta = a \sin[mx - kt] + a \sin[(m + \delta m)x - (k + \delta k)t]$$

where δm and δk are supposed to be small quantities. This expression can be written as:

$$\eta = 2a \cos \tfrac{1}{2}(\delta m x - \delta k t) \sin\left[\left(m + \frac{\delta m}{2}\right)x - \left(k + \frac{\delta k}{2}\right)t\right]$$

Since δm and δk are small:

$$\eta \cong 2a \cos \tfrac{1}{2}(\delta m x - \delta k t) \sin(mx - kt)$$

It is seen that the wave $2a \sin(mx - kt)$ is modulated by the term:

$$\cos \tfrac{1}{2}(\delta m x - \delta k t) = \cos \tfrac{1}{2}\, \delta m\left(x - \frac{\delta k}{\delta m} t\right)$$

which travels at a speed $U = \delta k/\delta m$. The wave velocity $C = k/m$, and $\delta k = \delta(mC)$, $U = \delta(mC)/\delta m$. If these values are taken at the limit when δm tends toward dm, U can be written as $U = d(mC)/dm$. This is just $U = C + m(dC/dm)$.

Since $m = 2\pi/L$, $dm = -2\pi\, dL/L^2$, and

$$U = C - L\frac{dC}{dL}$$

When the values $C = (gT/2\pi) \tanh(2\pi d/L)$ and $L = CT$ are inserted this becomes

$$U = \frac{C}{2}\left(1 + \frac{2md}{\sinh 2md}\right)$$

It is interesting to note that this expression is the same as the expression for the rate of energy propagation U_E. This stems from the fact that in the case of a wave train, there is no energy passing through the node of the wave train where the wave amplitude is zero. So, the velocity of energy propagation is that of a wave train or of a group of waves. For this reason, the group velocity U is equal to the rate of energy propagation. However, it is important to note that this statement holds true only in the case of a linear wave. The values of U/C_0 and C/C_0 where C_0 is the deep water wave celerity ($C_0 = gT/2\pi$) is shown on Fig. 16-11. The energy flux can now be expressed as a function of the group velocity U:

$$F_{av} = \tfrac{1}{8}\rho g H^2 U$$

16-5.4 Physical Interpretation

A physical interpretation of the wave energy flux can be given. In deep water, the particle velocity describes a circle as seen in Section 16-3.3. The particle velocity remains the same in absolute value along the circle. Consider a vertical plane AB passing through the circle (Fig. 16-12). The kinetic energy flux in this plane is zero. This energy, $\rho V^2/2$, is the same at point A and at point B. The only difference is that the direction is opposite. The potential energy is different at A from its value at B, and only the potential energy is transmitted through the plane AB. Since the average value of potential and kinetic energy are equal, there are only two possibilities. Either one-half the wave energy is transmitted with phase velocity or the total energy is transmitted at one-half the phase velocity, which is the deepwater group velocity.

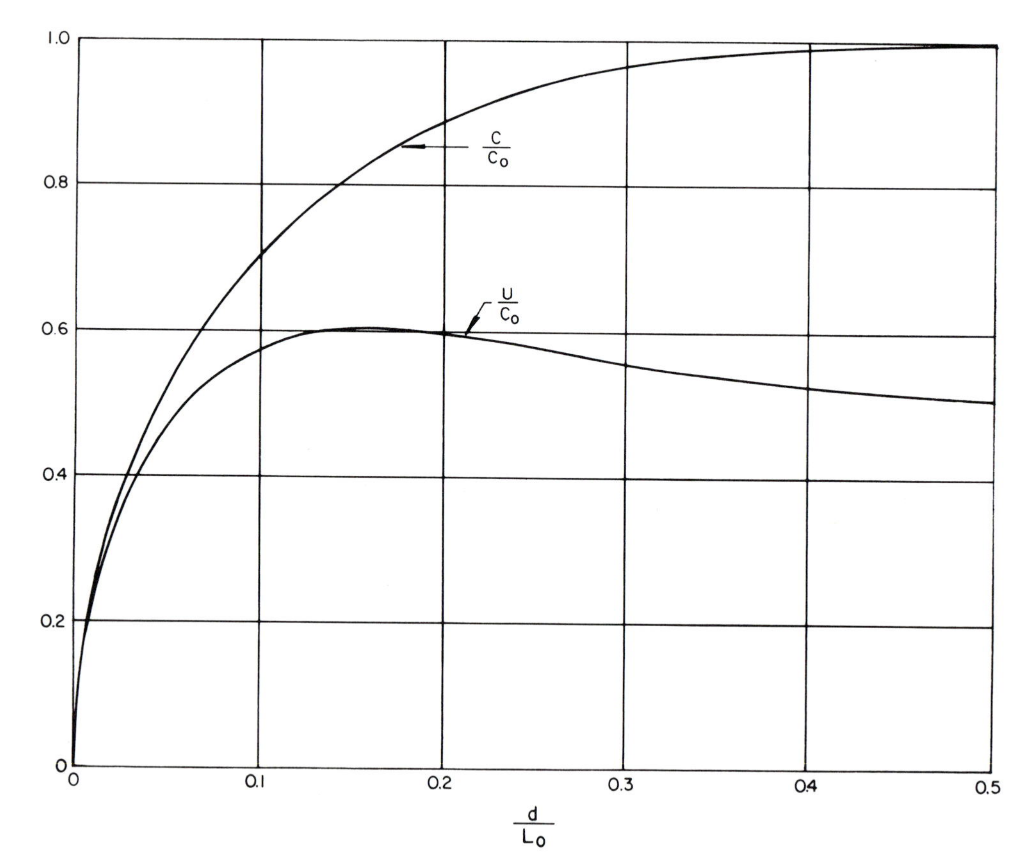

Figure 16-11

Relative wave velocity and group velocity vs depth.

Figure 16-12 *The potential energy only is transmitted by the wave.*

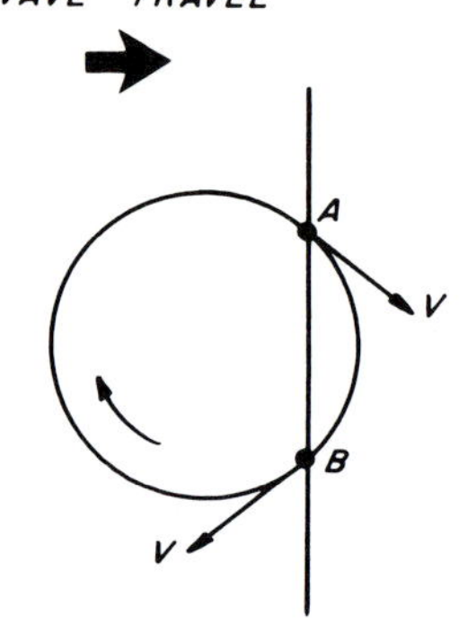

16-6 Wave Transformation

16-6.1 Wave Shoaling

In the case of a wave traveling over a very gentle slope, it is assumed that the wave motion is almost the same as if the bottom were horizontal. In a word, the flow pattern deformation due to the bottom slope is neglected. It is then assumed that the flux of transmitted energy is a constant. In this case the wave height $2a$ at a given depth d is known as a function of the deep water wave height $2a_0$ from the formula

$$F_{\text{av}}|_d = F_{\text{av}}|_{d \to \infty}$$

Thus,

$$\tfrac{1}{4}\rho g a^2 \frac{g}{k} \tanh md\left[1 + \frac{2md}{\sinh 2md}\right] = \tfrac{1}{4}\rho g a_0^2 \frac{g}{k}$$

and

$$\frac{2a}{2a_0} = \frac{H}{H_0} = \frac{1}{\{\tanh md[1 + (2md/\sinh 2md)]\}^{1/2}}$$

$$= \left(\frac{U_0}{U}\right)^{1/2} = K_s$$

where K_s is the shoaling coefficient. Inserting the expression for $L/L_0 = \tanh(2\pi d/L)$ where L_0 is the deep water wave length $L_0 = gT^2/2\pi$, it is then possible to calculate the value of H/H_0 as a function of d/L_0 only. The results of such calculation are shown in Fig. 16-13.

16-6.2 Bottom Friction and Wave Damping

At a given location such as $x = 0$ the velocity at the bed is

$$u_b = \left.\frac{\partial \phi}{\partial x}\right|_{z=-d} = \frac{ka}{\sinh md} \cos kt$$

The friction stress τ for a turbulent boundary layer is given by

$$\tau = \rho f u_b |u_b|$$

where f is the friction factor ($f \cong 0.015$). The average work done by this friction stress per unit area or as rate of energy dissipation is

$$D_f = \overline{\tau_b u_b} = \frac{1}{T}\int_0^T \rho f u_b^2 |u_b| \, dt$$

Figure 16-13 *Relative wave height variation with depth.*

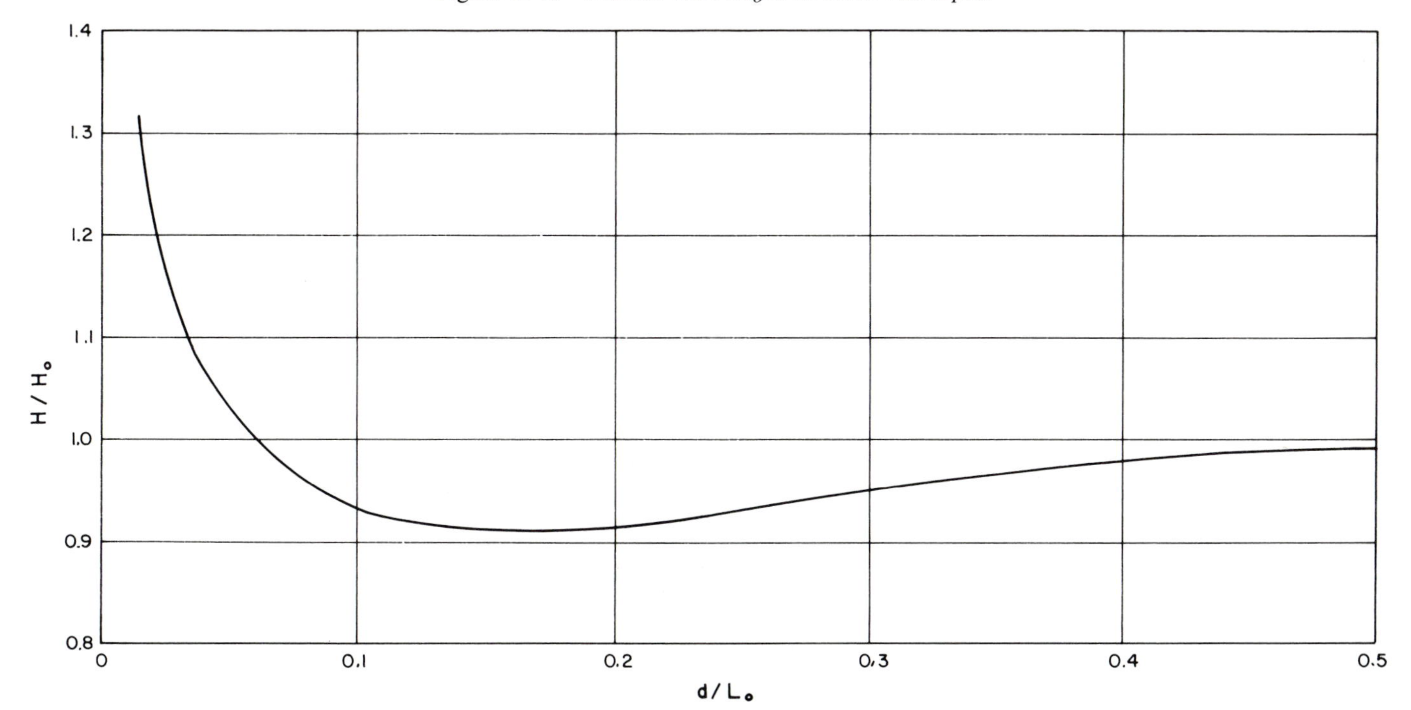

Since $H = 2a$ and $k = 2\pi/T$ then,

$$D_f = \rho f \frac{1}{T} \int_0^T \frac{(\pi H)^3}{T^3 \sinh^3 md} (\cos kt)^2 |\cos kt| \, dt$$

i.e.,

$$D_f = \tfrac{4}{3}\rho f \pi^2 \frac{H^3}{T^3 \sinh^3 md}$$

The variation of energy flux equals the rate of energy dissipation, i.e.,

$$\frac{d}{dx}(F_{av}) = -D_f$$

or, $H/H_0 = K_s K_f$. The damping factor K_f is

$$K_f = \left(1 - \frac{8}{\rho g} \int_{-\infty}^{d(x)} D_f \, dx\right)$$

which is solved numerically over complex topography.

16-6.3 Wave Refraction

16-6.3.1 A wave orthogonal is the locus of points which define the minimum time of travel for wave propagation between two points. Wave orthogonals are usually normal to wave crests. They indicate approximately the direction of propagation of the wave energy. When the wave arrives at an angle with the bottom contours and refracts, the distance between wave orthogonals b varies by a factor of

$$\frac{b_2}{b_1} = \frac{\cos \alpha_2}{\cos \alpha_1}$$

α is the angle of the bottom contours with the wave crest. Subscripts 1 and 2 refer to the two referenced water depths on both sides of the bottom contour. When the variation of wave height along a wave crest is small (which is not the case of wave diffraction) it is generally assumed that the wave energy flux between wave orthogonals is constant, i.e., $F_{av} b$ = constant, and

$$\tfrac{1}{8}\rho g H^2 U b = \text{constant}$$

Hence, the wave height varies with a refraction coefficient equal to the square root of the inverse ratio of the distance between wave orthogonals. The refraction coefficient K_r is $(b_0/b)^{1/2}$, where b_0 is the distance between wave orthogonals in deep water. The combined effects of shoaling, damping, and refraction are obtained by multiplying the shoaling coefficient K_s, the damping coefficient K_f and the refraction coefficient K_r, so that

$$H = K_s K_f K_r H_0$$

Wave refraction diagrams giving distance between wave orthogonals are obtained by graphical methods or by computer. Refraction effects remain small as long as the water depth is larger than $L/3$. Therefore, wave refraction calculations (and scale models) need to be done only over bottom topography shallower than such a depth.

16-6.3.2 The first method for drawing a wave refraction diagram consists of calculating the wave crests first (Fig. 16-14). The wave crests are obtained by drawing the envelope of circles from a preceding wave crest, the radius of the circles being proportional to the local values of the wavelength, or one of its multiples.

The second method consists of drawing the orthogonals directly by application of the Snell's law of wave refraction:

$$C_2 \sin \alpha_2 = C_1 \sin \alpha_1$$

where the subscripts 1 and 2 correspond to two successive values of the water depth.

Not only is it possible to calculate the refraction coefficient from the relative distance of two wave orthogonals but it is also possible to determine it from the curvature of the orthogonals.

16-6.3.3 In many cases refraction diagrams provide a reasonably accurate measure of the changes waves undergo on approaching a coast. Quite often they provide the only measure of these changes available. However, the accuracy of data determined from refraction diagrams is limited by the validity of the theory of their construction and the accuracy of the depth data on which they are based. The orthogonal direction change is derived for the simplest case of straight parallel contours and, although little error

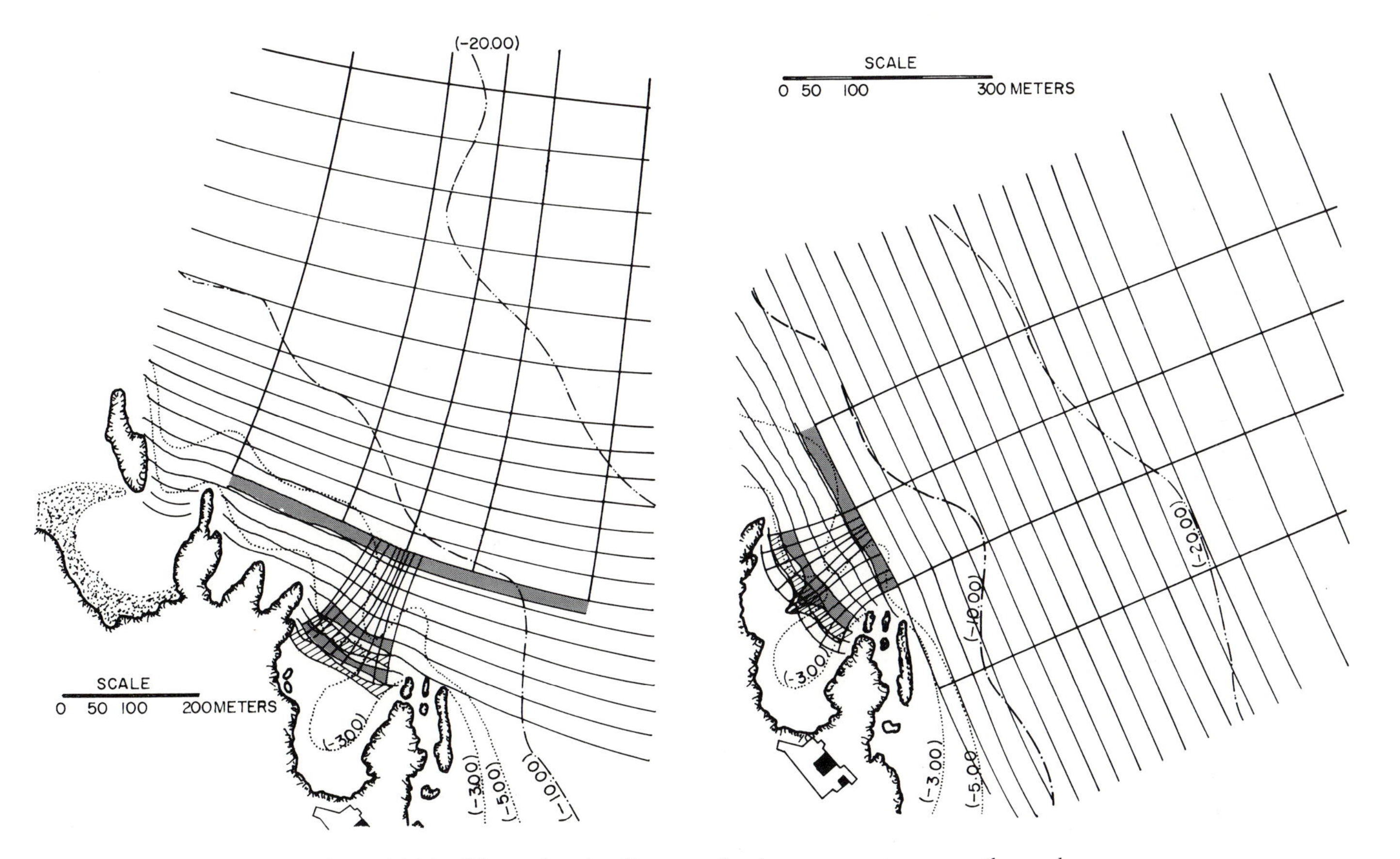

Figure 16-14 *Wave refraction diagrams showing wave crests, wave orthogonals, and relative wave heights along wave crests (shaded).*

is introduced by bringing orthogonals over relatively simple hydrography, it is difficult to carry an orthogonal accurately on shore over complex bottom features. Moreover, the equation is derived for small waves moving over relatively flat slopes. Although no strict limits have been set, accuracy as far as height changes are concerned cannot be expected where bottom slopes are steeper than 1/10. Model tests have indicated that direction changes occur nearly as predicted even over a vertical discontinuity. Nevertheless, a small wave reflection takes place. A third limitation is inherent in the assumption that no energy travels laterally along a wave crest. No strict limits have been set, but the accuracy of wave heights derived from orthogonals that bend sharply is questionable. A further problem is the interpretation of crossed orthogonals. Such cases would seem to lead to the formation of high, short-crested waves with breakers, and caustic, doubling wave crests due to nonlinear effects.

In short, refraction coefficients which are quite different from unity, say $K_r < 0.5$ and $K_r > 2$, must be doubted.

When they do occur as a result of one of the plotting procedures mentioned, other effects such as diffraction may be occurring. Nonlinear effects also influence wave diffraction, as can be seen by the doubling of wave crests at the ends of breakwaters.

16-7 Wave Thrust or Radiation Stress

16-7.1 Definitions

The concept of specific force has been defined in Section 14-2.1 as the sum of the momentum flux through a cross section A and the pressure force acting on this section A per unit weight of fluid; i.e.,

$$\left[\frac{V^2}{g} + \frac{p}{\rho g}\right]A$$

A similar definition is introduced in the field of water waves, the concept of "radiation stress." Radiation stress is, by definition, the average value of the sum $(p_w + \rho V_n^2)$ with respect to time, integrated along a vertical plane of unit width. The velocity component perpendicular to the considered plane is V_n, p_w is the pressure fluctuation due to water waves around the hydrostatic pressure from the still water level, i.e., $p_w = p - \rho g z$ where p is the total pressure, and ρV_n^2 is the momentum flux. When the plane is vertical and parallel to the wave crest ($V = u$), the radiation stress is

$$S_{xx} = \frac{1}{T}\int_0^T \int_{-d}^{\eta} (p_w + \rho u^2)\, dz\, dt$$

which is still called the "wave thrust."

16-7.2 Application to Water Waves

If one inserts the expression of p_w obtained from the linear wave theory, the corresponding integral has an average value equal to zero. However the second-order term yields

$$\tfrac{1}{16}\rho g H^2 \frac{4\pi d/L}{\sinh(4\pi d/L)}$$

Inserting the expression for u (at a first order of approximation), yields

$$\frac{\rho}{T}\int_0^T \int_d^{\eta} \left(\frac{H}{2} k \frac{\cosh m(d+z)}{\sinh md} \sin(kt - mx)\right)^2 dt\, dz$$

$$= \tfrac{1}{16}\rho g H^2\left[1 + \frac{4\pi d/L}{\sinh(4\pi d/L)}\right]$$

Note that these two expressions are of the same order in H^2. Therefore, the wave thrust is

$$S_{xx} = \tfrac{1}{16}\rho g H^2\left[1 + 2\frac{4\pi d/L}{\sinh(4\pi d/L)}\right]$$

which gives $S_{xx} = \tfrac{3}{16}\rho g H^2$ in shallow water and $S_{xx} = \tfrac{1}{16}\rho g H^2$ in deep water.

If we consider a wave traveling over a gently sloping bed from deep water into shallow water to the point at which the wave height becomes equal to the deep water wave height, as seen in Section 16-6.1 the wave thrust has to increase by a factor of 3. Therefore, the momentum balance requires an external force. This external force is obtained by differences in the hydrostatic pressure. Hence, the midwater level is set down below the still water level, as demonstrated in the following section.

16-7.3 Wave Set-Up/Wave Set-Down; Surf Zone Circulation

Let us consider the momentum balance in a slice of water bounded by the free surface $z = \eta$, a gently sloping bottom $z = -d(x)$, and two vertical planes parallel to the wave crest at $x = x_0$, $x = x_0 + dx$. The average value of $\bar{\eta} = 1/T \int \eta\, dt$ is defined and gives the variation of the midwater level with respect to the still water level. The total forces exerted on the planes $x = x_0$, $x = z_0 + dx$ are the sum of hydrostatic pressure and wave thrust. Thus,

$$F_{x0} = +\left[S_{xx} + \tfrac{1}{2}\rho g(d + \bar{\eta})^2\right]$$

$$F_{x0+dx} = -\left\{S_{xx} + \frac{d}{dx}(S_{xx})\,dx + \frac{1}{2}\rho g \times \left[(d + \bar{\eta})^2 + \frac{d}{dx}(d + \bar{\eta})^2\, dx\right]\right\}$$

Another external force is due to bottom pressure (since the bottom is not horizontal). It has a horizontal component,

$$F_b = -\left[\rho g(d + \bar{\eta})\frac{d(d)}{dx}\,dx\right]$$

By application of the momentum theorem (Chapter 12), the momentum balance yields

$$F_{x_0} + F_{x_0+dx} + F_b \cong \frac{dS_{xx}}{dx} + \rho g(d + \bar{\eta})\frac{d\bar{\eta}}{dx} \cong 0$$

or, since $\bar{\eta} \ll d$,

$$\frac{d\bar{\eta}}{dx} \cong -\frac{1}{\rho g d}\frac{dS_{xx}}{dx}$$

Since S_{xx} increases when the wave proceeds from deep water to shallow water, it is seen that $\bar{\eta}$ decreases, i.e., the midwater level is set down (Fig. 16-15).

After breaking inception, the wave energy is dissipated; therefore S_{xx} decreases and $\bar{\eta}$ increases. This is the wave set-up. The wave set-up is the elevation of the midwater level which takes place in the surf zone. If one assumes that all the wave energy is dissipated in the surf zone, the total force exerted on the beach (or on a wave absorber) is equal to the wave thrust in deep water, i.e.,

$$F = \tfrac{1}{16}\rho g H^2$$

Therefore, it is seen that dissipating the wave energy does not nullify the average wave force. If the wave is reflected instead of being dissipated, the wave force will be subjected to large fluctuation but its average value, at a first order of approximation, is zero.

The concept of radiation stress also explains the occurrence of surf beats, i.e., the oscillations of water level on the

Figure 16-15 *Effects of radiation stresses in the surf zone.*

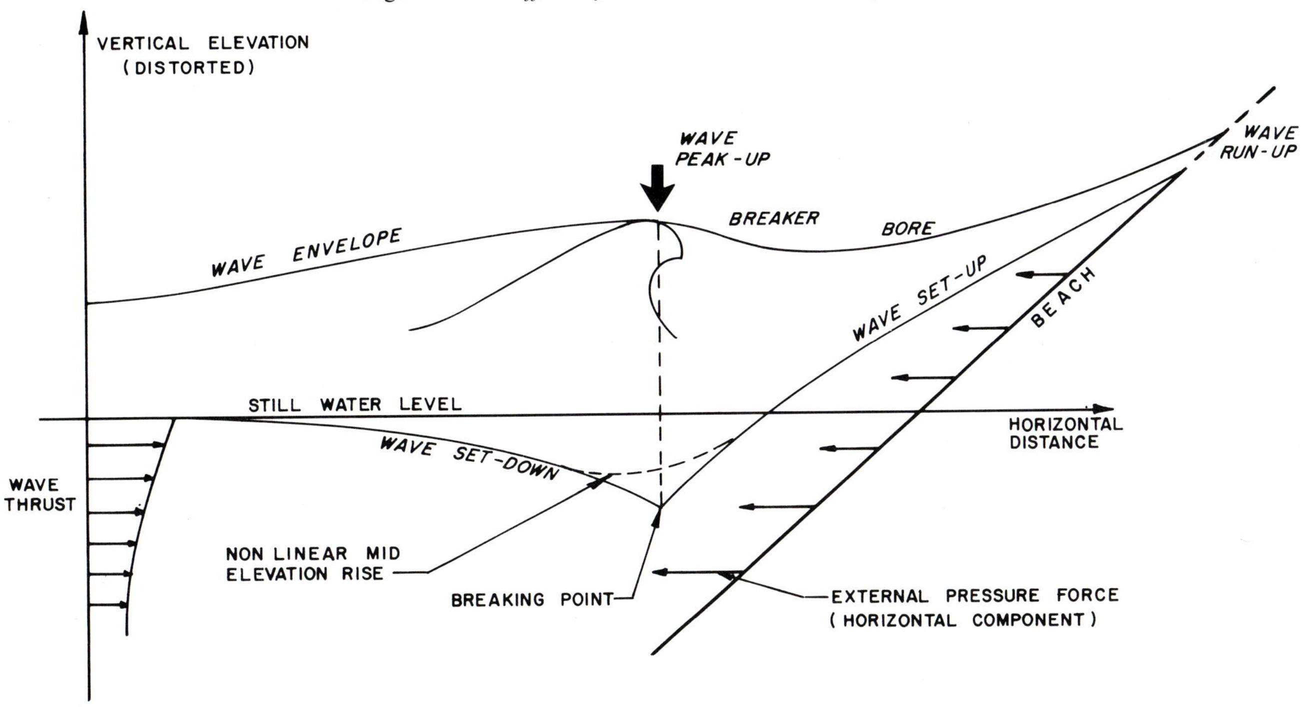

shore with a period equal to several times the wave period, due to variation of wave heights within wave groups.

If the breaking waves approach the shore at an angle, the wave thrust has a component parallel to the coast which cannot be balanced by pressure variations. Therefore a longshore current is generated. The piling of the water due to the wave thrust acting towards a beach is unstable and causes rip currents. The rip current has a momentum flux in opposite direction to the wave thrust. Therefore, the longshore current, instead of being parallel to the coast, tends to meander by interaction with the movable sediments forming the beaches. Transverse resonance effects (edge wave) enhance and/or trigger the phenomena.

16-8 A Formulary for the Small Amplitude Wave Theory

Figure 16-16 gives the values sinh $(2\pi d/L)$, cosh $(2\pi d/L)$, and tanh $(2\pi d/L)$ as functions of d/L. Tables 16-2 and 16-3 summarize a number of formulas for linear periodic waves in deep water, in an intermediate water depth, and in shallow water. Table 16-2 presents a set of formulas for progressive waves, Table 16-3 for standing waves.

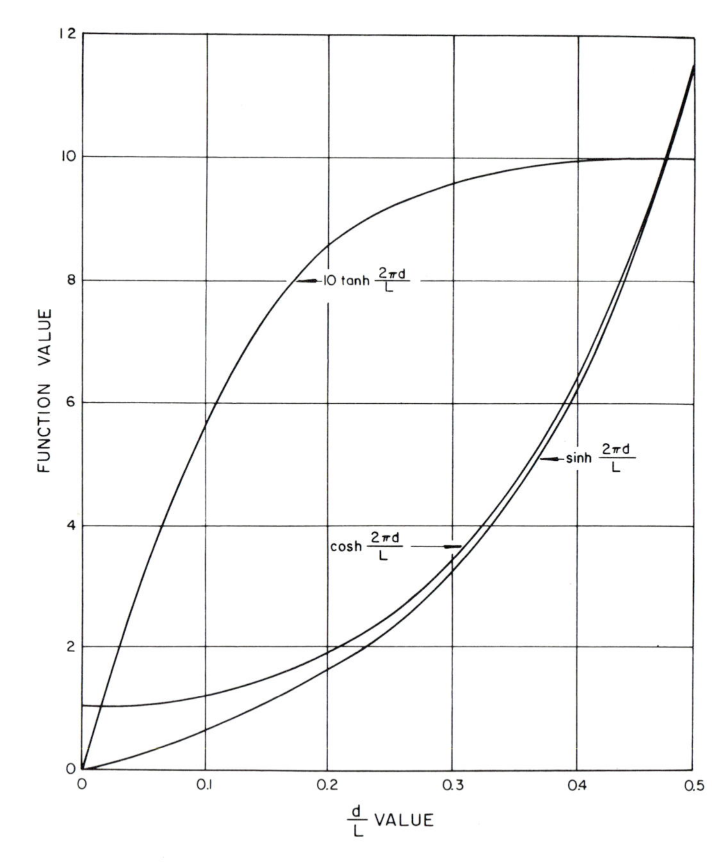

Figure 16-16 *Hyperbolic functions vs relative depth.*

PROBLEMS

16.1 Determine the streamlines, the equipotential lines, paths, and lines of equipressure (isobars) in a linear periodic progressive wave.

16.2 Determine the streamlines, the equipotential lines, paths, and lines of equipressure in a linear periodic standing wave.

16.3 Demonstrate by applying the linear long wave theory that the free surface of two-dimensional fundamental motion of a seiche in a basin in the form of a parabola $\{d(x) = d_0[1 - (x^2/a^2)]\}$ is a straight line. Demonstrate that the free surface of the first harmonic motion is a parabola.

16.4 Calculate the value of the wavelength L as a function of the water depth d in the case of a wave period $T = 8$ sec, 10 sec, and 12 sec. Determine the minimum value of d/L when L can be considered as the deep-water wavelength, and the maximum value of d/L permitting the shallow water approximation within 5% of accuracy.

16.5 Demonstrate that the velocity potential function for a linear periodic progressive wave in deep water is

$$\phi = -a\frac{k}{m}e^{-kz}\cos(kt - mx)$$

Table 16-2 *Periodic progressive wave, first approximation: linear theory*

	Deep water $d/L > 0.5$	Intermediate water depth $0.5 > d/L > 0.05$	Shallow water $d/L < 0.05$
Wave velocity	$C = \left[\frac{gL}{2\pi}\right]^{1/2} = \frac{gT}{2\pi} = \frac{g}{k}$	$C = \left[\frac{gL}{2\pi}\tanh\frac{2\pi d}{L}\right]^{1/2} = \frac{gT}{2\pi}\tanh\frac{2\pi d}{L}$	$C = (gd)^{1/2}$
Wave length	$L = \frac{gT^2}{2\pi}$ ($L_{ft} = 5.12\ T^2$ sec)	$L = \frac{gT^2}{2\pi}\tanh\frac{2\pi d}{L}$	$L = T(gd)^{1/2}$
	$k^2 = gm$	$k^2 = gm \tanh md$	$k^2 = m^2 gd$
Group velocity	$U = \frac{C}{2} = \frac{1}{2}\frac{gT}{2\pi}$	$U = \frac{C}{2}\left[1 + \frac{2md}{\sinh 2md}\right]$	$U = C = (gd)^{1/2}$
Potential function	$\phi = -\frac{H}{2}\frac{k}{m}\exp(mz)\cos(kt - mx)$	$\phi = -\frac{H}{2}\frac{k}{m}\frac{\cosh m(d+z)}{\sinh md}\cos(kt - mx)$	$\phi = -\frac{H}{2}\frac{gT}{2\pi}\cos(kt - mx)$
	$\phi = -\frac{H}{2}\frac{g}{k}\exp(mz)\cos(kt - mx)$	$\phi = -\frac{H}{2}\frac{g}{k}\frac{\cosh m(d+z)}{\cosh md}\cos(kt - mx)$	$\phi = -\frac{H}{2}\frac{g}{k}\cos(kt - mx)$
Velocity components	$u = \frac{2\pi}{T}\frac{H}{2}\exp(mz)\sin(kt - mx)$	$u = \frac{H}{2}k\frac{\cosh m(d+z)}{\sinh md}\sin(kt - mx)$	$u = \frac{H}{2}(g/d)^{1/2}\sin(kt - mx)$
	$w = \frac{2\pi}{T}\frac{H}{2}\exp(mz)\cos(kt - mx)$	$w = \frac{H}{2}k\frac{\sinh m(d+z)}{\sinh md}\cos(kt - mx)$	$w = 0$
Orbits	$x - x_0 = -\frac{H}{2}\exp(mz_0)\cos(kt - mx_0)$	$x - x_0 = -\frac{H}{2}\frac{\cosh m(d+z_0)}{\sinh md}\cos(kt - mx_0)$	$x - x_0 = -\frac{H}{2}\frac{T}{2\pi} \times (g/d)^{1/2}\cos(kt - mx_0)$
	$z - z_0 = \frac{H}{2}\exp(mz_0)\sin(kt - mx_0)$	$z - z_0 = \frac{H}{2}\frac{\sinh m(d+z_0)}{\sinh md}\sin(kt - mx_0)$	$z - z_0 = 0$
	Circles of radius: $R = \frac{H}{2}e^{+mz_0}$	Ellipses of: Semimajor axis: $\frac{H}{2}\frac{\cosh m(d+z_0)}{\sinh md}$ Semiminor axis: $\frac{H}{2}\frac{\sinh m(d+z_0)}{\sinh md}$	Straight lines horizontal amplitude: $\frac{H}{2}\frac{T}{2\pi}\left(\frac{g}{d}\right)^{1/2}$
Free surface	$\eta = \frac{H}{2}\sin(kt - mx)$	$\eta = \frac{H}{2}\sin(kt - mx)$	$\eta = \frac{H}{2}\sin(kt - mx)$
Pressure	$p = -\rho gz + \rho g\frac{H}{2}\exp(mz)\sin(kt - mx)$	(1)* $p = -\rho gz + \rho g\frac{H}{2}\frac{\cosh m(d+z)}{\cosh md}\sin(kt - mx)$ (2)* $p = -\rho gz_0 - \rho g\frac{H}{2}\frac{\sinh mz_0}{\sinh md\cosh md}\sin(kt - mx_0)$	$p = -\rho gz + \rho g\frac{H}{2}\sin(kt - mx)$
Energy per wave length per unit of length of crest	$E = \frac{1}{2}\rho gL\left(\frac{H}{2}\right)^2, E_K = E_P = \frac{1}{2}E$	$E = \frac{1}{2}\rho gL\left(\frac{H}{2}\right)^2, E_K = E_P = \frac{1}{2}E$	$E = \frac{1}{2}\rho gL\left(\frac{H}{2}\right)^2, E_K = E_P = \frac{1}{2}E$
Average energy flux per unit length of crest	$\frac{1}{4}\rho g\left(\frac{H}{2}\right)^2\frac{gT}{2\pi}$	$\frac{1}{4}\rho g\left(\frac{H}{2}\right)^2 C\left[1 + \frac{2md}{\sinh 2md}\right]$	$\frac{1}{2}\rho g\left(\frac{H}{2}\right)^2\sqrt{gd}$

*(1) Euler; (2) Lagrange.

Table 16-3 *Standing wave, first approximation: linear theory*

	Deep water $d/L > 0.5$	*Intermediate water depth* $0.5 > d/L > 0.05$	*Shallow water* $d/L < 0.05$
Potential function	$\phi = -\frac{H}{2}\frac{k}{m} e^{+mz} \cos mx \sin kt$	$\phi = -\frac{H}{2}\frac{k}{m}\frac{\cosh m(d+z)}{\sinh md} \cos mx \sin kt$	$\phi = -\frac{H}{2}\frac{g}{k} \cos mx \sin kt$
	$\phi = -\frac{H}{2}\frac{g}{k} e^{+mz} \cos mx \sin kt$	$\phi = -\frac{H}{2}\frac{g}{k}\frac{\cosh m(d+z)}{\cosh md} \cos mx \sin kt$	
Velocity components	$u = -\frac{H}{2} k e^{+mz} \sin mx \sin kt$	$u = -\frac{H}{2} k \frac{\cosh m(d+z)}{\sinh md} \sin mx \sin kt$	$u = -\frac{H}{2} (g/d)^{1/2} \sin mx \sin kt$
	$w = \frac{H}{2} k e^{+mz} \cos mx \sin kt$	$w = \frac{H}{2} k \frac{\sinh m(d+z)}{\sinh md} \cos mx \sin kt$	$w = 0$
Orbits	$x - x_0 = \frac{H}{2} e^{+mz_0} \sin mx_0 \cos kt$	$x - x_0 = \frac{H}{2}\frac{\cosh m(d+z_0)}{\sinh md} \sin mx_0 \cos kt$	$x - x_0 = \frac{H}{2}\frac{T}{2\pi}(g/d)^{1/2} \sin mx_0 \cos kt$
	$z - z_0 = -\frac{H}{2} e^{+mz_0} \cos mx_0 \cos kt$	$z - z_0 = -\frac{H}{2}\frac{\sinh m(d+z_0)}{\sinh md} \cos mx_0 \cos kt$	$z - z_0 = 0$
	Straight lines: $\frac{x - x_0}{z - z_0} = -\tan mx_0$	Straight lines: $\frac{x - x_0}{z - z_0} = -\frac{\tan mx_0}{\tanh m(d + z_0)}$	Horizontal lines of amplitude: $\frac{H}{2}\frac{T}{2\pi}(g/d)^{1/2} \sin mx_0$
Free surface	$\eta = -\frac{H}{2} \cos mx \cos kt$	$\eta = -\frac{H}{2} \cos mx \cos kt$	$\eta = -\frac{H}{2} \cos mx \cos kt$
Pressure	$p = -\rho gz - \rho g \frac{H}{2} e^{+mz} \cos mx \cos kt$	$p = -\rho gz - \rho g \frac{H}{2}\frac{\cosh m(d+z)}{\cosh md} \cos mx \cos kt$	$p = -\rho gz - \rho g \frac{H}{2} \cos mx \cos k$
Energy per wave length	$E_K = \frac{1}{4}\rho g\left(\frac{H}{2}\right)^2 L \sin^2 kt$	$E_K = \frac{1}{4}\rho g\left(\frac{H}{2}\right)^2 L \sin^2 kt$	$E_K = \frac{1}{4}\rho g\left(\frac{H}{2}\right)^2 L \sin^2 kt$
	$E_p = \frac{1}{4}\rho g\left(\frac{H}{2}\right)^2 L \cos^2 kt$	$E_p = \frac{1}{4}\rho g\left(\frac{H}{2}\right)^2 L \cos^2 kt$	$E_p = \frac{1}{4}\rho g\left(\frac{H}{2}\right)^2 L \cos^2 kt$
	$E = \frac{1}{4}\rho g L\left(\frac{H}{2}\right)^2$	$E = \frac{1}{4}\rho g L\left(\frac{H}{2}\right)^2$	$E = \frac{1}{4}\rho g L\left(\frac{H}{2}\right)^2$

and give the expression for the pressure $p(x,z,t)$ and the free surface. Demonstrate why the criterion $d/L \leq 0.5$ is generally considered as the limit of validity of the above expression for ϕ. Explain why the wave refraction diagram must start at $d/L > 0.3$.

16.6 Establish the expression for $p(x,z,t)$ for a two-dimensional periodic linear progressive wave in intermediate water depth.

16.7 Consider the relationship:

$$k^2 = mg \tanh md$$

and demonstrate the expression:

$$C = \left(\frac{gL}{2\pi} \tanh \frac{2\pi d}{L}\right)^{1/2}$$

16.8 Demonstrate that the particle path in a two-dimensional periodic linear standing wave is a straight line and sketch the corresponding flow pattern.

16.9 Determine the periods of free oscillation (fundamental and first ten harmonics) of a two-dimensional basin 10 ft long (3.048 m) and 2 ft deep (0.609 m). Determine the three longest periods of free oscillation of a rectangular basin 10 ft × 8 ft (3.048 m × 2.438 m) and 2 ft deep (0.609 m).

16.10 Determine the error on the value of C corresponding to the deep water approximation ($C = gT/2\pi$) when $d/L = 1/3$. Determine the minimum value of d/L which gives the same error for the shallow water approximation: $C = (gd)^{1/2}$. Determine the common value of d/L such that the error for the shallow water approximation equals the error for the deep water approximation.

16.11 The pressure $p(t)$ due to a periodic wave traveling in intermediate water depth is recorded at a fixed location on the sea bottom. Determine the operation which permits one to determine the wave height from the knowledge of this pressure fluctuation.

16.12 Establish the value of the pressure on a vertical wall (first approximation) and determine a corresponding method of calculation for determining the stability of a vertical breakwater.

16.13 Consider a straight shoreline with parallel bottom contours and a periodic wave arriving at an angle with the deep water bottom contours. Demonstrate that at any contour one has

$$\frac{\sin \alpha}{C} = \frac{\sin \alpha_0}{C_0} = \text{constant}$$

by application of the law of wave refraction. C is the water wave velocity, α is the angle of wave crest with the parallel bottom contours, and subscript 0 refers to infinite depth.

16.14 It will be assumed that the principle of conservation of transmitted energy between wave orthogonals is valid. Moreover, it can be demonstrated that the wave breaks over a horizontal bottom when the wave steepness

$$\frac{H}{L} \rightarrow 0.14 \tanh \frac{2\pi d}{L}$$

It will be assumed that this breaking criterion is still valid in the case of a wave breaking at an angle on a gently sloping beach. Now, consider a periodic wave arriving from deep water at an angle α_0 with the bottom contours, these bottom contours being straight and parallel and defining a very gently sloping beach. Establish a method for calculating the angle of breaking wave crest α_b with the shoreline, the depth of breaking d_b and the wave height of breaking H_b as a function of α_0, the deep water wave height H_0, and the wave period T.

16.15

1. Establish the expression for the maximum pressure fluctuation Δp in a periodic progressive wave as a function of depth z.
2. If a wave recorder is located at a depth $z = z_0$ what is the pressure multiplication factor K to apply to the amplitude in order to obtain the wave height.
3. Let Δp recorded be 1.33 psi (0.0933 kg/cm^2), $T =$ 8 sec, $z_0 = -25$ ft (7.62 m), $d = 96$ ft. (29.26 m). Calculate the wave height.

16.16 Establish the expression of the pressure fluctuation due to an incident periodic wave of height H at the toe of a vertical wall causing total wave reflection.

16.17 In a two-dimensional periodic wave, the pressure difference across the free surface due to capillary effect is

$$\Delta p = -A \frac{\partial^2 \eta}{\partial x^2}$$

where η is the free surface elevation and x the horizontal distance; $A = 75$ dyn/cm for water at 0°C; $A = 490$ dyn/cm for mercury (ρ mercury $= 13.56\ \rho$ water).

1. Establish the free surface boundary condition (linear approximation).
2. Give the expression for the phase velocity as a function of period and water depth.
3. Draw the curve for phase velocity as function of the wave period for deep water.

16.18 A wave filter is composed of wire mesh dropped into the fluid flow. Such a filter creates negligible flow disturbances. However, it introduces an internal friction force **F** proportional to the average velocity such that $\mathbf{F} = -K\mathbf{V}$. The average flow motion will be considered as irrotational. Establish the free surface condition which should be used instead of the Cauchy–Poisson condition for the free surface in the case where the void coefficient of the filter is close to unity and in the case where it has a finite value ε.

Chapter 17

Finite Amplitude Waves

17-1 Introduction to the Field of Finite Amplitude Wave Theory

17-1.1 Type of Power Series

17-1.1.1 This chapter deals with the problem of periodic two-dimensional finite amplitude waves, in which nonlinear effects are partly taken into account by the use of power series. The subject matter has been introduced briefly in Chapter 15 in the general introduction to water waves. The purpose of this chapter is to present more advanced considerations, which form the background for this subject matter. However, the mathematical details are beyond the scope of this book, so that only the main results will be presented.

It has been shown that solutions can be expressed as power series in terms of a quantity which is small compared to the other dimensions. The small quantity used here is H/L for small L/d; it is the most significant parameter in deep water. H/d is used for large L/d; it is most significant parameter in shallow water.

In the first case (development in terms of H/L), the first term of the power series is obtained by application of the linear theory. These are the Stokes (or "Stokesian") wave theories. Many series are expressed in terms of $2a/L$, where a is a parameter such that

$$\sum_1^n \left(\frac{2a}{L}\right)^n = \frac{H}{L} \qquad \left(a = \frac{H}{2} \text{ at the first order}\right)$$

while others are expressed in terms of H/L directly.

In the second case, the first term of the series is already a solution of nonlinear equations. They are called cnoidal wave theories because they are mathematically defined by the so-called cnoidal functions. The definition of these functions, as well as the development of these theories, are beyond the scope of this book. These are the essential basic principles.

17-1.1.2 In order to illustrate this method, the following example is presented: There exist in the American literature at least five different irrotational Stokesian wave theories at a fifth order of approximation; all of them are based on the same assumption, but the mathematical formulations are different. The free surface profiles are given for three of them ($\theta = mx - kt$):

$$\frac{\eta}{H} = a \sum_{1}^{n} r_n \cos n\theta \tag{17-1}$$

$$\begin{aligned} m\eta = mA \cos \theta &+ [B_{22} + B_{24}(mA)^2](mA)^2 \cos 2\theta \\ &+ [B_{33} + B_{35}(mA)^2](mA)^3 \cos 3\theta \\ &+ B_{44}(mA)^4 \cos 4\theta + B_{55}(mA)^5 \cos 5\theta \end{aligned} \tag{17-2}$$

where A is related to $H/2$ and the B coefficients are complex functions of d/L.

$$\begin{aligned} \frac{\eta}{A_0} = [1 &+ \beta_{13}(mA_0)^2 + \beta_{15}(mA_0)^4] \cos \theta \\ &+ [\beta_{22} + \beta_{24}(mA_0)^2]mA_0 \cos 2\theta \\ &+ [\beta_{33} + \beta_{35}(mA_0)^2](mA_0)^2 \cos 3\theta \\ &+ \beta_{44}(mA_0)^3 \cos 4\theta + \beta_{55}(mA_0)^4 \cos 5\theta \end{aligned} \tag{17-3}$$

The β coefficients are complex functions of d/L. This example of the mathematical representation for steady-state periodic waves over a constant depth illustrates the complexity of a systematic comparison between wave theories supposedly developed with the same aim. The complexity increases with the number of possible assumptions which can be used initially for developing a wave theory.

17-1.1.3 Consider further the case of irrotational waves. The values of the wave characteristics depend upon the number of terms chosen for the power series expansion, either in terms of wave steepness H/L (Stokesian solution) or in terms of relative height H/d (cnoidal-type solution). The Stokesian power series solution is not uniformly convergent, and the validity of the solution is lost when the relative depth d/L tends to a small value (say, $d/L < 0.1$ for a fifth-order solution), since the coefficient functions of d/L tend to infinity.

The same occurs in the case of the cnoidal wave solution. There is no unique cnoidal theory; rather, the literature contains several theories which may not be identical. Since all cnoidal representations are truncated series, the order of approximation is important because high-order terms are generally significant.

There are two types of cnoidal theories. The oldest is intuitive in nature, while the newer theories are straightforward and more rigorous. All are irrotational. The primary intuitive theory is that of Korteweg and de Vries (1895). The first and second terms of the series are deduced but no scheme is presented for extension to higher-order terms. The terms which are found are unique. More rigorous theories have been developed. All are based on a perturbation expansion. Unfortunately, even though rigor prevails, the newer theories also diverge very rapidly. It seems then that the high-order theory is not necessarily better than its lower-order counterpart.

17-1.2 Vorticity and Mass Transport

17-1.2.1 The problem consists initially of solving a problem satisfying continuity, momentum, and boundary conditions. The motion is assumed to be periodic and the wave profile is assumed to be that of a steady state. However, these assumptions are not sufficient for solving the nonlinear problem. Two more conditions are necessary. This leads to discussion of the problem of rotationality and mass transport (as they are related) and the arbitrariness of water wave theories.

The arbitrariness in the calculation of wave motion is inherent to the arbitrariness which prevails in the assumptions which are used in the calculation of the mass transportation. The wave motion can be determined by assuming that there is no mass transport at all. These are the closed orbit theories, such as the exact solution of Gerstner (1809) in deep water and the power series solutions of Boussinesq for shallow water. As a result of this assumption, the motion is found to be rotational and the vorticity is in the opposite direction to the particle rotation, i.e., in opposite direction to what should be expected physically under the influence of a shearing stress due to wind blowing in the wave direction (Fig. 17-1).

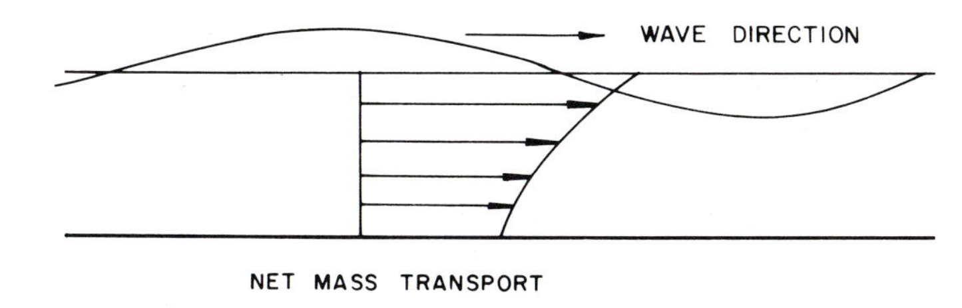

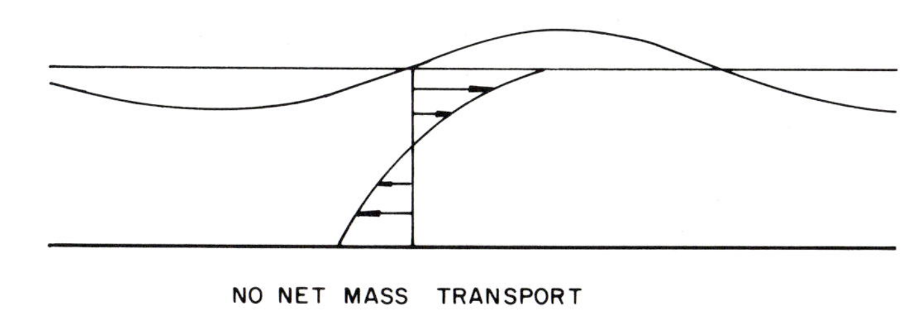

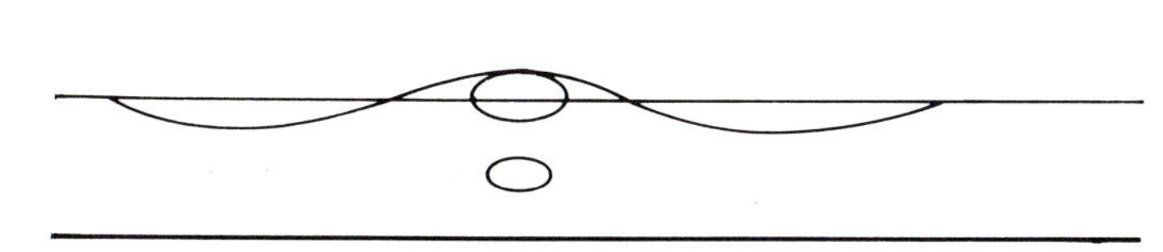

Figure 17-1 *Different kinds of assumptions on mass transportation in periodic progressive waves.*

The wave motion can also be assumed to be irrotational, in which case a mass transport distribution is found as a result of nonlinearity. These are the Stokesian wave theories which include Stokes (1847), Levi-Civita (1925), Struik (1926), and Nekrassov (1951).

Even though there is a given mass transport distribution which is a function of the vertical coordinate, the integrating constant is often determined by assuming that the average mass transport $\overline{U}$ is nil for the sake of continuity, i.e., a steady flow is superimposed such that

$$\overline{U} = \int_{-d}^{\eta} U(z)\, dz = 0$$

where $U(z)$ is the mass transport.

17-1.2.2 If one uses a power series solution in terms of wave steepness H/L to the second order of approximation, one finds that the general relationship between mass transport U and vorticity 2η is given by

$$2\eta = \frac{H^2}{4}\left[\frac{\partial U}{\partial z} - \frac{m^2 k \sinh 2m(d+z)}{\sinh^2 md}\right]$$

where it is recalled that H is the wave height, $m = 2\pi/L$, L is the wave length, $k = 2\pi/T$, T is the wave period, d is the water depth, and z is the vertical coordinate.

In the case where $U(z) = 0$ (Boussinesq wave)

$$2\eta(z)\Big|_{U=0} = -\frac{H^2}{4}\left[\frac{m^2 k \sinh 2m(d+z)}{\sinh^2 md}\right]$$

In the Stokesian wave theories ($\eta = 0$) where one assumes that the integrating constant is such that $\overline{U} = 0$, one has

$$U(z)\Big|_{\eta=0} = \left[\frac{mk}{2\sinh^2 md}\right]\left[\cosh 2m(d+z) - \frac{\sinh 2md}{2md}\right]$$

In the more general case where both U and η are different from zero, one can write

$$\eta(z) = \mu[\eta(z)|_{U=0}]$$

Figure 17-2 *Mass transport as a function of vorticity.*

WAVE DIRECTION →

S.W.L. → x

η	μ
>0	>0
0	0
<0	$-1<, <0$
<0	$= -1$
<0	< -1

which gives (assuming $\overline{U} = 0$)

$$U(z) = (\mu + 1)[U(z)|_{\eta=0}]$$

where μ is an arbitrary coefficient.

The case where $\mu = -1$ corresponds to the Boussinesq (closed-orbit) solution. $\mu = 0$ corresponds to the irrotational theory of Stokes. If $\mu > 0$, the average vorticity is in the same direction as the orbit direction, such as found when a strong wind blows in the wave direction. The case where $\mu < -1$ gives a negative vorticity and a negative mass transport at the free surface which can be due to a wind blowing locally in opposite direction to the wave propagation (a frequent nearshore occurrence). In the case where $-1 < \mu < 0$, the vorticity is in the opposite direction and the mass transport is smaller than in the irrotational case. These facts are schematically represented in Fig. 17-2.

Due to viscous forces, the mass transport at the bottom is always in the wave direction. Numerous experiments confirm this result, and the following value of the bottom mass transport velocity can be derived:

$$U_b = \tfrac{5}{16}H^2 \frac{mk}{\sinh^2 md}$$

17-1.2.3 Finally, it is recalled that the mathematical representation of water wave motions are found as a solution of a set of basic equations and assumptions. These basic equations are the equations of continuity and momentum, the solutions of which must satisfy given boundary conditions. The first assumption is that of rotationality or mass transport as they are related. However, another condition is required. For example, for progressive monochromatic waves, the solution sought represents a steady state such that the potential function $\phi = f(x - Ct)$, where C is a constant equal to the wave velocity. In this case, the solution is unique. Although the steady-state solutions are of the same form, C is still undetermined, and for the determination of C, another condition is required. For example, it can be assumed that the average horizontal velocity over a wave period at a given location is zero, but the mass transport is then necessarily minimum. Another condition is generally preferred. This condition consists of assuming that the average momentum over a wavelength is zero. In this case, another expression for C is found which results in a different mass transport. Thus, it is further realized that the calculation of wave theories is subject to arbitrariness because of different assumptions that lead to different values of C.

17-1.3 The Essential Characteristics of Some Two-Dimensional Periodic Progressive Irrotational Wave Theories

A brief review of some classical wave theories is now presented.

17-1.3.1 The *linear theory of Airy* in Eulerian coordinates gives the essential characteristics of the wave pattern in a simple formulation. The free surface is sinusoidal, particle paths are elliptic and follow a closed orbit (zero mass transport), and lines of equipressure are also sinusoidal. The terms in $(H/L)^2$ are neglected. However, nonlinear effects such as the wave set-down can be determined from this theory (Section 16-7).

The linear theory of Airy in Lagrangian coordinates also gives elliptic particle paths, but the free surface and lines of equipressure are now trochoidal (as in the deep water wave theory of Gertsner). (A trochoid is the locus described by a point within a circle rolling on a straight line.)

The *linear long wave theory* is the same as the theory of Airy where it is assumed that d/L is small. As a consequence, the formulas are simplified considerably. The pressure is hydrostatic and the horizontal velocity distribution is uniform. The wave velocity is simply $(gd)^{1/2}$.

17-1.3.2 The *theory of Stokes* at a second order of approximation is characterized by the sum of two sinusoidal components of period T and $T/2$, respectively. As a result, the wave crest becomes peaked and the troughs become flatter. The wave profile can even be characterized by the appearance of a hump in the middle of the wave trough. Similarly, the elliptical particle path is deformed and tends to hump under the crest and flatten under the trough.

In this theory as in all the following wave theories, there is mass transport as a result of irrotationality and nonlinearity. However, phase velocity, wave length and group velocity are the same as in the linear theories. The terms in $(H/L)^3$ are neglected.

The theory of Stokes at a third order of approximation is characterized by the sum of three sinusoidal terms of period T, $T/2$, and $T/3$, respectively. The same logical results are found. Phase and group velocity exhibit nonlinear corrections. The coefficients of H/L which are functions of d/L tend to infinity when d/L tends to zero so that the theory cannot be used in very shallow water. (The series is nonuniformly convergent.) The terms in $(H/L)^4$ are neglected.

The theory of Stokes at a fifth order of approximation is the sum of five sinusoidal terms. The coefficients of $(H/L)^n$ are functions of d/L and tend to large values for $n > 3$ even sooner than in the case of the third-order theory, i.e., for larger values of d/L. Consequently, the fifth-order wave theory is less valid than the third-order wave theory for small values of d/L and cannot be used when $d/L < 0.1$. The terms in $(H/L)^6$ are neglected.

17-1.3.3 The *theory of Keulegan and Patterson* belongs to the cnoidal family of water wave theories. It follows the same physical approach as the theory of Korteweg and de Vries. From a purely mathematical view point, there are some inconsistencies as some third-order terms are included while some other second-order terms are neglected; however, it gives relatively good results. The horizontal velocity component varies with depth, while the pressure is hydrostatic.

The *cnoidal wave theory of Laitone* is mathematically rigorous. At a first order of approximation, the vertical distribution of horizontal velocity is uniform. There is no mass transport. The terms in $(H/d)^2$ are neglected.

The theory of Laitone at a second order of approximation gives a nonuniform velocity distribution. There is mass transport. The vertical distribution of mass transport velocity is uniform. The second-order term becomes larger than the first-order term as H/d increases. (H/d is not necessarily a small parameter whereas H/L is always.) The series is nonuniformly convergent. The terms in $(H/d)^3$ are neglected. The results of this theory diverge significantly from experimental results.

17-1.3.4 The *solitary wave theory* of Boussinesq is the result of a purely intuitive approach. The vertical component of velocity is initially assumed to be linearly distributed from the bottom (equal to zero) to the free surface (equal to the linearized free surface velocity $\partial\eta/\partial t$). The vertical distribution of horizontal velocity is assumed to be uniform.

A correction due to path curvature (vertical acceleration) is added to the hydrostatic pressure. The equations of motion are linearized vertically but remain nonlinear horizontally, i.e., convective inertia terms where the vertical component of velocity appears are neglected, but the product $u(\partial u/\partial x)$ remains. The solution is then exact. As in any solitary wave theory, η has always a positive value and there is mass transport equal to the volume of the wave above the still water level. The terms in $(H/d)^2$ are neglected. The wave velocity is $[g(d + H)]^{1/2}$. The solitary wave theory of McCowan is more rigorous and satisfies the kinematic free surface boundary condition exactly. It corresponds to a higher-order solution than the theory of Boussinesq. The vertical distribution of horizontal velocity is nonuniform. The terms in $(H/d)^3$ are neglected.

17-1.3.5 The *stream function theory* of Reid and Dean is also irrotational ($\nabla^2\psi = 0$) even though the use of the stream function allows the study of rotational motion. It is actually a numerical method requiring the use of a computer. The solution is presented as a series of harmonic functions with unknown coefficients determined numerically in such a way that the deviation from the Bernoulli constant at the free surface is minimum.

17-1.3.6 The *formulation of Goda* is based on experimental results and assumes that the particle velocity at the crest of a limit wave is equal to the phase velocity. Accordingly an empirical correction parameter is introduced in the linear wave theory.

17-1.4 *The Theory of Gertsner*

The rotational deep water wave theory of Gerstner is exact. It is a closed orbits theory with the rotation in opposite direction to the particle path. Its interest is mostly historical, since it dates to 1807, although it was also developed independently by Rankine at a later date. It gives a simple geometric representation of the wave motion defined by the coordinates (x,z) of a given particle, such as (see Fig. 17-3; note the z coordinate system is positive downward):

$$x = \alpha - ae^{m\beta} \sin (kt - m\alpha) \qquad z = \beta - ae^{m\beta} \cos (kt - m\alpha)$$

$$m = \frac{2\pi}{L} \qquad k = \frac{2\pi}{T}$$

where (α,β) are two parameters which define the mean position of the considered particle. The particle describes circles of radius $r = ae^{m\beta}$. The wavelength is $L = gT^2/2\pi$. The free surface and the lines of equipressure are trochoids. The limit wave is a cycloid of steepness $H/L = 0.31$. Figure 17-3 illustrates some essential characteristics given by this theory. The free surface corresponds to $\beta = 0$ and is defined by the parametric equation

$$x = \alpha - a \sin (kt - m\alpha)$$
$$z = -a \cos (kt - m\alpha)$$

The pressure on a particle whose mean position is β is independent of time and is given by the expression

$$p = \rho g \beta + \frac{a^2}{2} k^2 e^{2m\beta} + \text{constant}$$

Figure 17-3 *Gertsner wave.*

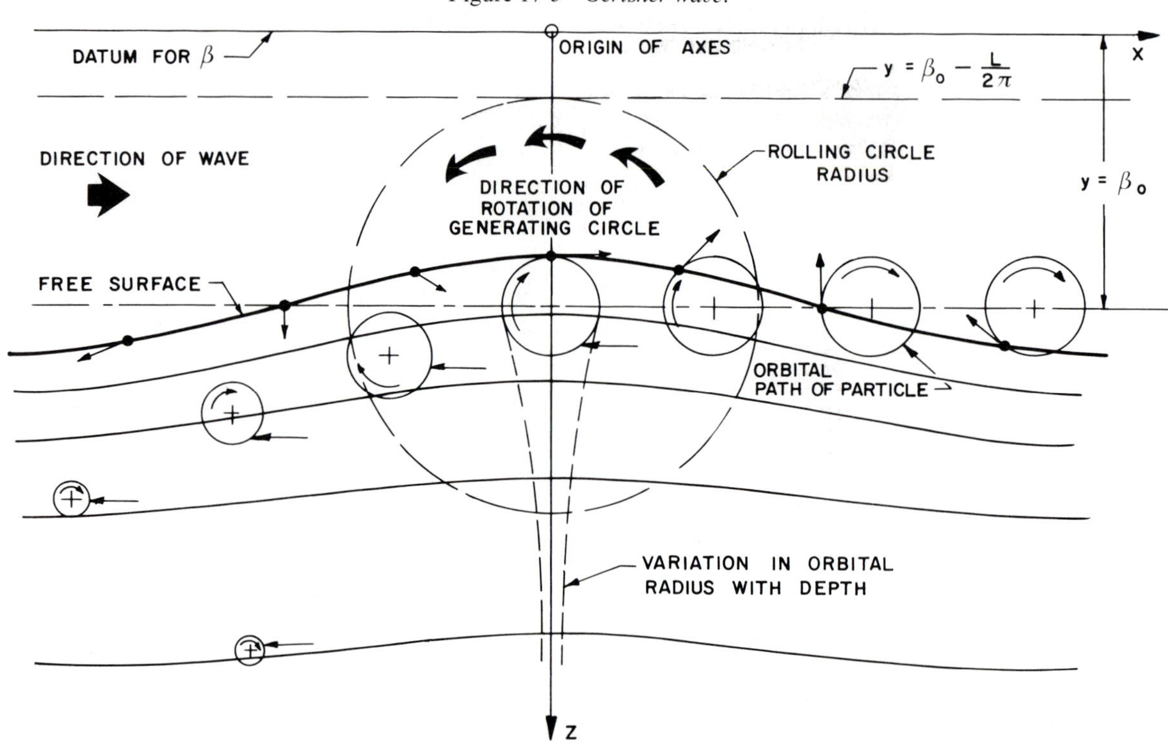

The pressure is constant if β is constant, and thus all particles at the same mean level have the same pressure as they move about; this pressure is the same as the pressure on that particle in its still position. The vorticity is

$$2\eta = \frac{2m^2ka^2e^{2m\beta}}{1 - m^2e^{2m\beta}}$$

Finally, the energy in a wavelength is

$$E = \tfrac{1}{2}\rho gLa^2\left(1 - \frac{m^2a^2}{2}\right)$$

17-1.5 Determination of the Domain of Validity of Finite Amplitude Wave Theories

17-1.5.1 Knowledge of the features of these wave theories is essential to the engineer interested, for example, in calculating the forces on submerged structures such as pilings. For this purpose, he needs reliable expressions for the velocity field within the wave. Upon choosing a suitable design wave (that is, mean water depth, wave height, and wave period), he then proceeds to select a theory to describe that wave. This choice is not easily made, since he must evaluate the applicability of at least a dozen theories. It may be hoped, for example, that a theory developed to a fifth order of approximation is more accurate than its lower-order counterparts, which may or may not actually be the case. Unfortunately, the range of validities of the various wave theories is not well defined. (An attempt to classify them has been presented in Fig. 15-7.) Also, only a few experimental results are available. Some are presented here.

In this comparison with experiments, one can choose the horizontal particle velocity under the crest as an important feature, since, in applications, velocity is generally most critical and the velocity under the crest is the greatest attained at any depth. Furthermore, a theory which prescribes the velocity field well is constrained to be good for other features, such as accelerations and pressures, a priori. Velocity profiles under the wave crest are shown in Figs. 17-4 and 17-5. A comparison of various free surface profiles given by theories and experiments is also presented in Fig. 17-6. All these comparisons are done for relatively shallow water waves of large relative amplitude. It is seen that the linear theory is always good at the bottom where the boundary condition is linear and the convective effects are relatively small. The linear theory loses its validity toward the free surface. However, in general, the nonlinear theories are not better than the linear theory. Even though the numerical method of Reid and Dean is not shown in this figure, it has been proven to be as good as or better than the other theories when compared to laboratory experiments. The empirical formulation of Goda is still the best. Due to its simplicity, it appears that an empirical relationship of this kind is the best method to recommend from an engineering standpoint. This leads us to discuss the selection of the "best" theory for practical purposes.

17-1.5.2 In a narrow two-dimensional wave tank, viscous effects in the boundary layer near the bottom induce vorticity which diffuses upward. Also, the fact that the wave tank is limited causes a return flow which nullifies the total mass transport. (If the wave tank is large, transverse effects take place, and the return flow meander.) The theoretical approach which takes into account viscous effects gives a mass transport which is relatively well verified by laboratory experiments. Figure 17-7 shows qualitatively the results of such theory.

17-1.5.3 Under storm condition, the wind stress induces a large mass transport and vorticity in the wave motion. This mass transport remains in a thin upper layer at the beginning of the storm and slowly diffuses downward under the influence of the turbulent fluctuations (mostly due to white caps). At the same time, the mass transport near the surface, which is in the main direction of the wave travel at the beginning of the storm, tends to be subjected to the Coriolis effect. Therefore the mass transport near the free surface is at a small angle with the wave direction and has a direction which changes with depth. Eventually the mass transport results in a storm surge which, in

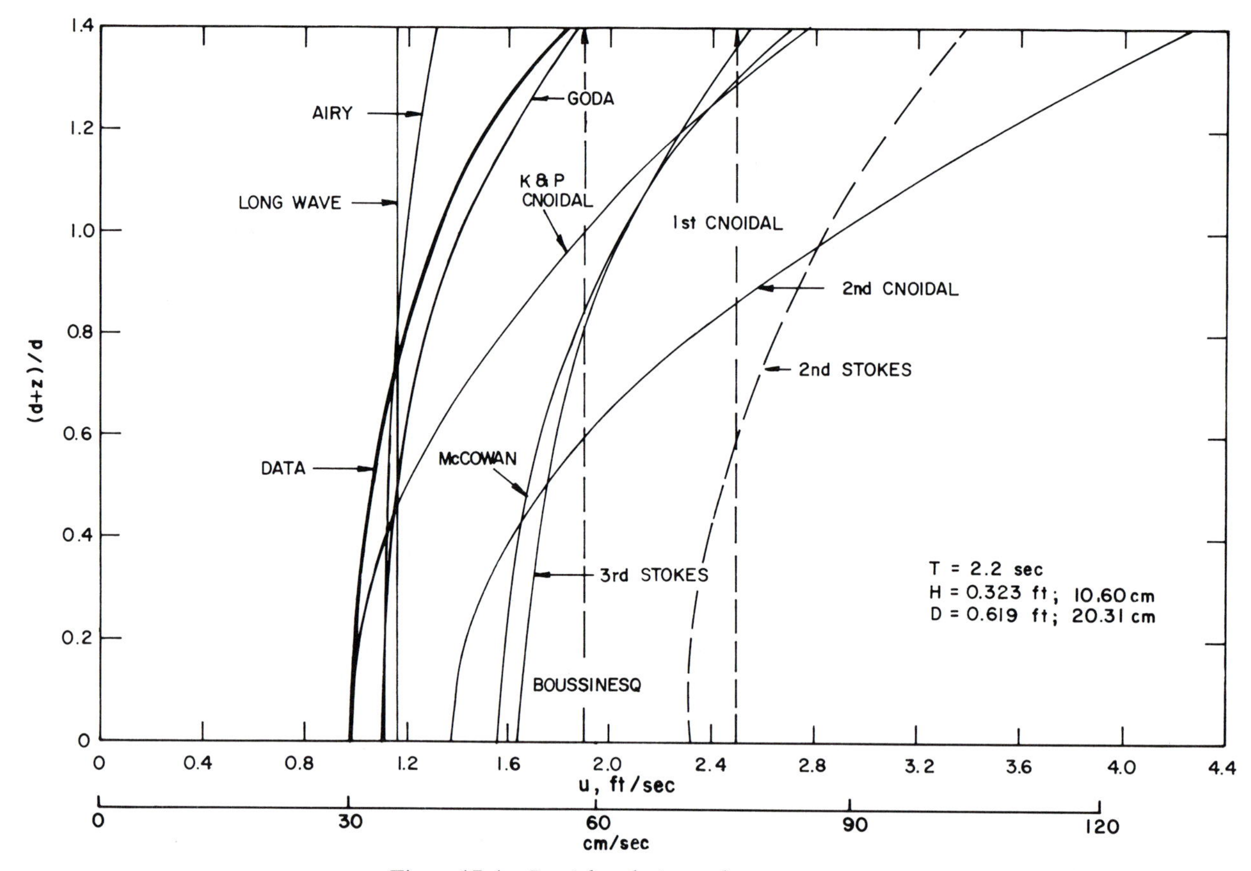

Figure 17-4 *Particle velocity under a wave crest.*

addition to being influenced by Coriolis acceleration, is also subjected to coastal boundary effects. The storm surge calculation method is based on the nonlinear long wave equations (see Section 18-1.2.4) independently of water wave effects. Therefore, it is seen that storm surge calculation and nonlinear wave theories are two separate approaches for what is actually a unique phenomenum resulting from air–sea interaction in shallow water. Actually, it is a fallacy to use irrotational or closed-orbit nonlinear wave theory for calculating wave forces on piles, for example. The wind-induced mass transport near the free surface could be as high as 80% of the particle velocity given by the linear theory. The nonlinear corrections are meaningless for an engineer confronted with decision on what his structure should look like. Failure of Texas towers may be attributed to the neglect of wind-induced mass transport in water waves.

The correct theory is a rotational wave theory with mass transport function of wind forces, duration of wind, and fetch (length of water surface over which the wind exerts its action), as done in wave hindcast methods. The description of such a method is beyond the scope of this book. However, a possible approach consists of finding a stream function solution of $\nabla^2\psi = f(\psi)$.

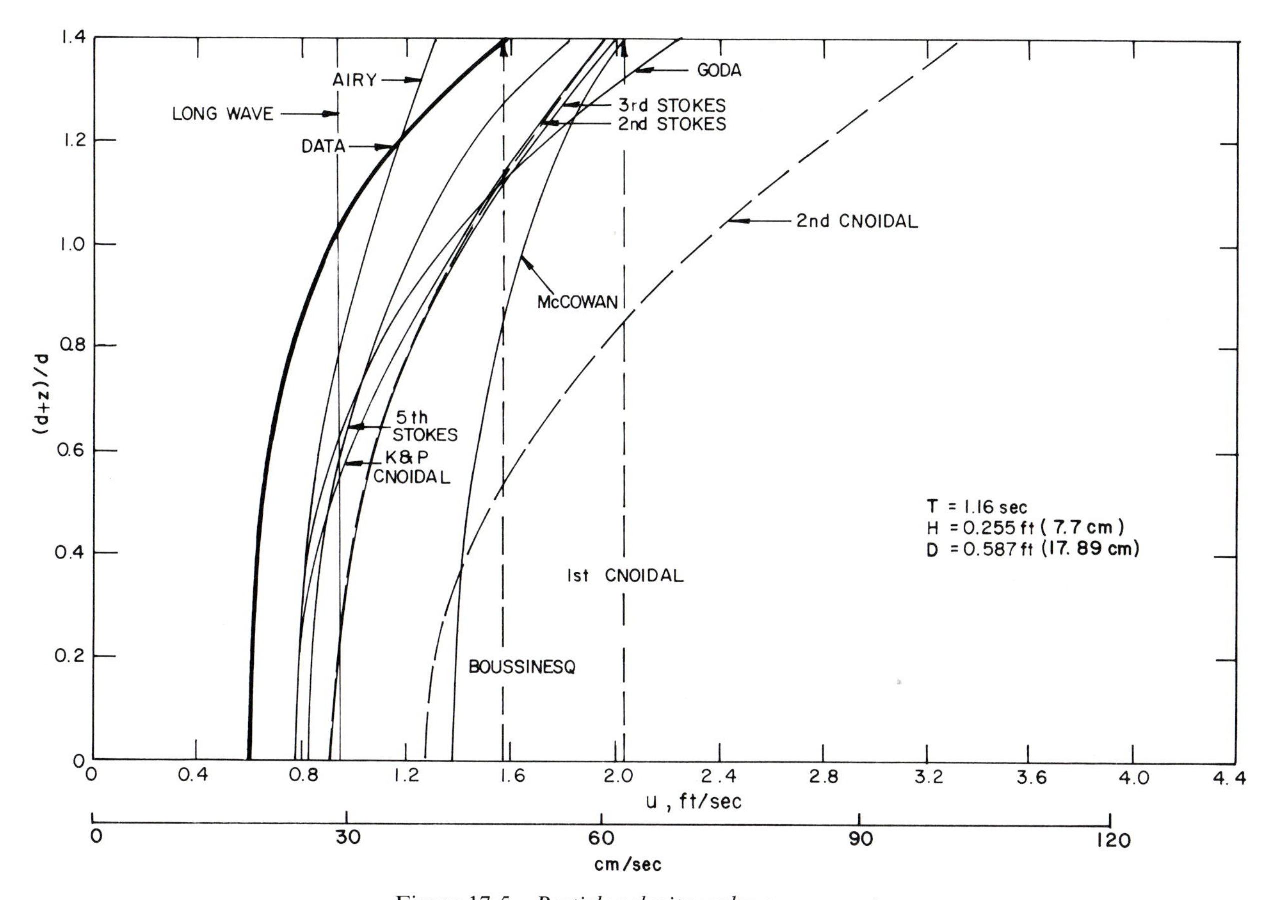

Figure 17-5 *Particle velocity under a wave crest.*

17-1.6 Limit Wave Steepness

Although it is beyond the purpose of this book to deeply analyze limit wave steepness and the phenomenon of wave breaking, the relationship between the rate of rotationality and the limit wave steepness is worthwhile mentioning.

It has been mentioned (Section 15-1.2.7) that wave-breaking inception will occur when the wave profile reaches a limit wave steepness $H/L|_{max}$. This limit steepness is theoretically 0.142 for a deep water irrotational periodic wave. Rotationality at the crest in the direction of the wave travel such as that due to a generating wind will reduce the limit wave steepness to a smaller value (see Fig. 17-8). A deep sea wave steepness larger than 0.10 is rarely encountered.

Rotationality in the opposite direction will theoretically increase the limit wave steepness. Such a case can be observed near the coasts when the wave travels in the opposite direction to a wind blowing offshore. At the limit, according to the closed-orbit Gertsner theory, the maximum limit steepness is 0.31, but the rotationality at the crest

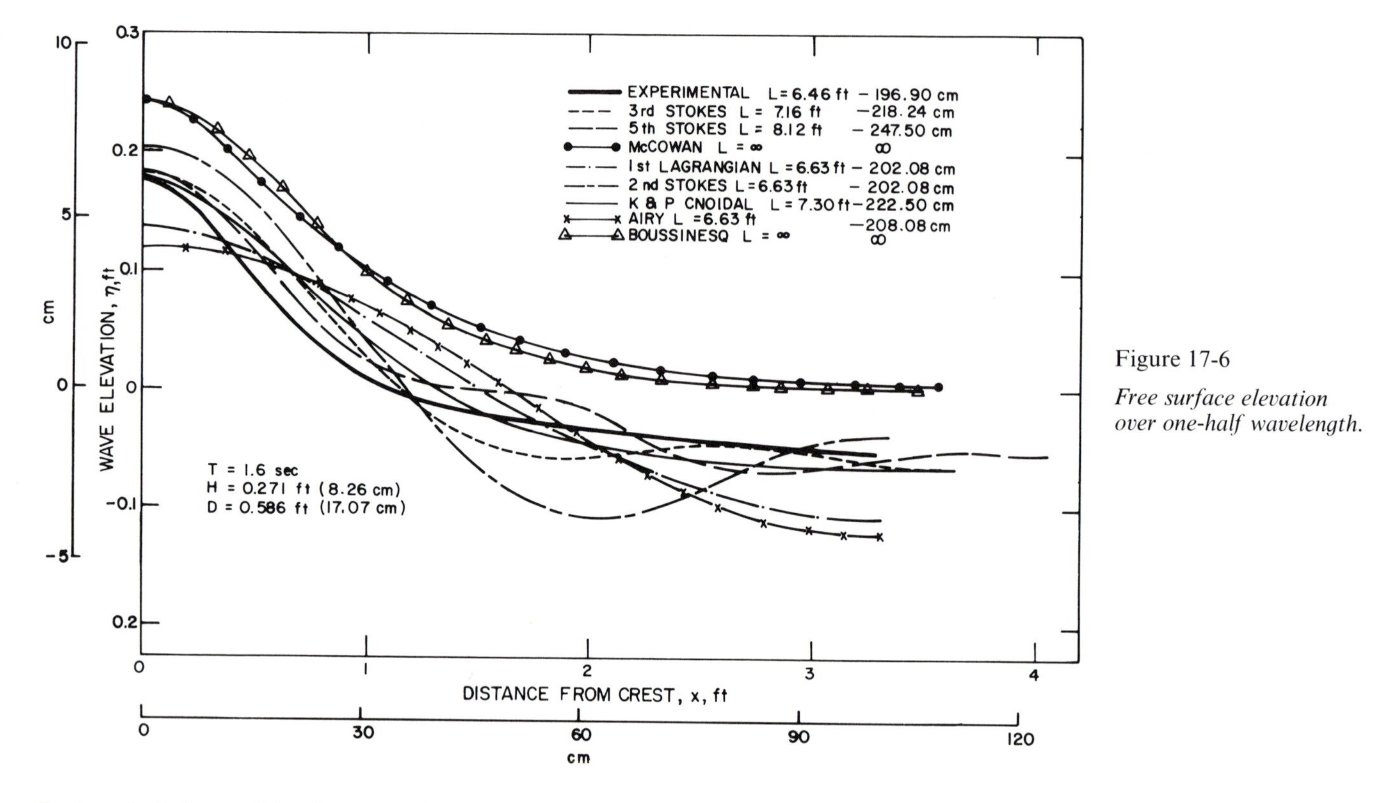

Figure 17-6

Free surface elevation over one-half wavelength.

is then infinity and in the opposite direction to the wave travel. It is evident that this result of the Gertsner theory has no physical significance.

It is seen how important it would be to establish a general rotational wave theory and to relate the rotationality and mass transport to the wind action and bottom friction. The effect of viscous friction at the bottom has already been subjected to investigation to some extent. However, a general theory of irregular waves with an arbitrary rotationality or mass transportation and quadratic bottom friction, valid for any wave height, any wave period, and any water depth, would be preferred. Then it will remain to calibrate some parameters as functions of wind stress, wind duration, and fetch length.

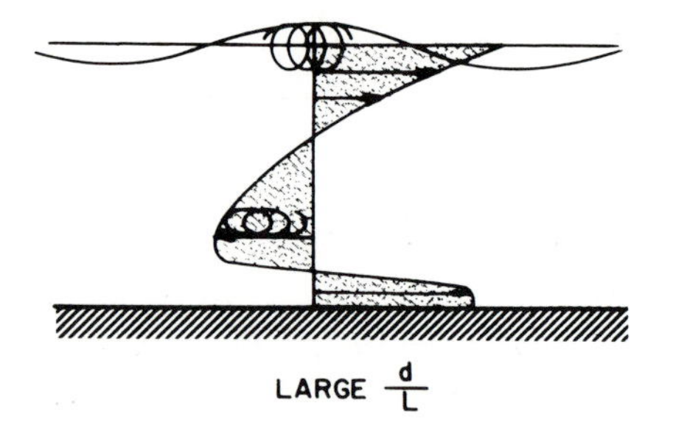

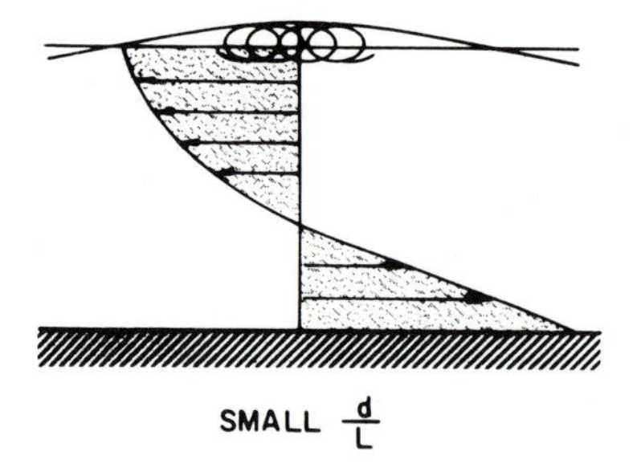

Figure 17-7

Periodic progressive wave with mass transportation effect.

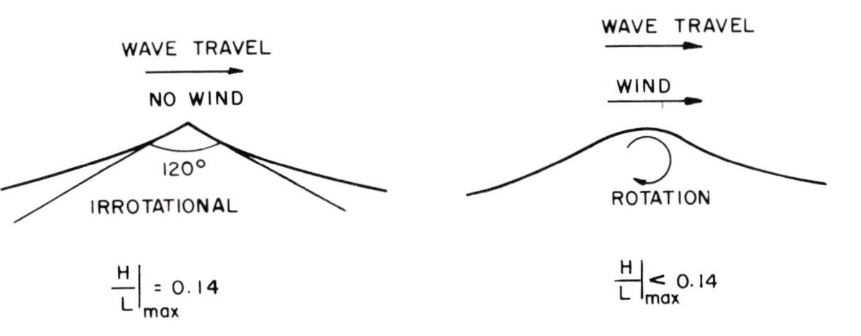

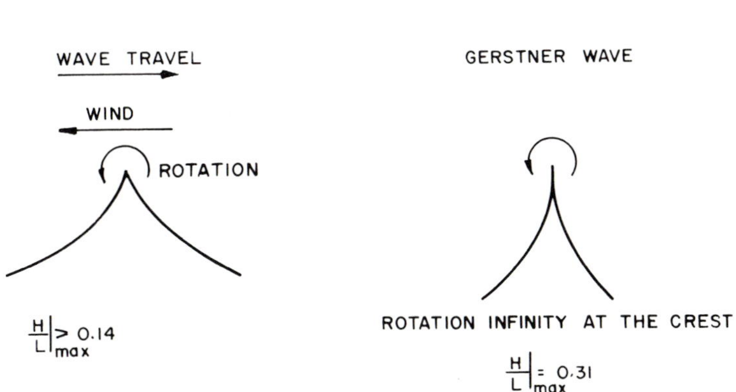

Figure 17-8
Rotationality and wave limit steepness.

17-2 Irrotational Waves: Methods of Calculation and Results

17-2.1 General Process of Calculation

It is assumed that the two unknowns, the velocity potential ϕ and the free surface elevation η, may be transformed into the following power series expressions with respect to a parameter a. This parameter is chosen arbitrarily but has the dimension of a length, usually taken to be a half-wave height.

$$\phi = a\phi_1 + a^2\phi_2 + a^3\phi_3 + \cdots + a^n\phi_n + \cdots$$
$$\eta = a\eta_1 + a^2\eta_2 + a^3\eta_3 + \cdots + a^n\eta_n + \cdots$$

When this expression is substituted in the continuity relationship $\nabla^2\phi = 0$, it is seen that each of these terms ϕ_n is an independent solution of the Laplace equation $\nabla^2\phi_n = 0$. Each ϕ_n also satisfies the fixed boundary condition $\partial\phi_n/\partial n = 0$ and the free surface condition.

The free surface conditions are expressed in terms of their values at the still water level ($z = 0$). Developing $\phi(x,y,0 + \eta,t)$ in terms of power series in η gives

$$\phi(x,y,\eta,t) = \phi(x,y,0,t) + \eta\left[\frac{\partial\phi(x,y,0,t)}{\partial z}\right] + \cdots$$

or

$$[\phi]_{z=\eta} = [\phi]_{z=0} + \eta\left[\frac{\partial\phi}{\partial z}\right]_{z=0} + \frac{\eta^2}{2!}\left[\frac{\partial^2\phi}{\partial z^2}\right]_{z=0} + \cdots$$

When this expression for ϕ is introduced into the free surface conditions given in Section 16-1.3.2, one obtains the kinematic condition

$$\frac{\partial}{\partial z}\left(\phi + \eta\frac{\partial\phi}{\partial z} + \cdots\right) = -\frac{\partial\eta}{\partial t} + \frac{\partial\eta}{\partial x}\frac{\partial}{\partial x}\left(\phi + \eta\frac{\partial\phi}{\partial z} + \cdots\right)$$
$$+ \frac{\partial\eta}{\partial y}\frac{\partial}{\partial y}\left(\phi + \eta\frac{\partial}{\partial z} + \cdots\right)$$

and the dynamic condition

$$-\frac{\partial}{\partial t}\left[\phi + \eta\frac{\partial\phi}{\partial z} + \cdots\right] + \frac{1}{2}\left[\left\{\frac{\partial}{\partial x}\left(\phi + \eta\frac{\partial\phi}{\partial z} + \cdots\right)\right\}^2 + \left\{\frac{\partial}{\partial y}\left(\phi + \eta\frac{\partial\phi}{\partial z} + \cdots\right)\right\}^2\right] + g\eta = 0$$

Now the problem is completely formulated. When the expressions for ϕ and η are substituted in the above free surface conditions, calculations can be completed.

17-2.2 Method of Solutions

These relationships must be verified for any value of a since a is arbitrarily taken. Grouping together the terms of common powers of a, it is found, respectively, that

$$a\left(\frac{\partial\phi_1}{\partial z} + \frac{\partial\eta_1}{\partial t}\right) + a^2\left(\frac{\partial\phi_2}{\partial z} + \frac{\partial\eta_2}{\partial t} + \eta_1\frac{\partial^2\phi_1}{\partial z^2} - \frac{\partial\phi_1}{\partial x}\frac{\partial\eta_1}{\partial x} - \frac{\partial\phi_1}{\partial y}\frac{\partial\eta_1}{\partial y}\right) + a^3(\cdots) + \cdots = 0$$

and

$$a\left(-\frac{\partial\phi_1}{\partial t} + g\eta_1\right) + a^2\left\{g\eta_2 - \frac{\partial\phi_2}{\partial t} - \eta_1\frac{\partial^2\phi_1}{\partial t\partial z} + \frac{1}{2}\left[\left(\frac{\partial\phi_1}{\partial x}\right)^2 + \left(\frac{\partial\phi_1}{\partial y}\right)^2 + \left(\frac{\partial\phi_1}{\partial z}\right)^2\right]\right\} + a^3(\cdots) + \cdots = 0$$

This leads to the following equations which are independent of the value of a

$$\frac{\partial\phi_1}{\partial z} + \frac{\partial\eta_1}{\partial t} = 0$$

$$\frac{\partial\phi_2}{\partial z} + \frac{\partial\eta_2}{\partial t} + \eta_1\frac{\partial^2\phi_1}{\partial z^2} - \frac{\partial\phi_1}{\partial x}\cdot\frac{\partial\eta_1}{\partial x} - \frac{\partial\phi_1}{\partial y}\frac{\partial\eta_1}{\partial y} = 0$$

$$\cdots\cdots\cdots\cdots\cdots\cdots$$

$$\frac{\partial\phi_n}{\partial z} + \frac{\partial\eta_n}{\partial t} = f(\phi_{n-1},\eta_{n-1})$$

and

$$-\frac{\partial\phi_1}{\partial t} + g\eta_1 = 0$$

$$-\frac{\partial\phi_2}{\partial t} + g\eta_2 - \eta_1\frac{\partial^2\phi_1}{\partial t\partial z} + \frac{1}{2}\left[\left(\frac{\partial\phi_1}{\partial x}\right)^2 + \cdots\right] = 0$$

$$\cdots\cdots\cdots\cdots\cdots\cdots$$

$$-\frac{\partial\phi_n}{\partial t} + g\eta_n = f'(\phi_{n-1}, \eta_{n-1})$$

Taken in pairs, the above equations may be solved for $\phi_n|_{z=0}$ and η_n when $\phi_{n-1}|_{z=0}$ and η_{n-1} are known, using $\nabla^2\phi_n = 0$ and $(\partial\phi_n/\partial z)|_{z=-d} = 0$. In particular, the two equations which are linear in a give

$$a\left(\frac{\partial\phi_1}{\partial z} + \frac{\partial\eta_1}{\partial t}\right) = 0 \quad \text{and} \quad a\left(-\frac{\partial\phi_1}{\partial t} + g\eta_1\right) = 0$$

which leads to

$$\left[\frac{\partial^2\phi_1}{\partial t^2} + g\frac{\partial\phi_1}{\partial z}\right]_{z=0} = 0$$

This may be compared with the Cauchy–Poisson condition, which was previously developed. The linear motion will then be defined by $\phi = a\phi_1$, and $\eta = a\eta_1$.

In practice, the study of nonlinear problems requires very long and tedious calculations. Sometimes, this may be reduced in the case of two-dimensional motion by using the function $W = \phi + i\psi$ where ψ is the stream function and $i = (-1)^{1/2}$.

17-2.3 The Bernoulli Equation and the Rayleigh Principle

In the case of periodic progressive wave moving in the OX direction at a celerity C, the general solution for ϕ is

$$\phi = P(z)F(x - Ct)$$

and

$$\frac{\partial\phi}{\partial t} = -CP(z)F'(x - Ct)$$

also

$$\frac{\partial \phi}{\partial x} = P(z)F'(x - Ct)$$

Hence it is seen that

$$\frac{\partial \phi}{\partial t} = -C\frac{\partial \phi}{\partial x} = +Cu$$

Introducing these relationships, it is seen that the Bernoulli equation in the case of progressive waves takes the form

$$-Cu + \tfrac{1}{2}(u^2 + w^2) + \frac{p}{\rho} + gz = 0$$

After some transformations, this becomes:

$$C^2 - 2Cu + u^2 + w^2 + \frac{2p}{\rho} + 2gz = C^2$$

C^2 being a constant, and p being also a constant at the free surface, the dynamic condition at the free surface becomes after division by C^2

$$\left(\frac{u - C}{C}\right)^2 + \frac{w^2}{C^2} + \frac{2g\eta}{C^2} = \text{constant}$$

The constant can be taken equal to unity. It is seen that the motion could be considered as a steady motion in a new system of relative coordinates. In this system of coordinates, the origin of the OX axis moves at the wave celerity C (the Rayleigh principle). The free surface and the bottom are streamlines in this moving system of coordinates.

17-2.4 Stokesian Theory at a Second Order of Approximation

Some mathematical results for a periodic irrotational progressive wave at a second order of approximation are presented in Table 17-1.

These have been obtained by a method similar to the one briefly outlined in the previous sections according to the assumptions described in Section 17-1.2.

Table 17-1 *Periodic progressive wave: second order of approximation*

	Deep water	*Intermediate water depth*
Wave velocity	$C = \left(\frac{gL}{2\pi}\right)^{1/2} = \frac{gT}{2\pi} = \frac{g}{k}$	$C = \left[\frac{gL}{2\pi}\tanh\frac{2\pi d}{L}\right]^{1/2}$
Third order	$C = \left\{\frac{gL}{2\pi}\left[1 + \left(\frac{H}{2}\frac{2\pi}{L}\right)^2\right]\right\}^{1/2}$	
Potential function	$\phi = -\frac{H}{2}Ce^{+mz}\sin(kt - mx)$	$\phi = -\frac{H}{2}C\frac{\cosh m(d+z)}{\sinh md}\sin(kt - mx) - \frac{3}{8}m\left(\frac{H}{2}\right)^2 C\frac{\cosh 2m(d+z)}{\sinh^4 md}\sin 2(kt - mx)$
Free surface	$\eta = \frac{H}{2}\cos(kt - mx) + \frac{1}{2}m\left(\frac{H}{2}\right)^2\cos 2(kt - mx)$	$\eta = \frac{H}{2}\cos(kt - mx) + m\left(\frac{H}{2}\right)^2\frac{\cosh md(\cosh 2md + 2)}{4\sinh^3 md}\cos 2(kt - mx)$
		$= \frac{H}{2}\cos(kt - mx) + m\left(\frac{H}{2}\right)^2\frac{1}{\tanh md}\left(1 + \frac{3}{2\sinh^2 md}\right)\cos 2(kt - mx)$
Midlevel elevation	$\Delta = \frac{m}{2}\left(\frac{H}{2}\right)^2$	$\Delta = m\left(\frac{H}{2}\right)^2\frac{1}{\tanh md}\left(1 + \frac{3}{2\sinh^2 md}\right)$
Mass transport (nonclosed orbits)	$U = m^2\left(\frac{H}{2}\right)^2 Ce^{+2mz}$	(1) No return flow: $U = m^2\left(\frac{H}{2}\right)^2 C\frac{\cosh^2 m(d+z)}{2\sinh^2 md}$
		(2) Average mass transport nil: $U = m^2\left(\frac{H}{2}\right)^2 C\frac{1}{2\sinh^2 md}\left(\cosh 2m(d+z) - \frac{\sinh 2md}{2md}\right)$

17-3 Differences between Water Waves and Unsteady Flow through Porous Media

17-3.1 A Review of the Basic Assumptions

Before concluding the study of the finite amplitude wave theories, it is particularly interesting to establish a parallel between unsteady flow through porous media and irrotational water waves and to point out the essential differences.

It is recalled that the study of the average flow through a porous medium can generally be performed by neglecting all inertial forces, local and convective (see Section 9-2.1.3). Pressure and body forces always balance friction forces. In the case of water waves, pressure and body forces balance inertial forces and the friction forces are neglected. As a consequence of this first equality between pressure, body force, and friction forces, a "water wave effect" for a flow through a porous medium is impossible. For example, consider a hump of the free surface elevation $\eta(x,t)$ and $h(x,t)$ with a zero velocity at time $t = 0$ such as shown on Fig. 17-9.

In the case of water waves, the potential energy is replaced by kinetic energy as $\eta(x,t)$ tends to zero and the kinetic energy is in turn changed into potential energy. The free surface elevation oscillates around the still water level (SWL) and the initial perturbation generates a water wave.

In the case of flow through porous medium, the potential energy is at any time dissipated by the friction forces. The free surface elevation tends slowly toward the still water level. A case of practical interest where the two phenomena can be observed jointly is when a wave (tidal wave or wind wave) oscillates along a pervious ground or a pervious quay. The water table motion induced by the waves at sea is damped very rapidly with distance.

These facts can, of course, be demonstrated and calculated exactly. A short parallel between the most typical equations for these two kinds of motion is now presented.

17-3.2 Dynamic Conditions

Both problems consist in defining a potential function ϕ satisfying the Laplace equation $\nabla^2\phi = 0$.

It is recalled (see Section 9-2.2.3) that the potential function for a flow through porous medium is defined in the case of two-dimensional motion by

$$u = -K\frac{\partial\phi}{\partial x}$$

$$w = -K\frac{\partial\phi}{\partial z}$$

where the average double bar $\bar{\bar{u}}$, $\bar{\bar{w}}$ is eliminated for the sake of simplicity. It is recalled that under such a condition,

$$\phi = \frac{p}{\rho g} + z$$

In particular, $\phi = h$ at the free surface where $z = h(x,t)$ by definition. Thus the free surface equation is

$$h(x,t) = \phi(x,z,t)|_{z=h}$$

or again

$$h(x,t) = \phi[x,h(x,t),t]$$

Although the word "dynamic" would now be misused, it has to be noted that this equation corresponds to the

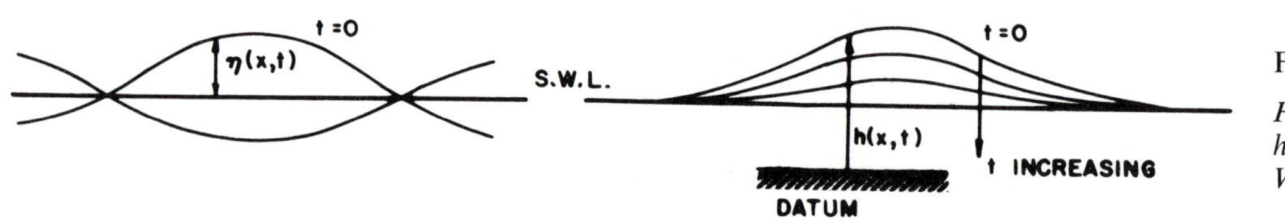

Figure 17-9

Flow through porous medium $h(x, t) \rightarrow$ *still water level (SWL). Water waves:* $\eta(x, t)$ *oscillates.*

free surface dynamic condition for water waves:

$$-\frac{\partial\phi}{\partial t}+\frac{1}{2}\left[\left(\frac{\partial\phi}{\partial x}\right)^2+\left(\frac{\partial\phi}{\partial z}\right)^2\right]+g\eta=f(t)$$

The basic difference is that in one case h is proportional to ϕ, and in the other case η is proportional to $\partial\phi/\partial t$ due to local inertia.

17-3.3 Kinematic Condition

It is recalled (see Section 16-1.3) that the kinematic condition for water waves is in two dimensions

$$\left(\frac{\partial\phi}{\partial z}=-\frac{\partial\eta}{\partial t}+\frac{\partial\phi}{\partial x}\frac{\partial\eta}{\partial x}\right)_{z=\eta}$$

A similar equation does exist for flow through porous medium. However, due to the fact that the water fills only the voids, and due to the change of definition of the potential function, this equation has to be slightly modified as a function of the void coefficient ε and the coefficient of permeability K as follows.

Let us consider an element $d\sigma$ of the free surface at time t (see Fig. 17-10) and at time $t + dt$. The volume of fluid within $ABCD$ is equal to the discharge through AB times the interval of time dt. On one hand the volume of water in $ABCD$ is $\varepsilon\, d\sigma\, dn$, where ε is the void coefficient and $dn = (\partial h/\partial t)\, dt \cos\alpha$. On the other hand, the fluid discharge through AB during the time dt is

$$[u \sin\alpha\, d\sigma + w \cos\alpha\, d\sigma]\, dt$$

By inserting $\tan\alpha = -(\partial h/\partial x)$ and replacing u and w by $-K(\partial\phi/\partial x)$ and $-K(\partial\phi/\partial z)$ respectively and then dividing by $\cos\alpha\, dt$, one obtains for continuity

$$\frac{\partial\phi}{\partial z}=-\frac{\varepsilon}{K}\frac{\partial h}{\partial t}+\frac{\partial\phi}{\partial x}\frac{\partial h}{\partial x}$$

This is the kinematic condition for flow through a porous media.

At the free surface $z = h(x,t) = \phi[x,h(x,t),t]$. Differentiating h with respect to t and x successively (* means

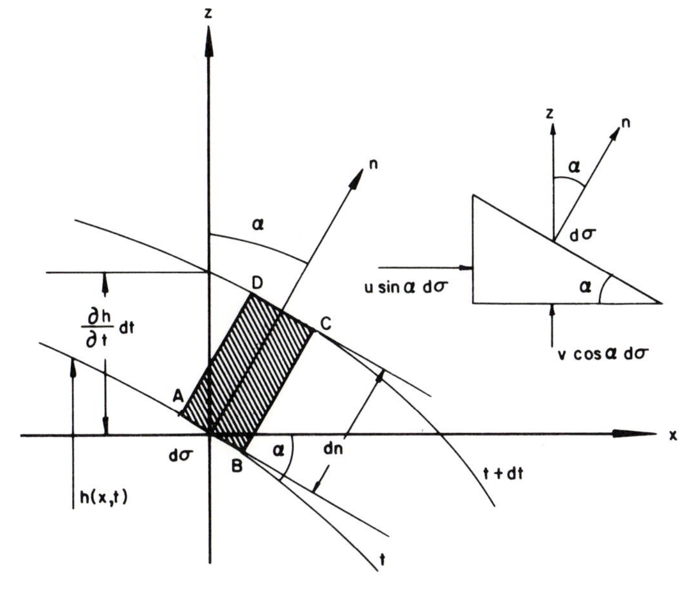

Figure 17-10 *Notation for the free surface equations.*

t or x), gives

$$\frac{\partial h}{\partial *}=\frac{\partial\phi}{\partial *}+\frac{\partial\phi}{\partial h}\frac{\partial h}{\partial *}$$

This can be written as

$$\frac{\partial\phi}{\partial x}=\left(1-\frac{\partial\phi}{\partial z}\right)\frac{\partial h}{\partial x}$$

and

$$\frac{\partial\phi}{\partial t}=\left(1-\frac{\partial\phi}{\partial z}\right)\frac{\partial h}{\partial t}$$

One can eliminate h from the kinematic condition of the previous section by substitution of the above two expressions. This gives

$$\frac{\varepsilon}{K}\frac{\partial\phi}{\partial t}=\left(\frac{\partial\phi}{\partial x}\right)^2+\left(\frac{\partial\phi}{\partial z}\right)^2-\frac{\partial\phi}{\partial z}$$

which can be written as

$$\frac{\varepsilon}{K}\frac{\partial \phi}{\partial t} = \frac{\partial}{\partial x}\left(\phi \frac{\partial \phi}{\partial x}\right) + \frac{\partial}{\partial z}\left(\phi \frac{\partial \phi}{\partial z}\right) - \frac{\partial \phi}{\partial z}$$

since $\nabla^2 \phi = 0$. When the vertical component of motion $\partial \phi/\partial z$ is negligible, this becomes

$$\frac{\varepsilon}{K}\frac{\partial \phi}{\partial t} = \frac{\partial}{\partial x}\left(\phi \frac{\partial \phi}{\partial x}\right)$$

When $\phi = h$ is used this becomes the *Dupuit approximation.* If the variations in h are small with respect to h, ($h = d + \eta$), the equation becomes

$$\frac{\varepsilon}{Kd}\frac{\partial \eta}{\partial t} = \frac{\partial^2 \eta}{\partial x^2}$$

which is the heat or *diffusion equation* of mathematical physics.

Table 17-2 summarizes the main equations for the formulation of a two-dimensional wave and flow through porous medium.

17-3.4 *Form of Solutions*

It is now possible to substantiate mathematically the fact that a water wave effect is impossible in the case of flow through a porous medium.

In the case of water waves, η is proportional to $\partial \phi/\partial t$ or $\partial \phi/\partial z$ is proportional to $\partial^2 \phi/\partial t^2$. In other words, when

$$\phi = f(x,z)e^{i\sigma t}$$

it is seen that $\partial \phi/\partial z$ is proportional to $\sigma^2 \phi$, which is real. The periodic water wave solution does exist.

In the case of flow through a porous medium, a wave

Table 17-2 *Equations for two-dimensional waves and flow through porous mediums*

	Irrotational water waves	*Flow through porous medium*
Definition of the potential function	$u = -\frac{\partial \phi}{\partial x}$	$u = -K\frac{\partial \phi}{\partial x}$
	$w = -\frac{\partial \phi}{\partial z}$	$w = -K\frac{\partial \phi}{\partial z}$
Expression of the potential function	$\phi = f(x,z,t)$	$\phi = \frac{p}{\rho g} + z$
Kinematic condition at the free surface	$\frac{\partial \phi}{\partial z} = -\frac{\partial \eta}{\partial t} + \frac{\partial \phi}{\partial x}\frac{\partial \eta}{\partial x}$	$\frac{\partial \phi}{\partial z} = -\frac{\varepsilon}{K}\frac{\partial h}{\partial t} + \frac{\partial \phi}{\partial x}\frac{\partial h}{\partial x}$
Dynamic condition at the free surface	$-\frac{\partial \phi}{\partial t} + \frac{1}{2}\left[\left(\frac{\partial \phi}{\partial x}\right)^2 + \left(\frac{\partial \phi}{\partial z}\right)^2\right] + g\eta = f(t)$	$\phi = h$
Kinematic and dynamic condition gives	$\frac{\partial^2 \phi}{\partial t^2} + g\frac{\partial \phi}{\partial z} = 0$ (linearized)	$\frac{\varepsilon}{K}\frac{\partial \phi}{\partial t} = \left[\left(\frac{\partial \phi}{\partial x}\right)^2 + \left(\frac{\partial \phi}{\partial z}\right)^2\right] - \frac{\partial \phi}{\partial z}$ or $\frac{\varepsilon}{K}\frac{\partial h}{\partial t} = \frac{\partial}{\partial x}\left[h\frac{\partial h}{\partial x}\right]$ (Dupuit approximation)

effect is impossible since h is proportional to ϕ and $\partial\phi/\partial z$ is proportional to $\partial\phi/\partial t$.

Then a solution such as $\phi = f(x,z)e^{i\sigma t}$ for wave motion would lead to an imaginary relationship between h and ϕ unless σ is imaginary, i.e., $i\sigma$ is then real, which means that the motion is exponential in time instead of oscillatory.

Figure 17-9 illustrates these considerations physically.

PROBLEMS

17.1 The super elevation of the midwater level in front of a vertical wall causing total wave reflection, (clapotis) at a second order of approximation, is

$$\Delta = 2m\left(\frac{H}{2}\right)^2 \frac{1}{\tanh md}\left[1 + \frac{3}{4\sinh^2 md} - \frac{1}{4\cosh^2 md}\right]$$

Let us consider the incident wave defined by wave height $H = 11$ ft (3.35 m), wave period $T = 6$ sec, water depth to the still water level $d = 26$ ft (7.92 m). Calculate successively:

1. The value for Δ.
2. The maximum and minimum water level elevation, η_{max}, η_{min}.
3. The pressure fluctuation Δp at the bottom (using the linear theory)
4. The maximum horizontal force on the wall and the resulting location. (It will be assumed that the pressure on the wall is linearly distributed between the free surface and the bottom.)

17.2 Draw free surface lines of equipressure, particle paths, streamlines according to the Gerstner wave theory in the case where $T = 12$ sec and wave height $2a = 14$ ft (4.26 m).

Chapter 18

The Long Wave Theory

18-1 Basic Equations

18-1.1 The Continuity Equations

18-1.1.1 Integrating the continuity equation div $\mathbf{V} = 0$ along a vertical from the bottom, $z = -d(x,y)$, to the free surface, $\eta(x,y,t)$, gives

$$\int_{-d}^{\eta} \left[\frac{\partial u}{\partial x} + \frac{\partial v}{\partial y} + \frac{\partial w}{\partial z} \right] dz = 0$$

One has

$$\int_{-d}^{\eta} \frac{\partial u}{\partial x}\, dz = (d + \eta) \frac{\partial \bar{u}}{\partial x} \qquad \int_{-d}^{\eta} \frac{\partial v}{\partial y}\, dz = (d + \eta) \frac{\partial \bar{v}}{\partial y}$$

$$\int_{-d}^{\eta} \frac{\partial w}{\partial z}\, dz = w|^{\eta}_{-d}$$

and

$$w|_{z=\eta} = \frac{d\eta}{dt} = \frac{\partial \eta}{\partial t} + u_s \frac{\partial \eta}{\partial x} + v_s \frac{\partial \eta}{\partial y}$$

$$w|_{z=-d} = u_b \frac{\partial (d)}{\partial x} + v_b \frac{\partial (d)}{\partial y}$$

u_s and v_s are the free surface horizontal velocity components, u_b and v_b are the bottom components; $\partial(d)/\partial x$ and $\partial(d)/\partial y$ are the bottom slopes in the X and Y directions which permits us to relate the vertical components w_b to the horizontal components. Assuming that the horizontal velocity distribution is uniform (i.e., $u = \bar{u} = u_s = u_b$, and $v = \bar{v} = v_s = v_b$), and adding these terms gives:

$$\frac{\partial \eta}{\partial t} + \frac{\partial [u(d + \eta)]}{\partial x} + \frac{\partial [v(d + \eta)]}{\partial y} = 0$$

When $v = 0$, the equation found in Section 3-3.1 is recognized.

18-1.1.2 In the study of tsunami wave generation, i.e., during earthquake, the water depth d is time-dependent due to ground vertical displacement. It is then sufficient to replace $\partial\eta/\partial t$ by $\partial(\eta + d)/\partial t$ in the above expression to obtain the continuity equation. (The momentum equation—neglecting friction—remains the same.)

18-1.1.3 In a river or an estuary with gentle variation of width $l(x)$ the continuity equation is ($A = hl$)

$$\frac{\partial A}{\partial t} + \frac{\partial(AV)}{\partial x} = 0$$

where V is the average velocity in cross section A. Developing this equation and neglecting $\partial l/\partial t$ (i.e., $(\partial l/\partial x) \cong (dl/dx)$) gives

$$\frac{\partial h}{\partial t} + \frac{\partial(hV)}{\partial x} + \frac{hV}{l}\frac{dl}{dx} = 0$$

The first two terms are easily recognized, and the last term is the correction due to a slight change of width dl/dx.

18-1.2 The Momentum Equation

18-1.2.1 The momentum equation for long waves can be established directly, by application of the generalized Bernoulli equation with a local inertia force, or from the Eulerian equation. This last approach is used.

Some approximations similar to the ones which were done in the study of boundary layer (Section 4-5.2.4) need to be done. It is assumed that w is small so that the terms $\partial w/\partial t$, $w(\partial w/\partial z)$, can be neglected. Also, the variation of w in the X, Y direction is very small, so that $u(\partial w/\partial x)$, $v(\partial w/\partial y)$ are also neglected. Then the momentum equation along the Z axis is reduced to $(\partial p^*/\partial z) = 0$, i.e.,

$$p^* = \rho g(-z + \eta)(+p_a)$$

The pressure is hydrostatic.

18-1.2.2 Let us now consider the momentum equation along the X axis. For simplicity, one considers the motion to be unidimensional. (In the long wave theory, the vertical component is not considered as one dimensional.) Therefore,

$$\rho\left(\frac{\partial u}{\partial t} + u\frac{\partial u}{\partial x} + w\frac{\partial u}{\partial z}\right) = -\frac{\partial p^*}{\partial x}$$

Adding the zero quantity $u[(\partial u/\partial x) + (\partial w/\partial z)]$, replacing p^* by its hydrostatic expression as seen in the previous section, and integrating gives

$$\rho\int_{-d}^{\eta}\left(\frac{\partial u}{\partial t} + \frac{\partial u^2}{\partial x} + \frac{\partial uw}{\partial z}\right)dz = -\frac{\partial}{\partial x}\int_{-d}^{\eta}\rho g(-z + \eta)\,dz$$

If it is assumed that u is uniform along a vertical, then

$$(d + \eta)\frac{\partial u}{\partial t} + (d + \eta)\frac{\partial u^2}{\partial x} + uw|_{-d}^{\eta} = -g\frac{\partial}{\partial x}\left(-\frac{z^2}{2} + \eta z\right)\Big|_{-d}^{\eta}$$

If w is eliminated and the continuity equation is used, then dividing by $(d + \eta)$ and using algebra will give

$$\frac{\partial u}{\partial t} + u\frac{\partial u}{\partial x} = -g\frac{\partial\eta}{\partial x}$$

18-1.2.3 In the two-dimensional case, a similar calculation gives:

$$\frac{\partial u}{\partial t} + u\frac{\partial u}{\partial x} + v\frac{\partial u}{\partial y} = -g\frac{\partial\eta}{\partial x}$$

$$\frac{\partial v}{\partial t} + u\frac{\partial v}{\partial x} + v\frac{\partial v}{\partial y} = -g\frac{\partial\eta}{\partial y}$$

which, associated with the continuity equation forms the two-dimensional long wave equation.

18-1.2.4 Adding the Coriolis components $-\Omega v$ and Ωu along the X and Y axes, respectively, a quadratic friction stress at the bottom, a quadratic shearing force due to wind at the free surface, and an atmospheric pressure gradient to the hydrostatic term, the equations of storm surge are obtained. The local inertia term is sometimes neglected and the storm surge is then considered as a succession of quasi steady motion. Adding also the Coriolis components, friction forces and body forces due to the sun-moon attraction, the equations of tidal motion are obtained. In the study of oceanic tide, the ground displacement due to earth tide (earth deformation) may also need to be taken into account.

18-1.2.5 For simplicity, consider the one-dimensional motion. Since the wave motion at the bottom is far from being negligible, the shearing force due to friction has in practice a relatively great importance in long wave theory. The long wave equation should be written (Fig. 18-1):

$$\frac{\partial u}{\partial t} + u\frac{\partial u}{\partial x} = -g\frac{\partial \eta}{\partial x} - \frac{\tau}{\rho(d + \eta)}$$

In general, τ will be assumed to be quadratic, i.e., $\tau = \rho f u^2$. The coefficient f is assumed to be independent of the variation of u with respect to time. It is assumed that f is the same as if the motion were steady. Consequently, f can be expressed in terms of the Chezy coefficient C_h: $f = g/C_h^2$. The long wave equation now becomes (see Section 14-1.1):

$$\frac{\partial u}{\partial t} + u\frac{\partial u}{\partial x} = -g\frac{\partial \eta}{\partial x} - \frac{g}{C_h^2}\frac{u|u|}{(d + \eta)}$$

It has been seen that the friction coefficients such as f, C_h, or n are used for gradually varied flow. Most often, they are also considered valid for unsteady motion, such as those due to flood waves or tidal motion in an estuary.

However, one must realize that the practical use of such empirical formulation is only due to the lack of further information on the friction factor for unsteady motion. The application of the Chezy coefficient to unsteady motion is valid provided the velocity distribution is influenced by the friction up to the free surface as in the case of a steady flow. This occurs when the expression for the boundary layer thickness, which increases with the wave period, becomes much larger than the actual depth h. Still, density effects in a marine estuary have a nonnegligible influence on the distribution of friction forces. They can only be known by analyzing field records.

18-1.2.6 When the zero quantity $g[d(d)/dx] \pm gS$ where S is the bottom slope (which can be positive or negative depending upon the motion direction with respect to the X-axis) is added to the long wave equation, $(d + \eta)$ is replaced by the total depth h, and u is replaced by V, the following results:

$$\frac{\partial V}{\partial t} + V\frac{\partial V}{\partial x} + g\frac{\partial h}{\partial x} = \pm gS - \frac{g}{C_h^2}\frac{V|V|}{h}$$

Such an equation is known as the *equation of Barré de St. Venant*, which is used in the study of flood waves or tidal estuary in connection with the continuity equation given in Section 18-1.1.3.

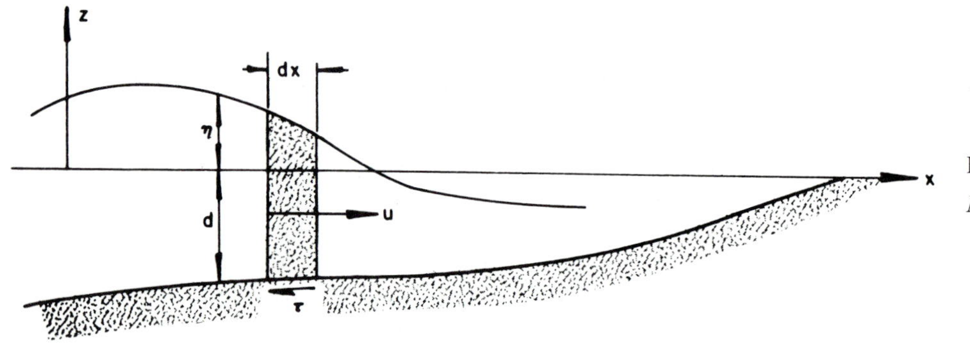

Figure 18-1

Notation for long wave theory.

18-1.3 The Insertion of Path Curvatures Effect: Boussinesq Equation

Consider again the Eulerian equation along a vertical axis,

$$\frac{dw}{dt} = -\frac{1}{\rho}\frac{\partial p}{\partial z} - g$$

in which one assumes that $dw/dt \cong \partial w/\partial t$. The nonlinear terms $u(\partial w/\partial z)$ and $w(\partial w/\partial z)$ are neglected, but the linear term $\partial w/\partial t$ is nonnegligible.

The pressure distribution is no longer hydrostatic. The vertical acceleration $\partial w/\partial t$, due to path curvatures, is going to modify this simple approximation as follows: On a horizontal bottom, the vertical component of velocity at the bottom w_b is nil. At the free surface w is equal to the velocity of the free surface itself $d\eta/dt = \partial\eta/\partial t + u(\partial\eta/\partial x)$ (see Section 16-1.3.2). The nonlinear term $u(\partial\eta/\partial x)$ will also be neglected in such a way that $w_s \cong \partial\eta/\partial t$. Consequently $w(z)$ increases from $w_b = 0$ to $w_s = \partial\eta/\partial t$. In the most general case, one can always say that $w(z)$ is given by a power series such as:

$$w(z) = \frac{\partial\eta}{\partial t}\sum_1^n A_n\left[\frac{z}{d+\eta}\right]^n$$

where $A_1 = 1$ ($z = 0$ at the bottom). The first term of this series ($n = 1$) is

$$w(z) = \frac{z}{d+\eta}\frac{\partial\eta}{\partial t}$$

It shows that w is linearly distributed from the bottom to the free surface. Moreover, if η is assumed small in comparison with d, neglecting the nonlinear term

$$-\frac{z}{(d+\eta)^2}\left(\frac{\partial\eta}{\partial t}\right)^2$$

yields

$$\frac{\partial w}{\partial t} = \frac{z}{d+\eta}\frac{\partial^2\eta}{\partial t^2}$$

Inserting this value in the Eulerian equation where the nonlinear terms are neglected gives

$$\frac{z}{d+\eta}\frac{\partial^2\eta}{\partial t^2} = -\frac{\partial}{\partial z}\left(\frac{p}{\rho} + gz\right)$$

When this is integrated in the vertical direction from a point z to the free surface, a distance $(d + \eta)$ from the origin, the result is:

$$\frac{p(z)}{\rho} = g[d + \eta - z] + \int_z^{d+\eta}\frac{z}{d+\eta}\frac{\partial^2\eta}{\partial t^2}\,dz$$

i.e.,

$$\frac{p(z)}{\rho} = g[d + \eta - z] + \frac{\partial^2\eta}{\partial t^2}\frac{(d+\eta)^2 - z^2}{2(d+\eta)}$$

The first term is the hydrostatic pressure, the second term is the correction due to vertical acceleration. Differentiating with respect to x and inserting $\partial p/\partial x$ in the Eulerian equation along a horizontal axis gives, when the small terms are neglected,

$$\frac{du}{dt} = -g\frac{\partial\eta}{\partial x} - \frac{\partial^3\eta}{\partial t^2\partial x}\left[\frac{(d+\eta)^2 - z^2}{2(d+\eta)}\right]$$

Now, averaging with respect to the vertical and neglecting some terms due to the fact that η is small by comparison with d gives

$$\frac{d\bar{u}}{dt} = \frac{1}{d+\eta}\int_{d+\eta}^{0}\left[g\frac{\partial\eta}{\partial x} + \frac{d^3\eta}{\partial t^2\partial x}\frac{(d+\eta)^2 - z^2}{2(d+\eta)}\right]dz$$

i.e., integrating and developing $d\bar{u}/dt$ leads finally to

$$\frac{\partial\bar{u}}{\partial t} + \bar{u}\frac{\partial\bar{u}}{\partial x} + g\frac{\partial\eta}{\partial x} + \frac{d+\eta}{3}\frac{\partial^3\eta}{\partial t^2\partial x} = 0$$

The first three terms are easily recognized from the nonlinear long wave equation. The last term is an approximate correction due to flow curvature. Such an equation is known as the Boussinesq equation.

18-2 The Linear Long Wave Theory

18-2.1 Basic Assumptions

If one neglects the convective inertia and friction terms in the momentum equation, one obtains

$$\frac{\partial u}{\partial t} = -g\frac{\partial \eta}{\partial x} \qquad \frac{\partial v}{\partial t} = -g\frac{\partial \eta}{\partial y}$$

Similarly, if the nonlinear terms $\partial(u\eta)/\partial x$ and $\partial(v\eta)/\partial y$ are neglected in the continuity equation, one has

$$\frac{\partial \eta}{\partial t} + \frac{\partial(ud)}{\partial x} + \frac{\partial(vd)}{\partial y} = 0$$

This set of equations characterizes the linearized long wave theory which is valid provided $\eta \ll d$ and $\eta/L(L/d)^3 \ll 1$.

In the case of near two-dimensional motion v is neglected and these equations can be written:

$$\frac{\partial u}{\partial t} = -g\frac{\partial \eta}{\partial x}$$

$$\frac{\partial}{\partial x}(Au) + l\frac{\partial \eta}{\partial t} = 0$$

where A is the cross section perpendicular to the velocity vector u and l is the width of the container at the free surface. This system of equations has a great number of exact solutions for various geometrical bottom topography.

18-2.2 The Linear Long Wave Equations

For example, in the two-dimensional case, with the water depth d as a constant, $d = A/l$. Differentiating the momentum equation with respect to x and the continuity equation with respect to t, and eliminating u gives

$$\frac{\partial^2 \eta}{\partial t^2} - gd\frac{\partial^2 \eta}{\partial x^2} = 0$$

Similarly differentiating the momentum equation with respect to t and the continuity equation with respect to x and eliminating η yields

$$\frac{\partial^2 u}{\partial t^2} - gd\frac{\partial^2 u}{\partial x^2} = 0$$

18-2.3 Harmonic Solution—Seiche

It can easily be verified that the solution can be that of a progressive wave such as

$$\eta = H\cos(mx - kt)$$

$$u = H\left(\frac{g}{d}\right)^{1/2}\cos(mx - kt)$$

with $m = 2\pi/L$, $k = 2\pi/T$, and $L = T(gd)^{1/2}$, or that of a standing wave (seiche)

$$\eta = H\cos mx\cos kt$$

$$u = H\left(\frac{g}{d}\right)^{1/2}\sin mx\sin kt$$

It can also easily be verified that the linear long wave theory is the limit case of the small amplitude wave theory when $d/L \to 0$ (see Section 16-3).

18-3 The Numerical Methods of Solution

18-3.1 The Versatility and Limits of Validity of Numerical Methods of Calculation

18-3.1.1 The nonlinear long wave equations are mostly treated by numerical methods and computer. For this purpose, one transforms the set of differential equations (continuity and momentum) into a finite difference scheme. The calculation consists of proceeding step by step. The values of $\eta(x,t)$ and $u(x,t)$ at a given time t_1 and at a given location x_1 are calculated from the knowledge of their values a small interval away. For example, consider the simple linearized long wave equation which is demonstrated in Section 18-2.1,

$$\frac{\partial u}{\partial t} = -g\frac{\partial \eta}{\partial x}$$

Consider the points 1, 2, 3, 4 in a T-X diagram separated by intervals Δx and Δt, respectively (see Fig. 18-2). Knowing the value of η at points 1, 2, 3, the value of u at point 4 is obtained from the equation:

$$\frac{\Delta u}{\Delta t} = -g \frac{\Delta \eta}{\Delta x}$$

The known values are used to find u_4:

$$\frac{u_4 - u_2}{\Delta t} = -g \frac{\eta_3 - \eta_1}{2\,\Delta x}$$

Then one proceeds step by step calculating successively u and η for the entire diagram. The history of the wave profile is so obtained as a function of time and distance.

It is easily seen that the great advantage of the long wave theory treated by numerical analysis is its versatility. Numerical methods are particularly convenient over complex boundary conditions, whereas the search for analytical solutions is beyond the scope of the best analysts. For example, the long wave theory may be applied in a river with variable cross section. Also, terms for bottom friction, wind stress on the free surface, and gradients due to complex pressure distribution can easily be taken into account. The term for bottom friction is particularly important for flood waves and tidal waves. Wind stress and free surface pressure gradients can also be taken into account numerically in the study of storm surges.

18-3.1.2 The error and limit of validity of numerical procedures is now discussed. The transformation of a differential equation into finite difference involves a systematic error. Indeed, it is known that by developing a differential term into finite difference by a Taylor expansion yields

$$\frac{\partial F}{\partial x} = \frac{\Delta F}{\Delta x} - \frac{\Delta x}{2} f''(x) \cdots - \frac{\Delta x^{n-1}}{n!} f^{(n)}(x) \cdots$$

The first task of any numerical computation is to insure that by taking $\partial F/\partial x = \Delta F/\Delta x$ the cumulative error due to the neglect of higher-order terms does not exceed the desired accuracy. Such a study involves the search for "stability criteria," which results in a relationship between intervals (generally space Δx and time Δt). In the case of high order derivative terms, stability criteria may not exist. The high order derivative terms have then to be replaced by a first-order derivative term of another variable, so the number of unknowns increases. For example, $\partial^2\eta/\partial t^2$ may be replaced by $\partial a/\partial t$, a being equal to $\partial \eta/\partial t$. Also, the choice of the interval is conditioned by the cumulative error, cost of the computing time, and the "round-off error." The round-off error is due to the fact that any numerical calculus is necessarily done with a limited number of figures or "digits." For example, most calculation done on computers is done with 8 digits, sometimes 16 digits if one used "double precision," or even more. But the increasing cost of computing time offsets this advantage.

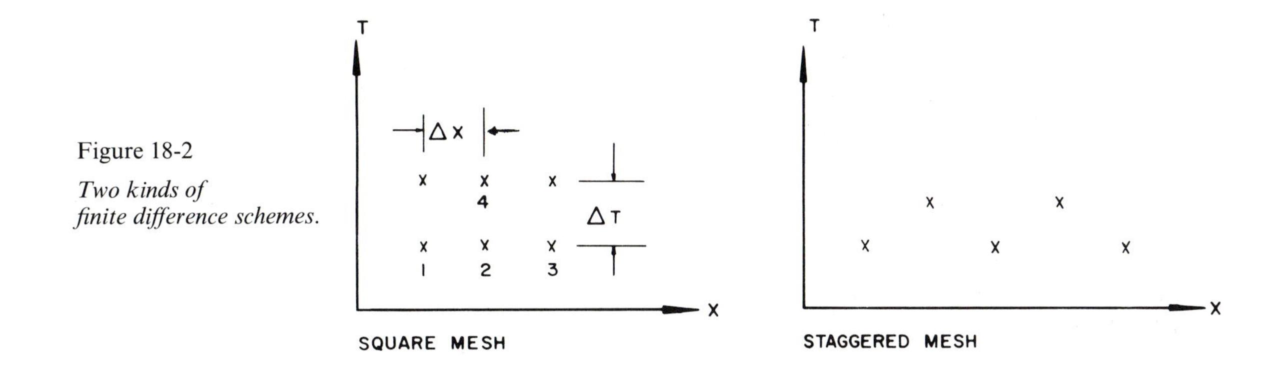

Figure 18-2
Two kinds of finite difference schemes.

In brief, the numerical implementation of the long wave theory always involves an inherent error, which increases with time and/or distance.

The study of the propagation of a bore, or an undulated wave, or the wave created by the breaking of a dam over a long distance is unreliable even if the stability criteria is satisfied because of the cumulative effect of the error. However, the study of a tidal wave in an estuary, or even of a flood wave with gentle variation of depth, is possible. Similarly, the propagation of a breaking wave (bore) over a steep beach, i.e., over a short distance because of the steep slope, may give reliable results.

18-3.2 The Method of Characteristics

18-3.2.1 The characteristic equations are now established. Consider the long wave equations in the following form:

Momentum: $$\frac{\partial u}{\partial t} + u\frac{\partial u}{\partial x} = -g\frac{\partial \eta}{\partial x}$$

Continuity: $$\frac{\partial \eta}{\partial t} + \frac{\partial[u(d + \eta)]}{\partial x} = 0$$

Then consider the zero quantity:

$$-g\left[\frac{d(d)}{dx} + S\right] = 0$$

When this is added to the right side of the momentum equation, the result is

$$\frac{\partial u}{\partial t} + u\frac{\partial u}{\partial x} + \frac{\partial[g(d + n)]}{\partial x} = -gS$$

Since $\partial(d)/\partial t = 0$, it can be added to the continuity equation. After the continuity equation is multiplied by the constant g the result is

$$\frac{\partial[g(d + \eta)]}{\partial t} + \frac{\partial[ug(d + \eta)]}{\partial x} = 0$$

Define $c = [g(d + \eta)]^{1/2}$ which has the dimension of a velocity. Inserting $c^2 = g(d + \eta)$, into the above equations and then differentiating so that $\partial c^2/\partial^* = c\partial(2c)/\partial^*$, and dividing the second equation by c yields

$$\frac{\partial u}{\partial t} + u\frac{\partial u}{\partial x} + c\frac{\partial 2c}{\partial x} = -gS$$

$$\frac{\partial 2c}{\partial t} + c\frac{\partial u}{\partial x} + u\frac{\partial 2c}{\partial x} = 0$$

Adding and subtracting these two equations gives

$$\frac{\partial}{\partial t}(u \pm 2c) + (u \pm c)\frac{\partial}{\partial x}(u \pm 2c) = -gS$$

It is now recalled that the total derivative of an expression $A(x,t)$ with respect to time is

$$\frac{dA(x,t)}{dt} = \frac{\partial A}{\partial t} + \frac{\partial A}{\partial x}\frac{dx}{dt}$$

Thus, the left-hand side of the above equation is the total derivative

$$\frac{d}{dt}(u \pm 2c)$$

provided that

$$\frac{dx}{dt} = u \pm c$$

This means that along a line of slope $dx/dt = u \pm c$, the relationship

$$\frac{d}{dt}(u \pm 2c) = -gS$$

will apply (see Fig. 18-3).

The lines of slope $dx/dt = u + c$ are called the advancing or positive characteristics. The lines of slope $dx/dt = u - c$ are called the receding or negative characteristics.

It is seen that from a given point (x,t) in the diagram, two lines of slopes $u + c$ and $u - c$ can be drawn.

18-3.2.2 An example, the "dam-break problem," is now presented: In the case of a horizontal bottom $S = 0$ and

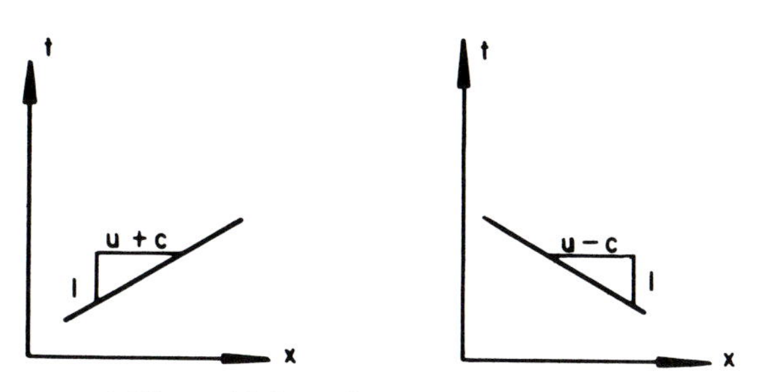

Figure 18-3 *Advancing and receding characteristics.*

$(d/dt)(u \pm 2c) = 0$ i.e., $u \pm 2c =$ constant values K_1 and K_2 along lines of slope $dx/dt = u \pm c$. Consider the case of a vertical wall of water which is suddenly released as in the case of a dam breaking (see Fig. 18-4). It is seen that when $t = 0, u = 0$, and $c = (gd)^{1/2}$, where d is the water depth and the constants K_1 and K_2 are

$$u + 2c = K_1 = 2(gd)^{1/2}$$
$$u - 2c = K_2 = -2(gd)^{1/2}$$

At time $t = t_0$, the water wall collapses. At the downstream water tip: $\eta = -d$ and $c = 0$. Hence the speed of the water tip is $dx/dt = u(+0)$ and

$$(u + 2c)|_{t=0} = (u + 2c)|_{t=t}$$

becomes $0 + 2(gd)^{1/2} = u$.

The water tip travels at a speed $2(gd)^{1/2}$. The speed of the rarefaction wave traveling upstream where $\eta = 0$ and $u = 0$ is $dx/dt = 0 + c$, i.e.,

$$(0 + 2c)|_{t=0} = (0 + 2c)|_{t=t}$$

and

$$c = \frac{dx}{dt} = (gd)^{1/2}$$

18-3.2.3 In general, the variation of u and η (or c) are gentle enough in order that a finite interval method can be applied. The basic characteristic equations are then written for finite intervals Δx, Δt:

$$\Delta(u \pm 2c) = -gS\,\Delta t \quad \text{along} \quad \frac{\Delta x}{\Delta t} = u \pm c$$

The time history of the wave evolution can then be determined step by step as follows. The values of $u(x,t_1)$ and $\eta(x,t_1)$ are given for a wave at a time $t = t_1$ (see Fig. 18-5).

The values of u and c are calculated at regular intervals Δx. The characteristic line of slope $\Delta x/\Delta t = u_1 + c_1$ is drawn from point 1 and the characteristic line of slope $\Delta x/\Delta t = u_2 - c_2$ is drawn from point 2.

Their intersection at point 3 defines x_3, t_3 graphically. Then, by applying the characteristic equation along these lines, the values of u_3 and c_3 (and consequently η_3) are found from the two equations

$$u_3 + 2c_3 = u_1 + 2c_1 - gS_1(t_3 - t_1)$$
$$u_3 - 2c_3 = u_2 - 2c_2 - gS_2(t_3 - t_2)$$

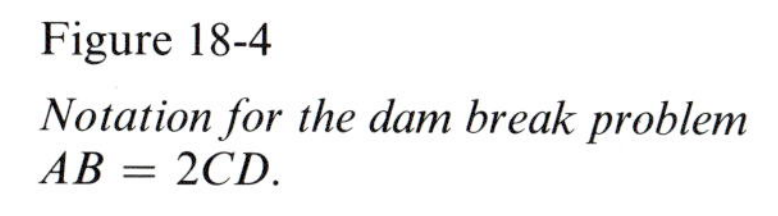
Figure 18-4

Notation for the dam break problem $AB = 2CD$.

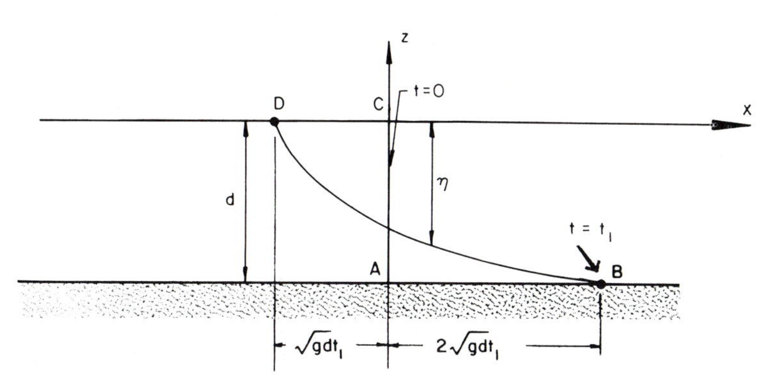

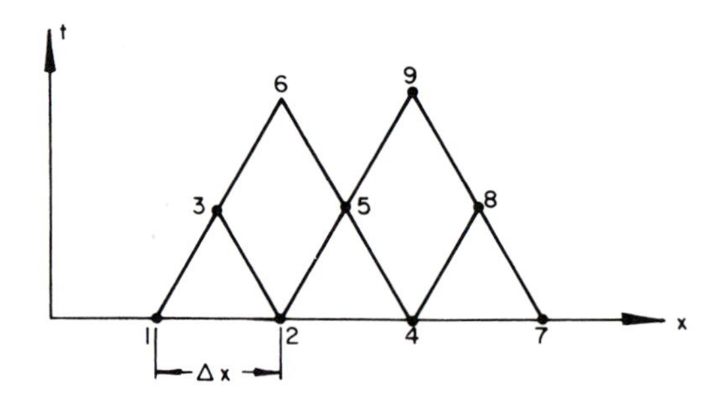

Figure 18-5 *Application of the method of characteristics.*

In the particular case of the Fig. 18-5, $t_2 = t_1$. Similarly, u_5 and c_5 are found from the points 2 and 4, u_6 and c_6 from the points 3 and 5, and so on.

18-3.2.4 By this process of calculation, it is seen that the state of a wave at a given point 1 (x_1, t_1) (see Fig. 18-6) has an influence upon the state of the wave at any other point between the characteristic lines issued from that point. Such characteristic lines define the domain of influence of this point. Similarly, the state of the wave at a given point 3 depends solely upon the state of the wave under the two characteristic lines crossing at point 3. Such a domain is the domain of dependence of point 3.

Any disturbance arising over a finite distance $(x_1 - x_2)$ will have an influence on the water behavior within a domain defined by a negative characteristic issued from x_1 and a positive characteristic from x_2.

18-3.2.5 In the case of tidal estuary, the bottom friction and variation of cross section must be taken into account. The momentum equation can easily be modified for taking into account the bottom friction as in Section 18-1.2.6. The continuity equation can also be modified for the variation of cross section as in Section 18-1.1.3. The same calculations done previously lead to

$$\Delta(u \pm 2c) = -\left[gS + \frac{g}{C_h^2}\left(\frac{u}{c}\right)^2 \pm \frac{uh}{l}\frac{\Delta l}{\Delta x}\right]\Delta t$$

in which case the method of characteristics will apply similarly.

18-3.3 Tidal Bore

18-3.3.1 When two positive characteristics $dx/dt = u + c$ cross each other (see Fig. 18-7), one set of values can be found from points a and b: u_2 and c_2; another set of values can be found from points a and c: u_1 and c_1.

The two values for u and the two values for c at the same location indicate a discontinuity or a vertical wall of water. There is bore inception.

18-3.3.2 The line $dx/dt = W$, where W is the speed of the bore, is a line of discontinuity for the mesh formed by the positive and negative characteristics. At the locus of the bore there are five unknowns to be determined, namely: u_1, η_1 on the low side of the bore, u_2, η_2 on the high side of the bore, and the bore velocity itself, W. Since $W > u_1 + c_1$, u_1 and η_1 are determined by application of the method of

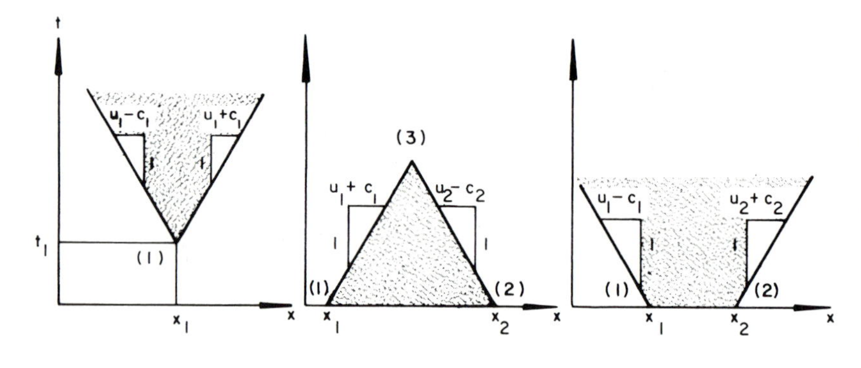

Figure 18-6 *Domain of influence and domain of dependence.*

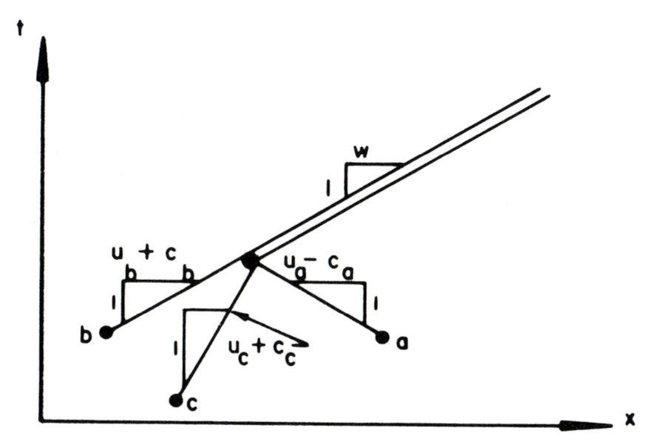

Figure 18-7 *Bore inception.*

characteristics directly from the points a and c (see Fig. 18-8). u_2, η_2 and W are determined from the three following equations:

1. The momentum equation for a moving hydraulic jump (see Section 12-4.6) which after some transformations gives:

$$\tfrac{1}{2}\rho g(h_2^2 - h_1^2) = \rho h_1(u_2 - u_1)(W - u_1)$$

2. The continuity equation for a moving hydraulic jump:

$$u_2 h_2 = W(h_2 - h_1) + u_1 h_1$$

Figure 18-8 *Bore propagation.*

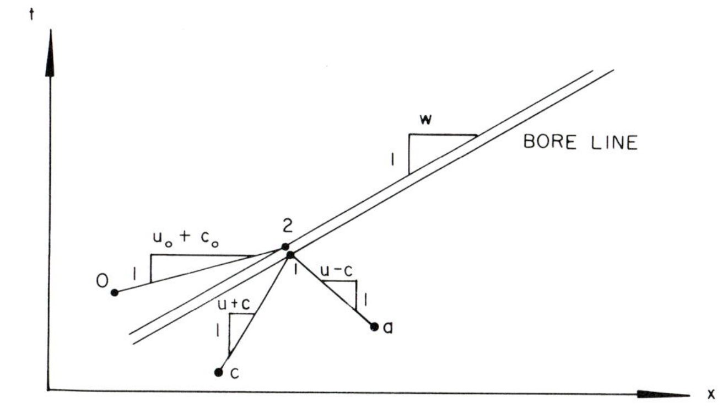

3. Since $u_2 + c_2$ always exceeds W, the positive characteristic equation which merges with the bore line at the considered point (x,t) is $[c_2 = (gh_2)^{1/2}]$:

$$u_2 + 2c_2 = u_0 + 2c_0 - gS(t_2 - t_0)$$

The wave profile on the high side of the bore is determined afterward by making use of u_2 and c_2 by applying the method of characteristics in a straightforward manner.

The problem of a limit solitary wave (see Section 18-4.2) travelling on an horizontal bottom and reaching a 1/10 bottom slope is treated on Fig. 18-9, as an example of applicability of the method of characteristics. The successive wave profiles are presented on Fig. 18-10 and are obtained by interpolating between the values of η at the crossing of characteristic lines by lines defined by t = constant.

18-3.4 The Direct Approach to Numerical Solutions

The application of the method of characteristics requires the solution of four unknowns at each point of crossing lines, namely, x, t, u, and η. In the case where x and t are specified *a priori*, only two unknowns remain: u and η. For this purpose the long wave equations can be treated directly by a finite difference process. The intervals Δx and Δt, i.e., the locations of x and t, are specified independently from the characteristic lines. For this purpose, a square mesh or a staggered mesh can be used (see Fig. 18-2).

18-3.4.1 In the case of the square mesh method, u and η at point 4 can be deduced from the value of u and η at points 1 and 3 directly as follows (see Fig. 18-11). From the continuity equation one has $\{(2) = \tfrac{1}{2}[(1) + (3)]\}$:

$$\eta_4 = \eta_2 + \frac{\Delta t}{2\,\Delta x}[u(d + \eta)|_3 - u(d + \eta)|_1]$$

and from the momentum equation one has:

$$u_4 = u_2 - \frac{\Delta t}{2\,\Delta x}[g(\eta_3 - \eta_1) + u_2(u_3 - u_1)]$$

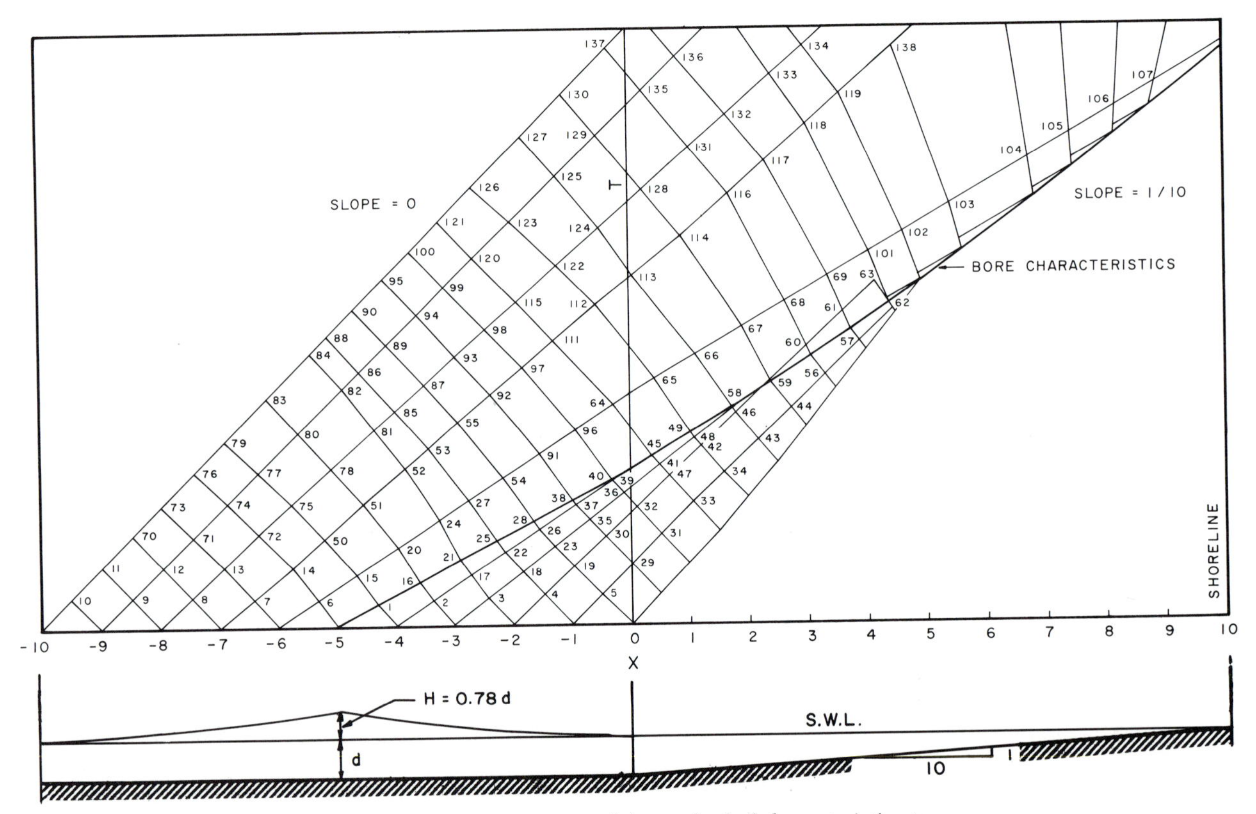

Figure 18-9 *Application of the method of characteristics to a wave breaking on a slope.*

The calculation will proceed by calculating u_5 and η_5 from points 2, 3, 6, and so on.

18-3.5 *The Stability Criterion*

The error made in transforming the differential equation into a finite difference equation may cause a cumulative error which may blow up as $t = \sum \Delta t$ increases. The finite difference method is stable provided the point (x,t) under consideration is within the domain of dependence formed by the intersection of the characteristics lines from the points which are used for its determination.

Consequently, since in general $(u + c)_1 > (u - c)_3$ (see Fig. 18-11), the criterion for stability is $\Delta t < (\Delta x/u + c)$.

The variation of u and c with respect to x and t also has to be gentle enough in order that the finite difference terms have a value close to the differential.

18-4 On Some Exact Steady-State Solutions

In general, an unsteady-state solution is a typical characteristic of the long wave theories, i.e., the wave profile changes its shape as the wave proceeds. This fact is inherent in the long wave equations. However, the insertion of bottom friction and a vertical acceleration term, even approximated, into the long wave equations permits

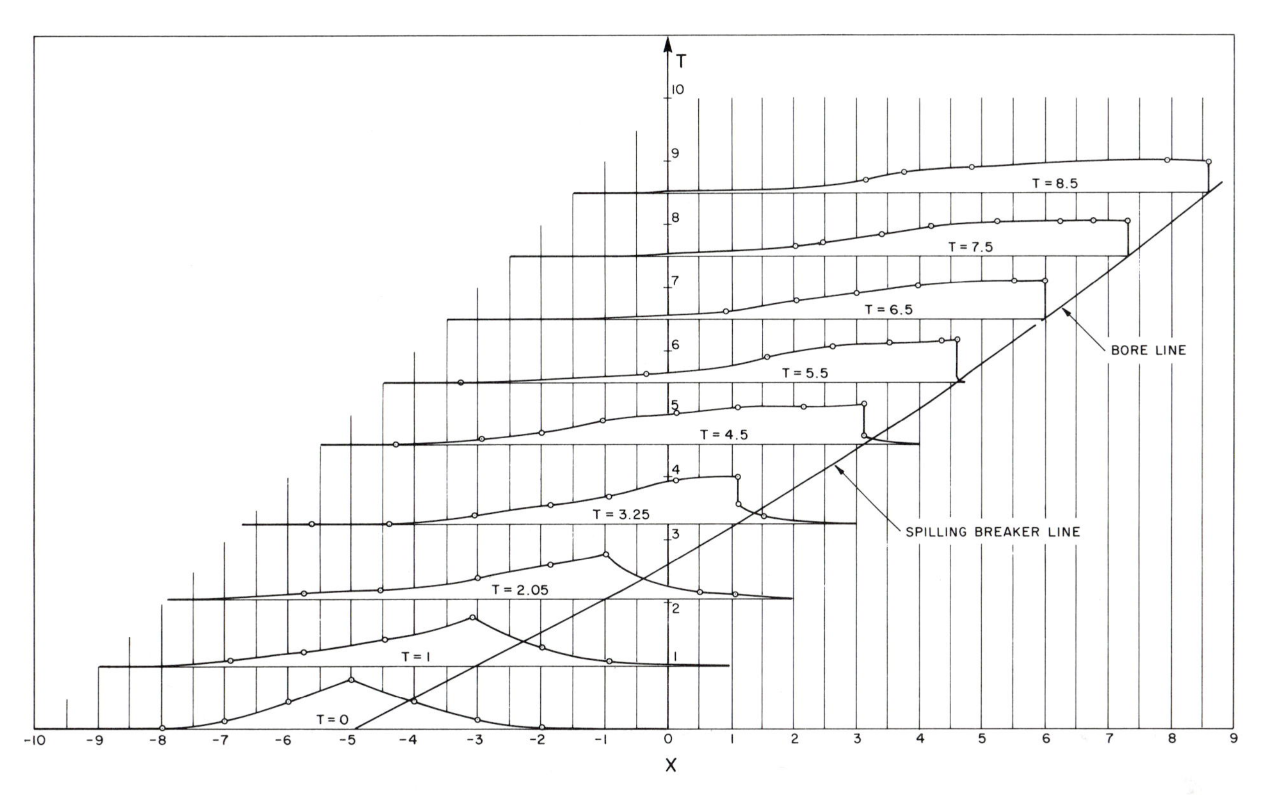

Figure 18-10 *Time-history of the wave profiles breaking on a "one-tenth" bottom slope.*

the finding of some special steady state solutions. They are the "monoclinal" wave, the solitary wave, and the cnoidal wave. Only the first two cases are analyzed in this book.

It is recalled that the assumption of a steady state solution results in the search of a function such as

$$\left.\begin{matrix}\eta\\ u\end{matrix}\right\} = f(x - Ct)$$

where C is a constant. Consequently,

$$\frac{\partial}{\partial t} = -C\frac{\partial}{\partial x}$$

Figure 18-11 *Notation for application of the square mesh method.*

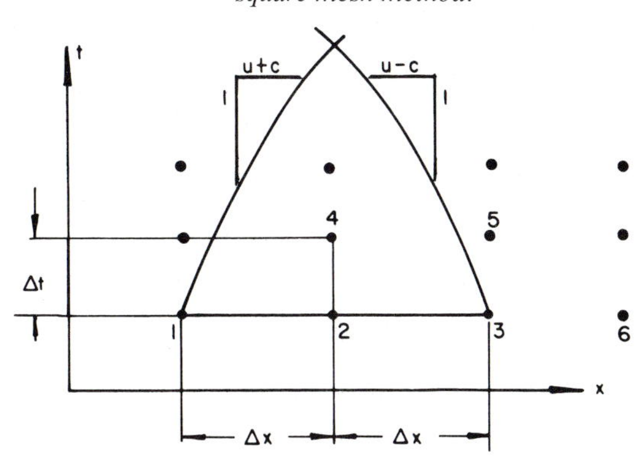

18-4.1 *Monoclinal Waves*

Consider the long wave equation with a bottom friction term such as $\tau = \rho f u^2$ (see Section 18-1.2.6):

$$\begin{cases} \dfrac{\partial h}{\partial t} + \dfrac{\partial(hu)}{\partial x} = 0 \\ \dfrac{\partial u}{\partial t} + u\dfrac{\partial u}{\partial x} + g\dfrac{\partial h}{\partial x} = gS - \dfrac{g}{C_h^2}\dfrac{u|u|}{h} \end{cases}$$

Inserting the relationship $\partial/\partial t = -C(\partial/\partial x)$ in the continuity equation gives $(u - C)(\partial h/\partial x) + h(\partial u/\partial x) = 0$ which integrates to $(u - C)h = A$(constant). The momentum equation becomes

$$(u - C)\frac{\partial u}{\partial x} + g\frac{\partial h}{\partial x} = g\left(S - \frac{u^2}{C_h^2 h}\right)$$

Eliminating $\partial u/\partial x$ between these two equations gives:

$$\left[g - \frac{(u - C)^2}{h}\right]\frac{\partial h}{\partial x} = g\left[S - \frac{u^2}{C_h^2 h}\right]$$

By using the value $u - C = A/h$ and then solving for $\partial h/\partial x$, it is found that:

$$\frac{\partial h}{\partial x} = \frac{S - [(Ch + A)^2/C_h^2 h^3]}{1 - A^2/gh^3}$$

This differential equation has a number of solutions, however, some are without physical significance. The solution depends upon whether the water depth is larger or smaller than the critical depth, as shown by Fig. 18-12. These waves are called monoclinal waves or uniformly progressive flow. This theory is particularly suitable for the study of flood waves in rivers.

18-4.2 *Solitary Wave Theory*

Consider the long wave equation for a horizontal bottom with a vertical acceleration term as it has been established in Section 18-1.3:

$$\frac{\partial u}{\partial t} + u\frac{\partial u}{\partial x} + g\frac{\partial \eta}{\partial x} + \frac{d + \eta}{3}\frac{\partial^3 \eta}{\partial t^2 \partial x} = 0$$

$$\frac{\partial \eta}{\partial t} + \frac{\partial[u(d + \eta)]}{\partial x} = 0$$

Inserting the relationship $\partial/\partial t = -C(\partial/\partial x)$ gives

$$\frac{\partial}{\partial x}\left[-Cu + \frac{u^2}{2} + g\eta + \frac{C^2(d + \eta)}{3}\frac{\partial^2 \eta}{\partial x^2}\right] = 0$$

and

$$\frac{\partial}{\partial x}[-C\eta + (d + \eta)u] = 0$$

The quantities between brackets are independent of x and consequently equal to constants. It is seen that these constants are nil since u and η tend to zero when $x \to \infty$.

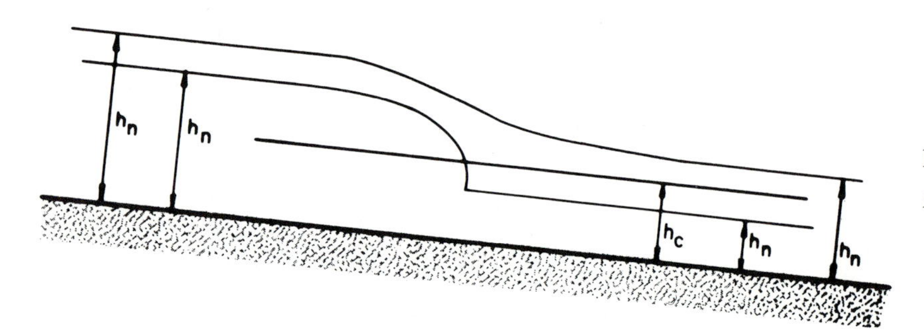

Figure 18-12
Monoclinal waves.

From the second equation one obtains $u = C\eta/d + \eta$. When this is substituted into the first equation, the results are

$$\frac{C^2\eta}{d+\eta} = \frac{C^2\eta^2}{2(d+\eta)^2} + g\eta + \frac{C^2(d+\eta)}{3}\frac{\partial^2\eta}{\partial x^2}$$

By considering η small with respect to d and developing the square root as $(1+\alpha)^{1/2} \cong 1 + \frac{1}{2}\alpha$, the above equation can be solved for C such that

$$C = \left[g(d+\eta)\left(1 + \frac{\eta}{2d} + \frac{d^2}{3\eta}\frac{\partial^2\eta}{\partial x^2}\right)\right]^{1/2}$$

When the free surface curvature is negligible, the value for C becomes

$$C \cong (gd)^{1/2}\left(1 + \frac{3}{4}\frac{\eta}{d}\right)$$

which can also be obtained directly by application of the momentum theorem.

In this case, since one has assumed C to be a constant and since the wave profile remains unchanged with time, one must have:

$$\frac{3\eta}{2d} + \frac{d^2}{3\eta}\frac{\partial^2\eta}{\partial x^2} = \text{constant} = \frac{H}{d}$$

H is a constant, which is specified in the following. This equation can be integrated as follows:

$$\frac{\partial^2\eta}{\partial x^2} = \frac{3\eta}{2d^3}[2H - 3\eta]$$

Since $\partial\eta/\partial x = 0$ when $\eta \to 0$,

$$2\frac{\partial\eta}{\partial x}\frac{\partial^2\eta}{\partial x^2}dx = \frac{3\eta}{d^3}(2H - 3\eta)\,d\eta$$

or

$$\left[\frac{d\eta}{dx}\right]^2 = \frac{3\eta^2}{d^3}(H - \eta)$$

It is seen that $\eta = H$ gives $d\eta/dx = 0$ and corresponds to the top of the wave; consequently, H is the wave height.

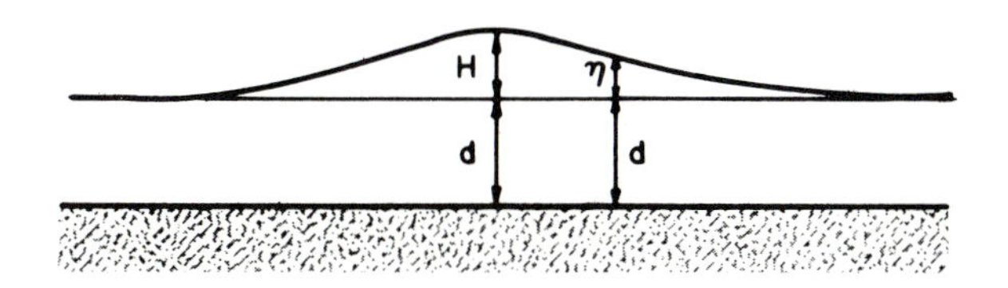

Figure 18-13 *Solitary wave.*

The equation can still be written:

$$\frac{d\eta}{dx} = \left(\frac{3}{d^3}\right)^{1/2}\eta(H - \eta)^{1/2}$$

After separation of the variables, it can be integrated in the form

$$\int_0^\eta \frac{d\eta}{\eta(H-\eta)^{1/2}} = \int_0^x \left(\frac{3}{d^3}\right)^{1/2} dx$$

which gives for the wave profile:

$$\eta = \frac{H}{\cosh^2[(3H/d)^{1/2}(x/2d)]}$$

as shown in Fig. 18-13.

PROBLEMS

18.1 Demonstrate that the two-dimensional linear long wave equation over a horizontal bottom is

$$gd\left(\frac{\partial^2\phi}{\partial x^2} + \frac{\partial^2\phi}{\partial y^2}\right) = \frac{\partial^2\phi}{\partial t^2}$$

18.2 Demonstrate that the two-dimensional linear long wave equation over a slow varying complex bottom topography for periodic motion is

$$\frac{\partial}{\partial x}\left[d\frac{\partial\phi}{\partial x}\right] + \frac{\partial}{\partial y}\left[d\frac{\partial\phi}{\partial y}\right] = -\frac{k^2}{g}\phi$$

where k is a wave number and $d = d(x,y)$ is the water depth.

18.3 Consider a square mesh of mesh size Δ in a X, Y system of coordinates. Demonstrate that the linear long wave equation for periodic motion for a finite difference scheme follows the relationship ($m = 2\pi/L$):

$$-4\phi_{i,j} + \phi_{i,j+1} + \phi_{i+1,j} + \phi_{i,j-1} + \phi_{i-1,j} + \frac{(\Delta m)^2}{gd}\phi_{i,j} = 0$$

18.4 Let us consider the linear long wave motion defined by the free surface equation, such as: $m = 2\pi/L$, $k = 2\pi/T$

$$\eta(x,y,t) = \cos kt \cos mx + \sin kt \cos my$$

This defines a motion obtained by the superposition of two standing waves at right angles.

1. Establish the equation of the lines of constant wave amplitude.
2. Determine the points of maximum amplitude.
3. Is there any amphidromic point (point of zero amplitude)?
   Describe the motion around it.
4. Draw a sketch of the curve of constant amplitude within a square defined by $0 < x/y < L/2$.

18.5 Establish the equation of motion for periodic linear long wave in a corner defined by a depth $d = cst$, and width $b = b_0 x$.

18.6 Consider the energy diagram as shown on Fig. 14-5. Indicate the modification which needs to be made to such a figure in the case of unsteady motion.

18.7 Demonstrate that the long wave equations can be transformed as $u + 2c$ = constant and $u - 2c$ = constant along a line of slope

$$\frac{dx}{dt} = u \pm c \pm \frac{gS}{(\partial/\partial x)(u \pm 2c)}$$

18.8 Transform the differential equations which are used in the method of characteristics for the study of the propagation of two-dimensional long waves into a finite difference system corresponding to intervals Δx, Δt forming a square mesh. Do the same in the case of a staggered mesh.

18.9 Demonstrate that the equation of Barré de St. Venant can be written:

$$\alpha\frac{\partial^2\phi}{\partial x^2} + 2\beta\frac{\partial^2\phi}{\partial x\partial t} + \gamma\frac{\partial^2\phi}{\partial t^2} = F$$

where

$$\alpha = \left[Q^2 - \frac{gA^3}{l}\right]$$

$$\beta = QA$$

$$\gamma = A^3$$

$$F = A^3 g\left[\frac{A}{l^2}\frac{dl}{dx} - S - \frac{lQ|Q|}{C_h^2 A^3}\right]$$

Give the definition for ϕ.

REFERENCES FOR PART THREE

Basset, A. B., *A Treatise on Hydrodynamics*, Vols. I and II. Reprinted by Dover Publications, Inc., New York, 1961.

Beach Erosion Board, A Summary of the Theory of Oscillatory Waves. Technical Report No. 2, U.S. Government Printing Office, 1942.

Biesel, F., Equations générales au second ordre de la houle irrégulière. *La Houille Blanche*, May 1952.

Birkhoff, G., *Hydrodynamics*. Princeton University Press, 1950.

Bouasse, H., *Houles, rides, seiches et marées*. Librairie Delagrave, Paris, 1924, pp. 92–145.

Boussinesq, J., Théorie de l'intumescence liquide appelée onde solitaire ou de translation se propagant dans un canal rectangulaire. Institut de France, Académie des Sciences, Comptes Rendus, June 19, 1871, p. 755.

Boussinesq, J., Essai sur la théorie des eaux courantes. Institut de France, Académie des Sciences, *Mémoires présentés par divers savants*, *23*, 1877.

Broer, L. J. F., On the propagation of energy in linear conservative waves. *Appl. Sci. Res.*, *A2*: 447–468, 1951

Carr, J. H., and Stelzriede, M. E., Diffraction of Water Waves by Breakwaters in Gravity Waves. U. S. National Bureau of Standards, Circular 521, pp. 109–125.

Chappelear, J. E., Shallow water waves. *J. of Geophys. Res. 67*, 1962.

Coulson, C. A., *Waves*, 4th ed. Oliver and Boyd, Edinburgh, 1947.

Danel, Pierre, On the Limiting Clapotis. U. S. National Bureau of Standards, Gravity Waves, NBS Circular 521, 1952, pp. 35–38.

Dean, R. G., Stream Function Wave Theory—Validity and Application. Specialty Conference on Coastal Engineering, ASCE, 1965.

Druet, C., Nomographic chart for determination of the monochromatic wave type in the region of foundation of a designed hydrotechnical structure. Paper S, II-I, 21st International Navigation Congress, Stockholm, 1965, pp. 183–201.

Dubreuil-Jacotin, J. L., Sur les ondes de type permanent dans les liquides hétérogènes. *Rendiconti della Accademia Nazionale dei Lincei*, Ser. 6, *15*: 814–819, 1932.

Dubreuil-Jacotin, J. L., Sur la determination rigoureuse des ondes permanentes périodiques d'ampleur finie, *Journal de Mathematiques Pures et Appliquées*, Ser 9, *13*: 217–291, 1934.

Eckart, C., The Propagation of Gravity Waves from Deep to Shallow Water. U. S. National Bureau of Standards, Gravity Waves, NBS Circular 521, 1952, pp. 165–173.

Freeman, J. C., and Le Méhauté, B., Wave breakers on a beach and surges on a dry bed. *Proc. Amer. Soc. Civ. Eng.*, *90*: 87–216, 1964.

Friedrichs, K. O., On the derivation of the shallow water theory, Appendix to the formation of breakers and bores by J. J. Stoker. *Comm. Pure Appl. Math.* 1: 81–85, 1948.

Gerstner, F., Theorie der Wellen, *Annalen der Physik*, *32*, 1809.

Goda, Y., Wave Forces on a Vertical Circular Cylinder. Report No. 8, Port and Harbor Technical Research Inst., Japan, 1964.

Ippen, A. T. (editor), *Estuary and Coastline Hydrodynamics*. McGraw-Hill, New York, 1966.

Johnson, J. W., O'Brien, M. P., and Isaacs, J. D., Graphical Construction of Wave Refraction Diagrams, Hydrographic Office, Navy Department, Publication No. 605, 1948.

Keller, J. B., The solitary wave and periodic waves in shallow water. *Comm. Appl. Math.*, December 1948.

Keulegan, G. H., and Patterson, G. W., Mathematical theory of Irrotational translation waves. *U.S. NBS J. Res. 24*, 1940.

Kinsman, B., *Wind Waves*. Prentice-Hall, Englewood Cliffs, New Jersey, 1965.

Koh, R. C. Y., and Le Méhauté, B., Wave shoaling. *J. Geophys. Res.*, *71*: 2005–2012, 1966.

Kotschin, N. J., Kibel, I. A., *et al.*, *Theoretische Hydrodynamik*. Adad. Verlag, Berlin, 1964.

Korteweg, D. J., and de Vries, G., On the change of form of long waves advancing in a rectangular canal on a new type of long stationary waves. *London, Dublin and Edinburgh, Philosophical Magazine*, Ser. 5, *39*: 422, 1895.

Kravtchenko, J., and Daubert, A., La Houle à trajectoires fermées en profondeur finie. *La Houille Blanche*, *12*, 1957.

Laitone, E. V., The second approximation to cnoidal and solitary waves, *J. Fluid Mech.*, *9*, 1960.

Lamb, H., *Hydrodynamics*. Dover Publications, New York, 1945; Cambridge University Press, 1932.

Le Méhauté, B., Mass transport in cnoidal waves. *J. Geophys. Res.*, *73*: 5973–5979, 1968.

Le Méhauté, B., Divoky, D., and Lin. A., Shallow water waves: A comparison of theories and experiments. *Proc. Amer. Soc. Civ. Eng.* (11th Conf. Coastal Eng.) *1*, 86–107, 1968.

Le Méhauté, B., and Webb, L., Periodic gravity waves over a gentle slope at a third order of approximation. *Proc. Amer. Soc. Civ. Eng.* (8th Cong. Coastal Eng.) *1*: 23–40, 1964.

Le Méhauté, B., *Theory of Explosion-Generated Water Waves*. Advances in Hydroscience, Vol. 7, Academic Press, New York, 1971.

Le Méhauté, B., Theory of wave agitation in a harbor. *Trans. Amer. Soc. Civ. Eng.*, *127*: 369–383, 1962.

Longuet-Higgins, M. S., and Stewart, R. W., Radiation stress and mass transport in gravity waves. *J. Fluid Mech.*, *13*: 481–509, 1962.

Longuet-Higgins, M. S., Mass transport in water waves. *Phil. Trans. Roy. Soc.*, *A 245*: 535–581, 1953.

Mei, C. C., and Le Méhauté, B., Notes on the equations of long waves over an uneven bottom. *J. Geophys. Res.*, *71*: 393–400, 1966.

McGowan, J., On the solitary wave. *London, Dublin and Edinburgh, Philosophical Magazine*, Ser. 5, *32*: 45, 1891; On the highest wave of permanent type. Ser. 5, *38*: 351, 1894.

McNown, J. S., Waves and Seiches in Idealized Ports. U. S. National Bureau of Standards, Gravity Waves, NBS. Circular 521, 1952.

Miche, A., Mouvements ondulatoires de la mer en profondeur constante ou décroissante. *Annales des ponts et chaussées*, 1944, pp. 25–78, 131–164, 270–292, 369–406.

Mitchell, J. H., The wave resistance of a ship. *London, Dublin and Edinburgh, Philosophical Magazine*, Ser. 5, *45*, 1898.

Milne-Thompson, L. M., *Theoretical Hydrodynamics.* Macmillan, London, 1938.

Munk, W. H., The solitary wave theory and its application to surf problems. *Annals New York Acad. Sci.*, *51*: 376–424, 1949.

Munk, W. H., and Arthur, R. S., Wave Intensity Along a Refracted Ray. U. S. National Bureau of Standards, Gravity Waves, NBS. Circular 521, 1952.

Pierson, W. J., Jr., The Interpretation of Crossed Orthogonals in Wave Refraction Phenomena. Technical Memorandum No. 21, Beach Erosion Board, Corps of Engineers.

Putnam, J. A., and Arthur, R. S., Diffraction of water waves by breakwaters. *Trans. Am. Geophys. Union*, *29*: 1948.

Rouse, H., *Fluid Mechanics for Hydraulic Engineers.* McGraw-Hill, New York, 1938.

Stokes, G. G., On the theory of oscillatory waves. *Trans. Cambridge Philosoph. Soc.*, *8*, 1847; and *Suppl.*, *Scientific Papers*, *1*.

Struik, D. J., Détermination rigoureuse des ondes irrotationnelles periodiques dans un canal à profondeur finie. *Mathematische Annalen*, *95*: 595–634, 1926.

Suquet, F., Remarks on graphical computation of wave refraction. International Association for Hydraulic Research, Grenoble, 1949.

Sverdrup, H. U., A Study of Progressive Oscillatory Waves in Water; A Summary of the Theory of Water Waves. Technical Reports No. 1, 2, Beach Erosion Board, Office of Chief Engineering, U. S. War Department.

Thomas, H. A., *The Propagation of Stable Wave Configurations in Steep Channels.* Carnegie Institute of Technology, Pittsburgh, Pa.

Thomas, H. A., *Hydraulics of Flood Movements in Rivers.* Carnegie Institute of Technology, Pittsburgh, Pa., 1937.

Thomas, H. A., Propagation of waves in steep prismatic conduits. Proceedings of Hydraulics Conference, University of Iowa, Studies in Engineering, Bulletin 20, 1940.

Ursell, F., The long-wave paradox in the theory of gravity waves. *Proc. Cambridge Philosoph. Soc.*, *49*, 1953.

Van Dorn, W. G., Tsunami. *Advan. Hydrosci.*, *3*: 1–48, 1965.

Wehausen, J. V., and Laitone, E. V., Surface waves, In *Handbuch der Physik*, Springer-Verlag, Berlin, 1960.

Wiegel, R. L., *Oceanographical Engineering.* Prentice-Hall, 1964.

Whitham, G. B., Mass momentum and energy flux in water waves. *J. Fluid Mech.*, *12*: 135–147, 1962.

Appendix A

Wave Motion as a Random Process

A-1 Introduction to a Real Sea State

A-1.1 A New Approach to Water Waves

A brief introduction to the hydrodynamics and mathematics of periodic and other analytically defined waves in a heavy fluid bounded by a free surface has been presented in Part Three. The developments along these lines have been continued for well over a century.

Any observation of a real sea reveals a continuously changing random pattern of bumps and hollows. Waves of different lengths travel at different speeds combining and recombining in constantly changing patterns even if they are all unidirectional. When multidirectional waves are present the patterns created are even more complex. It was thought for a long time, even by such men as Rayleigh and Stokes, that this apparently chaotic process was beyond adequate mathematical description. The best that could be done appeared to be the choice of a mean wave height and a mean wavelength, followed by the application of the classical wave theories. To the mathematician the classical wave theories themselves were fascinating and the apparent difficulties of dealing with a real sea did not concern him.

In comparatively recent times an approach to the understanding of real seas has been developed. This approach is based on the combination of statistics, Fourier analysis, and hydrodynamics. The theories of statistics have to be used to determine stable parameters for describing a random sea state; then Fourier analysis is used to break down the random process into harmonic components whose behavior can be analyzed by using classical hydrodynamic theories of wave motions.

Recent progress owes a great deal to the development of research on the statistical analysis of random noise by communications engineers. The study of sea waves has developed as a combination of time series analysis and statistical geometry governed by the laws of hydrodynamics. This section presents an introduction to this nondeterministic aspect of waves. Probability concepts and spectrum theory will be discussed.

A-1.2 Some Definitions and Stability Parameters

A part of a typical wave record is shown in Fig. A-1. The concepts of surface ordinates, "wave height," "wave periods," maxima and minima, and envelopes are illustrated. Some of the statistics of the various parameters which can be obtained from such a wave record as Fig. A-1 will be discussed. Clearly, if the wave record is very long it is impractical to keep it in its original form. Methods of condensing the gross details of the wave record are required whereas inevitably, after such a process, much detail will be lost.

The condensations of the real sea state need to have the property of stability. The term stability is used to describe a characteristic which does not change too much if the observation is repeated. For example, suppose two wave recorders were placed in the open sea at a distance of, say, 200 ft apart. The water surface-time history of the two records taken at the same time would be completely different. The sea surface records themselves are unstable. On the other hand, such things as the average wave height, mean square water surface fluctuation, etc. of the two records would be very close to being equal (so long as the records were of reasonable length). Such statistical properties are said to be stable.

Probability densities and distributions of sea surface parameters, together with the spectral (or variance–frequency) distribution of the sea surface, have been found to be concise and useful properties of this process. The spectrum is a form of probability distribution and has very desirable stability characteristics. The spectrum retains much information on wave amplitudes and "periods" but loses all information on phase position. Probability distributions, on the other hand, lose all information on wave periods if "wave height" probabilities are computed or vice versa.

The description of sea states in terms of spectra has already been used extensively by naval architects to predict ship performance. Spectral and other probability concepts are used by mechanical and aeronautical engineers to handle vibration problems. Many branches of geophysics, such as turbulence, seismology, and tide analysis, use the basic communications theory to describe the processes and analyze data.

Figure A-1 *A typical wave record at sea.*

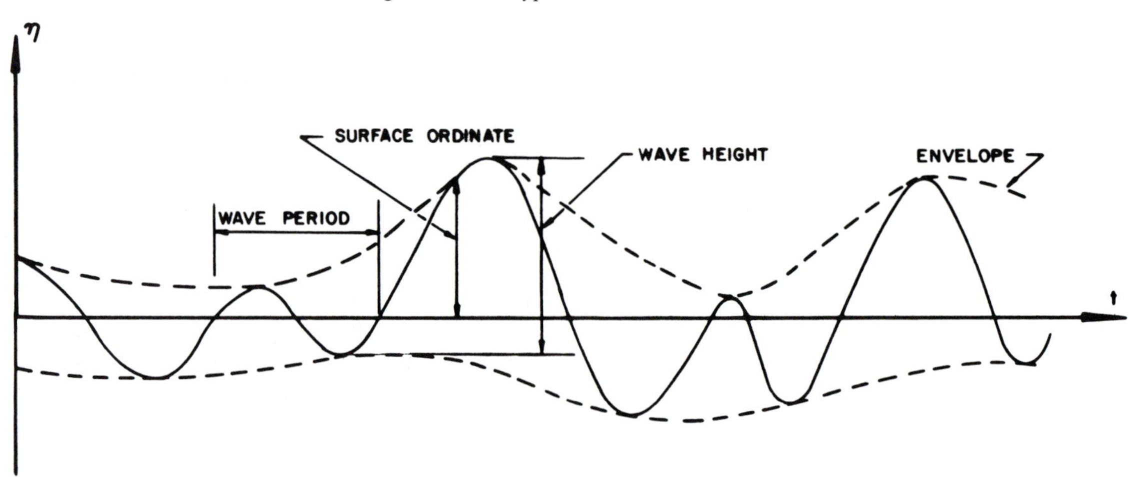

A summary of the most important probability distributions and an exposition of spectral representation of waves as random processes is now presented.

A-2 Harmonic Analysis

A-2.1 The Concept of Representation of Sinusoidal Waves in the Frequency Domain

Consider a sinusoidal wave profile given by

$$\eta = A \cos (mx + kt + \varepsilon)$$

where A is the amplitude, $m = 2\pi/L$ and $k = 2\pi/T$, where L and T are wavelength and period, respectively, and ε is a phase angle. If this wave is observed at a point as a function of time and the origin is chosen such that the initial phase position is zero, it can be written $\eta = A \cos kt$. This wave can be described as a wave of amplitude A and frequency k (in radians). It can be represented as in Fig. A-2, which is referred to as an amplitude spectrum.

Suppose a wave of arbitrary shape, but having a period T (frequency $2\pi/T$) is to be represented on an amplitude spectrum such as Fig. A-2. Recall that almost any periodic function can be represented as a Fourier series of fundamental period T and harmonics, so that

$$\eta = \sum A_n \cos (nkt + \varepsilon_n)$$

This wave would have an amplitude frequency spectrum appearing as in Fig. A-3.

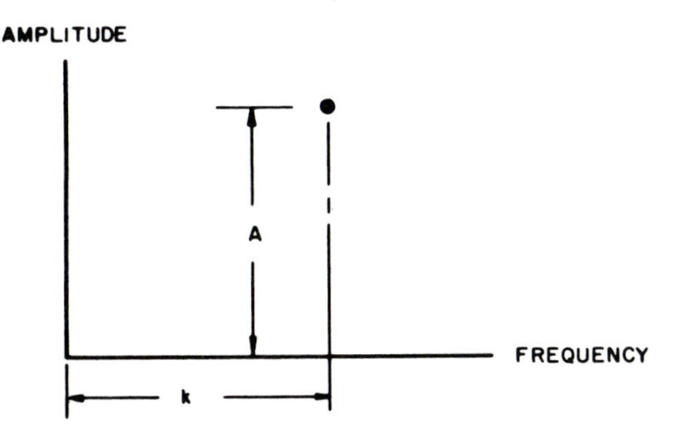

Figure A-2 *Amplitude spectrum of a sinusoidal wave.*

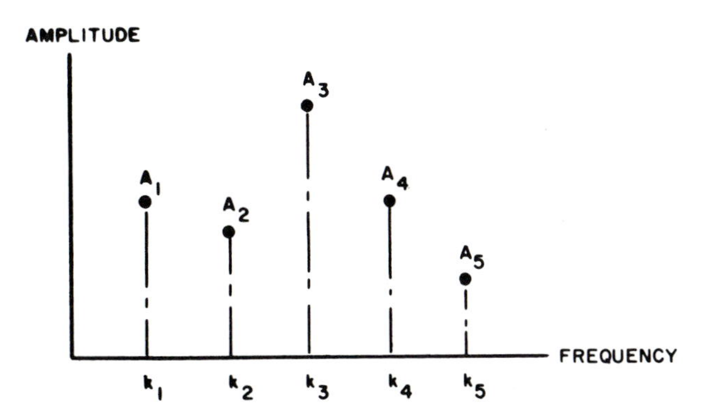

Figure A-3 *Amplitude spectrum of a periodic wave of arbitrary shape.*

If a water surface is described by a combination of a number of waves having different periods (frequencies) it could be represented as a diagram such as Fig. A-3, having amplitude and frequency components at many different places.

A-2.2 Review of Fourier Analysis

A-2.2.1 The immediate temptation is to take a wave record such as Fig. A-4, assume that it can be represented in the interval $0 < t < T_1$ by a Fourier series:

$$\eta(t) = \frac{A_0}{2} + \sum \left[A_n \cos \frac{2\pi nt}{T_1} + B_n \sin \frac{2\pi nt}{T_1} \right]$$

and evaluate A_n and B_n. However, the limits on the integer n have not yet been assigned. They should be $1 < n < \infty$, but n would obviously have to be chosen much smaller than the upper limit. It is hoped that for some comparatively large but finite value of n the series of the above equation will approach the observed water surface of Fig. A-4.

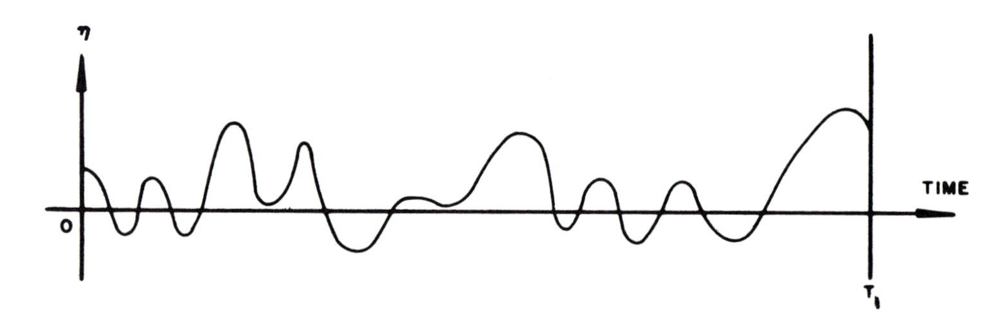

Figure A-4
A typical wave record at sea.

In order to determine the coefficients A_n and B_n in the right-hand side of the above equation, it is recalled that

$$\int_0^{T_1} \cos\frac{2\pi nt}{T_1}\sin\frac{2\pi mt}{T_1}\,dt = 0 \qquad \text{for all } m \text{ and } n$$

$$\int_0^{T_1} \cos\frac{2\pi nt}{T_1}\cos\frac{2\pi mt}{T_1}\,dt = 0 \qquad \text{for } m \neq n$$

$$= \frac{T_1}{2} \qquad \text{for } m = n$$

$$\int_0^{T_1} \sin\frac{2\pi nt}{T_1}\sin\frac{2\pi mt}{T_1}\,dt = 0 \qquad \text{for } m \neq n$$

$$= \frac{T_1}{2} \qquad \text{for } m = n$$

To use these properties multiply both sides of the previous equation for $\eta(t)$ by cos $[(2\pi mt)/T_1]$ and integrate between 0 and T_1.

$$\int_0^{T_1} \eta(t)\cos\frac{2\pi mt}{T_1}\,dt = \frac{A_0}{2}\int_0^{T_1}\cos\frac{2\pi mt}{T_1}\,dt$$

$$+ \sum_n A_n \int_0^{T_1}\cos\frac{2\pi nt}{T_1}\cos\frac{2\pi mt}{T_1}\,dt$$

$$+ \sum_n B_n \int_0^{T_1}\sin\frac{2\pi nt}{T_1}\cos\frac{2\pi mt}{T_1}\,dt$$

$$= \frac{A_m T_1}{2} \qquad (n = m)$$

It is now found that

$$A_n = \frac{2}{T_1}\int_0^{T_1}\eta(t)\cos\frac{2\pi mt}{T_1}\,dt$$

and similarly by multiplying by sin $[(2\pi m)/T_1]$

$$B_n = \frac{2}{T_1}\int_0^{T_1}\eta(t)\sin\frac{2\pi mt}{T_1}\,dt$$

A-2.2.2 Sufficient and necessary conditions that the function $\eta(t)$ can be represented by this set of equations cannot be stated precisely and fully within the scope of this book. (The reader is referred to many standard text-books on Fourier analysis.) It is sufficient for the present purpose to write one of the necessary conditions, which is

$$\int_0^{T_1} |\eta(t)|\,dt \neq \infty$$

This condition will be satisfied for wave records with T_1 finite.

The coefficients A_n and B_n can be represented on two figures such as Fig. A-3 (one for A_n and one for B_n). The points for each coefficient will occur at multiples of $k_n = (2\pi)/T_1$ where T_1 is the length of the record.

In order for the Fourier coefficients A_n and B_n to be representative of a sea state for all time, the record length T_n must be very long. The limits on n will be very large and the practical details of such a Fourier representation are seen to be enormous. A comparatively short record of the sea will yield values for A_n and B_n which would only be representative of that particular record at that particular time and place. A slightly different record taken, say, a few minutes later or a few feet away would yield completely different A_n and B_n. This simplified Fourier analysis is unstable, hence a much more stable description of the sea state must be found.

A-2.3 Fourier Analysis in Exponential Form

The expressions for the Fourier series representation given in the previous section will be more easily handled in their exponential form. The transformation is as follows:

$$\cos nkt = \tfrac{1}{2}(e^{inkt} + e^{-inkt})$$

$$\sin nkt = \frac{1}{2i}(e^{inkt} - e^{-inkt})$$

The previous expression for $\eta(t)$ given at the beginning of the section may now be written:

$$\eta(t) = \frac{A_0}{2} + \frac{1}{2}\sum_{1}^{\infty}(A_n - iB_n)e^{inkt} + \frac{1}{2}\sum_{1}^{\infty}(A_n + iB_n)e^{-inkt}$$

It follows from the definition of A_n and B_n that

$$A_n = A_{-n} \text{ since } \cos(-nkt) = \cos nkt$$

and

$$B_n = -B_{-n} \text{ since } \sin(-nkt) = -\sin nkt$$

The last term of the above equation can be written as

$$\frac{1}{2}\sum_{1}^{\infty}(A_n + iB_n)e^{-inkt} \equiv \frac{1}{2}\sum_{-1}^{-\infty}(A_n - iB_n)e^{inkt}$$

Inserting this equation into the previous one gives

$$\eta(t) = \sum_{-\infty}^{\infty} A_n' e^{inkt}$$

where A_n' is given by

$$A_n' = \tfrac{1}{2}(A_n - iB_n) \quad \text{and} \quad A_0' = \tfrac{1}{2}A_0$$

Hence

$$A_n' = \frac{1}{T_1}\int_0^{T_1} \eta(t)e^{-inkt}\,dt \qquad \text{for } n = 0, \pm 1, \pm 2 \ldots$$

It is seen that A_n' is a complex function of frequency (nk) for a given record of $\eta(t)$ of length T_1 containing information on phase and amplitude of the components making up that particular record. This operation is called a Fourier transform.

A-2.4 Random Functions

There still remains the problem of deriving a more stable description of a sea state and a method to handle longer wave records. As the record length tends to infinity, the condition $\int_0^{T_1} |\eta(t)|\,dt \neq \infty$ will obviously be violated. The Fourier transform giving the value of A_n' will no longer be valid. Some method has to be devised to handle a random process under the general concept of harmonic analysis. A random phenomenon is one in which the fluctuations of the quantity under observation as a function of time cannot be precisely predicted. No two water level records will ever be identical. They will, however, have certain identifiable statistical properties. In a random sea where the variety of wave forms is infinite, characterization by wave form is contrary to the inherent feature of the process. Characteristics that are common to all possible samples of that sea state are required.

A-2.5 Autocorrelation

Consider an observer who records a water surface elevation at time $t = t_1$ at a fixed point. What can he say about the water surface elevation at time $t = t_1 + \Delta t$ where $\Delta t = 0.1$ sec, 1.0 sec, 10 sec, or 100 sec? If the observer is watching a sinusoidal wave train he can say quite a lot about the times $t_1 + \Delta t$ but if the process is random he cannot be sure of the future. The best he can do is to give some estimate of the expected value of the sea surface elevation. (The expected value is defined as the average from an infinite number of observations.) The problem is to correlate the water surface elevation at time t with its value at time $t + \Delta t$. The correlation function

$$R(\tau) = \lim_{T_1 \to \infty} \frac{1}{T_1} \int_0^{T_1} \eta(t)\eta(t + \tau)\, dt$$

predicts the expected (average) value of the product of two values of the water surface which are separated in time at the same place by τ. To confirm that $\eta(t)$ and $\eta(t + \tau)$ are taken from the same record, the expression for $R(\tau)$ is called the *autocorrelation function* of $\eta(t)$. (Occasionally the cross correlation between two different signals may be required.)

Some properties of the autocorrelation as defined by $R(\tau)$ are developed:

1. The autocorrelation function is even, i.e., $R(\tau) = R(-\tau)$, or

$$\lim_{T_1 \to \infty} \frac{1}{T_1} \int_0^{T_1} \eta(t)\eta(t + \tau)\, dt = \lim_{T_1 \to \infty} \frac{1}{T_1} \int_0^{T_1} \eta(t)\eta(t - \tau)\, dt$$

2. The value of the autocorrelation at $\tau = 0$ is the mean square water surface fluctuation since

$$R(\tau = 0) = \lim_{T_1 \to \infty} \frac{1}{T_1} \int_0^{T_1} \eta^2(t)\, dt$$

3. The value of the autocorrelation as $\tau \to \infty$, if the observed phenomenon contains no periodic or drifting components, is zero. The demonstration of this property is beyond the scope of this book. It can be noted, however, that if a process is random the correlation between an observation at time t and an observation taken at time τ later would tend to become infinitesimally small as the time τ became large.

A-2.6 Autocorrelation to Energy (or Variance) Density Spectrum

A-2.6.1 The concept of a spectrum (amplitude and phase) in the frequency domain can be introduced. The idea of a density spectrum is similar to the idea of a probability density. The energy density spectrum really expresses the rate of change of variance or derivative of the signal as a function of frequency. However, the square of the amplitude is used instead of the amplitude. The density spectrum corresponding to a simple sinusoidal wave (Fig. A-2) would appear as a spike of infinite height at the frequency $(2\pi)/T$ since the variation of the square amplitude as a function of frequency is infinity. However, the spike has by definition a finite area:

$$\int_{-\infty}^{\infty} S(k)\, dk = \tfrac{1}{2}A^2$$

Such a function is referred to as a *Dirac delta function.* A periodic wave of arbitrary shape which has the amplitude spectrum of Fig. A-3 will show a comb of Dirac delta functions, each one of infinite height at the frequencies $k_1, k_2, k_3, \ldots$ but having finite areas equal to $A_1^2/2$, $A_2^2/2$, $A_3^2/2$, etc.

A-2.6.2 A random function which contains no periodic components will have no infinite density spikes in the frequency domain. Consider the Fourier transform of the autocorrelation function

$$\Phi(k) = \frac{1}{2\pi} \int_{-\infty}^{\infty} R(\tau) e^{-ik\tau}\, d\tau$$

Then by definition,

$$R(\tau) = \int_{-\infty}^{\infty} \Phi(k) e^{ik\tau}\, dk$$

When $\tau = 0$,

$$R(0) = \int_{-\infty}^{\infty} \Phi(k)\, dk = \lim_{T_1 \to \infty} \frac{1}{T} \int_0^{T_1} \eta(t)\eta(t)\, dt = \overline{\eta^2(t)}$$

It is seen that the total area under the curve $\Phi(k)$ is the mean square value of $\eta(t)$. The area of any element $\Phi(k)\, dk$ represents the mean square contribution of the variance of $\eta(t)$ in the interval $\pm\frac{1}{2}\delta k$. $\Phi(k)$ is the *variance density frequency spectrum* of $\eta(t)$. It is often miscalled the *energy spectrum* or the *power spectrum* or simply the spectrum of the process $\eta(t)$. It can be shown that the variance spectrum may be called the energy spectrum for deep water linear waves. It is never a "power" spectrum as used in the above definition, but it is always a variance spectrum for any water surface record—linear or nonlinear.

The reason for the term energy spectrum follows from the definition,

$$\int_{-\infty}^{\infty} \Phi(k)\, dk = \overline{\eta^2}(t)$$

If $\eta(t)$ is written as,

$$\eta(t) = \sum_{n=1}^{\infty} A_n \cos(knt + \varepsilon_n)$$

then upon squaring and averaging the right-hand side of the above equation

$$\overline{\eta^2} = \frac{1}{2} \sum A_n^2$$

It is now recalled that the average energy per unit surface area can be written as $E_{\text{av}} = \frac{1}{8}\rho g H^2$, where H is the wave height ($2A_n$). A wave of height $2A_n$ and frequency k_n in the confused sea contributes $\frac{1}{2}\rho g A_n^2$ to the energy of that sea. It is seen from the above equations that $\rho g \int \Phi(k)\, dk$ represents the energy in the sea (for linear waves). The description of a sea in terms of its variance spectrum $\Phi(k)\, dk$ is a purely statistical definition. If physical meaning such as "energy spectrum" is used, it should be born in mind that the "variance spectrum" and "energy spectrum" are only equivalent (after a factor of ρg is inserted) for deep water small amplitude wave motions.

All the operations described above are generally performed with high speed digital computers using fast fourier transform (FFT) techniques. Detailed discussions of such procedures are beyond the scope of this book.

A-2.6.3 Advantage can be taken of the even property of $R(\tau)$ to reduce the Fourier transform to the cosine transform.

$$\Phi(k) = \frac{1}{2\pi} \int_{-\infty}^{\infty} R(\tau) \cos k\tau\, d\tau$$

and from the symmetry $\Phi(k) = \Phi(-k)$ (by inspection) an alternative definition for the spectrum is

$$S(k) = \frac{1}{\pi} \int_{0}^{\infty} R(\tau) \cos k\tau\, d\tau$$

If one uses the frequency spectrum in cycles/sec (Hz), $f = (k/2\pi)$, the relationship is simply:

$$S(f) = 2 \int_{0}^{\infty} R(\tau) \cos 2\pi f\tau\, d\tau$$

One advantage of this definition is that the factor $1/(2\pi)$ in the Fourier transform is not required. It is also pointed out that in order to transform a spectrum $S(k)$ to the equivalent $S(f)$ it must be recalled that this is a density spectrum and the identity $S(f)\, df \equiv S(k)\, dk$ must be maintained so that $S(k) = S(f)/2\pi$. Some authors have also advocated the use of "period" spectrum defined by

$$S(T)\, dT \equiv S(f)\, df$$

where $f = 1/T$. This has some advantage in giving a "feel" for the effects of wave period but in fact there is no reason why the same "feel" cannot be developed for frequency (f or $k = 2\pi f$) once familiarity is achieved.

The spectrum and the autocorrelation function have many desirable properties when used to describe random processes such as sea states. They (particularly the spectrum) have stable statistical properties in that sample records taken from the same sea state yield closely reproducible spectra. One further obvious usefulness is the condensation of say a 30-min record of water surface into 50–100 points in a frequency domain. Attempts have been made to fit empirical curves to wave spectra.

A typical spectrum of sea-surface phenomena is shown in Fig. A-5. This figure illustrates the complete range of the wave spectrum from infratidal effects to capillary waves.

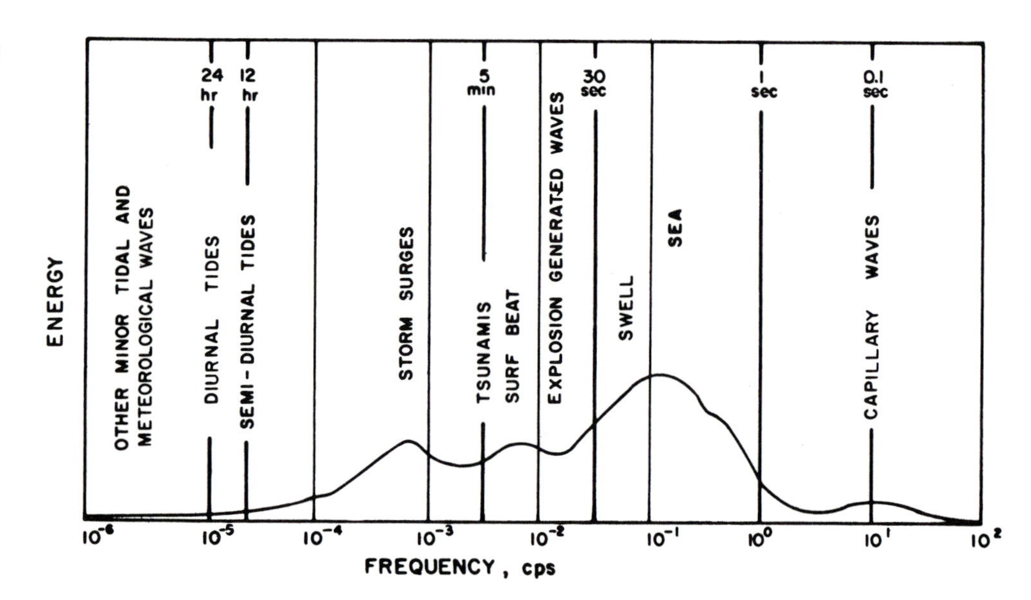

Figure A-5

Energy spectrum at sea.

A-3 Probability for Wave Motions

A-3.1 The Concepts of Probability Distributions and Probability Density

A probability distribution is normally defined when the proportion of values of the considered variable which are less than a particular value are plotted against that value. The most familiar shape for such a distribution is the typical "S" shape, as is shown in Fig. A-6.

The derivative of the probability distribution is called the probability density. The area under the probability density which lies between two values a, b defines the probability that the result of the event being observed lies between a and b. The total probability of the event having all possible outcomes must be unity.

Suppose the water surface elevation of the sea is being considered. Then

$$\int_{-\infty}^{\infty} p(\eta)\, d\eta = 1$$

where $p(\eta)$ is the probability density of the water surface elevation η. The probability distribution of η, say $P(\eta)$, would be (see Fig. A-6)

$$P(\eta) = \int_{-\infty}^{\eta} p(\eta)\, d\eta$$

A-3.2 The Probability Density of the Water Surface Ordinates

There is no theoretical justification for the form of the probability density of a sea state. It is usually assumed to be a Gaussian (normal) distribution with a zero mean value (measured from the still water level), i.e.,

$$p(\eta)\, d\eta = \frac{1}{(2\pi\overline{\eta^2})^{1/2}} \exp\left(\frac{-\eta^2}{2\overline{\eta^2}}\right) d\eta$$

where $\overline{\eta^2}$ is the mean square surface fluctuation.

The form of this equation may be varied. If η is measured in units of $(\overline{\eta^2})^{1/2}$ $(=k_\eta)$ then the above equation becomes

$$p\left(\frac{\eta}{k_\eta}\right) d\left(\frac{\eta}{k_\eta}\right) = \frac{1}{(2\pi)^{1/2}} \exp\left(-\frac{1}{2}\left(\frac{\eta}{k_\eta}\right)^2\right) d\left(\frac{\eta}{k_\eta}\right)$$

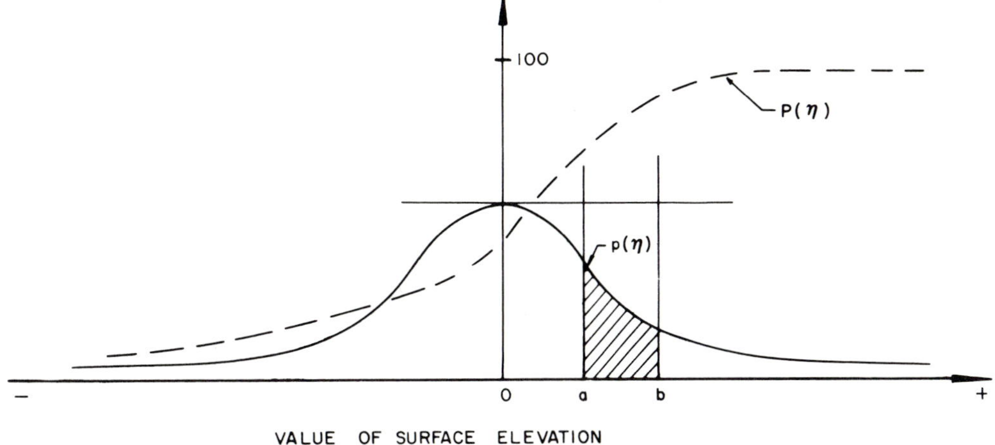

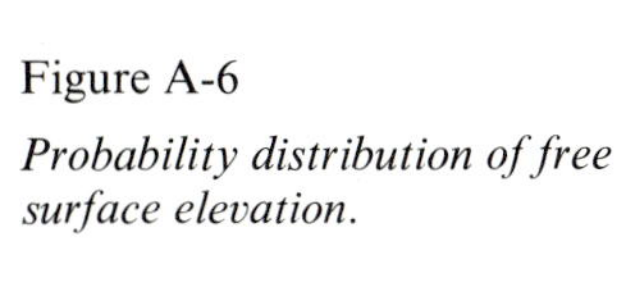
Figure A-6

Probability distribution of free surface elevation.

It is noted here that $\overline{\eta^2} = k_\eta^2$ is the mean square surface fluctuation and is equal to the area under the spectrum. Some authors use different definitions for the total area under the spectrum involving factors of 2, 4, or sometimes 8 or 16 for various reasons. Mathematically speaking, there is a preference for retaining

$$\int_0^\infty S(f)\, df = \overline{\eta^2} = k_\eta^2$$

and this definition is normal for spectral applications in most other fields. In other words, the total area under the spectrum is the variance of the process being studied.

A-3.3 Probability for Wave Heights

A one-dimensional wave train is assumed to be described by the function

$$\eta(t) = \sum A_n \cos (k_n t + \varepsilon_n)$$

where the range of k_n is distributed over the frequencies contained in the spectrum, the ε_n's are arbitrary phase positions and the A_n's are governed by the spectrum as a function of k_n.

If the spectrum is spread over a narrow range of frequencies such that the midfrequency is k_m and the actual values of k_n vary from this only by a small amount, the above equation can be written:

$$\begin{aligned}\eta(t) &= \sum A_n \cos (k_n t - k_m t + \varepsilon_n + k_m t) \\ &= A_c \cos k_m t + A_s \sin k_m t\end{aligned}$$

where

$$\begin{aligned}A_c &= \sum A_n \cos (k_n t - k_m t + \varepsilon_n) \\ A_s &= -\sum A_n \sin (k_n t - k_m t + \varepsilon_n)\end{aligned}$$

In the equation for $\eta(t)$ the assumption of a narrow range of k_n means that A_c and A_s vary very slowly with time.

Now let $R = (A_c^2 + A_s^2)^{1/2}$. By the central limit theorem of probability the values of A_c and A_s given above will be normally distributed so long as the sums are taken over a sufficiently large number of terms and

$$\overline{A_c^2} = \overline{A_s^2} = \overline{\eta^2} = \int_0^\infty S(f)\, df$$

Therefore, the probability that A_c, A_s lies within the element $dA_c\, dA_s$ is given by the following probability density function

$$p(A_c, A_s)\, dA_c\, dA_s = \frac{1}{2\pi\overline{\eta^2}} \exp\left[-\frac{A_c^2 + A_s^2}{2\overline{\eta^2}}\right] dA_c\, dA_s$$

so long as the variables A_c and A_s are statistically independent. This follows from the fact that $\overline{A_c A_s} = 0$. Now, put

$$A_c = R\cos\alpha \qquad A_s = -R\sin\alpha$$

$p(A_c,A_s)$ can be transformed by means of the identity $p(A_c,A_s)\,dA_c\,dA_s \equiv p(R,\alpha)R\,dR\,d\alpha$ to

$$p(R,\alpha)R\,dR\,d\alpha = \frac{1}{2\pi}\frac{R}{\overline{\eta^2}}e^{-R^2/2\overline{\eta^2}}\,d\alpha\,dR$$

The variables can be separated to give

$$\begin{aligned} p(R,\alpha)\,dR\,d\alpha &= p(R)\,dR\cdot p(\alpha)\,d\alpha \\ &= \frac{R}{\overline{\eta^2}}e^{-R^2/2\overline{\eta^2}}\,dR\cdot\frac{1}{2\pi}\,d\alpha \end{aligned}$$

so that

$$p(R)\,dR = \frac{R}{\overline{\eta^2}}e^{-R^2/2\overline{\eta^2}}\,dR$$

$$p(\alpha)\;d\alpha = \frac{1}{2\pi}\,d\alpha$$

In fact R can be considered as the wave amplitude $H/2$ and α can be considered as the wave phase position ε_n. The last equation indicates that the phase position of the waves has a uniform distribution between zero and 2π. $p(R)\,dR$ can be further simplified since for a sine wave the mean square value is equal to one-half of the square of the amplitude.

Therefore, if $H = 2R$, $\overline{\eta^2} = \frac{1}{2}\overline{R^2} = \frac{1}{8}\overline{H^2}$ so that

$$p(H)\;dH = \frac{2H}{\overline{H^2}}e^{-H^2/\overline{H^2}}\;dH$$

which is the well known Rayleigh distribution.

In spite of the many apparent assumptions in the derivation of this equation, it has been found to be an extremely good fit for observed wave height distributions in wind generated seas.

A-3.4 Probability for Wave Periods

There is at present no simple development of the probability distribution of wave periods from the spectrum. The probability of wave periods (based on zero crossings or on the crossing of $\eta(t)$ with the still water level) should be related to the autocorrelation function in some manner but the exact form has not been developed. In order to introduce some estimate of the "mean period" based on the variance density spectrum, the concept of spectrum moments is introduced. The moment of the spectrum of order n is given by

$$M_n = \int_0^\infty f^n S(f)\,df$$

This equation is in direct analogy to the moment of a probability density $\int_{-\infty}^{\infty} x^n p(x)\,dx$. The expected (average) value for the time between successive zero crossings of a process having the spectrum $S(f)$ is given by

$$\frac{1}{2}\bar{\bar{T}} = \frac{1}{2}\left(\frac{M_0}{M_2}\right)^{1/2} = \frac{1}{2}\left(\frac{\int_0^\infty S(f)\,df}{\int_0^\infty f^2 S(f)\,df}\right)^{1/2}$$

where $\bar{\bar{T}}$ is defined as the mean apparent wave period (the wave period will be twice the expected time between successive zero crossings). The proof of the above equation can be found in more advanced textbooks and is beyond the scope of the current chapter. In a similar manner the expected period between successive maxima and minima (points of zero gradient, $\partial\eta/\partial t = 0$) is found to be given by the ratio $\frac{1}{2}(M_2/M_4)^{1/2}$, and this is seen to be quite different from $\bar{\bar{T}}/2$ for a general spectrum shape $S(f)$. The two are identical if $S(f)$ is considered as a delta function having only one frequency.

The relationship of $\bar{\bar{T}}$ to the autocorrelation function is best illustrated by recalling a well-known statistical theorem. "The moment of a probability distribution of order n is given by the nth derivative of the characteristic function of the probability density at the origin." The characteristic function of a probability density is its Fourier transform. The Fourier transform of a spectrum is the autocorrelation function (see Section A-2.6) so that M_0, M_2, and M_4 are simply the zero, second, and fourth derivatives of the autocorrelation function at $\tau = 0$.

A-3.5 Probability for Subsurface Velocity and Accelerations

The probability distributions of subsurface velocities and accelerations in the case of small wave motions (linear assumption) are usually assumed to have a normal distribution. In order to define a normal distribution it is only necessary to be able to define the variance. This can be done for velocities and accelerations in terms of the surface variance spectrum and the use of hydrodynamic potential theory.

The water surface is described by the sum of an infinite number of sinusoids having the variance spectral distribution $S(f)$. For the case of any linear random process it can be demonstrated that the effect of a linear operation on the process can be described in terms of the spectrum of the process multiplied by the square of the generating transform. A sea state is described by

$$\eta(t) = \sum A_n \sin(2\pi f_n t + \varepsilon_n)$$

The corresponding subsurface velocities and accelerations are (see Section 16-3.4)

$$u(t) = \sum 2\pi f_n A_n \frac{\cosh m_n(d+z)}{\sinh m_n d} \sin(2\pi f_n t + \varepsilon_n)$$

and

$$\frac{\partial u(t)}{\partial t} = \sum 4\pi^2 f_n^2 A_n \frac{\cosh m_n(d+z)}{\sinh m_n d} \cos(2\pi f_n t + \varepsilon_n)$$

The square of the modulus of the operating transform for the process $u(t)$ is seen to be,

$$\left[2\pi f_n \frac{\cosh m_n(d+z)}{\sinh m_n d}\right]^2$$

Thus,

$$S_u(f)\,df = \left[\frac{2\pi f \cosh m(d+z)}{\sinh md}\right]^2 S_\eta(f)\,df$$

where $S_u(f)$ is the velocity spectrum at elevation z and $S_\eta(f)$ is the surface spectrum.

By definition of the spectrum,

$$\overline{u^2} = \int_0^\infty S_u(f)\,df$$

$$= \int_0^\infty \left[\frac{2\pi f \cosh m(d+z)}{\sinh md}\right]^2 S_\eta(f)\,df$$

It must be recalled in the transformation function that m is a function of f given by the wave equation (see Section 16-3.2)

$$(2\pi f)^2 = mg \tanh md$$

Once $\overline{u^2}$ is determined, the probability density distribution is given by

$$p(u)\,du = \frac{1}{(2\pi\overline{u^2})^{1/2}} \exp\left(-\frac{u^2}{\overline{u^2}}\right) du$$

In a similar manner the probability density of the acceleration and subsurface pressures can be derived.

A-4 Discussion of Nonlinear Problems

A-4.1 Effects of Nonlinearity on Probability Distributions

It is well known and readily observable that high waves tend to have shorter peaked crests and longer flatter troughs than the simple sinusoidal representation would indicate. Since the mean water level must remain the same it would be expected that the probability density of the surface ordinates will not be the same for large sea states as for lower ones. The positive ordinates will be larger but less frequent whereas negative ordinates will be smaller but more frequent. A skewness is introduced into the probability density (see Fig. A-7).

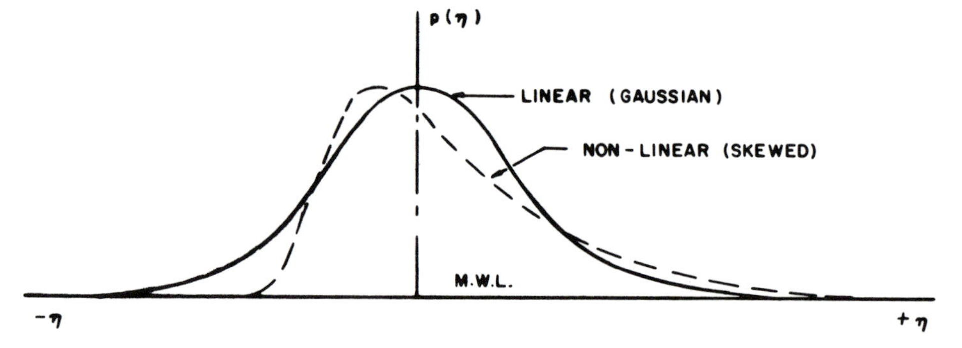

Figure A-7

An illustration of nonlinear effects on the probability distribution of free surface elevation.

The effect of skewness in water surface ordinate probability distributions is reflected to some extent in the probability distributions of wave heights. However this effect is not very marked and does not appear to be very important. It must be recalled that as the crests become more peaked the troughs become flatter. As far as wave heights are concerned these effects are compensatory and wave height probabilities for high seas are still very close to the Rayleigh distribution.

The effects on wave period distributions of large seas has not received much study. In random seas the concept of "wave period" does not really exist. The only thing which may be mentioned is that the times between successive up–down crossings of the mean water level will be shorter than the times between successive down–up crossings. The estimates of average time between successive zero crossings and successive maxima and minima as presented in the section on wave period probability will not be very reliable.

A-4.2 Effects of Nonlinearities on Spectra and Spectral Operations

Nonlinear problems can appear in wave spectrum studies in many forms. Clearly the mechanical process of harmonic analysis as described in Section A-2 is not affected by nonlinear problems. It is the interpretation of the resultant spectrum which is difficult. The high-frequency part of the spectrum is an aggregate of the small ripples mixed with the sea and the harmonics of some of the larger low-frequency wave components. It would be preferable to separate the harmonics since they travel at the phase speed of the "fundamental" wave, whereas the shorter waves travel much slower.

The spectrum destroys all details of phase position and also all differences between "crests" and "troughs." The normal wave spectrum is obviously an insufficient descriptor of a high (nonlinear) sea state.

One of the major advantages of the spectral description of a sea state became apparent in the section on subsurface velocities and accelerations. The velocity, acceleration, pressure, etc. spectra are easily determined from the water surface spectrum. The transfer functions for nonlinear sea states will not be so simple. One other form of nonlinearity arises in operations on the spectrum. For example, the prediction of wave force spectra from sea surface spectra when the function is of the form $f = au^2 + b(\partial u/\partial t)$ is one involving a nonlinear operation on the water surface spectrum. These problems are not yet solved satisfactorily.

PROBLEMS

A.1 Sketch the autocorrelation functions for a sine wave and a random noise. Suggest a method for detecting a sinusoidal component in a random process.

A.2 Sketch the variance density spectrum of a periodic wave. Why is the description of a periodic motion in density spectrum form not very useful?

A.3 If $f(x) = x^2$ and $p(x) = [1/(2\pi)^{1/2}] \exp -(\frac{1}{2}x^2)$, determine $p[f(x)]$ and sketch.

A.4 Show that the autocorrelation function of any stationary random variable is even.

A.5 In the determination of the spectrum of a random process, the autocorrelation is first computed followed by a Fourier transformation in the form of a cosine transformation. What would you expect to find if a sine transform on the autocorrelation was performed? Why?

A.6 A random variable x has the exponential probability density

$$p(x) = a \exp(-b|x|)$$

where a and b are constants. Determine the relationship between a and b and the probability distribution function $P(x)$. Sketch $p(x)$ and $P(x)$.

A.7 On the assumption of linear wave motion, derive the function for the variance of the subsurface bottom pressure in water depth d for a sea state given by

$$\eta(t) = \sum A_n \cos(\pi f_n t + \varepsilon_n)$$

A.8 The probability density distribution of wave heights in a confused sea is given by

$$p(H)\,dH = \frac{H}{2\overline{\overline{H^2}}} \exp\left(\frac{-H^2}{\overline{H^2}}\right) dH$$

Establish the relationships between the most probable wave height, the average wave height, the significant wave height (the mean of the highest one-third of the wave), and the highest wave (probability 0.01).

REFERENCES FOR APPENDIX A

Blackman, R. B., and Tukey, J. W., *The Measurement of Power Spectra*. Dover Publications, New York, 1958.

Cosley, J. W., and Tukey, J. W., An algorithm for the machine calculation of complex Fourier series. *Math. Computation*, Vol. *19*: 90, 297, 1965.

Jenkins, G. M., and Watts, G. W., *Spectral Analysis and Its Applications*. Holden-Day, San Francisco, Calif., 1969.

Kinsman, B., *Wind Waves*. Prentice-Hall, Inc., Englewood-Cliffs, New Jersey, 1965.

Lee, Y. W., *Statistical Theory of Communication*, Wiley, 1961.

National Academy of Sciences, Ocean wave spectra. Proceedings of a Conference, Easten, Maryland, May 1–4, Prentice-Hall, Englewood-Cliffs, New Jersey, 1961.

Appendix B

Similitude and Scale Model Technology

B-1 Introduction to Scale Model Technology

B-1.1 Definition and Advantages of Scale Models

Scale model technology is based on similitude or similarity between two phenomena at different scales. Similitude is a science which is aimed at establishing general functional relationships between various parameters based on dimensional analysis. Scale model technology is an engineering art in which a compromise is made with the law of similitude, to build a practical tool with which engineering problems can be solved.

If problems are too complicated to be analyzed by application of theoretical methods, the answer is scale model technology. The advantage of scale model vs theory is threefold:

1. Complex boundary conditions cannot be analyzed by analytical means. Numerical solutions and high-speed computers extend the power of analytical methods used. But scale models are the best analog computer.
2. Nonlinear effects are still the greatest source of mathematical difficulty. It is true that many problems can often be linearized, but the great advantage of scale model study is the reproduction of not only the linear forces in similitude, but also the nonlinear convective inertial force $\rho\nabla(V^2/2)$. The convective rotational force $\rho\mathbf{V} \times \mathbf{curl}\,\mathbf{V}$ may not always be reproduced in similitude, as the distribution of rotationality $\mathbf{curl}\,\mathbf{V}$ is often related to the distribution of friction forces, which cannot be reproduced in similitude exactly. This scale effect is often negligible, as will be seen in the following.
3. Short fully turbulent flow are statistically in similitude.

In general, a scale model does not reproduce all the aspects of the phenomenon under investigation, but a few effects which are of interest in the research. A knowledge and understanding of the physical laws which govern the effects studied permits us to determine the relative influence of "scale effects."

B-1.2 Economic Considerations

The scale is chosen as a compromise between economics on the one hand and the technical requirements for similitude on the other hand. From the economical viewpoint, the cost of scale model experiments increases approximately as the cube of the scale λ^3, since the area of the model increases as λ^2, and the theoretical duration of the test increases as $\lambda^{1/2}$, but the actual duration increases faster. Of course, this rule of thumb has many exceptions and is not valid if the scale of the model is too small. (In the case of a very small model, the cost of the operation actually increases as the difficulty of carrying out reliable measurements increases.) The "best" scale is not the largest possible one. As a matter of fact, the scales of models must primarily be dictated by economical considerations. Actually, the decisive elements rarely present themselves in a precise manner. It usually is almost impossible to evaluate, even roughly, the "economic expectation" of a model and particularly the variation in this economy with respect to the accuracy of the tests. Hence, the determination of the scales of optimum reduction will necessarily be an art rather than a science, but the value of the results obtained will nevertheless primarily depend upon the overall knowledge available to the designer. Among other things, it is imperative that he be acquainted with the laws of similitude because not only will they enable him to estimate the degree of precision that he may expect from the model, but, above all, they will render it possible to have a sound concept of the latter and to achieve the optimum precision compatible with a given sum available for research.

The first stage of the study of similitude is not concerned with a detailed discussion of the equations, but is essentially an examination of the problem as a whole and of its physical character. For example, it is impossible to form a sound concept of a model without *a priori* having an idea of the causes and origin of the phenomenon studied. The overall examination of the phenomena involved enables us, a priori, to eliminate basically erroneous methods of approaching the problem. It renders it possible to submit a broad outline of the boundaries of the rational conditions. In many cases, it also will enable us to determine the principal stages of the study and occasionally, the number and type of the models required. The detailed study of the equations of the phenomena will then define the approximations, scales, distortions, etc., and eventually, the number of models required, as well as the rules of similitude to be applied to each of them. For example, in the case of a conventional study of the protection of a port against wave action, it may happen that the installations as a whole are studied on a scale of 1/150, some particularly important structures (such as those at the entrance) on a scale of 1/75, and the stability of the individual structures on a scale of 1/40.

In summary, the problem of similitude is far from simple, and there are no standard solutions. The engineer has at his disposal a number of complex means which he must be able to handle with skill. The better he is acquainted with the tools, the more effective and economical the solution will be. From the technical viewpoint, the problem consists of examining when scale effects become important. This leads us to consider the forces which always need to be in similitude. Each separate case needs to be examined in detail. However, some general rules can be presented. For all practical purposes, a scale model must fulfill the following conditions:

1. It must be exact; i.e., it must reproduce with exactness the natural phenomenon under study.
2. It must be consistent; i.e., it must always give the same results under the same conditions.
3. It must be sensitive, or more exactly, its sensitivity has to be imposed by the needs of the reproduction of the phenomenon under investigation.
4. It must be economical, of reasonable size, and completed within a reasonable time interval.

B-1.3 General Considerations on the Concept of Similitude and Scale Model Technology

B-1.3.1 The rules of similitude can be obtained by three different approaches: dimensional analysis, inspectional analysis, or the most general method, which consists of deducing the conditions of similitude from the Navier–Stokes equation and the equations of elasticity. In textbooks on hydraulics, similitude is usually presented as a natural consequence of dimensional analysis. Then the similitudes of Froude, Reynolds, Mach, Weber, Cauchy, etc. are presented. But no scale model has ever been built according to an equality of Weber number, or even of Reynolds number. Practically, a Reynolds similitude does not exist in scale model technology; but the similitude of head loss, a function of the Reynolds number, is sometimes adjusted.

Because of the inherent inadequacies of dimensional analysis, the engineer will often rather deduce the laws of similitude by "inspectional analysis." Knowledge and understanding of the phenomena under study is necessary for deducing the rules of similitude, for neglecting phenomena of secondary importance, and for deciding the relative importance of scale effects. Inspectional analysis is also required for the interpretation of scale model results. The problem may need to be mathematically formulated in a differential form, but does not need to be integrated. The law of similitude can be deduced from the law of motion under a differential form.

B-1.3.2 The dynamics of a system obeys the general law

$$I + G + P + F + E + C = 0$$

where I is the inertial force, G the gravity force, P the pressure force, F the friction force, E the elastic force, and C the capillary force. These forces could be expressed vectorially, or on projection axis, in differential form, or in an integrated form. On the scale model, one has:

$$I' + G' + P' + F' + E' + C' = 0$$

and for similitude

$$I/I' = G/G' = \cdots = C/C'$$

Once all these factors are mathematically expressed, the conditions of similitude are obtained. It is then realized that complete similitude is impossible. Hence, it is important to know what can be neglected and what are the scale effects. This is the inspectional analysis method.

B-1.3.3 It is then important to distinguish between conditions of similitude and criteria of similitude. The *conditions of similitude* are an ensemble of formulas deduced from the physical laws governing the phenomena under investigation; e.g., the similitude condition governing sediment transport is obtained by an analysis of the mechanics of sediment transport. They have an absolute definition which cannot be changed unless an improvement in the knowledge of the physical law is obtained. They are not chosen by the experimenter, but are imposed on him. Unfortunately, it is known that in the field of sediment transport, many phenomena still remain to be analyzed or clarified. Thus, the conditions of similitude are not as well defined as they should be. A choice of what is important will have to be based again on the knowledge of these laws obtained by "inspectional analysis."

In performing a model study, an experimenter must specify certain criteria, such as model wave conditions and fineness of model bottom features. The *criteria of similitude* is a free choice of the experimenter to a very large extent. For example, sea states vary from day to day and from hour to hour. The experimenter will choose, for the sake of simplicity as well as practical necessity, a characteristic wave condition and will only be able to reproduce simplified storm and swell conditions on the scale model. He will choose the wave direction and the wave amplitude and the duration guided by his knowledge of natural conditions. In particular, the wave generator will generate waves at an angle which corresponds to the dominant direction of storm wave energy. Even though he is guided by his knowledge of wave statistics, his final choice will be determined by a trial and error method which permits him to reproduce, for example, a wave agitation pattern or the same bottom evolution as observed in the prototype. This faithful reproduction of bottom evolution determines the choice of the wave characteristics, rather than the strict conditions of similitude of wave motion.

Other criteria of similitude will be to what extent he wants to reproduce the fineness of the bottom topography, a typical tide cycle, a typical flood wave, the currents and their variations with time, and so on. In summary, the criteria of similitude are specified by the experimenter as reasonable approximations for simplification of model operation.

B-2 Short Model vs Long Model

B-2.1 Froude Similitude: Nondissipative Short Models

B-2.1.1 There are two kinds of scale models for which the law of similitude are deduced: short models and long models. The short models are considered either as non-dissipative or as fully turbulent. The nondissipative short models are the ones where the flow pattern is essentially governed by inertia force and gravity only in the case of free surface flow, or inertia forces and pressure gradient only in the case of flow under pressure.

The ratio of inertial force (dimensionally equal to ρV^2), to gravity or pressure force (dimensionally equal to $\rho g L$) is a Froude number $V^2/(gL)$, which is to be the same on a scale model as on a prototype. The Froude similitude stems from the fact that the gravity acceleration is the same on the model as in the prototype. An example of such a scale model is the flow over a spillway, when boundary layer effects are negligible.

The inertial force and the gravity and pressure forces are always present in water wave problems, and the Froude similitude applies when viscous effects are negligible.

B-2.1.2 From an equality of Froude numbers,

$$\left.\frac{V^2}{gL}\right|_p = \left.\frac{V^2}{gL}\right|_m$$

(p refers to prototype, m to model) one deduces

$$\left(\frac{V_m}{V_p}\right)^2 = \frac{L_m}{L_p} = \lambda$$

where λ is the geometric scale. Thus the ratio of scale velocities is $\lambda^{1/2}$. The ratio for time scale is also $\lambda^{1/2}$, since

$$\frac{Tm}{Tp} = \frac{L_m/V_m}{L_p/V_p} = \frac{\lambda}{\lambda^{1/2}}$$

The ratio of discharges is like the ratio of areas λ^2 times ratio of velocities $\lambda^{1/2}$; i.e., $\lambda^{5/2}$. The ratio of powers is discharge times height, i.e., $\lambda^{7/2}$. The ratio of angular velocities or frequencies $1/T$ is $\lambda^{-1/2}$, etc. Since the ratio of ρg is unity, the ratio of pressure is λ, and the ratio or scale of forces is (area $\times$ pressure) λ^3. The ratio of powers defined by force times velocity is $\lambda^{7/2}$. The ratio of shearing stress τ is also λ like the pressure.

B-2.2 Generalized Froude Similitude: Fully Turbulent Short Models

B-2.2.1 Let us now consider the viscous forces. It is known that the ratio of inertial force to viscous force is dimensionally represented by a Reynolds number, VL/ν, where ν is the kinematic viscosity. An equality of Reynolds number and Froude number is possible only at scale unity. Hence, Froude similitude is possible in two cases only: (1) where viscous forces are negligible, as in a gravity wave prior to breaking or in a flow over a weir; or (2) where the flow is very turbulent and the flow pattern to be reproduced on scale model is short, as in a hydraulic jump or a breaking wave. Indeed, in the latter case, the dissipation of energy is mostly due to turbulent fluctuations and is not due to laminar viscous effect. While these viscous effects are linearly related to the velocity, the turbulent fluctuations are quadratic, i.e., proportional to the square of the average velocity V^2, as are the inertial forces. Thus, the ratio of dissipative forces to gravity force in a very turbulent flow is also a Froude number. This situation allows us to use the "generalized Froude similitude." Of course, the depth at which the air bubble penetrates in a breaker will be relatively larger in the field than in the model, as the size of the bubble (determined by capillary effects) is approximately the same in any air entrainment phenomenon. But the dissipative forces are in similitude and proportional to V^2, and the total amount of energy dissipated is in similitude ($\lambda^{7/2}$), even though the fine structure of the flow could be different.

B-2.2.2 This is evidenced when one considers, for example, the shearing stress (see Section 8-3.3):

$$\tau = \rho\overline{u'v'} + \mu\frac{\partial\bar{u}}{\partial y} = \rho l^2\left(\frac{\partial\bar{u}}{\partial y}\right)^2 + \mu\frac{\partial\bar{u}}{\partial y}$$

where $\bar{u}$ is an average velocity, μ the viscosity, ρ the density, l the Prandtl mixing length, and u' and v' the turbulent velocity components. In the case where $\tau_m = \lambda\tau_p$ and

$$\mu\frac{\partial\bar{u}}{\partial y} \ll \rho l^2\left(\frac{\partial\bar{u}}{\partial y}\right)^2$$

it can easily be seen that the shearing forces τ will be in similitude provided the mixing lengths are such that $l' = \lambda l$ in accordance with the similitude of Froude. This condition is fulfilled in the case of fully turbulent motion occurring over a short distance and presenting a large velocity gradient.

B-2.2.3 An example will illustrate these considerations. The rate of energy dissipated by a hydraulic jump or a moving bore, dE/dt, depends only upon the discharge Q and the depth of water before and after the jump (h_1 and h_2). One has seen (Section 14-5.1.2) that the rate of energy dissipation by an hydraulic jump is:

$$\frac{dE}{dt} = \rho g Q\frac{(h_1 - h_2)^3}{4h_1h_2}$$

independently of the scale. Therefore, the ratio

$$\frac{\left.\frac{dE}{dt}\right|_m}{\left.\frac{dE}{dt}\right|_p} = \lambda^{5/2}\frac{\lambda^3}{\lambda^2} = \lambda^{7/2}$$

which is in accordance with the Froude similitude, since the rate of energy dissipation is a power dimension. So a small hydraulic jump, defined by h_1, h_2, and Q, is in similitude with a large jump defined by $h_1 = \lambda h'_1$, $h_2 = \lambda h'_2$, and $Q = \lambda^{5/2}Q'$ at a scale λ. Indeed, the phenomenon is very turbulent. The average flow pattern is the same. The turbulent fluctuation may statistically be different, but the gross effect remains the same. It is the reason why stilling basins, for example, could be investigated by scale models. In the case of breaking waves, it is sufficient to remember that scale effects will be negligible for the phenomena under consideration, provided the height of breaker is larger than say, 5 cm. This statement is based more on experience than on theory.

B-2.2.4 From the most general viewpoint, it is known that the average fluid flow obeys the Reynolds equation

$$\rho\frac{\partial\overline{u_i}}{\partial t} + \rho\overline{u_j}\frac{\partial\overline{u_i}}{\partial x_j} = -\frac{\partial(\bar{p} + \rho g z)}{\partial x_i} + \mu\frac{\partial^2\overline{u_i}}{\partial x_i\,\partial x_j} - \frac{\partial}{\partial x_j}(\rho\overline{u'_i u'_j})$$

where $(\overline{u_i})$ is the average velocity and u'_i is the turbulent fluctuation velocity. It is known that $\bar{u}_i = u_i$ and $u'_i = 0$ in the case of a perfect fluid ($\mu = 0$) and in the case of a laminar flow. If this equation applies for the prototype, a similar equation should also apply for the scale model, and for similitude one should have

$$\frac{\left.\rho\frac{\partial\overline{u_i}}{\partial t}\right|_m}{\left.\rho\frac{\partial\overline{u_i}}{\partial t}\right|_p} = \frac{\left.\rho\overline{u_j}\frac{\partial\overline{u_i}}{\partial x_j}\right|_m}{\left.\rho\overline{u_j}\frac{\partial\overline{u_i}}{\partial x_j}\right|_p} = \cdots = \frac{\left.\frac{\partial(\rho g z)}{\partial x_i}\right|_m}{\left.\frac{\partial(\rho g z)}{\partial x_i}\right|_p} = \cdots = \frac{\rho\overline{u'_i u'_j}|_m}{\rho\overline{u'_i u'_j}|_p}$$

Since $\rho_m = \rho_p$, $g \equiv g$ and $z_m/z_p = \lambda$, one may deduce that the scale for pressure p is also λ. The scale for acceleration $\partial u_i/\partial t$ is unity, as is g. Consequently, the scale for time is $\lambda^{1/2}$, and the scale for velocity is

$$\frac{\overline{u_i}|_m}{\overline{u_i}|_p} = \lambda^{1/2}$$

It is seen also that a similitude of viscous force is impossible unless $\lambda =$ unity, and that a similitude of Reynolds stresses is possible only if $\overline{u_i'u_j'} \approx (\overline{u_i})^2$ which is the case of very turbulent flow on short structures. It is also seen that similitude is possible, as previously stated, in the case of a perfect fluid ($\mu = 0$, $\bar{u}_i = u_i$), where the only forces are inertia (linear or nonlinear) and gravity-pressure. Consequently, a first scale effect is due mostly to the fact that the viscous forces are not in similitude. The value of scale model experiments depends upon their relative importance, i.e., the thickness of the boundary layer and its effect on related phenomena such as separation and wakes. Fortunately, many fluid flows may be successfully considered as that of a perfect fluid ($\mu = 0$) or as very turbulent. The Froude similitude or generalized Froude similitude is then the rule for all short flow patterns.

B-2.2.5 In the case of flow under pressure, which could be subjected to cavitation the similitude parameter $\Delta p/(\rho V^2)$, may intervene requiring to reduce the atmospheric pressure p_a to a scale model partial vacuum.

B-2.3 Similitude of Head Loss; Long Models

B-2.3.1 It has been seen in elementary hydraulics that one has Froude similitude, defined by an equality of Froude numbers, and the Reynolds similitude, defined by an equality of Reynolds numbers. Actually, very few scale models are ever built according to an equality of Reynolds number, as the fluid velocity will have to be larger on the scale model as it is on the prototype (unless one uses different fluids).

For practical purposes, to the old teaching tradition which consists of presenting a parallel between Froude and Reynolds similitude, one will make a parallel between what we can call "short model" and "long model." In a short-scale model, viscous friction is unimportant as compared to gravity and inertia; therefore, it is governed by Froude similitude, as previously seen. Also, energy dissipation may result from a fully turbulent condition, as in the case of a hydraulic jump or a wave breaking on a beach. Boundary layer effects in both cases are unimportant. The dissipative forces are also proportional to the square of velocity like the inertial forces. This is the generalized Froude similitude seen in the previous section.

On the other hand, in a long model, friction has a definite influence on the flow pattern; therefore, in addition, a similitude of head loss is required. This head loss is a function of the Reynolds number, but is not determined by the so-called Reynolds similitude requiring an equality of Reynolds numbers. Therefore, similitude for long models requires, in addition to the Froude similitude, another condition which makes the similitude sometimes impossible as it is shown in the following section.

B-2.3.2 Indeed, the head loss ΔH in a pipeline is

$$\frac{\Delta H}{l} = f\left(\frac{UD}{\nu}\right)\frac{U^2}{2gD}$$

where l is the length of the pipe, D is its diameter, and U is the average velocity. For similitude of inertia and gravity, one must have

$$\left.\frac{U^2}{gD}\right|_m = \left.\frac{U^2}{gD}\right|_p$$

and for geometric similarity:

$$\left.\frac{\Delta H}{l}\right|_m = \left.\frac{\Delta H}{l}\right|_p$$

Therefore,

$$f\left(\left.\frac{UD}{\nu}\right|_m\right) = f\left(\left.\frac{UD}{\nu}\right|_p\right)$$

This cannot be achieved in the case of a smooth prototype pipeline. Indeed, the similitude for Reynolds numbers is

$$\left.\frac{UD}{\nu}\right|_m = \left.\frac{UD}{\nu}\right|_p \times \lambda^{3/2}$$

i.e., $(UD/\nu)|_m < (UD/\nu)|_p$, and $f_m > f_p$. The friction coefficient f_m can only be larger than f_p, as shown in Fig. 13-16 or in Section 13-4.3.2. In the case of a rocky rough gallery (not lined by concrete walls), the relative roughness is large and the boundary layer is turbulent. Then, it is possible to adjust by trial and error the scale model roughness in order to obtain the same friction coefficient. A "Froudian" discharge is then obtained, i.e., $Q_m = Q_p \lambda^{5/2}$, when the relative roughness

$$\left.\frac{k_s}{D}\right|_m = \left.\frac{k_s}{D}\right|_p$$

are theoretically identical, and the Reynolds number $(UD/\nu)|_m$ exceeds a given value, in such a way that the boundary layer is "completely rough," or again $k_s U^*/\nu > 70$ (see Fig. 13-15). But, in any case, the velocity distribution which is also a function of the Reynolds number (see Fig. 13-13) could never be in similitude exactly. It is a scale effect, which is most often acceptable. The same considerations apply to free surface flows.

B-2.3.3 Water wave scale models could, in general, be considered as short models, whether in Froude similitude (wave propagation) or in the generalized Froude similitude (wave breaking). The magnitude of long shore currents and location of rip currents may, to some extent, depend upon friction characteristics of the beaches, in which case the study of this phenomena would have to be considered as belonging to the categories of long models and, therefore, may not be studied on scale models. Nevertheless, most scale model studies have to deal with short coastal structures (like entrance of harbor); thus, the water motion is not too dependent upon the friction coefficients. The main dissipative mechanism is due to wave breaking. If viscous damping is too significant, as is the case when the wave has to travel a long distance in very shallow water, a correction coefficient, determined theoretically, can be applied. Very rarely do we have to be concerned with adjustments of roughness for similitude of energy dissipation, and coastal models can be considered as short models. On the other hand, as has been pointed out previously, a similitude of head loss is imperative for models of rivers and estuaries—these are long models. Despite this adjustment, the vertical velocity distribution which is a function of the Reynolds number can never be in similitude.

B-2.3.4 Ship resistance is due to the waves created by the hull displacement, and to the boundary layer drag and flow separation. The first phenomenon—the ship wave or Kelvin wave—would exist even in a perfect fluid, and it is a function of a Froude number only. From this standpoint, the ship is a short model and this effect could be investigated in similitude. But the boundary layer development along the hull is a long model, and since the hull is smooth, this force is not reproduced in similitude. One can only obtain an approximate effect, introducing a thin wire at the bow of the scale model, in order to cause some turbulence in the scale model boundary layer. Similarly, waves breaking on the large armor blocks covering a breakwater core create very violent and turbulent flow motion around these blocks. Therefore, it is to be expected that the corresponding pressure forces will be in similitude, and studies on stability of breakwater are done according to the law of the generalized Froude similitude (a short model). The study of the permeability of breakwater requires the reproduction of similitude of permeability and energy loss by distorting the size of the material of the core of the breakwater. From this standpoint, these are long models, since the stream tubes within the porous core are long.

B-2.4 On Distortion

B-2.4.1 A model is distorted when one of its scales (say the vertical scale μ) is different from the horizontal scale λ. $D = \mu/\lambda$ is the rate of distortion. A long model can also have two vertical scales, one for cross sections and another one for the bottom or energy slope, For example, the slope of a scale model river could be adjusted to satisfy a similitude condition for head losses.

B-2.4.2 A short model cannot be distorted as the flow pattern on the scale model would be completely different from what it is on the prototype. A long model could be distorted to some extent, without too much scale effect. It is actually a method frequently used to overcome the constraint imposed by the required equality of friction coefficient between model and prototype. Also, it permits reduction of the horizontal scale of the model (i.e., the cost) without sacrificing too much accuracy.

B-2.4.3 In general, scale models used in the study of water waves cannot be distorted. The wave velocity is

$$C = \frac{gT}{2\pi} \tanh \frac{2\pi d}{L}$$

The $\tanh|_m$ must be the same as the $\tanh|_p$ and this is only possible if

$$\left.\frac{d}{L}\right|_m = \left.\frac{d}{L}\right|_p$$

Then both time scale and wave velocity scale are like $\lambda^{1/2}$. Since the ratio of wavelengths L_m/L_p is given by the horizontal scale λ, it is deduced that the ratio of water depth d_m/d_p should also be λ. Then the wave pattern is in similitude. As such, water wave models could be considered as short models despite some viscous damping. There are some exceptions in the case of long waves and in the case of movable bed scale models.

B-2.4.4 In the case of long waves

$$\tanh \frac{2\pi d}{L} \to \frac{2\pi d}{L}$$

and

$$L = T(gd)^{1/2} \qquad C = (gd)^{1/2}$$

The model could then be distorted. The velocity scale is

$$\left(\frac{d_m}{d_p}\right)^{1/2} = \mu^{1/2}$$

and the time scale is

$$\frac{T_m}{T_p} = \frac{L_m/d_m^{1/2}}{L_p/d_p^{1/2}} = \frac{\lambda}{\mu^{1/2}}$$

This time scale also applies to a distorted tidal model and a distorted model of a river.

B-2.4.5 In the case of a movable bed scale model which is used to study sediment transport and bottom evolution of rivers, estuaries, and beaches, distortion is not an engineering trick for reducing the size of the model and the bottom friction, but is the extrapolation of a natural observed phenomenon. The method to obtain a satisfactory scale model is first to obey the law of nature, even though this law may not be fully understood. For example, a small river flowing in its own alluvium can be considered a distorted model of a large river. This means that the ratio depth to width of the small river is comparatively greater than the relative depth of the larger one. The ratios of depths μ and widths λ are approximately related by the *law of Lacey* ($\lambda^2 = \mu^3$), in accordance with the "regime theory." Similarly, a beach in a protected area has a relatively steep slope, whereas a beach in an exposed area tends to have a more gentle slope. The vertical scale being defined by the ratio of incident wave heights, a protected beach can be considered as a distorted scale model of an exposed beach. In both the case of the river and the beach, the choice of distortion becomes a stringent condition to be respected quantitatively. However, the natural law determined by statistical observation of natural phenomena needs to be modified in the case of a river model because scale models generally do not use the same material as the prototypes in order to satisfy other conditions of similitude. In the case of beaches, a natural distortion based on the use of sand both in the scale model and in the prototype is unavoidable. However, the use of sand is not always recommended because it leads to too large a distortion and, subsequently, to large scale effects. Lighter materials (pumice, coals, plastic) are preferred.

B-2.4.6 Similitude of wave refraction is also obtained when $d_m/d_p = \mu$ and $L_m/L_p = \mu$, which is obtained with $T_m/T_p = \mu^{1/2}$. However, the model could still be distorted. Such similitude is used in movable bed scale model of beaches.

It is pertinent to point out that since the only requirement of a movable bed model is a reproduction of bottom evolution, it is not necessary that this be achieved through exact similitude of water motion. Since the model is distorted, similitudes of wave refraction and wave breaking are only aimed at. These conditions are the most susceptible to producing satisfactory long shore current and sediment transport distribution. This is achieved by keeping the ratio of wave lengths and wave heights like the vertical scale μ. Based upon this condition, the following wave characteristics are preserved in the model: (1) wave steepness; (2) refraction pattern and angle of refraction with bottom contours; (3) breaking angle of wave crests with shorelines, if the distortion is not too large; and (4) breaking depth. Also, the scale for long shore current and mass transport velocities is approximately $\mu^{1/2}$. Therefore, the ratio of scales of wave particle velocity to current velocity is approximately unity. The study of movable scale model technology is beyond the scope of this book.

B-3 Scale Effects

B-3.1 Capillary Effects

The influence of capillary effects has to be minimized on scale models of water waves for two reasons:

1. The wave celerity is a function of surface tension.
2. Surface tension introduces a wave damping effect.

Let us consider the first problem and try to determine the suitable minimum wave period for scale experiments. It is known that the wave velocity (first order of approximation) is given by the following formula (Problem 16-17):

$$C^2 = \left(\frac{gL'}{2\pi} + \frac{A}{\rho}\frac{2\pi}{L'}\right)\tanh\frac{2\pi d}{L'}$$

where L' is the wave length ($C = L'T$), T the wave period, g the gravity acceleration, d the water depth, ρ the density, and A the surface tension (74 dyn/cm for air–water interface at 20°C). If L is the wavelength when A is assumed to be equal to 0, then

$$L' = L\left(1 + \frac{\Delta L}{L}\right)$$

After some straightforward calculation, one obtains the scale effect defined by $\Delta L/L$ from the following equations:

In the case where d/L is small,

$$\frac{\Delta L}{L} \approx \frac{1.55 \times 10^{-3}}{dT^2}$$

where d is in centimeters and T is in seconds. If $(2\pi d)/L$ is large,

$$\frac{\Delta L}{L} \approx \frac{T^8}{(0.105)^4}$$

Figure B-1 illustrates these results, where $\Delta L/L$ is given as a function of T and d. It is seen that $\Delta L/L$ is smaller than 1% when $T > 0.35$ sec and $d > 2$ cm. In practice, it can be stated that the lower limit of limiting scale effects is $T > 0.35$ sec.

Figure B-1 *Capillary effects.*

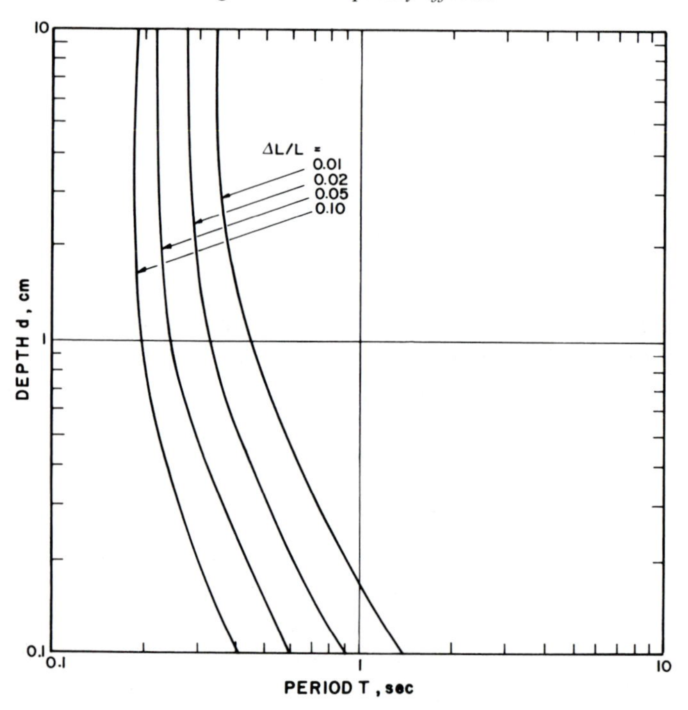

B-3.2 *Wave Damping*

Let us now consider the wave damping phenomenon by viscous effects. The wave damping phenomenon is strongly affected by scale; viscous damping is mostly due to boundary dissipation. The wave height H is a function of the horizontal distance x as

$$H = H_0 e^{-\Delta x}$$

where Δ is the damping coefficient.

In a rectangular tank of width b and depth d

$$\Delta = \Delta_b + \Delta_s$$

Δ_b is due to the solid walls and is

$$\Delta_b = \frac{2m}{b}\left(\frac{\nu}{2k}\right)^{1/2} \frac{mb + \sinh 2md}{2md + \sinh 2md}$$

where m is the wavenumber and equals $2\pi/L$, k is the frequency and equals $(2\pi)/T$, and ν is the kinematic viscosity. Δ_s is due to a contaminated film on the free surface which, *at the maximum*, can be

$$\Delta_s = \frac{2m}{b}\left(\frac{\nu}{2k}\right)^{1/2} \frac{mb \cosh^2 md}{2md + \sinh 2md}$$

From these formulas, it can be verified that the viscous damping is not negligible in many wave tanks where the water depth is smaller than 2 cm. This water depth should be considered as a reasonable lower limit for scale model studies.

B-3.3 *Density Effects*

One important scale effect is due to the fact that in the wave tank one usually uses fresh water instead of sea water. The slight difference in density, which is approximately 3%, changes the wave forces accordingly. Let us consider, for example, the case of a rockfill breakwater. The minimum weight W of the rocks or armor units covering the breakwater is proportional to

$$W \approx \frac{\rho_b H^3}{\left[\dfrac{(\rho_b \cos\alpha - \rho_s)\tan\lambda - \rho_b \sin\alpha}{\rho_s}\right]^3}$$

where α is the slope of the breakwater near the free surface and $\tan\lambda \cong 1$. ρ_b is the density of the armor, and ρ_s the density of sea water; λ is a friction angle for the material.

This formula can be approximated as

$$W \approx \frac{g\rho_b H^3}{\left(\dfrac{\rho_b - \rho_s}{\rho_s}\right)^3 K_\Delta \cot\alpha}$$

as presently used by engineers. K_Δ is a damage coefficient ≈ 3 for rocks. Therefore, for similitude:

$$\frac{W_m}{W_p} = \frac{H_m^3}{H_p^3} \frac{\left[\rho_b \Big/ \left(\dfrac{\rho_b - \rho}{\rho}\right)^3\right]_m}{\left[\rho_b \Big/ \left(\dfrac{\rho_b - \rho_s}{\rho_s}\right)^3\right]_p}$$

where ρ is the fresh water density. It is found that

$$\frac{\rho_{bp}}{\left(\dfrac{\rho_{bp} - \rho_s}{\rho_s}\right)^3} = \frac{\rho_{bm}}{\left(\dfrac{\rho_{bm} - \rho}{\rho}\right)^3}$$

If $\rho_{bm} = \rho_{bp}$, it is seen that the error on W due to the use of fresh water is about 10%.

B-3.4 *Wave Forces on Structures*

The important case of a cylindrical structure is now analyzed. One must distinguish between large piles and small piles, characterized by the value of the ratio of the diameter D to the wavelength $L: D/L$. In the first case (small D/L), the wave force is given by the Morison formula as a function of the horizontal component of velocity u of the incident wave as (see Section 13-5.2.8):

$$dF = \left(\rho C_D D \frac{u^2}{2} + \rho C_M \frac{\pi D^2}{4} \frac{\partial u}{\partial t}\right) dz$$

where C_D is a drag coefficient and C_M is the virtual mass coefficient. C_D and C_M are supposed to be constant parameters which are functions of the Reynolds number, $(uD)/\nu$, roughness of the pile, etc. Actually, C_D and C_M are time dependent functions (but these functions are still not well defined), and the wave velocity field u is not free, but modified by the presence of the pile. The coefficients C_D and C_M are obtained experimentally. The coefficients C_M and C_D must be the same on the model and on the prototype. For a perfect fluid, $C_M = 2$ and $C_D = 0$. In practice, a wake develops on the lee side and on the front side of the pile alternately, hence $C_M \neq 2$ and $C_D > 0$.

The wake effect does not change the value of C_M very much, but it strongly affects the drag coefficient C_D. The dependence of C_D upon the Reynolds number makes similitude possible only if $(uD)/\nu > 2 \times 10^5$ in the case of uniform flow, but actually u is time dependent and varies from positive value to negative value alternately. The effect of roughness remains small, except that it may change the inception of flow separation and hence influence the wake. From this standpoint, the development of the boundary layer around the pile is to be considered as a long model (even though the body itself is short). Therefore, wave forces on small piles cannot be studied accurately on scale models unless the ratio of inertial force to drag force is large. This ratio is represented by the Iversen modulus

$$\frac{\left.\dfrac{\partial u}{\partial t}\right|_{\max} D}{\left. u^2 \right|_{\max}}$$

which is actually an inverse Froude number where the gravity acceleration g is replaced by the particle acceleration $\partial u/\partial t$.

In the case of very large piles (large D/L) the problem is one of wave diffraction. The drag force is small, and in some cases (like large circular or ellipsoid cylinders), the problem can be linearized and treated completely analytically with good accuracy. The solution is given by a potential function. It is evident that such wave motions around very large piles ($D/L > 0.5$) are very well reproduced on scale models even at a small scale. Moreover, complex boundary conditions due to complex structural forms and nonlinear effects, particularly important in the case where the wave breaks on the structure, are also in similitude, whereas they cannot be analyzed by theory.

All these considerations being borne in mind, it is possible to establish the critical value of H/D for which inertial forces

$$(F_M)_{\max} = \rho C_M \frac{\pi D^2}{4} \int_{-d}^{\eta} \left(\frac{\partial u}{\partial t}\right)_{\max} dz$$

equal the drag force

$$(F_D)_{\max} = \rho C_D D \int_{-d}^{\eta} u_{\max}^2 \, dz$$

as a function of the wave height H and the relative depth D/L. u, at a first order of approximation, is equal to:

$$u = k \frac{H}{2} \frac{\cosh m(d + z)}{\sinh md} \cos (kt - mx)$$

The calculations have been performed for two values of the wave height, namely $H \to 0$ and $H = H_b$, such that $H_b/L_b = 0.14 \tanh (2\pi d/L_b)$ (limit wave steepness).

For the sake of simplicity, this work has been carried out by assuming $C_D = 1$ and $C_M = 2$ as the most realistic average values. The results of these calculations are presented in Fig. B-2. On the high side of these two curves, the drag forces dominate and there is no way of obtaining a satisfying scale model investigation except at scale unity. On the low side, the inertial force dominates in such a way that similitude is more valid.

From these considerations, it is seen that in the case of small D/L, the larger the scale, the better. However, large values of D/L can be studied satisfactorily even at a small scale.

This nomograph can also be used as a guide for forms of body other than circular pile. For example, if D is considered as the beam of a ship, it is seen that the study of ship behavior in water waves can be done in similitude.

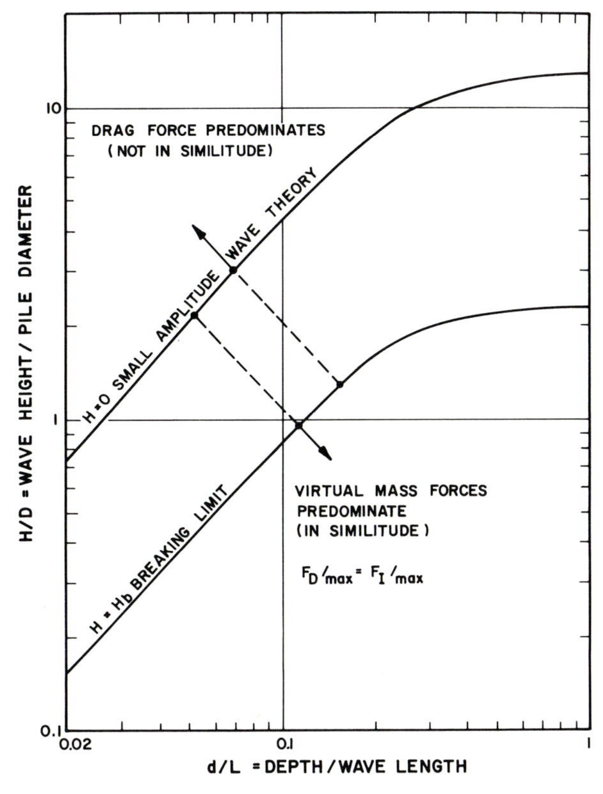

Figure B-2 *Forces on cylinders.*

Another significant criterion used in engineering practice is the "excursion ratio"—the ratio of the horizontal wave particle amplitude to pile diameter, which for the sake of similitude has to remain smaller than unity.

B-3.5 Special Effects

B-3.5.1 The Cauchy similitude, or similitude of elasticity, could be made compatible with the Froude similitude. This similitude is required to study ship mooring, for example, since the elastic forces due to mooring lines have to be in similitude with other forces. (Also, the solid friction of the ship against the fenders has to be made compatible. This is done by making the fenders very slippery, smooth, and oily.)

The Cauchy similitude is obtained either from the Boussinesq equation for elastic material or from the equation of the phenomena to be studied, through inspectional analysis. In general, it is sufficient that the Poisson ratio be the same on the model as in the prototype, and the modulus of elasticity E_m be such that $E_m = E_p \lambda$.

B-3.5.2 Similarly, in studies of stratified fluid (thermal outfall), the prevailing conditions of similitude are imposed by an equality of Richardson number and densimetric Froude number.

The Richardson number is dimensionally the dynamic ratio of the buoyancy to inertia force gradient and is a direct measure of the stability of a density stratified flow.

$$R_i = \frac{-\dfrac{g}{\rho}\dfrac{d\rho}{dz}}{\left(\dfrac{du}{dz}\right)^2}$$

ρ is the density, z a vertical coordinate, and u the horizontal velocity. The densimetric Froude number is

$$F_r = \frac{u^2}{\dfrac{\Delta\rho}{\rho} gd}$$

where d is a vertical distance (depth) and $\Delta\rho$ the variation of density about ambient. Then the ratio of Richardson numbers

$$\frac{R_{im}}{R_{ip}} = \frac{\Delta\rho_m}{\Delta\rho_p}\left(\frac{u_p}{u_m}\right)^2 \lambda = \left(\frac{F_{rp}}{F_{rm}}\right) = 1$$

Thus, preserving the densimetric Froude number also preserves the Richardson number.

One also finds a short model (near the outfall) which cannot be distorted and a long model for the study of the far field which could be distorted. Then the similitude of surface heat loss imposes some scale effects, the treatment of which is judged beyond the scope of the present book.

B-3.6 Other Scale Effects

Let us just mention:

1. The solid friction forces are generally relatively larger on scale model than on prototype (i.e., a concrete cap on rockfill breakwater or a ship against a fender).
2. The structural strength of material is generally not reproduced in similitude. For example, if the armor units covering breakwaters were in complete structural similitude, one should be able to crush them in the hand. Scale models of Dolosses or Tetrapods, scaled at 1/50, falling from a height of, say, 1 m, do not break, whereas their prototype counterparts falling from 50 m certainly do.
3. Air compression effects, such as due to the slamming of a wave against a vertical wall or buoyant tanks subjected to pressure variations cannot be investigated in similitude, unless the scale model is operated in a partial vacuum.
4. Air entrainments such as found in hydraulic jumps or in a breaking wave do not present the same concentration distribution at a small scale as in the prototype.
5. The wave diffraction pattern on a distorted model is not in similitude. Similitude of turbulent fluctuations and sediment transport by suspension does not exist. Only a similar bottom evolution could be attempted after a lengthy calibration of the scale model based on prototype observations.
6. Shock waves due to underwater explosions are not in similitude since the compressibility of water is not scalable. The same reasoning applies to slamming or impact phenomena. The free surface is "harder" on scale models.
7. The wave reflection coefficient of a smooth wall is smaller on a scale model than on the prototype. The opposite tends to apply in the case of wave absorbers made of ripraps (large rocks). This scale effect is corrected by adding a wire mesh on the scale model wave absorber.

B-4 Typical Scales Used in Engineering Practice

1. Breakwater stability/rockfill cofferdam: 1/30 to 1/50.
2. Wind–wave penetration in harbor: 1/100 to 1/150. (The water depth cannot be smaller than 2 cm and the wave period has to be larger than 0.5 sec)
3. Spillways, bottom outlet, water power structures: typically, 1/50 to 1/100.
4. River, estuary—distorted: typically, 1/100 vertical; 1/800 horizontal.
5. Beaches, shoreline processes—distorted: typically, 1/100 vertical; 1/300 horizontal.
6. Ship dynamics problems: typically, 1/100.

PROBLEM

B.1 One wants to reproduce a permeable rockfill cofferdam on a distorted scale model (scales λ, μ). One wants the discharge through the cofferdam to be in similitude with the discharge through a diversion tunnel. The material model has a size $\delta_m = D\delta_p$, where D is a coefficient of distortion. Give the expression which permits the calculation of D as a function of λ, μ, d_p, and u_p. One will assume that the flow through porous medium obeys the general law

$$\frac{\Delta H}{L} = f\left(\frac{U\delta}{\nu}\right)F(\varepsilon)\frac{U^2}{2g\delta}$$

$F(\varepsilon)$ is a function of the void coefficient which will be assumed to be the same on the model as on the prototype.

REFERENCES FOR APPENDIX B

Biesel, F. and Le Méhauté, Note on the similitude of scale models for studying seiche in harbors. *La Houille Blanche*, *3*: 392–407, July 1955.

Birkhoff, G., *Hydrodynamics*. Princeton University Press, Princeton, New Jersey, 1950.

Bridgman, P. W., *Dimensional Analysis*. Yale University Press, New Haven, Conn., 1937.

Langhaar, H. L., *Dimensional Analysis and Theory of Models*. Wiley, New York, 1951.

Le Méhauté, B., On Froude–Cauchy similitude. Santa Barbara, Proceedings of the Coastal Engineering Speciality Conference, American Society of Live Engineers, October 1965, pp. 327–346.

Le Méhauté, B., and Hwang, L-S, Harbor design: Scale model or computer? In *Topics of Ocean Engineering*, Vol. 2, Gulf Publishing Co., Houston, Texas, 1970, pp. 3–24.

Le Méhauté, B., A comparison of coastal and fluvial similitude. Coastal Engineering Conference, Washington, D.C., Sept., 1970.

Ledov, L. I., *Similarity and Dimensional Methods in Mechanics*, Academic Press, New York, 1959.

Van Dorn, W. G., Boundary dissipation of oscillatory waves. *J. Fluid Mech.*, *24*: 769–779, 1966.

Notation

A or ΔA	Cross section or element of cross section Also constant due to surface tension
C	Wave (or phase) velocity
C_D	Drag coefficient
C_M	Inertial coefficient
C_f	Resistance coefficient on a boundary due to shearing stress
C_h	Chezy coefficient
C_x	Drag coefficient
D	Diameter of a pipe
E	Modulus of elasticity; specific energy
F	Force
F'	Force on a body due to added mass
F_e, F_i	External forces, internal forces
H	Total head, sum of kinetic head, piezometric head, and pressure head ($H = V^2/2g + p/\varpi + z$) Also wave height
ΔH	Head loss
I	Specific force
K	Bulk modulus of elasticity Also hydraulic conductivity of a porous medium Also $AC_h(R_H)^{1/2}$: "conveyance" of a channel
L	Wavelength (wave theory)
M	Mass of a body
M'	Added mass
OX, OY	Horizontal axes (usually at the still water level)
OZ	Vertical axis (usually positive upward)
P	Wetted perimeter Probability distribution function
Q (or q)	Discharge
R	Total force on a boundary due to shearing stress Also radius of a cylinder or of a sphere Also the reflection coefficient in wave theory

R (or r) — Radius. Also the universal gas constant

R (or r), θ — Coordinates of a point in a cylindrical system of coordinates

R_H — Hydraulic radius

$\mathrm{R_e}$ — Reynolds number

$R(\tau)$ — Autocorrelation function

S — Wave spectrum. Also bottom slope $= \tan\theta$

S_{xx} — Radiation stress or wave thrust in the x direction

S (or ΔS) — Element of streamline

S_c — Critical bottom slope

S_f — Slope of the energy line

T (or ΔT) — Interval of time or period (wave theory)

$\bar{\bar{T}}$ — Average time interval between the crossing of free surface elevation with the still water level

U — Group velocity. Also average velocity in a pipe ($U = Q/A$). Also velocity of an immersed body in a fluid

U_E — Rate of propagation of energy. Also East–West velocity component

U_s — North–South velocity component

$U_R = \frac{\eta_0}{L}\left(\frac{L}{d}\right)^3$ — Ursell parameter

U_0 — Velocity outside the boundary layer

$\mathbf{V}$ — Velocity vector

$\bar{\mathbf{V}}(\bar{u},\bar{v},\bar{w})$ — Average velocity vector with respect to time and its components in a turbulent flow

$\bar{\bar{\mathbf{V}}}(\bar{\bar{u}},\bar{\bar{v}},\bar{\bar{w}})$ — Average velocity vector with respect to space and its components in a flow through porous medium

$\mathbf{V}'(u',v',w')$ — Velocity vector for the turbulent fluctuations and its components

W — Velocity of propagation of a bore. Also vertical velocity component in a spherical system of coordinates

X, Y, Z — Volume of body force (gravity) along OX, OY, OZ, respectively ($X = 0$, $Y = 0$, $Z = -\rho g z$)

$a = \frac{\partial u}{\partial x}$, $b = \frac{\partial v}{\partial y}$, $c = \frac{\partial w}{\partial z}$ — Coefficients of linear deformation along OX, OY, and OZ, respectively

d — Depth (wave theory)

e — Thickness of a pipe wall

$f = \frac{g}{C_h^2}$ — Friction coefficient

$f = \frac{1}{2}\left(\frac{\partial w}{\partial y} + \frac{\partial v}{\partial z}\right)$, $g = \frac{1}{2}\left(\frac{\partial u}{\partial z} + \frac{\partial w}{\partial x}\right)$, $h = \frac{1}{2}\left(\frac{\partial v}{\partial x} + \frac{\partial u}{\partial y}\right)$ — Coefficients of angular deformation along OX, OY, and OZ, respectively

g — Gravity acceleration

h — Depth (channel). Also water depth. Also coefficient of angular deformation

$h^* = \frac{h}{E}$ — Reduced water depth

h_c — Critical depth

h_n — Normal depth

h_1, h_2 — Conjugate water depths

k — Coefficient of permeability for porous medium. Also coefficient of von Kármán in the theory of turbulence, universal constant $\cong 0.4$. Also $2\pi/T$ in the wave theory and periodic motions where T is the period

$k = \dfrac{C_p}{C_v}$	C_p is the specific head at constant pressure and C_v the specific heat at constant volume
k_s	Characteristic number for roughness size in pipes and wall boundaries
l	Prandtl's mixing length Also width of a channel at the free surface
m	Mass; $2\pi/L$
n (or Δn)	Element perpendicular to an area or to a streamline Also element of an equipotential line
p	Pressure; probability density function
Δp	Pressure difference over a finite interval
p^*	$p + \rho g z$: pressure and gravity force
$\bar{p}$	Average pressure with respect to time in a turbulent flow
$\bar{\bar{p}}$	Average pressure with respect to space in a flow through porous medium
$q^* = g/E(2gE)^{1/2}$	Reduced discharge
r_0	Radius of a pipe
$s = \dfrac{kD}{v}$	Strouhal number
t, T	Time
u	Local velocity in a pipe or in a boundary layer
$u^* = (\tau_0/\rho)^{1/2}$	Shear velocity
u, v, w	Components of the velocity vector **V** along the three coordinate axes OX, OY, and OZ, respectively
v_r	Component of the velocity vector along a radius in a cylindrical system of coordinates
v_θ	Component of the velocity vector perpendicular to a radius in a cylindrical system of coordinates
x, y, z	Coordinates of a point along OX, OY, and OZ, respectively
Γ	Circulation of velocity
Φ	Dissipation function Also angle of latitude in spherical coordinates
Φ_m	Dissipation function due to viscous force
Φ_t	Dissipation function due to turbulent fluctuations
α	Correction factor for the kinetic energy term in a pipe Also ($\tan \alpha$) slope
α, β	Constant parameters in the Gerstner wave theory
δ	Diameter of a particle of a porous medium Also boundary layer thickness (general definition)
δ^*	Displacement thickness for a boundary layer
ε	Coefficient of Boussinesq for shearing stress due to turbulent exchange Also void coefficient for a porous medium
η	Elevation of the free surface around the still water level Also rotation in (x, z)
$\bar{\eta}$	Average elevation of the free surface around still water level or wave set-down, set-up
θ	$\tan \theta$: bottom slope Also momentum thickness for a boundary layer Also longitude in a spherical system of coordinates
λ	Second coefficient of viscosity for gas Also scale, horizontal
μ	Coefficient of viscosity Also scale, vertical
ν	Kinematic coefficient of viscosity ($\nu = \mu/\rho$)
$\xi = \dfrac{1}{2}\left(\dfrac{\partial w}{\partial y} - \dfrac{\partial v}{\partial z}\right)$, $\eta = \dfrac{1}{2}\left(\dfrac{\partial u}{\partial z} - \dfrac{\partial w}{\partial x}\right)$, $\zeta = \dfrac{1}{2}\left(\dfrac{\partial v}{\partial x} - \dfrac{\partial u}{\partial y}\right)$	Coefficients of rotation along OX, OY, and OZ, respectively

ρ	Density
σ	Normal stress
$[\sigma]$	Normal stress for a turbulent flow
τ	Shearing stress Also time interval
$[\tau]$	Shearing stress for a turbulent flow
τ_0	Shearing stress at the wall
ϕ	Potential function: $\mathbf{V} = -\mathbf{grad}\, \phi$ $(u = -\partial\phi/\partial x,\ v = -\partial\phi/\partial y,\ w = -\partial\phi/\partial z)$
$\Phi(k)$	Variance spectrum
ψ	Stream function: $u = \partial\psi/\partial y,\ v = -\partial\psi/\partial x$
ω	Angular rotation
ϖ	Specific weight ($\varpi = \rho g$)
$\dfrac{\partial A}{\partial *}$	Partial derivative (with respect to *)
$\dfrac{dA(x,y,z,t)}{dt}$	Total derivative (with respect to t) $= \partial A/\partial t + u\, \partial A/\partial x + v\, \partial A/\partial y + w\, \partial A/\partial z$
$\mathbf{A} \cdot \mathbf{B}$	Scalar product $= \lvert A\rvert \lvert B\rvert \cos$ (angle between **A** and **B**)
$\mathbf{A} \times \mathbf{B}$	Vector product $= \mathbf{1}\lvert A\rvert \lvert B\rvert \sin$ (angle between **A** and **B**), where **1** is a vector perpendicular to the plane AB
∇	$\mathbf{i}\, \partial/\partial x + \mathbf{j}\, \partial/\partial x + \mathbf{k}\, \partial/\partial z$ where **i**, **j**, **k** are unit vectors
grad A	Gradient of A, i.e., total variation of A with respect to space $\mathbf{grad}\, A = \mathbf{i}\dfrac{\partial A}{\partial x} + \mathbf{j}\dfrac{\partial A}{\partial y} + \mathbf{k}\dfrac{\partial A}{\partial z}$
div **A**	Divergence of **A**: scalar sum of $\partial A_x/\partial x + \partial A_y/\partial y + \partial A_z/\partial z$
curl A	Rotation of A. Vector of components 2ξ, 2η, 2ζ $\mathbf{curl\, A} = 2(\mathbf{i}\xi + \mathbf{j}\eta + \mathbf{k}\zeta)$ $\mathbf{curl\, A} = \begin{vmatrix} \mathbf{i} & \mathbf{j} & \mathbf{k} \\ \dfrac{\partial}{\partial x} & \dfrac{\partial}{\partial y} & \dfrac{\partial}{\partial z} \\ A_x & A_y & A_z \end{vmatrix}$
$\nabla^2 A = \text{div}\, \mathbf{grad}\, A$	Laplacian, scalar sum of $\dfrac{\partial^2 A}{\partial x^2} + \dfrac{\partial^2 A}{\partial y^2} + \dfrac{\partial^2 A}{\partial z^2} = \dfrac{\partial^2 A}{\partial r^2} + \dfrac{1}{r}\dfrac{\partial A}{\partial r} + \dfrac{1}{r^2}\dfrac{\partial^2 A}{\partial \theta^2} + \dfrac{\partial^2 A}{\partial z^2}$ When $\nabla^2 A = 0$, A is a harmonic function.

Answers to Selected Problems

Chapter 1

1.1 Streamlines:

$$y - y_0 = \frac{C}{A + Bt_0}(x - x_0)$$

Paths:

$$x - x_0 = \frac{A}{C}(y - y_0) + \frac{1}{2}\frac{B}{C^2}(y - y_0)^2$$

1.2 At a point (r,θ) within the circle:

$$u = k[R + r \sin (kt + \theta)]$$
$$w = kr \cos (kt + \theta)$$

The streamlines are circles of radius R_s, such that

$$R_s = [R^2 + r^2 + 2rR \sin \theta]^{1/2}$$

is centered at the point where the circle touches the plane ($t_0 = 0$).
When $\theta = 0$ and $k(t - t_0) = \alpha$, the paths are

$$\begin{cases} x - x_0 = R\alpha + r \sin \alpha \\ z - z_0 = R + r \cos \alpha \end{cases}$$

which are the parametric equations of a trochoid.

1.5 Streamlines are

$$\frac{K}{\cos mx} = \sinh m(d + z)$$

where K is a constant of the integration.
Paths: Calculate $(x - x_i)^2$, $(z - z_i)^2$ and add. Solution:

$$\left(\frac{x - x_i}{A}\right)^2 + \left(\frac{z - z_i}{B}\right)^2 = 1 \qquad \text{(ellipse)}$$

where

$$A = \frac{H}{2}\frac{\cosh m(d + z_0)}{\sinh md}$$

$$B = \frac{H}{2}\frac{\sinh m(d + z_0)}{\sinh md}$$

1.7 Moving body in still fluid:

$$F = y^3 - (x + Ut) = 0$$

Boundary condition:

$$-U - u + 3vy^2 = 0$$

Fixed body, fluid moving at velocity U at infinity:

$$y - x^{1/3} = 0$$

$$\frac{u}{v} = 3x^{2/3}$$

1.8

$$F = z^2 - A^2(x + Vt) = 0$$

$$w = -\frac{A^2}{2z}(V - u)$$

1.9

$$F = (x - u_s t)^2 + (y - v_s t)^2 + (z - w_s t)^2 - R^2 = 0$$

$$(u - u_s)(x - u_s t) + (v - v_s)(y - v_s t) + (w - w_s)(z - w_s t) = 0$$

1.10 The motion is steady with respect to the considered system co-ordinates. Then relative streamlines and paths are identical and defined by sinusoids of horizontal axis. The length L ($m = 2\pi/L$), and the amplitudes are

$$\frac{H}{2} e^{mz} \quad \text{in the case defined in Section 1-3.3.}$$

$$\frac{H}{2} \frac{\sinh m(d + z)}{\sinh md} \quad \text{in the case defined in Problem 1.5}$$

The free surface and bottom are streamlines.

Chapter 2

2.1

$$\frac{\partial u}{\partial x} = 0.1/t$$

2.3 $a = 0$, $h = -\zeta = 5/t$; all other coefficients are zero.

2.4

$$a = 0, h = -\zeta = \frac{1}{2}\left[\frac{\alpha y}{\mu} - \frac{e}{2\mu}\left(\alpha - \frac{2\mu V}{e^2}\right)\right]$$

If $\alpha = 0$, $V \neq 0$ the flow is created by the moving plane; if $\alpha \neq 0$, $V = 0$ the flow is due to a gradient of pressure dp/dx.

2.5 Cylindrical:

$$v_r = \frac{\partial \phi}{\partial r}, \; v_\theta = \frac{1}{r}\frac{\partial \phi}{\partial \theta}, \; w = \frac{\partial \phi}{\partial z}$$

Spherical:

$$v_r = \frac{\partial \phi}{\partial r}, \; v_\theta = \frac{1}{r}\frac{\partial \phi}{\partial \theta}, \; v_\phi = \frac{1}{r \sin\theta}\frac{\partial \phi}{\partial \Phi}$$

2.6

$$\frac{\partial v_\theta}{\partial r} + \frac{v_\theta}{r} - \frac{1}{r}\frac{\partial v_r}{\partial_\theta} = 0$$

2.7

$$v_r = 0$$

$$\frac{1}{2}\left(\frac{\partial v_\theta}{\partial r} + \frac{v_\theta}{r}\right) = \frac{1}{R_2^2 - R_1^2}[\omega_2 R_2^2 - \omega_1 R_1^2]$$

For irrotationality: ωR^2 = constant, i.e., $v \times R$ = constant.

Chapter 3

3.2 First case: div $\mathbf{v} = 0$ (incompressible)
Second case: div $\mathbf{v} = A$ (compressible)

3.3

$$\frac{\partial \rho}{\partial t} + \frac{1}{r}\frac{\partial}{\partial r}(\rho r v_r) + \frac{1}{r}\frac{\partial(\rho v_\theta)}{\partial \theta} + \frac{\partial(\rho v_z)}{\partial z} = 0$$

3.4

$$\frac{\partial^2 \phi}{\partial r^2} + \frac{1}{r}\frac{\partial \phi}{\partial r} + \frac{1}{r^2}\frac{\partial^2 \phi}{\partial \theta^2} + \frac{\partial^2 \phi}{\partial z^2} = 0$$

3.5

$$\text{div } \mathbf{v} = -wz$$

3.7

$$\frac{1}{r^2}\frac{\partial}{\partial r}(v_r r^2) + \frac{1}{r \sin\theta}\frac{\partial}{\partial \theta}(v_\theta \sin\theta) + \frac{1}{r \sin\theta}\frac{\partial v_\phi}{\partial \Phi} = 0$$

$$\frac{1}{r^2}\frac{\partial}{\partial r}\left(r^2 \frac{\partial \phi}{\partial r}\right) + \frac{1}{r^2 \sin\theta}\frac{\partial}{\partial \theta}\left(\sin\theta \frac{\partial \phi}{\partial \theta}\right) + \frac{1}{r^2 \sin^2\theta}\frac{\partial^2 \phi}{\partial \Phi^2} = 0$$

3.8 The change of volume between section x and $x + dx$ is

$$\frac{\pi D^2}{4}\frac{\partial u}{\partial x} dx\, dt$$

and is equal to the sum of the following:

1. The change in volume by compression is:

$$\frac{\pi D^2}{4}\frac{\partial p}{\partial t} dt \frac{dx}{K}$$

2. Since the increase in the stress in the pipe wall is equal to

$$\frac{\partial p}{\partial t} dt \frac{D}{eE}$$

the change of volume by pipe diameter variation is

$$\frac{\partial p}{\partial t} dt \frac{\pi D^2}{4}\frac{D}{eE} dx$$

3.9 Take an elementary volume

$$\text{vol} = h(a \sin\Phi\, \partial\theta\, a\, \partial\Phi)$$

The derivative of the volume due to variation of free surface elevation is

$$\frac{\partial \eta}{\partial t}(a^2 \sin \Phi\, \partial\theta\, \partial\Phi)$$

The discharge through two adjacent side elements are:

$$U_E ha \sin \Phi\, \partial\theta \qquad \text{and} \qquad U_S ha\, \partial\Phi$$

The discharge differences between opposite sides are:

$$\frac{\partial}{\partial \theta}(U_S ah\, \partial\Phi)\, d\theta$$

$$\frac{\partial}{\partial \Phi}(U_E ha \sin \Phi\, \partial\theta)\, d\Phi$$

Chapter 4

4.1 $$\frac{dV}{dt} = 1.72\, L/t_0^2$$

4.2 $$\frac{dT}{dt} = \frac{\pi}{3}\cos\frac{\pi t}{12} - \frac{1}{40} \quad \text{°F/hr, where } t \text{ is in hours.}$$

(If °F and miles are replaced by °C and kilometers, respectively, then one finds the same solution in °C/hr.)

4.4 1. $\eta = \left[\frac{141 - t}{32}\right]^2$ ft $= [0.172(141 - t)]^2$ cm (t in sec)

2. $\rho \frac{\partial u}{\partial t} = -0.16$ lb/ft^3 $= -0.00256$ g/cm^3

3. $\rho u \frac{\partial u}{\partial x} = 1.78\pi(2g\eta)^{1/2}$ lb/ft^3 $= 0.0284\pi(2g\eta)^{1/2}$ g/cm^3

4. |Local inertia| $\ll$ convective inertia at any time. (It can be found that they would be equal when $\eta \cong 0.014$ ft, $\eta \cong 0.43$ cm.)

5. $\eta \to 15.7$ ft when $t \to \infty$ or $y = 478$ cm when $t \to \infty$.

$$t = \frac{2A}{B^2}\left[B(\eta_0^{1/2} - \eta^{1/2}) + q_0 \log\frac{B\eta_0^{1/2} - q_0}{B\eta^{1/2} - q_0}\right]$$

where $B = (\pi\Phi^2/4)(2g)^{1/2}$

4.7 $$u_{max} = a\left(\frac{g}{d}\right)^{1/2} \text{ at } x = \frac{1}{2}$$

$$w_{max} = ka \text{ at } x = 0, l, \text{ and } z = d$$

$$\frac{\left|\rho u \frac{\partial u}{\partial x}\right|_{max}}{\left|\rho \frac{\partial u}{\partial t}\right|_{max}} = \frac{a}{d}$$

4.9 r-direction:

$$A_r = \frac{\partial v_r}{\partial t} + v_r\frac{\partial v_r}{\partial r} + \frac{v_\theta}{r}\frac{\partial v_r}{\partial \theta} + v_z\frac{\partial v_r}{\partial z} - \frac{v_\theta^2}{r}$$

θ-direction:

$$A_\theta = \frac{\partial v_\theta}{\partial t} + v_r\frac{\partial v_\theta}{\partial r} + \frac{v_\theta}{r}\frac{\partial v_\theta}{\partial \theta} + v_z\frac{\partial v_\theta}{\partial z} + \frac{v_r v_\theta}{r}$$

z-direction:

$$A_z = \frac{\partial v_z}{\partial t} + v_r\frac{\partial v_z}{\partial r} + \frac{v_\theta}{r}\frac{\partial v_z}{\partial \theta} + v_z\frac{\partial v_z}{\partial z}$$

4.10 1. $\frac{\partial V}{\partial t} + \frac{\partial(V^2/2)}{\partial x}$ and $\frac{V^2}{R}$

2. $\frac{\partial u}{\partial t} + \frac{\partial(V^2/2)}{\partial x}$, $\quad \frac{\partial v}{\partial t} + \frac{V^2}{R}$, $\quad \frac{\partial w}{\partial t}$

4.12 $$\frac{2\omega \cos \Phi U_E}{g} = 7.65 \times 10^{-6}$$

Chapter 5

5.1 $$-2\mu\left(\frac{\partial \xi}{\partial z} - \frac{\partial \zeta}{\partial x}\right), \qquad -2\mu\left(\frac{\partial \eta}{\partial x} - \frac{\partial \xi}{\partial y}\right)$$

5.2 $\mu\nabla^2\mathbf{V} = \mu\nabla^2\,\mathbf{grad}\,\phi = \mu\,\mathbf{grad}\,\nabla^2\phi \equiv 0$ and $\nabla^2\phi \equiv 0$ for continuity.

5.4 r-direction:

$$\mu\left[\frac{1}{r}\frac{\partial}{\partial r}\left(r\frac{\partial v_r}{\partial r}\right) + \frac{1}{r^2}\frac{\partial^2 v_r}{\partial \theta^2} + \frac{\partial^2 v_r}{\partial z^2} - \frac{v_r}{r^2} - \frac{2}{r^2}\frac{\partial v_\theta}{\partial \theta}\right]$$

θ-direction:

$$\mu\left[\frac{1}{r}\frac{\partial}{\partial r}\left(r\frac{\partial v_\theta}{\partial r}\right) + \frac{1}{r^2}\frac{\partial^2 v_\theta}{\partial \theta^2} + \frac{\partial^2 v_\theta}{\partial z^2} - \frac{v_\theta}{r^2} + \frac{2}{r^2}\frac{\partial v_r}{\partial \theta}\right]$$

z-direction:

$$\mu\left[\frac{1}{r}\frac{\partial}{\partial r}\left(r\frac{\partial v_z}{\partial r}\right) + \frac{1}{r^2}\frac{\partial^2 v_z}{\partial \theta^2} + \frac{\partial^2 v_z}{\partial z^2}\right]$$

5.5

$$\sigma_{rr} = 2\mu \frac{\partial v_r}{\partial r}, \sigma_{\theta\theta} = 2\mu\left(\frac{1}{r}\frac{\partial v_\theta}{\partial \theta} + \frac{v_r}{r}\right)$$

$$\sigma_{zz} = 2\mu \frac{\partial v_z}{\partial z}$$

$$\tau_{r\theta} = \mu\left[r \frac{\partial}{\partial r}\left(\frac{v_\theta}{r}\right) + \frac{1}{r}\frac{\partial v_r}{\partial \theta}\right]$$

$$\tau_{rz} = \mu\left[\frac{\partial v_r}{\partial r} + \frac{\partial v_r}{\partial z}\right]$$

$$\tau_{\theta z} = \mu\left[\frac{\partial v_\theta}{\partial z} + \frac{1}{r}\frac{\partial v_z}{\partial \theta}\right]$$

Chapter 6

6.4

$$u(z) = U\left(1 + \frac{\mu_1}{\mu_2}\frac{e_1}{e_2}\right)^{-1}\frac{z}{e_1} \quad \text{when} \quad z < e_1$$

and

$$u(z) = U\left\{1 - \left[1 - \left(1 + \frac{\mu_1}{\mu_2}\frac{e_2}{e_1}\right)^{-1}\right]\left[\frac{e_1 + e_2 - z}{e_2}\right]\right\} \quad \text{when} \quad z > e_1$$

6.5 1. $\partial u/\partial x = 0$, $v = w = 0$, $u = 0$ when $z = \mp h$

$$\frac{1}{\rho}\frac{\partial p}{\partial x} = v\frac{d^2u}{dz^2}$$

2. $u = (\rho gj/2\mu)(h^2 - z^2)$, $Q = (2\rho gj/3\mu)h^3$
3. $\bar{u} = \rho gjh^2/3\mu$, $u = (3\bar{u}/2h^2)(h^2 - z^2)$
4. $d^2u/dz^2 = -3\bar{u}/h^2$, $dp/dx = -3\mu\bar{u}/h^2$
5. $\xi = 0$, $\zeta = 0$, $\eta = \frac{1}{2}(\partial u/\partial z) = -\rho gjz/2\mu$
6. $\rho gjQ = 2(\rho g)^2j^2h^3/3\mu$, $j = 3\mu/2\rho gh^3$
7. $w = 0$

$$\frac{1}{\rho}\frac{\partial p}{\partial x} = v\,\nabla^2 u \cong v\frac{\partial^2 u}{\partial z^2}$$

$$\frac{1}{\rho}\frac{\partial p}{\partial y} = v\,\nabla^2 v \cong v\frac{\partial^2 u}{\partial z^2}$$

$\partial p/\partial x = -3\mu\bar{u}/h^2$, $\partial p/\partial y = -3\mu\bar{v}/h^2$ ($\bar{v} = v$ average), $\partial\bar{u}/\partial x + \partial\bar{v}/\partial y = 0$. Since $\partial^2 p/\partial x\,\partial y \equiv \partial^2 p/\partial y\,\partial x$, one must have $\partial\bar{u}/\partial y - \partial\bar{v}/\partial x = 0$. $\phi = -h^2p/3\mu$.

6.9

$$\rho\frac{d\eta}{dt} - \mu\,\nabla^2\eta = \rho\left(\xi\frac{\partial v}{\partial x} + \eta\frac{\partial v}{\partial y} + \zeta\frac{\partial v}{\partial z}\right)$$

$$\rho\frac{d\zeta}{dt} - \mu\,\nabla^2\zeta = \rho\left(\xi\frac{\partial w}{\partial x} + \eta\frac{\partial w}{\partial y} + \zeta\frac{\partial w}{\partial z}\right)$$

6.10

$$\frac{\partial p}{\partial x} = \frac{\mu}{r}\frac{d}{dr}\left(r\frac{du}{dr}\right)$$

$$u = \frac{\rho gj}{4\mu}(R^2 - r^2)$$

$$j = \frac{8vQ}{\pi gR^4}$$

6.11 After simplification, the equations of motion become:

$$\mu\frac{\partial^2 u}{\partial z^2} + 2\rho\omega v \sin\Phi = 0$$

$$\mu\frac{\partial^2 v}{\partial z^2} - 2\rho\omega u \sin\Phi = 0$$

$$\frac{\partial u}{\partial x} + \frac{\partial v}{\partial y} = 0$$

and the boundary conditions

$$\mu\frac{\partial u}{\partial z}\bigg|_{z=0} = 0, \ \mu\frac{\partial v}{\partial z}\bigg|_{z=0} = -\tau \qquad (u, v \to 0 \text{ when } z \to \infty)$$

Eliminating v yields

$$\frac{\partial^4 u}{\partial z^4} + \frac{4\rho^2\omega^2\sin^2\Phi}{\mu^2}u = 0$$

which gives, with $a = (\rho\omega\sin\Phi/\mu)^{1/2}$ and $U_0 = \tau/\mu a(2)^{1/2}$

$$u = U_0 e^{-az}\cos\left(\frac{\pi}{4} - az\right)$$

It is also found that

$$v = U_0 e^{-az}\sin\left(\frac{\pi}{4} - az\right)$$

At $z = 0$ $|\mathbf{V}| = U_0$ and is at 45° angle with the wind stress (on the right in the northern hemisphere and on the left in the southern hemisphere). The projection of the velocity vector is like a logarithmic spiral $\rho = U_0 e^{-\theta}$, and the angle θ of the velocity vector with the axis OX is $\theta = \pi/4 - az$, i.e., varies linearly with depth.

Chapter 9

9.1 *One layer:*

$$\frac{\Delta H}{\Delta L} = C_x\frac{V^2}{2g\delta}\frac{(1 - \varepsilon)^2}{\varepsilon^2}, \ C_x = \frac{24v}{V\delta}$$

$Q = 1.43 \times 10^{-2}\ \Delta H$ ft^3/sec ($1328\ \Delta H$ cm^3/sec)
$\Delta H < 0.76$ ft (23.16 cm) the Darcy law is valid
$R_e > 100$, for turbulence
From Fig. 9-5, $C_x = 1$
$\Delta H > 319$ ft (97.2 m) for turbulence

Two layers:

$$\Delta H_{\text{total}} = \frac{12\nu Q}{gA}\frac{(1-\varepsilon)^2}{\varepsilon^2}\sum\frac{\Delta l}{\delta^2}$$

$Q = 3.06 \times 10^{-3}\ \Delta H$ ft^3/sec ($2.842\ \Delta H$ cm^2/sec)
$\Delta H < 2.14$ ft (0.652 m); the Darcy law is valid
$\Delta H > 550$ ft (167.6 m) for turbulence

Three layers:
$Q = 4.15 \times 10^{-3}\ \Delta H$ ft^3/sec ($3.855\ \Delta H$ cm^3/sec)
$\Delta H < 1.58$ ft (0.48 m); the Darcy law is valid
$\Delta H > 424$ ft (129.2 m) for turbulence

9.2 $K = \lambda^{-3/4}$

Chapter 10

10.1 $p - p_\infty = \frac{1}{2}\rho U^2(1 - 4\sin^2\theta)$

10.2
$$2(u\zeta - w\xi) = -\frac{\partial}{\partial y}\left(\frac{V^2}{2} + \frac{p}{\rho} + gz\right)$$
$$2(v\xi - u\eta) = -\frac{\partial}{\partial z}\left(\frac{V^2}{2} + \frac{p}{\rho} + gz\right)$$

10.3 $v_r = 0$, $v_\theta = -1.5U\sin\theta$, $p - p_\infty = \frac{1}{2}\rho U^2(1 - \frac{9}{4}\sin^2\theta)$

10.4 The equation of streamlines yields the following equalities:

$$v\,dx = u\,dy,\ w\,dx = u\,dz,\ w\,dy = v\,dz$$

which gives an expression such as $u(\partial u/\partial x)\,dx = u\,du$. Add the equations along the three axis and simplify.

10.5
$$T = \frac{S_1 S_2}{S_1 + S_2}\frac{(h)^{1/2}}{0.3A(2g)^{1/2}}$$

10.6
$Q_{AM} = 3.05$ ft^3/sec (86,300 cm^3/sec)
$Q_{MB} = 1.85$ ft^3/sec (52,400 cm^3/sec)
$Q_{CN} = 3.2$ ft^3/sec (90,600 cm^3/sec)
$Q_{ND} = 4.4$ ft^3/sec (124,000 cm^3/sec)

10.7 2. $T = 2\pi(LF/gf)^{1/2}$
3. $z = (W_a - W_a')(Lf/gF)^{1/2}\sin 2\pi t/T$

10.9 4 unknowns:

$$z, Q_1, Q_2, Q_3$$

4 equations:

$$L\,dz = (Q_1 + Q_2 + Q_3)\,dt$$
$$z_1 = z + \frac{Q_1^2}{2gD^2} + K_{0-1}\frac{(Q_1 + Q_2 + Q_3)^2}{2g}$$
$$z_1 = z + \frac{Q_2^2}{2gD^2} + K_{0-2}\frac{(Q_2 + Q_3)^2}{2g} + \frac{L}{gD}\frac{d(Q_2 + Q_3)}{dt} + K_{0-1}\frac{(Q_1 + Q_2 + Q_3)^2}{2g}$$

and so on. K_{i-j} = coefficient for head loss between i and j.

Chapter 11

11.2
$$\rho\left(\frac{\partial^2\psi}{\partial y\,\partial t} + \frac{\partial\psi}{\partial y}\frac{\partial^2\psi}{\partial y\,\partial x} - \frac{\partial\psi}{\partial x}\frac{\partial^2\psi}{\partial y^2}\right) = -\frac{\partial p^*}{\partial x} + \mu\left(\frac{\partial}{\partial y}\nabla^2\psi\right)$$

and

$$\rho\left(-\frac{\partial^2\psi}{\partial x\,\partial t} - \frac{\partial^2\psi}{\partial x^2}\frac{\partial\psi}{\partial y} + \frac{\partial\psi}{\partial x}\frac{\partial^2\psi}{\partial x\,\partial y}\right) = -\frac{\partial p^*}{\partial y} - \mu\frac{\partial}{\partial x}\nabla^2\psi$$

11.4 $\psi = \mathbf{V}(y\cos\alpha - x\sin\alpha)$

11.5
$$\psi = Uy\left[1 - \frac{R^2}{x^2 + y^2}\right] = U\left(r - \frac{R^2}{r}\right)\sin\theta$$

11.6 $u = 0$; $v = 2x$; $\partial u/\partial y - \partial v/\partial x = -2$ (rotational); vorticity $2\zeta = -2$; $\partial u/\partial x + \partial v/\partial y = 0$ (incompressible). No equipotential line; it is a Couette flow between two parallel plates.

11.7 $\psi = \frac{1}{2}y^2$

11.9 They are circles of radii

$$r = 1.4R\ (50\%)$$
$$r = 3.1R\ (10\%)$$
$$r = 10R\ (1\%)$$

11.10
$$\phi = -\frac{Q}{2\pi}\ln r + \frac{K}{2\pi}\theta$$
$$\eta = \eta_\infty - (v_r^2 + v_\theta^2)\frac{1}{2g}$$
$$\eta = \eta_\infty - \frac{Q^2 + K^2}{8\pi^2 g}\frac{1}{r^2}$$

11.11 Equipotential lines:

$$r = e^{-(K/Q)(\theta - \theta_0)}$$

Streamlines:

$$r = e^{(Q/K)(\theta - \theta_0)}$$

$$\psi = \frac{Q}{2\pi}\theta - \frac{K}{2\pi}\ln r$$

11.12 Take the potential function for a source and a sink of same strength apart by a distance $2a$ such as

$$\phi = \frac{Q}{2\pi}(\ln r_2 - \ln r_1)$$

where r_1 and r_2 are measured from source and sink, respectively. Insert the relationships (see figure below):

$$r_1^2 = r^2 + a^2 - 2ar\cos\theta$$
$$r_2^2 = r^2 + a^2 + 2ar\cos\theta$$

Let $2a(Q/2\pi) = K$ and take the limit when a tends to zero.

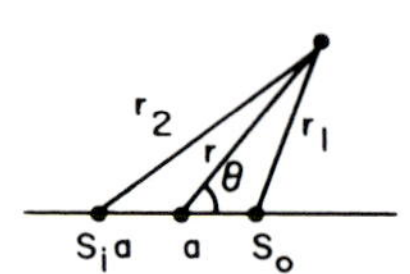

11.14 $v_\theta = -\partial\psi/\partial r$, $v_\theta =$ where $\sin\theta = \Gamma/4\pi RU$

$$p - p_\infty = \tfrac{1}{2}\rho U^2\left[1 - \left(-2\sin\theta + \frac{\Gamma}{2\pi RU}\right)^2\right]$$

Total force

$$X = -\int_0^{2\pi} pR\cos\theta\, d\theta = 0$$

$$Y = \frac{\rho U\Gamma}{\pi}\int_0^{2\pi}\sin^2\theta\, d\theta = -\rho U\Gamma$$

11.15

$$\phi = \frac{Q}{2\pi}\ln\frac{r_1}{r_2} + Ur\cos\theta$$

Shape:

$$\psi = 0 \quad \text{gives} \quad r = \frac{(Q/2\pi)(\theta_2 - \theta_1)}{U\sin\theta}$$

11.16

$$\left.\frac{\partial\psi}{\partial x}\right|_A = \frac{\psi_5 - \psi_1}{a}, \qquad \left.\frac{\partial\psi}{\partial x}\right|_B = \frac{\psi_1 - \psi_4}{a}$$

$$\frac{\partial^2\psi}{\partial x^2} = \frac{\left.\frac{\partial\psi}{\partial x}\right|_A - \left.\frac{\partial\psi}{\partial x}\right|_B}{a} = \frac{\psi_4 + \psi_5 - 2\psi_1}{a^2}$$

Similarly, $\partial^2\psi/\partial y^2$ is determined. Since $\nabla^2\psi = 0$, one finds ψ_1.

11.17 $x = C\cosh\phi\cos\psi$; $y = C\sinh\phi\sin\psi$. Equipotential lines ($\phi =$ constant):

$$\frac{x^2}{C^2\cosh^2\phi} + \frac{y^2}{C^2\sinh^2\phi} = 1$$

Streamlines ($\psi =$ constant):

$$\frac{x^2}{C^2\cos^2\psi} - \frac{y^2}{C^2\sin^2\psi} = 1$$

Foci: $(0,C)$ and $(0,-C)$.

Chapter 12

12.2

$$F = \frac{\rho Q^2}{\pi R_0^2}\left[\frac{3}{2} - \frac{1}{2}\left(\frac{R_0}{R}\right)^2 - \ln\frac{R}{R_0}\right]$$

12.3 The total force by momentum is $2\rho QV$ and it is found to be only ρQV by integration of pressure. The difference is due to the force acting at A and B.

12.6 Starting equations:

Energy:

$$\frac{V_1^2}{2g} + \frac{p_1}{\rho g} + h_1 = \frac{V_2^2}{2g} + \frac{p_2}{\rho g} + h_2 + \Delta H$$

Continuity: $V_1 h_1 = V_2 h_2$.

Momentum:

$$\left(\frac{V_1^2}{g} + \frac{p_1}{\rho g} + \frac{h_1}{2}\right)h_1 - \left(\frac{V_2^2}{g} + \frac{p_2}{\rho g} + \frac{h_2}{2}\right)h_2 = F$$

First case:

$$F = \frac{p_1}{\rho g}(h_2 - h_1)$$

(h can sometimes be neglected by comparison with $p/\rho g$)

Second case:

$$F = \int_1^2 \frac{p(x)}{\rho g} \sin \alpha \, dx$$

ΔH (head loss) is negligible so the value of the integral F is obtained from $p(x)$ by application of the generalized Bernoulli equation.

Third case:

$$\frac{p}{\rho g} = \frac{p_a}{\rho g}$$

The sum of external forces due to atmospheric pressure equals zero; so all the terms $p/\rho g$ disappear and $F = 0$.

12.7 $\rho Q V = \rho g z A$ (external force), and since $\rho Q V = \rho(2gz)AC_c$, $C_c = \frac{1}{2}$.

12.8 $$\rho g \frac{z^2}{2} - \rho g \frac{h^2}{2} = \rho q[V - f(\alpha)]$$

$$f(\alpha) \cong /2g(Z_w + H - h)]^{1/2} \sin \alpha$$

Inserting $q = f(H)$, the function $z = f(H)$ is obtained.

12.9 $F = \rho Q(V - U)$; power of the jet: $\rho Q(V^2/2)$. Transmitted power: $\rho Q(V - U)U$. In the case of the bucket, $F = 2\rho Q(V - U)$.

12.10 In the first case the pressure at the bottom of the vertical wall is $\rho g y_2$ while in the second case it is $\rho g(y_1 + h)$ depending upon the exact location of the jump with respect to the bottom drop.

12.15 (a) Given V_a wind velocity, D_1 the diameter of the windmill, A the cross section of the boat and, C_D its drag coefficient. The unknowns are D_2 the diameter of the propeller, V_b the velocity of the jet generated by the propeller. And the equations are (ρ_a is the density of the air and ρ_w that of the water).

1. Force exerted by the wind on the windmill = force exerted by the propeller + boat drag, i.e.,

$$\rho_a Q_a(V_a + V_b) = \rho_w Q_w(V_w - V_b) + \rho_w A C_D V^2$$

and

$$Q_a = V_a \frac{\pi D_1^2}{4}, \; Q_w = V_w \frac{\pi D_2^2}{4}$$

2. Power of the windmill = power of the propeller = power dissipated by the boat drag, i.e.,

$$\rho_a Q_a(V_a + V_b)^2 = \rho_w Q_w(V_w - V_b)^2 = \rho_w A C_D V_b^3$$

The boat can go upwind if D_2 is such that this system of equations is verified, and V_w and V_b are subsequently determined.

(b) Downwind, the force on the windmill is

$$\rho_a Q_a(V_a - V_b)$$

She cannot go faster than wind as the force on the windmill tends to zero when she tends toward the same velocity as the wind.

Chapter 13

13.1 $\delta^* = \delta/2, \delta/3$; $\theta = \delta/6, 2\delta/15$; $\delta^{**} = \delta/4 \cdots$

13.2 Since $\psi = (\nu x U_0)^{1/2} f(\eta)$, $\eta = y(U_0/\nu x)^{1/2}$

$$v = -\frac{\partial \psi}{\partial x} = \frac{1}{2}\left(\frac{\nu U_0}{x}\right)^{1/2}\left(\eta \frac{\partial f(\eta)}{\partial \eta} - f\right)$$

13.3 $A_0 = 0$, $A_1 = 0$, $A_3 = 0$, $A_4 = 0$, $A_5 = -\frac{1}{2}A_2^2$, $A_6 = 0$, $A_7 = 0$, $A_8 = -\frac{11}{4}A_2^3$, $A_9 = 0$, $A_{10} = 0$,

$$A_{11} = -\frac{1\,375}{8\;11!} A_2^4, \qquad A_{3n+2} = \left(-\frac{1}{2}\right)^n \frac{A_2^{n+1} C_n}{(3n+2)!}$$

13.4 $y = 0$, $u = 0$, $\partial^2 u/\partial y^2 = 0$; $y = \delta$, $u = U_0$, $\partial u/\partial y = 0$, $\partial^2 u/\partial y^2 = 0$.

$$a_0 = 0, a_2 = 0, a_1 = 2(U_0/\delta), a_3 = -2U_0/\delta^3, a_4 = U_0/\delta^4.$$

$$\tau_0 = \rho\nu \left.\frac{\partial u}{\partial y}\right|_{y=0} = \frac{2\rho\nu U_0}{\delta}$$

13.5 $$f''' - f'^2 + 1 = 0$$

Multiplying this equation by f'' and integrating,

$$\frac{df'}{d\eta} = (f' - 1)[\tfrac{2}{3}(f' + 2)]^{1/2}$$

i.e.,

$$\eta = (2)^{1/2}\left[\tanh^{-1}\frac{(2 + f)^{1/2}}{(3)^{1/2}} - \tanh^{-1}\left(\frac{2}{3}\right)^{1/2}\right]$$

Finally,

$$f' = \frac{u}{U} = 3\tanh^2\left(\frac{\eta}{(2)^{1/2}} + 1.146\right) - 2$$

13.6 From the continuity and momentum equations, one has

$$\frac{\partial u}{\partial t} + \frac{\partial u^2}{\partial x} + \frac{\partial uv}{\partial y} = -\frac{1}{\rho}\frac{\partial p}{\partial x} + \nu \frac{\partial^2 u}{\partial y^2}.$$

Outside the boundary $u \to U$ and $\partial^2 U/\partial y^2 \to 0$; so

$$\int_0^\delta \left(\frac{\partial u}{\partial t} + \frac{\partial u^2}{\partial x} + \frac{\partial uv}{\partial y}\right) dy = \int_0^\delta \left(\frac{\partial U}{\partial t} + U\frac{\partial U}{\partial x} + v\frac{\partial^2 u}{\partial y^2}\right) dy$$

Inserting $v = -\int_0^y (\partial u/\partial x)\, dy$, and rearranging,

$$\frac{\partial}{\partial t}\int_0^\delta (U-u)\, dy + \frac{\partial}{\partial x}\int_0^\delta [u(U-u)\, dy] + \frac{\partial U}{\partial x}\int_0^\delta (U-u)\, dy = \frac{\tau_0}{\rho}$$

Finally,

$$\frac{\tau_0}{\rho} = \frac{\partial}{\partial t}(U\delta^*) + \frac{\partial}{\partial x}(U^2\theta) + \delta^* U\frac{\partial U}{\partial x}$$

13.7 $$\frac{\partial u}{\partial t} = v\frac{\partial^2 u}{\partial y^2}, \qquad u = U_0\frac{\cosh k(h-y)}{\cosh kh}\cos kt$$

$$F = \mu\frac{\partial u}{\partial y}\bigg|_{x=0} = \mu k U_0 \tanh kh \cos kt$$

13.8 1. $\dfrac{d}{dx}\displaystyle\int_0^\delta u^2\, dy - U_0\frac{d}{dx}\int_0^\delta u\, dy = -v\frac{\partial u(0)}{\partial y} + g\delta$

2. $a_0 = 0,\ a_1 = 2,\ a_2 = -1$

$$\frac{8}{15}\frac{d}{dx}(2gx\delta) - \frac{2}{3}2g(x)^{1/2}\frac{d}{dx}[(x)^{1/2}\delta] = -\frac{2v}{\delta}(2gx)^{1/2} + g\delta$$

$[U_0 = (2gx)^{1/2}]$

3. $\beta = 3(2)^{1/2}[v/(g)^{1/2}]^{1/2}$

4. $x_0 = \left(\dfrac{3Q}{2(2g)^{1/2}\beta}\right)^{4/3}$

5. $y_0 = \beta\left(\dfrac{3Q}{2(2g)^{1/2}\beta}\right)^{1/3}$

6. $\dfrac{d}{dx}\displaystyle\int_0^\delta u^2\, dy = -v\frac{\partial u(0)}{\partial y} + g\delta$

7. $\dfrac{d\delta}{dx} = \dfrac{5g}{6Q^2}\left(\dfrac{3vQ}{g} - \delta^3\right)$

8. $$\frac{5g}{6Q^2}(x - x_0) = \frac{1}{6\delta^{*2}}\ln\left[\left(\frac{\delta^* - \delta_0}{\delta^* - \delta}\right)^2\left(\frac{\delta^{*2} + \delta^*\delta + \delta^2}{\delta^{*2} + \delta^*\delta_0 + \delta_0^2}\right)\right] + \frac{1}{(3)^{1/2}\delta^{*2}}\left[\tan^{-1}\frac{2\delta + \delta^*}{\partial^*(3)^{1/2}} - \tan^{-1}\frac{2\delta_0 + \delta^*}{\partial^*(3)^{1/2}}\right]$$

where $\delta^* = \left(\dfrac{3vQ}{g}\right)^{1/3}$

9. $\delta_3 = \left(\dfrac{3vQ}{g}\right)^{1/3}$

13.11 $$M' = \rho\int_0^{2\pi}\int_0^{\pi}\int_R^{\infty}\frac{v_r^2 + v_\theta^2}{U^2}r^2\sin\theta\, dr\, d\theta\, d\psi = \tfrac{2}{3}\rho\pi R^3$$

13.12 The drag force is maximum under the crest. The inertial force is maximum when the free surface elevation is at the still water level. The maximum total force at a given level occurs before the crest reaches the pile at a time which varies slightly with the vertical coordinates z. The maximum total force on the pile is obtained by numerical integration and by trial and error.

Chapter 14

14.3 $$R_H = \frac{A}{P} = \frac{dA}{dP} = \frac{b\, dh}{(db^2 + dh^2)^{1/2}}$$

By integrating and taking $b = R_H$ when $h = 0$, one finds:

$$h = R\{\ln[b + (b^2 - R^2)^{1/2}] - \ln R\}$$

Chapter 16

16.1 Streamlines:

$$\frac{k}{m}\frac{\sinh m(d+z)}{\sinh md}\cos mx = \text{constant}$$

Isobars:

$$z = -a\frac{\cosh m(d + z_0)}{\cosh md}\cos(kt - mx)$$

16.2 Streamlines:

$$\frac{k}{m}\frac{\sinh m(d+z)}{\sinh md}\sin mx = \text{constant}$$

Isobars:

$$z = -2a\frac{\cosh m(d + z_0)}{\cosh md}\cos mx \sin kt$$

16.3 Fundamental free surface elevation:

$$\eta = \frac{2x}{a^2}\sin\frac{2\pi}{T_1}t$$

Harmonic free surface elevation:

$$\eta = \frac{3x^2 - a^2}{a^3}\sin\frac{2\pi}{T_2}t \qquad \left(\frac{T_1}{T_2} = 3^{1/2}\right)$$

16.6
$$\frac{p}{\rho g} = -z + \frac{H}{2}\frac{\cosh m(d+z)}{\cosh md}\sin(kt - mx)$$

16.8
$$\frac{y - y_0}{x - x_0} = -\tanh m(d + z_0)\cot mx_0$$

16.9 $L = 2l/n$; $T_n = 2.494, 1.247, \ldots, 0.392$ sec

$$\frac{4}{T^2gd} = \left(\frac{r}{10}\right)^2 + \left(\frac{s}{8}\right)^2 \begin{cases} r = 0 & s = 1 \\ r = 1 & s = 0 \\ r = 1 & s = 1 \end{cases}$$

16.11
$$H = \frac{2(p_{\max} - \rho gd)}{\rho g}\cosh md, \left(\frac{2\pi}{T}\right)^2 = mg\tanh md$$

16.12 The maximum pressure on the vertical wall is approximated by a linear distribution between the elevation $d + H$ where the pressure is zero, and the bottom where the pressure is $\rho g\{d + H[\cosh(2\pi d/L)]^{-1}\}$. The underpressure acting on the vertical breakwater is assumed to be distributed between this latter expression and ρgd on the harbor side of the breakwater. The pressure on the harbor side is hydrostatic. The overturning momentum and bottom stress are then determined as in a gravity dam.

16.14 Distance between orthogonals:

$$\frac{b_0}{b_b} = \frac{\cos\alpha_0}{\cos\alpha_b}$$

Energy flux:
$$H_0^2 b_0 U_0 = H_b^2 b_b U_b$$

$$\frac{L_b}{L_0} = \frac{L_b}{H_b}\times\frac{H_b}{H_0}\times\frac{H_0}{L_0} \qquad \frac{L_b}{L_0} = \frac{\sin\alpha_b}{\sin\alpha_0} = \tanh\frac{2\pi d_b}{L_b}$$

Let
$$\alpha = \frac{2\pi d_b}{L_b}, t = \tanh\alpha, s = \sinh 2\alpha$$

Then
$$0.14t^{5/2}\left(1 + \frac{2\alpha}{s}\right)^{1/2} = \frac{H_0}{L_0}\left(\frac{1 - \sin^2\alpha_0}{1 - t^2\sin^2\alpha_0}\right)^{1/4}$$

which gives d_b/L_b as a function of H_0, L_0, and α_0. So L_b is obtained from $L_b/L_0 = t$ and d_b and finally α_b.

16.15 1. $\Delta p = \rho gH\dfrac{\cosh m(d+z)}{\cosh md}$

2. $K = \dfrac{\cosh md}{\cosh m(d + z_0)}$

3. $H = 6.5$ ft (1.98 m)

16.16 $\Delta p = \rho g\dfrac{H}{\cosh md}$

16.17 1. $p = p_a - A\dfrac{\partial^2\eta}{\partial x^2}, \dfrac{\partial\eta}{\partial t} \cong -\dfrac{\partial\phi}{\partial z}\Big|_{z=0},$

$$\frac{\partial p}{\partial t} = -A\frac{\partial^3\phi}{\partial z\,\partial x^2}\Big|_{z=0}$$

The free surface condition is:

$$-\frac{A}{\rho}\frac{\partial^3\phi}{\partial z\,\partial x^2} = g\frac{\partial\phi}{\partial z} + \frac{\partial^2\phi}{\partial t^2}$$

2. Insert ϕ from Table 16-2 into the free surface condition, since $d(\cosh a) = \sinh a\,da$, $d(\sinh a) = \cosh a\,da$. One obtains

$$C^2 = \left(\frac{L}{T}\right)^2 = \left(\frac{gL}{2\pi} + \frac{A}{\rho}\frac{2\pi}{L}\right)\tanh\frac{2\pi d}{L}$$

Water: $C_{\min} = 23.1$ cm and $L = 1.71$ cm; Mercury: $C_{\min} = 19.4$ cm and $L = 1.202$ cm

3. When T tends to zero, C tends to infinity. When T tends to infinity, C tends to $gT/2\pi$. The capillary effect becomes negligible when $T \gtrsim 0.3$ sec in water.

16.18 $\mathbf{V} = \mathbf{grad}\,\phi$

$$\frac{1}{\varepsilon}\frac{d\mathbf{V}}{dt} + \mathbf{grad}\left[\frac{p}{\rho} + gz\right] + K\mathbf{V} = 0$$

Linearizing:

$$\mathbf{grad}\left[\frac{1}{\varepsilon}\frac{\partial\phi}{\partial t} + \frac{p}{\rho} + gz + K\phi\right] = 0$$

At the free surface where $z = \eta$,

$$\frac{\partial\eta}{\partial t} = \frac{1}{\varepsilon}\frac{\partial\phi}{\partial z} \qquad \frac{1}{\varepsilon}\frac{\partial^2\phi}{\partial t^2} + \frac{g}{\varepsilon}\frac{\partial\phi}{\partial z} + K\frac{\partial\phi}{\partial t} = 0$$

Chapter 17

17.1 1. $\Delta = 4.24$ ft (1.29 m)
2. $\eta_{\max} = 15.24$ ft (4.65 m)
$\eta_{\min} = 6.76$ ft (2.06 m)
3. $\Delta p = 412$ lb/ft^2 (2009 kg/m^2)
4. 37,800 lb/ft run at 13.7 ft above the base (56.179 kg/m at 4.17 m)

Chapter 18

18.3

$$\frac{\partial\phi}{\partial x} \cong \frac{\phi_{i+1,j} - \phi_{i-1,j}}{2\Delta}$$

$$\frac{\partial^2\phi}{\partial x^2} \cong \frac{-2\phi_{i,j} + \phi_{i+1,j} + \phi_{i-1,j}}{\Delta^2}$$

The same is found for y; then they are inserted in the wave equation.

18.4 1. η can still be written $\eta = B\cos(kt - \psi)$ where $B\cos\psi = \cos mx$, $B\sin\psi = \cos my$, $B^2 = \cos^2 mx + \cos^2 my$, $\tan\psi = \cos my/\cos mx$. The curves of equal amplitude are defined by $B =$ constant.

2. $B^2 = 2$ at $\{x, y\} = n(L/2)$ $(n = 0, 1, 2, \ldots)$

3. Transform

$$x = x' + \frac{L}{4}$$

$$y = y' + \frac{L}{4}$$

Then, $B^2 = \sin^2 mx' + \sin^2 my'$. When x', $y' \to 0$, $B \to 0$ (the wave amplitude is zero). Near this point, the lines of equal amplitude are circles of equations

$$x'^2 + y'^2 = \left(\frac{B}{m}\right)^2$$

Also, the maximum amplitude on these circles is defined by $\cos(kt - \psi) = 1$; i.e., $kt = \psi$. The angle is defined by

$$\tan\psi = \frac{\sin my'}{\sin mx'} \cong \frac{y'}{x'}$$

The crest rotates in circles around x', $y' = 0$, with the wave period T. It is an amphidromic point.

18.5

$$b\frac{\partial^2\eta}{\partial t^2} = gd\frac{\partial}{\partial x}\left(b\frac{\partial\eta}{\partial x}\right)$$

Let $\eta = A(x)\sin nkt$ $(n = 1,2,3,\ldots)$. Then

$$\frac{d^2A}{dx^2} + \frac{1}{x}\frac{dA}{dx} + B^2A = 0$$

$$B^2 = \frac{4\pi^2n^2}{gd}$$

A solution is

$$A = J_0(Bx)$$

$$\eta = \eta_0 J_0(Bx)\sin 2\pi nt$$

18.7 The characteristic equation

$$\frac{d}{dt}(u \pm 2c) = -gS$$

can still be written:

$$\frac{\partial}{\partial t}(u \pm 2c) \pm \left[u \pm c \pm \frac{gS}{(\partial/\partial x)(u \pm 2c)}\right]\frac{\partial}{\partial x}(u \pm 2c) = 0$$

i.e.,

$$\frac{d}{dt}(u \pm 2c) = 0$$

along lines of the defined slopes.

18.9 Continuity:

$$\frac{\partial A}{\partial t} + \frac{\partial Q}{\partial x} = 0, \qquad Q = \frac{\partial\phi}{\partial t}, \qquad A = -\frac{\partial\phi}{\partial x}, \qquad d\phi = Q\,dt - A\,dx.$$

Then $u = Q/A$ and $h = A/l$ ($l =$ width) are expressed as functions of ϕ in the momentum equation:

$$\frac{\partial u}{\partial t} + u\frac{\partial u}{\partial x} = -gS - \frac{g}{C_h^2R_H}u|u|$$

Appendix A

A.1 The method for detecting periodic components consists of determining the autocorrelation function at an arbitrarily large lag time τ.

A.2 The density spectrum of a periodic function is in general an infinite series of delta functions. A periodic phenomenon is best described in terms of a simple Fourier series rather than a density spectrum. The relative amplitudes of the harmonics cannot be represented in a "density" domain.

A.3 For each value of x there is only one value of $p(x)$. It is required after the transformation of x to $y = f(x)\ (=x^2)$ that the probability $p(y)$ is also unique. In particular, $p(x)$ in the range $x - \partial x/2 < x < x + \partial x/2$ is now represented by $p(y)$ in the y plane in the range $y - \delta y/2 < y < y + \delta y/2$. This requirement is only met if the identity $p(x)\,dx \equiv p(y)\,dy$ is satisfied for all x and y. This yields the relationship $p(y) = p[f(x)] = p(x)(dx/dy)$. However, in this particular problem one further point arises. $p(y)$ can only exist for $y = x^2$ for positive values of y. In short, for each value of y, two values of x, $\pm x$ can satisfy the relationship $y = x^2$, Hence, $p(y)\,dy \equiv p(-x)\,dx + p(x)$ is the identity required and since $p(x)$ is an even function,

$$p(y)\,dy = 2p(x)\,dx \qquad \text{when } y = x^2$$

Therefore

$$p(y) = 2p(x)\frac{dx}{dy}$$

$$= \frac{1}{(2\pi y)^{1/2}} e^{-y/2} \qquad \text{for } y > 0$$

$$= 0 \qquad \text{for } y < 0$$

A.4

$$R(\tau) = \lim_{T\to\infty} \frac{1}{T}\int_{t=\theta}^{t=T} f(t)f(t+\tau)\,dt$$

Substitute $t = t - \tau$:

$$r(\tau) = \lim_{T\to\infty} \frac{1}{T}\int_{t=-\tau}^{t=T-\tau} f(t-\tau)f(t)\,dt$$

does not alter the average properties. Now the limits can be shifted without loss of generality since the process is a stationary one. Hence

$$R(\tau) = \lim_{T\to\infty} \frac{1}{T}\int_{0}^{T} f(t-\tau)f(t)\,dt$$

$$= R(-\tau) \qquad \text{(by definition).}$$

A.5 $R(\tau)$ can be represented by the Fourier series:

$$R(\tau) = a_0 + \sum_{n=-\infty}^{\infty} a_n \cos n\omega t + \sum_{n=-\infty}^{\infty} b_n \sin n\omega t$$

The sine transformation corresponds to the operation

$$\int R(\tau) \sin m\omega\tau \, d\tau$$

From the general theory of Fourier analysis, this integral only exists for any specified value of m when $R(\tau)$ has sinusoidal components containing the argument $m\omega\tau$. Since $R(\tau)$ is an even function, the assumed Fourier series

$$R(\tau) = a_0 + \sum a_n \cos n\omega t + \sum b_n \sin n\omega t$$

cannot contain any sinusoidal (asymmetric) components. Hence

$$R(\tau) = a_0 + \sum a_n \cos n\omega t$$

and it follows that the sine transformation is identically zero.

A.6 $b = 2a$,

$$P(x) = \tfrac{1}{2}e^{-2a|x|} \qquad \text{for } x < 0$$

$$= \tfrac{1}{2} + \tfrac{1}{2}[1 - e^{-2ax}] \qquad \text{for } x > 0$$

A.7

$$\eta(t) = \sum A_n \cos(2\pi f_n t + \varepsilon_n)$$

From the linear theory for periodic waves

$$P = \rho g\left[z + \frac{1}{\cosh kd}\right]$$

Therefore

$$S_p(f) = (\rho g)^2 \frac{1}{\cosh^2 kd} S_\eta(f)$$

$$\sigma = \int_0^\infty S_p(f)\,df = (\rho g)^2 \int_0^\infty \frac{df}{\cosh^2 kd} S_\eta(f)$$

A.8

H_{average}	$= 1.25 \times H$ mp,	Probability	$= 0.460$
$H_{\text{most probable}}$	$= 1 \times H$ mp		$= 0.606$
$H_{\text{significant}}$	$= 2 \times H$ mp		$= 0.135$
H_{maximum}	$= 3 \times H$ mp		$= 0.01$

Appendix B

B.1

$$V_m = V_p \mu^{1/2}, \qquad Q_m = Q_p \times \lambda\mu^{3/2}, \qquad \left.\frac{\Delta H}{L}\right|_m = \left.\frac{\Delta H}{L}\right|_p \times \frac{\mu}{\lambda}$$

$$\frac{\lambda}{D} f(R_p \mu^{1/2} D) = f(R_p)$$

where

$$R_p = \left.\frac{Ud}{v}\right|_p$$

and R_p is expressed as a function of $(\Delta H/L)|_p$

Index

A

B

F

G

H

I

Index

J

K

L

M

N

O

P

Q

R

S